Yashpal Singh Malik / Raj Kumar Singh / Mahendra Pal Yadav

Emerging and Transboundary Animal Viruses

新发和跨界动物病毒

主　编　〔印〕亚什帕尔·辛格·马利克
拉吉·库马尔·辛格
马亨德拉·帕尔·亚达夫

主　译　金宁一　马呜潇　李　昌

副主译　白杰英　郑　敏　尹革芬

天津出版传媒集团
天津科技翻译出版有限公司

著作权合同登记号:图字:02-2020-99

图书在版编目(CIP)数据

新发和跨界动物病毒 / (印)亚什帕尔·辛格·马利克(Yashpal Singh Malik),(印)拉吉·库马尔·辛格(Raj Kumar Singh),(印)马亨德拉·帕尔·亚达夫(Mahendra Pal Yadav)主编 ; 金宁一,马鸣潇,李昌主译. —天津:天津科技翻译出版有限公司,2024.4

书名原文: Emerging and Transboundary Animal Viruses

ISBN 978-7-5433-4425-9

Ⅰ.新… Ⅱ.①亚… ②拉… ③马… ④金… ⑤马… ⑥李… Ⅲ.①动物病毒 Ⅳ.①S852.65

中国国家版本馆 CIP 数据核字(2024)第 027373 号

授权单位: Springer Nature Singapore Pte Ltd
出　　版:天津科技翻译出版有限公司
出 版 人:刘子媛
地　　址:天津市南开区白堤路 244 号
邮政编码:300192
电　　话:(022)87894896
传　　真:(022)87893237
网　　址:www.tsttpc.com
印　　刷:天津新华印务有限公司
发　　行:全国新华书店
版本记录:889mm×1194mm　16 开本　17 印张　540 千字
2024 年 4 月第 1 版　2024 年 4 月第 1 次印刷
定价:138.00 元

(如发现印装问题,可与出版社调换)

译者名单

主　译　金宁一　马鸣潇　李　昌

副主译　白杰英　郑　敏　尹革芬

译　者　（按照姓氏汉语拼音排序）

白杰英　杜寿文　费东亮　高玉伟　郝鹏飞　贾雷立
金宁一　李　昌　李乐天　李　俏　李体远　李　霄
廖　明　刘　昊　刘立明　刘　全　刘　玮　鲁会军
马鸣潇　瞿孝云　任林柱　尚　超　田明尧　王　宏
王茂鹏　尹革芬　尹荣兰　张　赫　赵翠青　郑　敏
庄忻雨

主编简介

亚什帕尔·辛格·马利克　印度兽医研究所(IVRI)和印度农业研究委员会(ICAR)国家研究员。主要研究成果包括病毒病的流行病学、病毒-宿主互作、生物多样性、病毒特性和诊断。曾在明尼苏达州大学(美国圣保罗)接受分子病毒学高级培训;在加拿大渥太华大学安大略研究所(加拿大安大略省)和中国武汉市病毒学研究所学习病毒学。获得青年科学家印度微生物学家协会奖(2000),ICAR贾瓦哈拉尔·尼赫鲁奖(2001),北阿坎德邦科学技术委员会青年科学家奖 (2010)。被授予了多项杰出的荣誉会员资格, 即美国农业科学院会员资格(2010); 国家兽医科学会会员 (2010)。获科学与工业研究理事会高级研究奖学金(1997—2000);ICAR初级研究奖学金(1995—1997);学士学位学业优秀奖学金(1990—1995)。

任国际病毒学分类委员会双RNA病毒科,小RNA病毒科研究组的成员和世界病毒学协会管理委员会成员。开发了多项技术和诊断试剂盒,申请了2项国家专利,撰写著作5部,发表文章25篇,担任《免疫病理学杂志》主编,并在施普林格杂志《病毒病》编辑了“人类和动物感染的肠道病毒”特刊,在《当前的药物代谢杂志》编辑了“治疗进展和他们的生物医学观点”特刊,在《实验生物学和农业科学杂志》编辑了“疾病诊断与治疗进展的生物医学观点”特刊。

拉吉·库马尔·辛格　IVRI主任兼副校长。是兽医微生物学、生物技术、分子流行病学、诊断和疫苗学方面的著名科学家。曾任IVRI和穆科特瓦尔校区病毒学部负责人。申请国家专利10项(授权2项,申报8项),已开发8种减毒活疫苗/候选疫苗,26种诊断试验/化验/试剂盒。撰写著作2部,发表科研论文245篇,撰写书评52篇,主笔论文15篇,特邀评论/摘要24篇。获得了多项杰出奖项,包括ICAR拉菲阿哈默德基德瓦伊奖和团队研究奖、农业研究领导奖,获得粮农组织在美国加利福尼亚大学戴维斯分校设立的培训奖学金等。

马亨德拉·帕尔·亚达夫　萨达尔·瓦拉巴伊·帕特尔农业技术大学副校长,病毒学领域的著名科学家之一。主要研究贡献包括研制马流感、传染性喉气管炎和家禽大肠杆菌病本地疫苗,分离和鉴定动物病毒,以及开发多种动物疾病诊断试剂盒。担任过病毒学教授(1981—1982),在IVRI Mukteswar任教授和病毒司司长(1982—1987),动物卫生组织首席科学家和负责人(1987—1993),印度马研究中心主任(1993—2000)。任IVRI副所长、所长(2000—2006),后任萨达尔·瓦拉巴伊·帕特尔农业技术大学副校长(2006—2009)。

获得了许多奖项,包括总理勋章(1996);印度61骑兵兰斯勋章(1996);ICAR特别奖(1998);世界动物卫生组织国际荣誉奖(2000);马利卡IAAVR奖(2001);IAAVR杰出兽医奖(2002),以及印度病毒学协会颁发的卓越病毒学奖(2013)等。在兽医病毒学、细菌学、传染病学、疫苗学、生物技术和马传染病等领域都有研究。

编者名单

Alexander Malogolovkin Federal Research Center for Virology and Microbiology, Pokrov, Russia London School of Hygiene and Tropical Medicine, London, UK

Alexey Sereda Federal Research Center for Virology and Microbiology, Pokrov, Russia

Alyssa Kleymann Department of Biomedical Sciences, Ross University School of Veterinary Medicine, Basseterre, St. Kitts and Nevis, West Indies

Amrita Haikerwal Center for Advanced Research (CFAR), King George's Medical University (KGMU), Lucknow, India

Anamika Mishra ICAR-National Institute of High Security Animal Diseases,Bhopal, Madhya Pradesh, India

Anastasia N. Vlasova Food Animal Health Research Program, Department of Veterinary Preventive Medicine, CFAES, Ohio Agricultural Research and Development Center, College of Veterinary Medicine, The Ohio State University, Wooster, OH, USA

Ashwin Ashok Raut ICAR-National Institute of High Security Animal Diseases, Bhopal, Madhya Pradesh, India

Atul Pateriya ICAR-National Institute of High Security Animal Diseases, Bhopal, Madhya Pradesh, India

B. C. Bera ICAR-National Research Centre on Equines, Hisar, Haryana, India

B. N. Tripathi ICAR-National Research Centre on Equines, Hisar, Haryana, India

Balamurugan Vinayagamurthy ICAR-National Institute of Veterinary Epidemiology and Disease Information (ICAR-NIVEDI), Bengaluru, Karnataka, India Editors and Contributors

D. Senthil Kumar ICAR-National Institute of High Security Animal Diseases, Bhopal, Madhya Pradesh, India

Denis Kolbasov Federal Research Center for Virology and Microbiology, Pokrov, Russia

Dilip K. Sarma Veterinary Microbiology, Assam Agricultural University, Guwahati, Assam, India

Dinesh Kumar School of Bioengineering & Food Technology, Shoolini University of Biotechnology and Management Sciences, Solan, Himachal Pradesh, India

Diwakar D. Kulkarni ICAR-National Institute of High Security Animal Diseases, Bhopal, Madhya Pradesh, India

F. Singh ICAR-National Institute of High Security Animal Diseases, Bhopal,Madhya Pradesh, India

Fun –In Wang School of Veterinary Medicine, National Taiwan University, Taipei, Taiwan

Gheyath K. Nasrallah Department of Biomedical Science, Biomedical Research Center, College of Health Sciences, QA Health, Qatar University, Doha, Qatar

Govindaraj Gurrappa Naidu ICAR-National Institute of Veterinary Epidemiology and Disease Information (ICAR-NIVEDI), Bengaluru, Karnataka, India

Harsh Kumar School of Bioengineering & Food Technology, Shoolini University of Biotechnology and Management Sciences, Solan, Himachal Pradesh, India

Houssam Attoui Institut National de la Recherche Agronomique (INRA), National Institute for Agricultural Research, UMR1161 Virologie, INRA-ANSES-ENVA,Paris, France

K. Jung Food Animal Health Research Program, Department of Veterinary Preventive Medicine, Ohio Agricultural Research and Development Center, College of Food, Agricultural and Environmental Sciences, The Ohio State University, Wooster, OH, USA

Kuldeep Dhama ICAR-Indian Veterinary Research Institute, Izatnagar, Uttar Pradesh, India

Linda J. Saif Food Animal Health Research Program, Department of Veterinary Preventive Medicine, CFAES, Ohio Agricultural Research and Development Center, College of Veterinary Medicine, The Ohio State University, Wooster, OH, USA

Mahendra Pal Yadav Sardar Vallabh Bhai Patel University of Agriculture and Technology, Meerut, Uttar Pradesh, India; ICAR-Indian Veterinary Research Institute, Izatnagar, Uttar Pradesh, India

Manjunatha N. Belaganahalli One Health Institute, School of Veterinary Medicine,University of California, Davis, CA, USA

Nadia Touil Faculté de Médecine et de Pharmacie, Centre de Gé nomique des Pathologies Humaines (GENOPATH), Université Mohamed V, Rabat, Morocco; Laboratoire de Biosé curité et de Recherche, H?pital Militaire d'Instruction Med V de Rabat, Rabat, Morocco

Narender S. Maan Lala Lajpat Rai University of Veterinary and Animal Sciences, Hisar, Haryana, India

Nassim Kamar Department of Nephrology, Dialysis an Organ Transplantation, CHU Rangueil, INSERM U1043, IFR - BMT, University Paul Sabatier, Toulouse, France

Naveen Kumar ICAR-National Institute of High Security Animal Diseases,Bhopal, Madhya Pradesh, India

Nitin Virmani ICAR-National Research Centre on Equines, Hisar, Haryana, India

Nobumichi Kobayashi Department of Hygiene, Sapporo Medical University, Sapporo, Hokkaido, Japan

Parimal Roy Indian Council of Agricultural Research-National Institute of Veterinary Epidemiology and Disease Informatics (ICAR-NIVEDI), Bengaluru, Karnataka, India

Peter P. C. Mertens School of Veterinary Medicine and Science, University of Nottingham, Loughborough, UK; The Pirbright Institute, Woking, Surrey, UK

Pradeep Gandhale ICAR-National Institute of High Security Animal Diseases,Bhopal, Madhya Pradesh, India

Q. Wang Food Animal Health Research Program, Department of Veterinary Preventive Medicine, Ohio Agricultural Research and Development Center, College of Food, Agricultural and Environmental Sciences, The Ohio State University, Wooster, OH, USA;

Raj Kumar Singh ICAR-Indian Veterinary Research Institute, Izatnagar, Uttar Pradesh, India

S. Pavulraj ICAR-National Research Centre on Equines, Hisar, Haryana, India

S. B. Sudhakar ICAR-National Institute of High Security Animal Diseases,Bhopal, Madhya Pradesh, India

S. N. Langel Food Animal Health Research Program, Department of Veterinary Preventive Medicine, Ohio Agricultural Research and Development Center, College of Food, Agricultural and Environmental Sciences, The Ohio State University, Wooster, OH, USA

Shailendra K. Saxena Center for Advanced Research (CFAR), King George's Medical University (KGMU), Lucknow, India; CSIR-Centre for Cellular and Molecular Biology, Hyderabad, India

Souvik Ghosh Department of Biomedical Sciences, One Health Center for Zoonoses and Tropical Veterinary Medicine, Ross University School of Veterinary Medicine,Basseterre, St. Kitts and Nevis, West Indies

Sudipta Bhat ICAR-Indian Veterinary Research Institute, Izatnagar, Uttar Pradesh,India

Sushila Maan Lala Lajpat Rai University of Veterinary and Animal Sciences, Hisar, Haryana, India

Swatantra Kumar Department of Center for Advanced Research (CFAR), King George's Medical University (KGMU), Lucknow, India

Taruna Anand ICAR-National Research Centre on Equines, Hisar, Haryana, India

Tridib Kumar Rajkhowa Department of Veterinary Pathology, College of Veterinary Sciences and Animal Husbandry, Central Agricultural University, Aizawl, Mizoram, India

V. P. Singh ICAR-National Institute of High Security Animal Diseases, Bhopal, Madhya Pradesh, India

Yashpal Singh Malik ICAR-Indian Veterinary Research Institute, Izatnagar, Uttar Pradesh, India

项目资助

本书的出版获得中国医学科学院人兽共患病毒病防控关键技术研究创新单元(项目编号:2020-12M-5-001)和锦州医科大学人兽共患病防控协同创新中心(项目编号:182231506003)的资助。

中文版前言

在过去几十年里，一些新发传染病和跨界动物传染病呈现出世界范围内的流行和大流行趋势,其发病率和死亡率一直呈上升趋势。新发传染病和跨界传染病的病原体在自然界中可能一直存在,但由于各种原因,使其成为危害生态的新发病原体。这些病原体可在动物和人之间,野生动物、人类和家畜之间,或野生动物、家畜与人之间传播。纵观新发动物传染病和跨界动物传染病的发展过程，大部分新发传染病和跨界动物传染病都属于人兽共患病,其流行和大流行直接影响全球化与贸易、公共卫生、社会福利,可能危害生态,甚至被恐怖分子利用或用作生物战直接对国家和国际安全构成严重的威胁等。目前时有发生的裂谷热、西尼罗河热、SARS、亨德拉、H5N1 型禽流感、尼帕、寨卡和 H1N1 甲型猪流感等病毒性传染病,尤其是新冠病毒在全球范围内的大流行,更凸显出新发传染病和跨界动物传染病已对全球动物和人类公共卫生安全构成威胁。

正是基于新发和跨界动物传染病对人类和动物的重大威胁,来自世界各地的 50 多位世界知名病毒学领域专家、年轻学者共同撰写了这本书,这是近年来动物病毒学领域的一部经典名著,在国际上享有很高的学术声誉。本书共分为 15 章,内容涵盖了新发和跨界动物病毒对公共卫生的影响及一些重要的新发和跨界动物疫病，以及近年来在控制和管理方面取得的经验,包括非洲猪瘟病毒、猪瘟病毒、冠状病毒、细环病毒、捷申病毒、黄病毒、环状病毒、马流感病毒、施马伦贝格病毒、克里米亚-刚果出血热病毒、猪繁殖与呼吸综合征病毒、小反刍兽疫病毒、萨佩罗病毒和戊型肝炎病毒等,讨论了相关的控制策略,总结了诊断学的最新研究进展,介绍了常规疫苗和重组疫苗,以及近年来用于新型疫苗研发的分子生物学操作手段。

该书为新发和跨界动物疫病提供了精准的和最新的信息,内容全面、系统,对兽医专业人员、临床医生、公共卫生专家、研究人员、学生/学者、动物生产者等读者均具有重要的参考价值。引进出版该书对于推动我国动物病毒学的进一步发展,并与国际先进水平接轨具有非常重要的意义,可以提高相关从业人员的实践能力和理论水平,更好地服务公共卫生事业。

该书翻译过程中,恰逢新冠疫情,译者们克服了重重困难,历时两载有余,终于成稿。但由于水平有限,文字表述和理解上还存在一些疏漏或不够精准之处,译文中难免会有瑕疵和不妥,敬请读者给予批评指正。

前言

畜牧业是许多国家的经济支柱,新发动物病毒性疫病的出现和跨界(跨越国界)动物病毒性疫病对畜牧业及全球粮食和营养安全构成了严重威胁。在过去的几十年里,一些动物和(或)人类传染病的流行和大流行直接影响了经济、社会福利和公共卫生,凸显了新发疫病防控的重要性。目前世界多个地区都出现了新发疫病和其他新的病原体。此外,新发病毒性疫病/感染的出现,如裂谷热病毒、西尼罗河热病毒、SARS冠状病毒、亨德拉病毒、甲型禽流感(H5N1)病毒、尼帕病毒、寨卡病毒和甲型猪流感(H1N1)病毒,已对全球动物和公共卫生构成了威胁。因此,新发、再发和跨界病毒感染已成为研究人员和公共卫生工作者的关注热点。病毒性疫病增加的主要原因是缺乏安全、廉价的预防和治疗手段。此外,对病毒性疫病的控制及管理仍然具有挑战性,需要具备专业的检测和鉴别病原体的能力,如研发快速、灵敏、性价比高的检测试纸/试剂盒;建立区域和周边诊断实验室;对易感动物和接触动物进行疫病临床和血清监测;跨界疫病的边境管制;隔离设施、虫媒控制和限制动物离开疫区;以及其他有关的一般健康管制措施,例如,动物尸体处理、动物卫生措施及管理措施。

目前,有许多关于动物病原体研究的文章/新闻。然而,有关新出现、再出现和跨界动物病毒的图书较少。这方面图书对于研究者了解该领域的最新知识和发展趋势会有所帮助和参考。书中选用了几个例子来说明病毒性疫病对农业和畜牧业经济产生的危害,强调了动物病毒性疫病防控的重要性。本书针对最近已经出现或重新出现的动物病毒性疫病提供了精准的和最新的数据,这些疫病的出现是由于复杂的环境因素相互作用和那些不受国界限制、具有跨界特点的动物性病毒。最后,有数章内容给出了当今全球范围内关于新出现、再出现和跨界动物病毒的信息,重点介绍了分子水平的最新工具,特别是诊断、预防和治疗方面的新进展。

本书由来自世界各地的50多位作者共同编写,论述了影响经济、公共卫生的重要动物病毒或病毒性疫病。此外,书中重要的临床数据都以表格形式列出,各章末还附有参考文献,可供读者深入学习。本书共15章,对重要的动物病毒进行了介绍。第1章主要强调了新发疫病对公共卫生的重要性。本章综述了近年来在控制和管理新发疫病和跨界动物疫病方面取得的经验,以及成功开发更好的控制方法的限制因素、局限性以及未来的研究需要。第2章介绍了非洲猪瘟病毒的流行病学、免疫病理生物学和诊断方法,并简要概述了病毒疫苗研发的最新进展。第3章对典型的猪瘟病毒也进行了类似的概述。第4章讨论了猪传染性胃肠炎病毒、猪腹泻病毒和猪δ冠状病毒的研究进展。第5章对细环病毒进行了概述。第6章对猪捷申病毒进行了概述。

黄病毒在全球范围内都极易传播。第 7 章概述了动物在黄病毒生命周期中的重要作用,阐明了猪是流行性乙型脑炎病毒的扩增宿主,候鸟是西尼罗病毒的宿主。虫媒病毒是呼肠孤病毒科中最大的一个属,第 8 章重点讨论了蓝舌病病毒,阐明了虫媒病毒的分类关系、流行病学、复制机制和进化过程。第 9 章详细阐述了马流感病毒的全球现状。第 10 章对施马伦贝格病毒进行了概述。第 11 章论述了一种蜱传病毒,即克里米亚-刚果出血热病毒,其被归类为生物安全 4 级。第 12 章详细描述了对经济有重要影响的猪疫病病原——猪繁殖与呼吸综合征病毒。

第 13 章对小反刍兽疫病毒进行了概述。小反刍兽疫病毒是一种小型反刍动物麻疹病毒,属于副黏病毒科。小反刍兽疫病毒对家养和野生小反刍动物具有高度传染性,属于世界动物卫生组织报告传染病和具有重要经济意义的跨界动物病毒性疫病,称为"小反刍动物瘟疫"。第 14 章对萨佩罗病毒进行了概述。最后一章概述了戊型肝炎病毒的相关成果。由于恶劣的卫生条件和低质量的饮用水,戊型肝炎在发达国家和发展中国家都是非常突出的问题。

我们相信,基于专家们对重要动物病毒的深入研究和精心整理,本书将成为读者极好的信息来源。本书可供以下人员参考阅读:兽医专业人员,临床医生,公共卫生专家,动物养殖人员,病毒学、病毒性疫病、病毒性人兽共患病及其管理专业的师生,以及医药行业和生物医学专家。

首先感谢所有的编者,感谢他们的支持和辛勤工作,使这本书顺利出版。还要特别感谢所有审稿人,感谢他们的专业能力和对本书的严格审读。感谢 Springer Nature 为本书的出版提供平台,特别感谢 Springer Nature 生物医学图书副主编 Bhavik Sawhney 博士,感谢所有编辑的帮助和密切合作,最终使得该书顺利出版。

谨以此书献给所有杰出的病毒学专家，他们不仅具有发明力和创造力，而且还有着惊人的热情。感谢他们在编写过程中给予的莫大帮助，包括动物病毒病原体/疾病的重要方面，这些病原体/疾病都具有新发和跨界传播的特性。

目 录

第 1 章 新发和跨界动物病毒性疫病：展望与防控

Mahendra Pal Yadav, Raj Kumar Singh, Yashpal Singh Malik

1.1 引言

疾病是指动植物和人类机体所处的一种状态或某一正常生理功能受到损伤，通常表现出一些临床症状。动物和人类疾病可以根据不同的标准被划分为不同种类，例如，传染病与非传染病、感染性疾病与非感染性疾病、动物传染病与非动物传染病、急性病与慢性病等。传染病主要由病毒、细菌、支原体、真菌和立克次体等引起。在所有传染病中，病毒性传染病危害最为严重且最难控制，这主要是因为其传播速度快，并且不像其他传染病那样具有经济有效的抗病毒制剂。新发传染病是指在某一地区或某一宿主新出现的传染病，并迅速传播，但缺乏防控意识，缺少有效的诊断方法、经验或其他的因素，又被称为新发疫病。曾经发生过的疫病在具有易感宿主的地区再次出现，被称为再发疫病（Lederberg 等，1992；Daszak 等，2000），例如，结核，在许多国家由于采取有效措施，检测已为阴性，但是由于人类免疫缺陷病毒（human immunodeficiency virus, HIV）的感染蔓延，结核病会再次流行。印度和一些发展中国家，为了改善当地动物品种，从国外引进易感动物种群/品种进行杂交，而这种行为导致了动物寄生原虫感染再次出现，例如，进口家畜或其杂交品种中的巴贝斯虫和无浆虫。这些从国外进入的疫病，称为“外来病”。

传染病因其流行病学和病原学特点而具有动态性，因而令人担忧，主要表现为宿主、病原体和环境因素形成统一体，可感染家畜、野生动物和人类。宿主与环境间错综复杂的关系为疫病的发生创造了条件。新发疫病的“溢出”可从“家畜传染到野生动物”“野生动物传染到人类”或“家畜传染到人类”等各类跨界传播。促使不同新发疫病（emerging infectious disease, EID）跨界传播的因素包括：全球性旅行、城市化和对人类的生物医学操作导致人类 EID 的发生；农业集约化导致家畜 EID 的发生；野生动物迁徙导致野生动物 EID 的发生；人类对野生动物的迁徙地的入侵、接触与生态控制导致野生动物-人类共患 EID 的发生；侵占、新引进种群、“溢出”“回流”，以及科技和工业促进了家畜-人类共患 EID 发生。

1.2 新发疫病

“新发疫病”一词用来指生物群体中疫病动态的变化。新发传染病是指疫病最近传染给新的宿主，或发病率升高或传播范围扩大，或由病原体进化引起发病（Lederberg 等，1992；Daszak 等，2000）。这一定义涵盖了对医疗和兽医公共卫生构成重大威胁的一系列人和动物传染病。在世界动物卫生组织（World Organization for Animal Health, WOAH）列出的病毒性传染病中，牛瘟、小反刍兽疫（peste-des-petits ruminant, PPR）、口蹄疫（foot-and-mouth disease, FMD）、非洲猪瘟（African swine fever, ASF）、结节性皮肤病和裂谷热（rift valley fever, RVF）都已经发生变化。其中，牛瘟已经被彻底消灭了。从 20 世纪 90 年代到 2011 年，世界粮农组织（Food and Agricultural Organization, FAO）、WOAH、欧洲联盟和国际原子能机构采取了一系列措施，并进行指导和协调，使牛瘟得到了有效控制，这些措施主要包括泛非牛瘟运动、印度全国牛瘟根除计划和全国消灭牛瘟计划（Yadav，2011）、FAO 的全球消灭牛瘟计划，以及有该病流行国家的政府采取的措施。通过采取这些有效的措施，最终在 2011 年 6 月 28 日，FAO 具有历史意义地宣布全球消灭牛瘟。

1.3 跨界疫病

“外来疫病”和“跨界动物疫病”（transboundary animal disease, TAD）这两个词经常可互换使用。虽然所有 TAD 都是外来的，但并不是所有外来疫病都被列入 TAD 清单。TAD 是一种具有高度传染性和传播性的家畜流行病，不受国界限制，迅速传播到新的地区和区域，并造成严重的社会经济影响和危害公共卫生安全。几乎所有疫病都会对家畜、家禽、鱼类和其他动物，以及食品和其他产品的质量和数量造成不利影响，例如，兽皮、骨、纤维、羊毛和用于耕作、运输和牵拉的畜力。TAD 直接导致动物产品、生产力和收益下降，进而影响了人类的生活。在当前全球化快速增长的形势下，TAD 对经济和公众安全与健康构成了严重威胁，并导致相关产品和生产力大幅下降；扰乱了贸易、旅游、地方和国家经济；而且由于食品品质差和人兽共患病/感染而威胁着人类健康。因此，TAD 不仅对某个国家的经济和公共健康产生严重的不良影响，而且也会对全世界的经济和公共健康产生重大的不利影响。

1.3.1　EID 和 TAD 的病原体起源

EID 和 TAD 的病原体在自然界中可能一直存在，但由于某些因素发生改变，为疫病的发生创造了机会。这些病原体可在动物和人之间，野生动物、人类和家畜之间，或野生动物、家畜与人之间传播。然而，病原体在自然界中持续存在和传播主要是由人兽共患病的聚集、"溢出"和"回流"机制决定的。

1.3.2　跨界疫病的潜在威胁

由于 TAD 造成的重大经济影响、人兽共患的特性，以及在未来新发 TAD 的威胁日趋增加，TAD 对国家安全带来的风险已成为人们关注的焦点。近年来频频报道 TAD 中具有人兽共患特性的许多传染病对动物和人类健康造成了重大影响，例如，高致病性禽流感(highly pathogenic avian influenza，HPAI)、疯牛病、西尼罗热、裂谷热、严重急性呼吸综合征、亨德拉、尼帕、埃博拉、寨卡和克里米亚–刚果出血热等(Malik 和 Dhama 2015；Munjal 等，2017；Singh 等，2017，2019)。根据 FAO 的评估，近 1/3 的世界肉类贸易因为疯牛病、高致病性禽流感和其他动物疫病而被禁止。有证据表明，近年来自 TAD 的威胁有所增加。随着发展中国家人民收入的提高，对动物蛋白和产品(牛奶、肉、蛋、鸡和鱼)的需求增加，将来暴发动物疫病的风险也可能会进一步增大。为了满足人们对肉食的需求，动物的饲养量正在迅速增长。20 世纪 90 年代，东亚的家禽产量每年以 12%的速度增长，5~6 年就翻一番。类似于 TAD 发生，人类中出现了新的病毒性传染病，例如，埃博拉、严重急性呼吸综合征、寨卡病毒病、克里米亚–刚果出血热、尼帕病毒病和牛海绵状脑病，以及已有传染病表现出新的抗原性或新的生物型，例如，在欧洲家禽中出现的传染性法氏囊病强毒毒株，在美国出现的鸡新城疫强毒株(Riemenschneider，2005；Singh 等，2017)。通过虫媒传播的病原体，例如，蓝舌病病毒(bluetongue virus, BTV)、非洲马瘟病毒(African horse sickness virus, AHSV)、裂谷热病毒和西尼罗河热病毒，有以流行形式传播的可能。对于医学研究所提出的有关 TAD 控制的建议报告(Anonymous，2003)，Riemenschneider (2005) 已进行了评议。下文将简要讨论可能导致 TAD 威胁增加的一些问题。

1.3.2.1　全球化与贸易

当今世界，动物源性食品的数量增加，贸易速度加快，新的贸易路线和航空运输均增加了新的感染和疫病发生的风险。现在 24 小时内可以到达世界任何地方，这比大多数传染病的潜伏期还要短，因此，携带病原体的动物或人因缺乏临床症状而不能被及时发现。与加工食品相比，在贸易中新鲜商品的快速增长，更有可能把病原体带到世界上更远的地方。

1.3.2.2　满足不断增长动物蛋白需求和集约化动物生产体系

近几十年，为满足人们通过肉类和肉类产品、牛奶和奶制品、鸡蛋、鱼和鱼产品获得大量动物蛋白和其他营养物质的需求，采取了集约化生产体系和动物高密度饲养，增加了发展中国家和其他地方的收入。为了提高动物源性食品产量,常常在拥有大量人口且饲养方式比较落后的城市周边地区进行动物饲养。在这样的高产地区,疫病会以更快的速度暴发,影响更多的动物,造成重大的经济损失。

1.3.2.3 森林生态改变带来的影响

由于森林滥伐和在热带雨林放牧，使家畜暴露在森林的生态环境中，导致一些以前只在野生动物生态体系中传播的病原体和传播媒介进入或接触家畜,成为全新的病原体和传播媒介。家畜对这些病原体完全易感且无免疫力，同时针对这些新病原体缺乏特异的诊断试剂盒和疫苗，使得疫病传播更加迅速和严重,造成非常高的发病率、死亡率,以及严格的贸易管制和经济损失。

1.3.2.4 气候变化、全球变暖以及微生物进化

气候变化和全球变暖似乎正在改变降雨量和天气模式。北半球气温升高可能会引起蓝舌病、非洲马瘟(African horse sickness, AHS)、裂谷热和类似虫媒疫病的昆虫传播媒介的分布。BTV 有 27 个血清型,在世界多地都有发生。然而,直到最近欧洲才有报道。由于近几十年来欧洲的天气变得越来越热,血清 1 型 BTV 传入了欧洲。该血清 8 型 BTV 毒株与尼日利亚毒株亲缘关系最近。普遍认为，病原体入侵可能是由于进口被感染的动物园动物或被感染的蠓。由于非洲气候变化,在东非和西非都暴发了裂谷热。

1.3.2.5 TAD 严重威胁国家安全

上述讨论的许多因素都使得 TAD 对国家和国际安全构成严重的威胁。通常发展中国家受害最为严重。在其他因素中,发展中国家的兽医公共卫生服务通常远远落后于医疗公共卫生服务。此外,与人类疾病报告不同,动物疫病报告体系通常是被动报告而不是主动对疫病实施监测。其他一些因素也导致 TAD 带来更大的威胁:①农民对家畜流行病的严重威胁缺乏认识；②对外来疫病缺乏诊断设施,以及由于担心失去国内外市场,对类似 HPAI 的动物疫病少报，以期本国达到 WOAH 制定的无感染状态标准;③赔偿方案不健全。

1.4 新型病毒性疫病暴发的处置

当一种外来的病毒性疫病首次传入一个国家时，初期可能影响一种动物、少数动物或大量动物。为遏制疫情而采取的策略取决于病毒的性质、传播速度、媒介的作用、风险评估、沟通和管理、响应时间,以及国家关于疫病防控的法规。因此,有必要在个案的基础上制定对外来病和 TAD 的预防和控制战略计划。在印度,这类病毒性疫病包括 ASF、传染性胃肠炎(transmissible gastroenteritis, TGE),猪水疱病,裂谷热,AHS,西尼罗河热,东方马脑脊髓炎,西方马脑脊髓炎和委内瑞拉马脑脊髓炎,C 型、SAT 1 型、SAT 2 型和 SAT 3 型口蹄疫,尼帕,亨德拉,严重急性呼吸综合征和朊病毒病——牛海绵状脑病和痒病。

1.5 抗病毒感染的生物安全和生物安保措施

及时采取适当的生物安保措施是保护和改善动物健康的重要手段。在家畜、家禽和鱼类的管理中,要避免因无知和疏忽而违反生物安保,要及时采取生物安全和生物安保措施以最大限度地降低传染病(包括 EID 和 TAD)带来的风险。引起人兽共患病和动物其他传染病高发的重要原因是在家畜管理中违反生物安保，尤其是畜禽病毒性传染病更为突出。野生动物、动物和人类之间的密切接触，以及与人类密切相关的畜禽饲养均可促进病毒性传染病和其他传染病的传播，这些传染病有可能对世界各地的人口健康、经济和粮食安全构成潜在的威胁。对动物健康和公共卫生、国民经济和全球粮食和营养安全构成重大威胁的人兽共患

病的典型例子是新出现的病毒性传染病/感染，例如，裂谷热、西尼罗河热、严重急性呼吸综合征、亨德拉、禽流感(H5N1)、尼帕、寨卡和猪流感(H1N1)。由于生物安保方面存在漏洞，类似于口蹄疫的一些病毒性传染病在几十年都没有报告的国家的情况（包括英国在内的一些发达国家）也再次发生。

1.5.1　生物安全和生物安保

生物安全和生物安保是相互关联的术语，但在不同的语境中使用。世界卫生组织(World Health Organization，WHO)、FAO 和 WOAH 制定了指南。生物安全是指工作人员和处理生物制剂的人员的防护及防护设备，防止他们暴露在病原体中，意外接触病原体/毒素或意外释放。因此，生物安全是指用于防止个人、实验室和环境暴露于潜在疫病或生物危害所需的相关知识、技术和设备。不同于生物安全，生物安保在不同语境中有着不同的含义。生物安保包括防止微生物的相关用品溢出、被盗、丢失、转移或故意从实验室释放，防止某些生物体/毒素的进口。生物安保是一套预防措施，旨在降低故意传播传染病的风险，保障含有敏感生物材料和具有生物武器潜力的设施。简言之，生物安保就是生物风险管理。一旦一种疫病在全球范围内被根除，就由 WHO、FAO 和 WOAH 等国际机构根据本领域专家的建议，决定是否保留该病毒的野生毒株和疫苗毒株、是否储备备用疫苗及随后销毁疫苗库存的政策。

1.5.2　生物安保政策、措施和行动计划

生物安保政策、措施和行动计划包括风险评估、沟通和管理，在海港、旱港和农场进口动物的检疫；在国际和洲际边界设立检查站和防疫站，进行临床监测；在国际边境建立免疫带，规划和实施结构化疫病监测，包括临床监测和血清学监测。从农场到国家和国际各级都需要遵守生物研究安全和生物安全。处理最危险的 TAD 病原体要求必须在生物安全 3 级和生物安全 4 级实验室进行，以确保生物研究安全、生物安全和生物防护。为有效地管控 EID 和 TAD，需要严格执行合理的动物园卫生措施，例如，检疫啮齿动物和病媒控制，动物圈舍及其周边消毒，粪便、尿液、饲料及饲料废料妥善处置等。

1.5.2.1　国际边境的生物安全

每个国家都需要在其国际边境建立严格和万无一失的生物安全机制，以防止外来病原体/疫病随着家畜和其他动物及其产品的进口而传入。对于有可能受到 TAD 影响的国家，就有必要制定区域性生物安全计划以确保本区域的生物安全。如果周边接壤国家缺乏有效的生物安全措施，就不可能保证本国的生物安全。

不同的国家都面临着许多 TAD 的风险，包括炭疽热、鼠疫、马鼻疽、莱姆病、传染性马子宫炎、马流产沙门菌病、HPAI、FMD(SAT 1~3 型)、拉沙热、狂犬病、亨德拉和尼帕、西尼罗河、高致病性新城疫、兔出血病、牛海绵状脑病、AHS、马脑脊髓炎、马传染性贫血、鸡传染性贫血、马流行性感冒、水疱性口炎、裂谷热、恶性卡他热，以及其他绵羊、山羊和鹿等传染性贫血。必须采取生物安全措施，防止这些疫病通过国际贸易传入。WOAH 通过制定有效的标准来防止动物疫病在全球的传播，促进了动物和动物产品的安全贸易。为预防病原体在国内和国际间传播，需要制定与国际标准一致的生物安全措施。为检查来自境外的病毒和其他的病原体，必须在海港、机场和监管有疏漏的国际陆地边界设立充足的基础设施，包括检查站和检疫设施。为确保进口的畜禽产品无病，应配备训练有素的人力资源、完善的仪器和现场诊断检测/试剂盒。

1.5.2.2　国家层面的生物安全措施

国家层面的生物安全措施由“外部生物安全”和“内部生物安全”两部分组成。“外部生物安全”

指防止外来动物疫病及TAD,“内部生物安全”指国内区域的生物安全、片区的生物安全,以及农场级别的生物安全。印度已经制定了动物跨州流动的规定,但需要严格执行。现代检测体系可用于动物和动物产品的识别和跟踪，提供有关动物来源的信息,以及用于生产和食品安全的环境保护。

1.5.2.3 农业生物安全

为了有效预防和控制疫病，必须将生物安全融入农场的每一个操作环节。农场生物安全应包括“生物排除”(防止病原体进入畜群的措施)和“生物遏制”。后者主要针对病原体已传入农场,并在易感动物群中传播或进一步向其他农场扩散的情况进行处理。

农场层面执行严格的生物安全措施对预防疫病的扩散具有至关重要的作用。诸如农场的位置和布局、动物卫生措施的实施和农场的日常管理等重要问题,要制订切实可行的计划,灵活运用新的知识、概念和技术。针对不同的家畜品种及其生产体系,对于具体的感染风险或一般的疫病预防,都建议采取广泛的生物安全措施。针对牛、羊、猪、家禽和鱼类生产体系采用的生物安全措施已给出建议。

适用于所有物种和农场的一般生物安全措施和干预措施包括：

1. 保持畜群封闭,外购动物来源要清晰。

2. 尽量减少购买/转移/交换的动物数目,以及引进动物的畜群数量。

3. 避免从市场或经销商处购买动物。

4. 对于农场所引进或重新引进的动物,要进行适当的隔离和检测。

5. 不鼓励租用公牛或种马,并在繁殖季节后又将其归还等农业行为。

6. 避免引入健康状况不确定的生物材料。

7. 所有新引进动物都要有健康和预防接种记录。

新引进动物应在独立隔离设施中进行隔离/检疫2~8周，隔离期间应观察动物的发病/症状，并对重要疫病进行筛选，之后再与农场的其他家畜混群。建议在检疫隔离期间对重要传染病进行采样、实验室检测。还可以在解除隔离前2周内，接种本地区流行的疫病疫苗,从而提高免疫保护。

动物疫病可以从一个农场传播到另一个农场,导致动物发病、死亡,带来经济损失。参观禽畜饲养场、疫病实验室、禽鸟、啮齿动物、车辆、饲料和其他无生命体时,参观人员往往也是传染源。动物疫病除了对经济产生不利影响外，对环境和人类健康也会产生负面影响。最好的方法是实施有效的生物安全措施。实验室工作人员进入动物饲养场或实验室前后要求洗澡,进入后要穿胶靴、一次性工作服,戴头套和手套。实验室所有废水排放前都应该进行预处理，避免病原体对环境造成污染。进入农场的大门处要设有消毒池,供进出人员和车辆消毒，实验室人员和农场工人的衣服和鞋子必须单独存放。在疫情暴发期间,最大限度减少农场内的人员和动物流动，并且确保个人的健康和卫生是加强农业生物安全的最低准则。

1.6 早期疫病报告和补偿在动物健康管理中的关键作用

利用WOAH批准的诊断检测，进行及时、快速和准确的疫病报告是有效检测病原体和建立早期反应机制的必要手段,可避免疫病进一步传播。针对新的外来病，建议制订与技术指南配套的本地的标准操作规范、决策水平,并提供足够的资金和法律支持。据观察,对于扑杀病禽和接触病禽补偿的缺乏或不足,以及对销售、销售价格和出口的负面影响,使农场主未能及时报告动物疫病,最终导致了TAD的传播。鉴于对贸易的考虑,并不鼓励对高致病性禽流感和口蹄疫等进行预防接种。非疫病国家一般不愿意从实施动物预防接种的国家进口动物或动物产品。从公共卫生的角度来看,疫苗接种计划可能会控制一个国家病毒的传播,从而降低高致病性禽流感传播给人类的风险。然而，大多数国家更倾向于采取根除政策而不是接种疫苗,以便尽早宣布本国无疫病/感染,恢复出口。为了确保畜牧养殖户的配合,必须提供充足和及时的补偿,以减少他们因扑杀病禽和农场关闭期间遭受的损失。

未能及时报告疾病加速了 TAD 在国内和国际间的传播。据报道,在 1996 年向国际当局报告 HPAI(H5N1)禽流感病毒导致疾病/感染广泛传播之前,该病毒可能已经在受影响地区的家禽中传播了数月。一项对 2001 年英国暴发的口蹄疫疫情的事前研究表明,口蹄疫病毒(foot-and-mouth disease virus, FMDV)是在报告之前 3 周引入的,随后禁止牲畜流动,早期报告和禁止牲畜流动可使疾病的传播减少约 40%。

1.7　EID 和 TAD 控制

EID 和 TAD 对动物和人类健康、世界经济和全球环境福祉构成的威胁日趋严重。然而,很难预测疫病发生的数量,这些疫病很可能在一个国家或地区迅速扩散,威胁动物和人类的生命健康及该地区或国家的经济。全球对肉类需求的不断增长,尤其是在东南亚地区,使人口和动物数量达到了前所未有的水平。事实上,肉类产量增长最快的国家,其农场生物安全往往很差。这种情况为动物疫病跨界传播创造了巨大的空间,带来人类健康问题。其中一些问题在国际移民组织的报告中有详细的描述,之后 Riemenschneider(2005)对这些问题做了进一步的分析和讨论。有关 EID 和 TAD 控制步骤的建议包括早发现和早响应,防控措施,政府机构协调,国际协作,对生态、微生物进化和病毒传播的了解,扩建监测系统,疫病监测预警,以及跨领域合作关系等。

1.7.1　防控措施

在发展中国家,动物疫病的防控措施往往不尽如人意。向受影响的养殖户提供适当补偿等激励措施,有助于促进养殖户及时报告动物疫病。通过模拟演习来评估防范水平,将有助于建立对突然出现并在短时间内大面积传播的 EID 和 TAD 进行快速检测和响应的信心。对动物疫病暴发启动的进口禁令必须基于可靠的科学证据,以确保有关国家主动向国际机构 WOAH、FAO 和 WHO 报告疫病。国家兽医服务体系不健全导致无法及早确诊疫病,以及对其调查及时响应,许多地区的诊断服务已经恶化。持续的结构性的国家兽医服务升级方案必须被充分考虑,以便从提供诊断、免疫接种和治疗患病动物等的服务转变为提供检测和质量保证的服务。疫病监测、早期预警,以及应急预案作为国家兽医服务的重要组成部分,需要积极地开展。

1.7.2　政府机构协调

虽然公共卫生和国家安全属于国家政府的范畴,但政府机构放权及改进国际协调,充分尊重地方、区域或国家边境的权利,对有效处理 TAD 的威胁具有重要的作用。然而,政府在行政层面上给予支持也至关重要,以确保充分和及时地反应,通过控制家畜流动,关闭活体畜禽市场,分享诊断服务、专门知识、资金等方式避免疫病传播。

1.7.3　国际协作

FAO、WHO 和 WOAH 等国际机构的技术支持和指导是制定和及时实施控制和管理 EID 和 TAD 计划和模式的关键。FAO 于 1994 年通过跨界动物疫病应急中心的运作及跨界动植物病虫害紧急预防系统,建立了全球跨界动物疫病防控框架,用于早期预警和应对疫病的威胁,遵循动物-人类生态系统界面调查的合作研究方法。已证明这些机制在控制、预防和消灭疫病方面具有巨大的帮助和作用。

1.7.4　生态、微生物进化和传播

微生物的进化,特别是病毒的进化,是一个连续的过程。因此,有必要对新出现的传染病进行基础研究,包括病毒和其他微生物来源的研究,以便为导致新微生物出现的因素提供新的视角。疫病的生态学、社会因素、病毒和微生物的传播和扩

散、人类和动物种群因迁移和其他因素引起的生态和人口变化等因素协同作用，导致新发感染的发生。这些病毒和微生物传播的信号应该被视为预警信号。生物多样性应包括微生物和病毒，环境影响评估包括在发展规划中考虑到的卫生方面。

1.7.5 监控系统的建立

提供足够的资金建立国家、区域和国际一级实验室反应网络来加强监测，这一点非常重要。通过将公共领域和私人领域的实验室联合起来，预期这种网络可增强各级检测能力和防止自然或故意传播 EID (Anonymous,2003)。WHO 利用全球 100 多个实验室建立了监测网络，实现了 WHO 对流感病毒的持续调查，这是实验室监测 EID 和 TAD 网络的最好例子之一。这些实验室应该有多学科团队，包括兽医、医生、生态学家、昆虫学家、疫苗学家、流行病学家、分子生物学家和免疫学家，可能还有其他专家。

1.7.6 监测与预警

先进的疫病监测需要具备能够预测特定疫病可能在何时、何地发生的能力，以便进行更有针对性的监测。这些可操作的情报可能来自对气候条件、植被、野生动物种群统计、贸易模式或传播媒介种群统计和分布分析(Anonymous, 2003)。

1.7.7 抗原基因型与疫苗更新

引起疫病的微生物，尤其是病毒和细菌微生物，经常由于自然突变和免疫压力作用而改变它们的抗原构成。免疫压力则是在宿主有疫苗抗体的情况下，病原微生物野生株在机体内持续存活时产生。RNA 病毒的基因组分节，更容易发生重组、基因缺失等，导致抗原发生改变。人类和动物的 A 型流感病毒通过在人类、鸟类、猪和包括马在内的其他物种中传播，各种血清型的血凝素(haemagglutinin, HA) 和神经氨酸酶(neuraminidase, NA)基因在传播过程中发生互换，不断进化出新的变异株。随着抗原构成的改变，目前的疫苗毒株不能对新型病毒产生保护。类似的情况也发生在 FMD 病毒中，它有 7 个型和很多的亚型、支系和基因型，还有 PPR 病毒和新城疫病毒等。病毒或病原体的新抗原型还可通过进口畜禽从国外输入。因此，有必要通过结合目前可产生强大而持久免疫力的疫苗毒制/菌株制定升级疫苗的计划。这需要从暴发的疫病中分离野毒，建立毒种储备库，特别是从疫苗免疫失败的病例中分离野毒。家禽、马和人的流感疫苗，以及家畜的口蹄疫和脑脊液疫苗都要定期进行更新。

1.8 TAD 的教训

1.8.1 接种疫苗的优点

在预防兽医学中，免疫接种是一种有价值和经过充分检验的有效预防动物疫病的方法，通过免疫接种可促进动物健康和安全，并降低人类接触人兽共患病原体的风险。预防性免疫接种的实践、原则，以及免疫程序显著地减少了许多威胁生命健康的病毒性和细菌性疫病的流行。利益攸关方缺乏认识，无法获得价格适中的诊断方法和疫苗，向畜牧养殖户提供的兽医服务不佳，以及对其家畜不进行免疫接种等风险可能对农村畜牧养殖户的生计造成严重后果。广泛实施有效的免疫接种规划，可能会减少对抗生素的需求，而抗生素需求减少反过来又有助于降低耐药性菌株出现的风险。世界兽医协会认为，全球兽医行业有必要教育公众，特别是动物饲养员和生产者，让他们了解动物和人类接种疫苗的好处。

动物免疫接种的主要目的和动机是保护、改善和促进同类和食用动物的健康和安全，通过符合成本的方式增加家畜生产效益，防止传染病通过动物源性食品、密切接触和其他传播途径，从家畜和野生动物传播给人类。这些目标促进了利用不同的方法开发兽用疫苗，从粗糙但有效的全病原体制剂到分子设计的亚单位疫苗、基因工程生物制剂或嵌合体、定向抗原制剂和动物免疫注射用的裸 DNA。

1.8.2　疫苗研究

研究和开发疫苗最成功的成果是生产出一种满足市场需求，并适合在生产实际中应用的产品，进而实现其预期的目的。针对主要细菌、病毒、原虫和多细胞病原体的兽用疫苗已经成功生产出来，同时已经在现场成功应用，并且采用了新的技术。这些兽用疫苗不仅对动物健康和生产具有重大的影响，而且还通过增加食品安全供应，即牛奶、肉、蛋和鱼的供应，以及防止动物向人类传播传染病，为人类健康带来有利影响。兽医和医疗的研究人员与健康专业人士之间的持续互动将成为应用新技术、提供人类疾病动物模型和应对新出现传染病的主要动力。目前市面上有 100 多种不同的兽用疫苗（Meeusen 等，2007）。

1.8.2.1　多价疫苗

与单价疫苗相比，多价（二价、三价和多价）疫苗应优先规划用于控制流行中的一种以上疫病，以节省资金、时间和其他费用，并减轻执行机构的负担，如兽医和家畜保健工作人员的负担。

1.8.2.2　疫苗反应监测

相关机构应建立一套适用于现场的疫苗接种后血清学监测系统，寻找充足的血清转阳证据，按标准操作程序随机收集样本，应用区分自然感染与疫苗免疫动物（differentiation between infected and vaccinated animal, DIVA）检测可鉴别疫苗接种诱导的免疫反应和野毒感染诱导的免疫反应。

1.8.2.3　贸易风险分析

有关活禽及其产品在国际贸易中的疾病传播风险，世界贸易组织的《实施卫生与植物卫生措施协建议》（简称《SPS 协议》）（sanitary and phytosanitary, SPS）推动了风险分析的应用，即进口风险分析。

WOAH 制定的进口风险分析标准及其相关指南符合《SPS 协议》的要求。决策树分析采用核心建模方法，代表风险分析的关键步骤。应为产品 IRA 制订决策树，以评估每一步的风险等级。

风险评估和建模之间的相关性表明，两者应属于同一子部件。贸易有关进口风险分析必须包含评估结果，评估结果要符合《SPS 协议》要求。流行病学和经济模型相结合有助于寻求更好的风险应对方法。矩阵结合风险、暴露量，以及可能产生的后果，用于定性进口风险分析，但这种方法存在缺陷，需要更好的方法。WOAH 制订的进口风险分析标准及其相关指南表明，应考虑到贸易量。一些已发表的定性进口风险分析仅假设了目前的贸易水平和模式，但没有具体说明贸易额，这就限制了进口风险分析的使用，导致无法确定缓解措施（将风险降低到可接受的水平），无法确定是否符合对等原则和 SPS 协议。定性进口风险分析是否能满足《SPS 协议》中规定的所有标准尚不确定。尽管如此，相关部门还是根据现有范畴制定了目前的标准和指南，以便更好地为基于科学的决策服务。

1.9　无疫病区

现在从无疫病区和无疫病隔离区进行贸易和安全商品贸易的选择已具备促进国际贸易的积极机制。在印度，FMD 防控计划已经开始实施，目的是建立无 FMD 疫区。也可以为其他疫病创建类似的区域，例如，蓝舌病、羊痘、山羊痘、PPR 和其他重要疫病。隔离区生物安全是一个新的概念，即通过单一的一套生物安全措施管理隔离区的生物安全。设立区域/隔离区将促进家畜和家禽产品的国际贸易。在印度，中央政府和联邦政府制定了有关动物在这些区域和隔离区内迁徙的法律。

1.10　经济支撑是控制和管理 TAD 的关键

1.10.1　跨界疫病的经济负担

TAD 对动物健康和生产构成威胁，给各国经济造成巨大损失。最近暴发的牛海绵状脑病、FMD和HPAI,揭示了动物疫病对畜牧业、人类健康及其福祉构成真实且日益严重的全球威胁。TAD对活体动物及其产品的贸易产生了不利影响。2001年，英国暴发口蹄疫，损失估计超过90亿美元。Rushton等人(2005)评估,HPAI(H5N1)造成的泰国和越南经济损失占国内生产总值的0.5%~1.5%。这些疫病在印度造成的经济损失高达11.07亿INR/年(Tripathi等,2018)。

FAO和WOAH认为,在过去几十年里,平均每年出现一种新发动物疫病，而这些新发动物疫病中有3/4属于人兽共患病,可以传染给人类。例如,口蹄疫、蓝舌病、马流感、猪流感及HPAI,都对农民和工业造成了重大的经济损失。

1.10.2 充足的资金支持

用于动物卫生研发的资金支持并不容易得到充足的保障，特别是在发展中国家更不容易获得足够的财政支持。在这些国家的家畜饲养者大多是贫困人员，当地政府应该对这些项目发展给予支持,特别是对弱势的农民,应通过为他们提供诊断和疫苗的激励或补贴来促进猪、绵羊、山羊和家禽的饲养、提高产量低的牛的库存。为满足紧急需求，可以考虑通过公私合作筹集用于紧急疫病控制的风险基金,此外还可以考虑有利于农民的家畜健康保险政策。

1.11 同一个世界,同一个健康

对于重要的TAD,例如,禽流感、PPR和口蹄疫,建议FAO、WHO和WOAH本着“共同健康”的观念,监督跨国、区域或全球进行更充分的规划协调。“同一个世界，同一个健康”(One World One Health, OWOH)概念由FAO、WHO和WOAH提出,其根源在于包括人类、动物和病原体在内的生物之间的相互作用，而环境被视为一个独特的动态系统，其中各个组成部分间的健康都是相互联系和依赖的。如今,“同一种健康,同一种医学”(One Health One Medicine, OHOM)这一新的互相协调的观念反映了这种相互依存的生态系统的整体观点。OWOH可以被视为地方、国家和全球层面上的协作和多学科努力,以确保人类、动物和环境的最佳健康状态。对传染病的控制影响着人类历史的进程,被认为与健康的概念紧密相关。

1.12 TAD防控经验

1.12.1 禽流感

1996年出现的HPAI(H5N1)影响到亚洲、欧洲、非洲和北美等多个国家。该病毒对野生鸟类和家禽也造成了影响。人类因与被感染禽类密切接触而被传染的零星病例具有相当高的死亡率,增加了社会对“禽流感”大流行的担忧。禽流感病毒(avian influenza virus, AIV)对全球健康和经济构成持续威胁。H5N8亚型高致病性禽流感自2014年首次暴发以来,迅速传播到亚洲、欧洲和北美,这一特征在其他高致病性AIV中尚未观察到。然而，导致该病毒全球迁移的病理生物学特征尚不清楚。在候鸟中进行的模拟研究结果显示，支持AIV跨洲传播风险的病理生物学特征表明,H5N8和其他亚型AIV之间存在特征差异，例如,H5N6和H5N1的传播速度较慢。较低的反复感染和较低的死亡率及迁徙恢复率促进了候鸟种群的迁徙。尽管自2006年以来印度一直报告H5N1型AIV流行，但2017年首次在候鸟和家禽中报告H5N8型AIV。为探索人类和鸟类种群中两种病毒遗传特性、分布模式及全球传播假设机制的异同 (Bui等,2014)，研究者对人类和鸟类种群中H5N1亚型AIV和H7N9亚型AIV进行流行病学比较研究，结果表明H7N9亚型AIV比H5N1亚型AIV多样化的速度要快得多。对某些H7N9亚型AIV毒株的分析显示,它与经过基因修饰的可传播的H5N1亚型AIV有相似之处，都对人类呼吸道易感。H5N1亚型AIV和H7N9亚型AIV在人类和禽类中的流行病学差异研究，进一步引出了为何 H7N9 亚型 AIV 传播速度比 H5N1 型 AIV 的传播速度快的问题。

1.12.2　非洲猪瘟

ASF 是在许多国家出现的一种高传染性、致死性的新发疫病。虽然在 1921 年就有关于 ASF 的介绍，并且对非洲、欧洲和南美洲的 50 多个国家造成影响，但关于其发病机制、免疫逃避和流行病学等关键问题仍然尚不确定(Arias 等，2017)。由于疫苗的缺乏，与许多其他猪传染病相比，该病在猪群中的传播造成的卫生、社会和经济影响更大。目前，ASF 在撒哈拉以南非洲、撒丁岛、跨高加索地区、俄罗斯联邦和欧洲联盟的中东部国家都存在。ASF 已经传播至中国(2018 年 8 月首次报道)、保加利亚(2018 年 8 月首次报道)、比利时(2018 年 9 月首次报道)和越南(2019 年 2 月首次报道)，这说明 ASF 对全球养猪业的威胁日益严重(Netherton 等，2019)。目前还没有针对 ASF 获批的疫苗，导致 ASF 的威胁日益突出，所以这一领域还需要做进一步的研究，以期开发出针对 ASFV 的减毒活疫苗。通过利用宿主物种的基因修饰，有可能获得对猪瘟病毒和猪繁殖与呼吸综合征病毒(porcine reproductive and respiratory syndrome virus, PRRSV)具有抗性的猪(Burkard 等，2018)。基于此，基因修饰也可以作为一种增加宿主对非洲猪瘟病毒(African swine fever virus, ASFV)抗性的可行解决方案。将野猪属的疣猪或灌木猪的基因序列插入到家猪基因组中，繁殖的后代能够减少 ASFV 的复制和(或)ASFV 感染后所带来的疫病负担。然而，为了使猪具备完全抗 ASFV 的能力，可以尝试一种更有效的策略，例如，将宿主细胞上的病毒受体作为靶点，阻止病毒的进入和复制。

研究者已经根据病毒分离株的毒力对被 ASFV 感染猪的不同临床过程进行了描述，并且对降低毒力的 ASFV 分离株的基因组进行了测序，已经确定了与病毒表型相关的基因序列。针对巨噬细胞和猪的靶基因修饰、敲除和转基因病毒的检测有助于了解病毒毒力因子，以及病毒如何调节宿主反应。针对 ASF 在欧洲和亚洲迅速传播，且又缺乏疫苗的情况，ASF 防控重点应该是严格的海关和边境保护，以使 ASF 阴性国家避免 ASFV 感染/疫病。

针对 ASFV，需优先对有毒力基因，以及与宿主防御和免疫逃避相关的基因、多基因家族在抗原变异中的作用、免疫应答的逃避机制、决定病毒持续感染的因素及其引起的后果，以及 ASFV 与对 ASFV 感染有耐受性的非洲野猪间的相互作用进行研究。这些研究将为 ASF 的发病机制提供一个完整的认识。为了解不同的流行病学情况，需要阐明不同的宿主(包括野生动物宿主、病媒和环境因素)在疾病传播中的具体作用。为此，需要进一步调查被感染野猪促进 ASF 在北欧传播和持续存在的情况。ASF 曾对中国的生猪和猪肉生产构成了现实的威胁。受影响国家正计划通过增加肉用家禽的产量来弥补猪肉生产的损失。

野猪种群卫生控制方面的空白使得 ASF 控制变得困难。提高兽医、猎人和农民对 ASF 的认识应成为控制 ASF 的优先事项。为了便于对受影响地区进行监测，非损伤性取样方面需要取得进展。目前和未来的测试需要对非损伤性模式进行优化。今后研究重点应是提供可靠的血清学检测，以及替代原代细胞培养的细胞系。研制安全有效的预防 ASF 的疫苗对 ASF 控制和预防是极其有利的，虽然取得了一些进展，但这种疫苗仍然缺乏(Arias 等，2017)。

1.12.3　口蹄疫

继 2011 年牛瘟在全球范围内被成功根除之后，偶蹄动物 FMD 成为 WOAH 列出的另一个重要病毒性疫病，从 FMD 流行国家到无 FMD 国家/地区，该病对家畜和畜产品贸易均造成重大经济损失和不利影响。Knight-Jones 等人(2016)详细介绍了全球 FMD 研究的最新进展及差距，并概述了全球现状和研究需求。得出的结论强调了目前可用的疫苗和控制工具已使许多发达国家消灭FMD。然而,在许多发展中国家,FMD仍未得到控制。主要原因是生物安全措施是成功控制FMD的基础,但由于显而易见的原因,发达国家的FMD控制措施在发展中国家难以有效实施。在目前的情况下,对疫苗进行改进,使之对更广泛的FMDV毒株具有更持久的免疫保护和更低的生产成本,可能是增强我们控制FMD能力的最重要发展方向。虽然在解决当前灭活疫苗的关键技术方面还受到限制,但几种新的候选疫苗研制已经取得了令人鼓舞的进展,然而这些疫苗还没有商业化。虽然研制新疫苗至关重要,但使用目前的疫苗已多次有效地控制了FMD。此外,为了提高免疫力,应充分保证FMD疫苗的质量,并提供足够数量,以便根据适当的战略满足免疫的全覆盖。FMD疫苗的设计和完成方面的项目还需要提供更好的培训和支持。另一个研究领域是病毒的遗传和分子研究,以阐明宿主-病毒的相互作用。强大的工具和分析正在增进我们对FMDV进化、生态学和流行病学各个方面的理解。这进而将有利于FMD许多领域的研究,包括病毒学、疫苗和诊断研发等。此外,改进的遗传技术有可能揭示控制FMD的关键信息,例如,传播链、疫苗匹配和病毒循环水平。

世界上许多国家政府对控制FMD尤为重视。除了在欧洲和南北美洲建立的研究机构进行传统的FMDV防控研究外,在中国、印度和非洲也开展了引人注目的工作。南美洲和欧洲的经验表明,通过几十年的持续投入,FMD是可以控制的,即使曾经FMD近乎不可控制的地区也一样能够得到控制。但是,如果要改进和更广泛地对FMD进行控制,就必须在地方和国际层面持续投资以用于FMD研究。改进的DIVA诊断学方法提高了我们在疫苗接种的动物中检测出被野毒感染动物的能力。增强对动物及其产品被FMDV感染现状把控的信心,反过来会为更有效的疫病控制和贸易的国际标准,以及减少贸易限制开辟道路。严格的审批程序增加了诊断试剂和疫苗新技术进入市场的时间。然而,如果审批不严格,不合格的产品可能会被投放市场。因此,需要平衡这两个需求。通过必要改变对现有技术适当放宽,例如,改变疫苗品种,特别是在迫切需要的时候改变疫苗品种(Knight-Jones 等,2016)。

1.12.4 牛瘟:已被消灭的动物疫病

牛瘟,又称“牛疫”,曾对全球多个地区,特别是亚洲、非洲、欧洲和美洲的畜牧业和农业经济构成严重威胁。其周期性大流行造成了灾难性的家畜流行病和巨大的经济损失。通过大规模疫苗接种规划、动物园卫生措施、政策支持、国际合作等,牛瘟成功地在全球被消灭了。

新发病畜群的发病率和死亡率均可高达95%~100%,造成巨大的经济损失。在印度,被记录在案的大约有20万头牛死亡,20世纪50年代,每年受影响的牛达40万头,平均死亡率为50%。在整个人类历史中,牛瘟给社会、经济和生态造成了灾难性后果,甚至改变了各民族和国家的历史。在印度,牛瘟委员会证实了牛瘟的存在(Hallen 等,1871)。通过大规模疫苗接种和动物园的卫生措施,人类已经成功地战胜了这种疫病。2011年6月28日,FAO宣布在全球消灭牛瘟,这是继1980年人类根除病毒性传染病天花约30年后,全球首次消灭动物病毒性传染病(Yadav 等,2016)。在热带国家,获得充足有效的高质量疫苗、冻干疫苗及疫苗的冷链运输都受到限制,而缺乏结构化临床监测和血清检测的基础设施又限制了执行大规模疫苗接种的规划。在印度,根据疫病的流行病学情况和采取的战略,全国被划分为四个区域,重点在洲际和国际边界及牛和羊的迁徙路线上接种疫苗,建立免疫带,再加上严格的临床监测和血清检测,对防控牛瘟病毒感染具有很大帮助。来自FAO、WOAH、欧州联盟和国际原子能机构的财政支持和(或)技术指导是印度2006年实现无感染状态的驱动力。随着牛瘟的成功根除,全球畜牧业变得更加安全，从而提高了畜牧养殖的条件。

牛瘟的成功控制和根除是印度兽医服务的有益经验和里程碑，为现场兽医、研究人员、政策规划人员和捐助机构，以及其他利益相关方提供了能力建设和信心，以便在国家层面成功实施家畜疫病控制规划。印度成功根除牛瘟不仅促进了印度乳业的发展，而且在近十年也促进了肉类和其他乳制品的出口。今天，印度不仅在牛奶产量上位居世界第一，而且还是最大的水牛肉出口国。成本–效益分析显示，在牛瘟防治计划上每投入 1 美元，便为印度奶业带来约 20 美元的收益，以此增加牛奶、肉类产量及提高农业生产力（Uppal, 2011）。

1.13　展望

面对 EID 和 TAD 日趋严重的局面，多样化、动态和精心规划的疫病监测及其监测方法是任何国家健康畜牧业生产系统的可持续性发展和人民福祉的关键。为了战胜流行的、正在出现的、重新出现的 EID 和 TAD，需要强大的监测和精确检测系统，而且这些系统要具有灵活性、可行性和适应现场条件的能力。在这方面，诊断试纸条/芯片实验室检测仅需要一个小时。为选择用非损失性方法从不同动物物种和野生动物中取样，需要克服采样的障碍。必须坚持疫病报告的透明度，并向 WOAH 报告。由于动物和动物产品的贸易，世界贸易组织有向 WOAH 报告重大疫病的国际义务，而且世界贸易组织所有成员国必须遵守。现在该领域应用先进的诊断和分子检测工具，以确保快速检测和确认能够导致人类和动物疫病的病原体。这必须与国家级的疫病监测、检测联网相配套，以便建立用于传染病预测的早期预警系统（Saminathan 等，2016）。还需要优先考虑适当开发和应用新的有效、安全和廉价的疫苗，以及疫苗供应系统，并采用新型疫苗的接种规划和免疫调节及有效治疗模式，这将有助于及时制定针对病毒和其他病原体传染病的预防和控制战略。此外，应遵守良好的管理措施、标准的生物安全和生物研究安全措施/做法、适当的卫生、动物园卫生检疫措施。

此外，还需要按照世界贸易组织的 SPS 协议设想，现场控制和检查病原体的传播和适当的贸易限制。及时执行这些概念和战略，以及适当加强资金支持各种多层面研究和发展方案，需要有长远的眼光和全面的解决办法。这些措施将极大地减少疫病的发生和暴发，减轻传染病造成的经济负担，促进畜禽健康繁殖和生产，促进畜禽产业的可持续发展。大流行的威胁下降和对公共卫生关注度的提高，最终使得在“同一个健康”的理念下，整个社会的经济和福利得到改善。发展中国家需要优先关注人工智能、全球定位系统、遥感和可溯性在疫病的监测和管理方面的应用。同样，最新的基因编辑技术、基因突变技术、纳米技术、电子鼻技术等，都应该用于高效的疫病诊断和药物传递。在制定畜禽养殖政策时，既要提高畜禽的生产性能，又要兼顾畜禽后代的健康。可利用现代技术在本地品种中培育具有抗病能力（完全或部分）的家畜和家禽。

参考文献

Anon (2003) Institute of Medicine (IOM) report

Anonymous (2017–18) Annual report, ICAR-Directorate of foot-and-mouth disease, Mukteswar, India, pp 1–88

Anonymous (2018–19) Annual report, Department of Animal Husbandry, Dairying and Fisheries, Ministry of Agriculture and Farmers Welfare, Govt. of India, New Delhi, pp 61–64. http://dahd.nic.in

Arias M, Jurado C, Gallardo C, Fernandez-Pinero J, Sanchez-Vizcaino JM (2017) Gaps in African swine fever: analysis and priorities. Transbound Emerg Dis 65(S1):235–247. https://doi.org/10.1111/tbed.12695

Bui C, Bethmont A, Chughtai AA, Gardner L, Sarkar S, Hassan S, Seale H, MacIntyre CR (2014)

A systematic review of the comparative epidemiology of avian and human influenza A H5N1 and H7N9—lessons and unanswered questions. Transbound Emerg Dis 63(6):602–620. https://doi.org/10.1111/tbed.12327

Burkard C, Opriessnig T, Mileham AJ, Stadejek T, Ait-Ali T, Lillico SG et al (2018) Pigs lacking the scavenger receptor cysteine-rich domain 5 of CD163 are resistant to porcine reproductive and respiratory syndrome virus 1 infection. J Virol 92:e00415-18. https://doi.org/10.1128/JVI.00415-1

Daszak P, Cunningham AA, Hyatt AD (2000) Emerging infectious diseases of wildlife—threats to biodiversity and human health. Science 287:443–449. https://doi.org/10.1111/tbed.12180

Hallen JHB, Mcleod K, Charles JG, Keer HC, Allijan MM (1871) The cattle plague commission report to government of India. Calcutta Publication, Calcutta, pp 1–999

Knight-Jones TJD, Robinson L, Charleston B, Rodriguez LL, Gay CG, Sumption KJ, Vosloo W (2016) Global foot-and-mouth disease research update and gap analysis: 1—overview of global status and research needs. Transbound Emerg Dis 63(S1):3–13. https://doi.org/10.1111/tbed.12528

Lederberg J, Shope RE, Oaks SC Jr (eds) (1992) Emerging infections: microbial threats to health in the United States. National Academy of Sciences, Washington, DC

Malik YS, Dhama K (2015) Zika virus—an imminent risk to the world. J Immunol Immunopathol 17(2):57–59. https://doi.org/10.5958/0973-9149.2015.00019.2

Meeusen ENT, Walker J, Peters A, Pastoret P-P, Jungersen G (2007) Current status of veterinary vaccines. Clin Microbiol Rev 20(3):489–510. https://doi.org/10.1128/CMR.00005-07

Munjal A, Khandia R, Dhama K, Sachan S, Karthik K, Tiwari R, Malik YS, Kumar D, Singh RK, Iqbal HMN, Joshi S (2017) Advances in developing therapies to Combat Zika virus: current knowledge and future perspectives. Front Microbiol 8:2677. https://doi.org/10.3389/fmicb.2017.01469

Netherton CL, Cornell S, CTO B, Dixon LK (2019) The genetics of life and death: virus-host interactions underpinning resistance to African swine fever, a viral hemorrhagic disease. Front Genet 10:402. https://doi.org/10.3389/fgene.2019.00402

Peeler EL, Reese RA, Thrush MA (2015) Animal disease import risk analysis—a review of current methods and practice: open access article. Transbound Emerg Dis. 62:480–490. https://doi.org/10.1111/tbed.12180

Riemenschneider CH (2005) Avian influenza and other transboundary animal diseases, Director, Liaison Office for North America, Food and Agriculture Organization of the United Nations. Presentation at "Health in Foreign Policy Forum 2005", Washington, DC, 4 Feb 2005

Rushton J, Viscarra R, Guerne Bleich E, McLeod A (2005) Impact of avian influenza outbreaks in the poultry sectors of five South East Asian countries (Cambodia, Indonesia, Lao PDR, Thailand, Viet Nam) outbreak costs, responses and potential long term control. World's Poultry Sci. J. 61(1):491–514

Saminathan M, Rana R, Ramakrishnan MA, Karthik K, Malik YS, Dhama K (2016) Prevalence, diagnosis, management and control of important diseases of ruminants with special reference to Indian scenario. J Exp Biol Agric Sci 4(3S):3338–3367. https://doi.org/10.18006/2016.4(3s).338.367

Singh RK, Dhama K, Malik YS, Ramakrishnan MA, Karthik K, Tiwari R, Khandia R, Munjal A, Saminathan M, Sachan S, Desingu PA, Kattoor JJ, Iqbal HMN, Joshi SK (2017) Ebola virus—epidemiology, diagnosis and control: threat to humans, lessons learnt and preparedness plans—an update on its 40 year's journey. Vet Quart 37(1):98–135. https://doi.org/10.1080/01652176.2017.1309474

Singh RK, Dhama K, Chakraborty S, Tiwari R, Natesan S, Khandia R, Munjal A, Vora KS, Latheef SK, Karthikh K, Malik YS, Singh R, Chaicumpaj W, Mourya DT (2019) Nipah virus: epidemiology, pathology, immunobiology and advances in diagnosis, vaccine designing and control strategies—a comprehensive review. Vet Quart 39(1):26–55. https://doi.org/10.1080/01652176.2019.1580827

Tripathi BN, Kumar N, Barua S (2018) Peste des Petits ruminants: sheep and goat plague. Today and Tomorrow's Printers and Publishers, New Delhi, pp 1–180

Uppal PK (2011) FAO sponsored final project report "National testimonies" under the Global Rinderpest Eradication Programme (GREP)-(GCP/GLO/302/EC), pp 1–134

Yadav MP (2011) FAO sponsored final project report on "Laboratory contributions for rinderpest eradication in India" under the Global Rinderpest Eradication Programme (GREP)-(GCP/GLO/302/EC), pp 1–58

Yadav MP, Uppal PK, Rao JR (2016) Animal sciences. In: Singh RB (ed) 100 Years of agricultural sciences in India. NAAS, New Delhi, pp 158–258

第 2 章 非洲猪瘟病毒

Alexander Malogolovkin, Alexey Sereda, Denis Kolbasov

2.1 引言

ASF被认为是对猪群危害性最为严重的一种疾病，对全世界的家养猪和野猪都构成了威胁。ASF一旦传入新的地区，其死亡率高达100%。近期，在许多非洲国家、高加索地区、俄罗斯联邦，以及东欧国家的猪场和野生动物中暴发了ASF。令人担忧的是，最近关于东南亚ASF流行病学情况的报告表明，ASF已发生大面积流行。据记载，2019年，在柬埔寨、缅甸、越南和老挝暴发过破坏力极强的ASF，而为了有效控制该病的传播，数百万头猪被扑杀(Gogin等，2013；Nurmoja等，2017a；Oganesyan等，2013；Okoth等，2013；Owolodun等，2010；Pejsak等，2014)。

Montgomery报道在非洲首次发现了ASF(De Kock等，1940；Edgar等，1952)。已经证实ASF在家猪中存在，并表现出了与猪瘟[又称经典猪瘟(classical swine fever, CSF)]类似的临床症状。Hess、Hay、DeTray、Plowright和Malmquist进行了一些后续的研究，讲述了病毒的分离，以及ASFV基本生物学特性、传播和发病机制(Anderson 1986；Bool等，1970；Hammond和Detray，1955；Pan和Hess，1985；Pan等，1980；Parker等，1969)。

ASF是一种新发的跨界疫病。非洲以外的许多国家都暴发过ASF，例如，葡萄牙、西班牙、法国、荷兰、意大利、苏联、巴西、古巴和海地(Boinas等，2011；Caporale等，1988；Costard等，2013；Korennoy等，2017；Lyra，2006；Terpstra和Wensvoort，1986)。记载ASF暴发的地区可能距离其流行地区很远，距离暴发地区1000km。在西班牙，ASF持续存在30多年并且严重地影响了该国的养猪产业，该国执行了严格的控制政策、有效的监管机制，以及对病毒诊断和流行病学的全面的研究，ASF被成功地消灭(Arias等，2001；Pastor等，1989；Sanchez-Vizcaino等，1981)。

自从1978年起，ASF在意大利撒丁区就已经有记载，一直困扰着当地的养猪业者和兽医(Jurado等，2018；Mur等，2018)。事实上，在乔治亚州通报了ASF暴发情况以前，早在2007年就已经有了关于ASF的现代历史记载(Costard等，2009；Onashvili等，2012)。自那时起，ASF迅速扩散到高加索地区和俄罗斯联邦，首先感染野猪群，随后传播到家猪群。2017年，俄罗斯联邦兽医及植物卫生检疫监督局在报告中称，在俄罗斯暴发ASF超过1000例，导致46个地区超过80万头猪死亡。猪肉产量从2007年的1119吨下降到2017年的608吨，减产近50%(Kovalev，2017)。尽管有ASF的流行，但在此期间，高度工业化的养猪场的生猪产量每年都在增加。

本章将对ASF在俄罗斯的流行病学，以及从ASF地方性流行10多年中吸取的教训进行介绍。ASFV是ASF的病原体。ASFV是唯一一种能够同时感染蜱和猪，且能够复制的DNA虫媒病毒(Alonso等，2018)。到目前为止，由于ASFV极端的宿主范围及其复杂的病毒基因组结构，使得ASFV成为非洲猪瘟科的唯一成员。接下来，本章将讨论ASFV在不同宿主中的传播特性和免疫发病机制。ASFV可在被感染宿主的单核细胞(单核细胞/巨噬细胞)中进行复制，并且具有复杂的多功能免疫逃避系统(Reis等，2017a)，使得ASFV成为“完美的杀手”，至今仍然是不能防御的病原体。目前还没有针对ASFV的安全、有效的疫苗，但一些研究团队对ASFV疫苗研究的报告结果令人非常鼓舞(Arias等，2018；Dixon等，2013；Rock，2017)。本章的疫苗部分总结了ASFV疫苗研发的最新进展和知识空白。

在此，我们将引导读者了解ASF流行病学和控制方面所面临的最新挑战和解决方案，并讨论ASFV疫苗研究的突出问题。读者从中将找到有关ASFV生物学特性、预防、控制和发病机制等最新的专业参考资料。虽然参考文献没有完全被列出，但我们非常感谢本章涉及的所有对ASF研究做出宝贵贡献的研究者。

2.2 结构与分类

ASFV是非洲猪瘟病毒科非洲猪瘟病毒属的唯一成员。ASFV是一种大型且复杂的dsDNA虫媒病毒（Alonso等，2018）。病毒基因组长165~194kbp。病毒粒子具有多层膜和核蛋白核心结构。表层膜（囊膜）由不同形式的脂质和糖基化蛋白组成。病毒颗粒直径170~190nm。ASFV基因组结构与核质巨DNA病毒的其他成员一样，由单分子双链线性DNA组成。ASFV基因组有两个末端，通过末端反向重复序列以flip-flop形式形成共价封闭。目前GenBank中已公开21个ASFV完整基因组序列。关于ASFV基因组结构组成和复制的更多信息可在国际病毒分类委员会（International Committee on Taxonomy of Viruses, ICTV）网站上查阅。

ASFV在自然宿主的单核吞噬细胞、巨噬细胞和特定的网状细胞中可以有效复制。在体外，ASFV可在单核细胞/巨噬细胞中生长，并且可适应于内皮细胞系生长。一些研究表明，ASFV对内皮细胞的适应可能导致表型减弱（Carlson等，2016；O'Donnell等，2016）。

ASFV的抗原多样性在非洲东部最具代表性，ASFV可通过不同的传播周期进行传播。基于ASFV核心衣壳蛋白P72（B646L）的核苷酸测序，迄今为止已经确定了23种基因型（Achenbach等，2017）。历史上在欧洲和加勒比等国家和地区暴发的ASF是由基因Ⅰ型ASFV引起。最近在高加索地区、俄罗斯和东欧等国家和地区流行的ASF来源于2007年传入格鲁吉亚的基因Ⅱ型ASFV。

为了追溯流行区的病毒来源和分布，现已提出新的ASFV分型标记物（Gallardo等，2014；Goller等，2015）。ASFV中心可变区（CVR，B602L）和基因间隔区（I73R-I329R）完善了ASFV的基因分型。2012年至2018年，在分离到的ASFV分离株中发现了多个ASFV IGR变异株。

另外，根据血细胞吸附抑制试验和体内交叉保护试验，将ASFV分离株分为若干血清型。到目前为止，已经确定了8种血清型，但也可能存在更多血清型（Malogolovkin等，2015a；Sereda等，1994；Sereda和Balyshev，2011）。最近，已经在CD2v（EP420R，血凝素）和C型凝集素样蛋白（EP153R）中发现了血清型特异性的遗传特征（Malogolovkin等，2015b）。这种方法可能会使人们对ASFV遗传多样性和抗原多样性之间的关系有更多了解。

ASFV毒株不同可能会引起急性、温和型和慢性疾病。已经在ASFV基因组中发现了几种毒力因子。ASFV分离株可能会丢失多基因家族MGF360/530中的一些成员（Borca等，2018；O'Donnell等，2016）或者导致表型减弱的MGF110（Zani等，2018）。敲除干扰素抑制剂基因或CD2v的重组ASFV毒株（Abrams等，2013；Monteagudo等，2017；Neilan等，2002）对家猪的致病性有所减弱。使用不同的病毒模型，对CD2v蛋白的毒力和防护作用进行研究，但获得的结果具有争议（Burmakina等，2016；Monteagudo等，2017）。

ASFV较为独特，可吸附被感染巨噬细胞周围的红细胞（图2.1）。最初，该现象用于CSF和ASF的鉴别诊断。之后，ASFV CD2v蛋白被鉴定是病毒血凝素（Galindo等，2000；Rodríguez等，1993）。有趣的是，一些ASFV存在截断的或不规则的CD2v（EP402R），因此还不能证明其有吸附能力。一些非吸附型ASFV毒株表型为弱毒，已被用作疫苗研究的模型（King等，2011；Sanchez-Cordon等，2017）。

2.3 流行病学

最近已经发表了有关欧洲和非洲ASF流行病学的全面综述（Bosch等，2017；Brown等，2018；Cisek等，2016；Gogin等，2013；Mur等，2012）。在

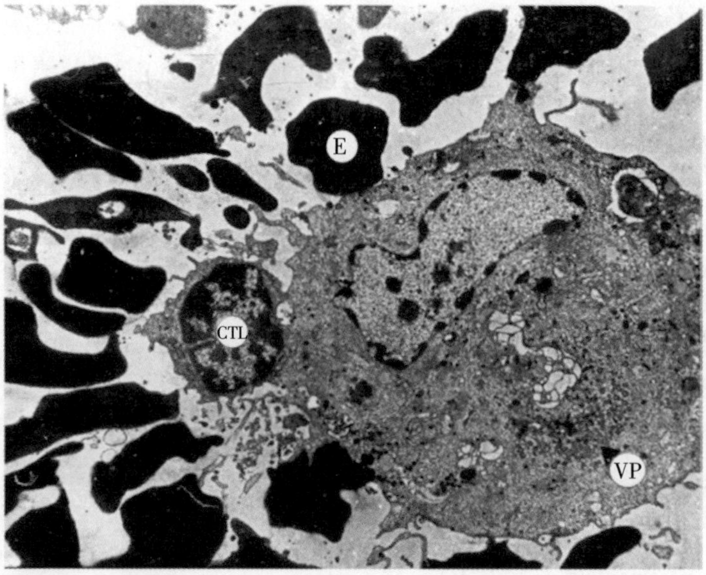

图 2.1 感染 ASFV 的巨噬细胞被红细胞(E)包围,并被细胞毒性 T 淋巴细胞(cytotoxic T lymphocyte, CTL),以及病毒蛋白(viral protein,VP)-病毒质(超薄切片电子显微镜检查)吸附。

这里重点描述俄罗斯 ASF 流行病学的状况及 10 年来的流行态势。

自 2007 年 12 月,ASF 首次传入北高加索地区以来,ASFV 就一直在俄罗斯存在,并且在过去 10 年中,ASF 从俄罗斯传播至整个东欧, 直到在波罗的海国家中流行, 对家猪和野猪均造成影响(Gogin 等,2013)。流行的 ASF 由基因Ⅱ型 ASFV 引起,致家猪和野猪的死亡率高达 100%(Malogolovkin 等,2012)。ASFV Armenia/2008 毒株的早期数据显示,Armenia/2008 毒株引起的家猪和野猪死亡率较高,但该毒株具有中度传染性(Gabriel 等,2011;Pietschmann 等,2015)。然而,波罗的海各州血清阳性野猪的存活数量有所增加(Nurmoja 等,2017a,b)。

俄罗斯联邦由 85 个联邦主体组成,并且通过兽医立法,每个联邦主体都要负责 ASF 的控制和预防。ASF 发病情况取决于地区当局的能力和资源。每个联邦主体都有不同结构的猪养殖部门,如果庭院养殖占比很高, 就会使该地区存在传入 ASF 的高风险,并且难以控制。

自 ASFV 首次传入俄罗斯联邦境内以来的 10 年时间(2007 年 12 月至 2017 年),已报告 ASF 疫情暴发 1274 起。其中 50%以上发生在小型私人控股农场或庭院养殖猪,7%来自工业养猪场,约 40%来自野猪(图 2.2)。尽管 ASF 病例数量几乎每年都在增加,但在过去 12 年中,俄罗斯境内猪的种群数量已经提高到了 7 000 000 头(Karaulov 等,2018)。

ASF 在俄罗斯境内已经传播了 10 年,要收集庭院养猪场猪群的准确及最新信息仍比较困难。由于受这种情况的限制, 不受控制的动物的流通对 ASF 的流行起到了关键作用(Súnchez-Vizcaíno 等,2012)。由于庭院养猪数量未知,所以无法通过兽医服务来控制庭院猪场猪的健康状况。一些 ASF 病例仅在饲养者报告疑似疾病后才被检测出;大多数情况下,首次传入的病毒发现的较晚,以至于无法及时确认首例病例并采取相应控制措施。一个地区的兽医服务体系薄弱,以及采取应对疫情暴发措施较晚,均导致 ASF 传播到邻近地区。

根据国家法规,接到 ASF 报告后,必须禁止在疑似猪场内猪的所有活动。通过实验室检测如果确诊为 ASF, 地区主管部门必须在半径 5~

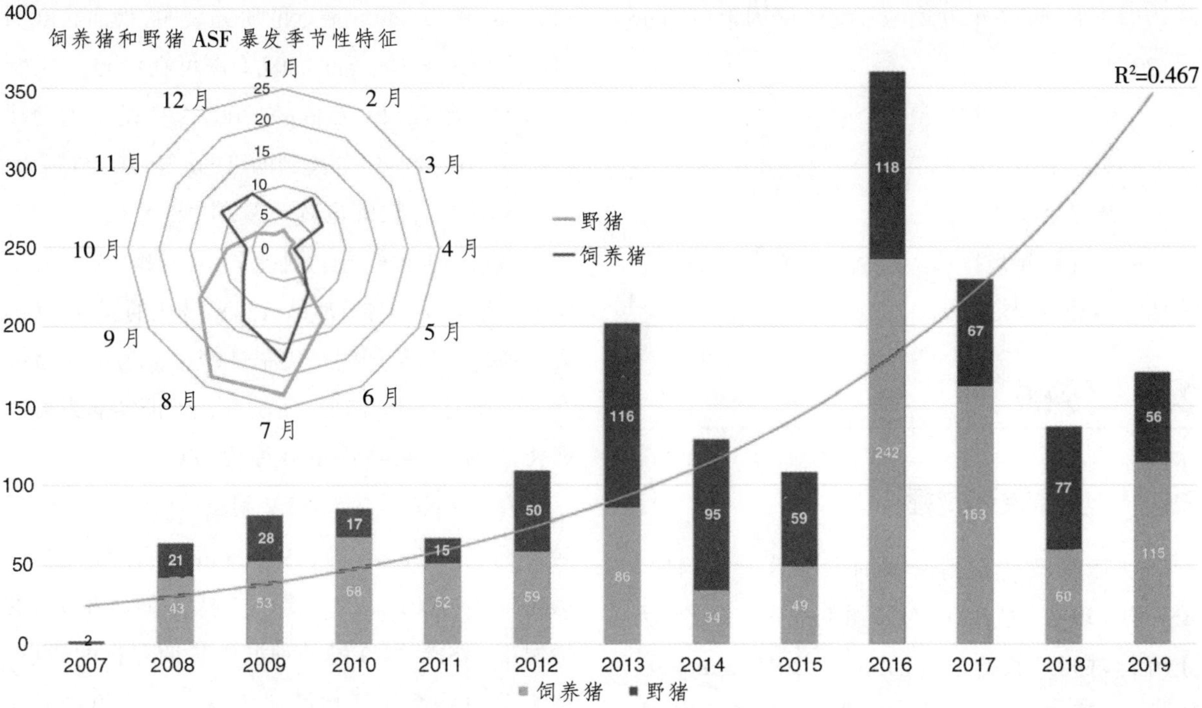

图 2.2 2007—2017 年俄罗斯 ASF 疫情暴发的累计数量及其季节性特征。

100km 应尽快实施扑杀。地区主管部门负责划定控制和监测区域。因为扑杀造成的极大损失和高额的补偿费用,所以扑杀策略的缺点在于,生猪养殖者很有可能会迟报,以及地区主管部门对于扩大控制区域的积极性较低。

虽然 ASF 已经流行了 10 多年,但影响 ASF 传播的主要风险因素没有变化,即感染/病猪进入新地区、猪肉产品被 ASFV 污染、疫情缓报,以及生猪养殖者、兽医和屠宰者之间的合作薄弱(Kolbasov 等,2018b)。

很明显,白猪是导致 ASF 在野猪间短距离传播的原因,但这种机制不能解释 ASF 在很短的时间内就能传播数千公里这一问题。野猪种群可分为:自然栖息地的野猪和在竞技场内用于狩猎的野猪。不幸的是,狩猎者和管理者并不顾动物的健康(Kolbasov 等,2018b)。如果在死亡的野猪中检测到 ASFV,则需在当地兽医的监督下,在适当的地点对死亡动物的尸体进行焚化,并对该地区施行被动监测。最初,ASFV 的流行主要有两个驱动因素:社会经济和人类行为,而野猪传播中所起的作用仍未完全明了。许多实例证明,以“跳跃”为特征的空间传播模式是由猪和猪肉产品的非法流动所致(Kolbasov 等,2018a)。

ASFV 传入家猪群的一个主要因素是较低的或不适当的生物安全措施:约 80%的 ASF 疫情是在圈养猪场中暴发。其中大多数 ASF 疫情与非法贸易及被感染猪不受控制的流通有关。众所周知,野猪在通过行政边界将病毒传入新的地区中起着至关重要的作用。同时,还未证实病毒传播至商业养猪场与感染 ASFV 的野猪有关。此外,有许多证据表明,在森林里非法处理家猪的尸体,随后在野猪种群中检测到了 ASFV。在家猪中,对 ASF 的控制相对容易。另外值得注意的是,蜱并未参与当前东欧和俄罗斯境内的 ASF 流行。

有趣的是,将 1977 年对 ASF 的流行病学调查结果与目前俄罗斯暴发 ASF 疫情中收集到的数据进行比较后发现(Korennoy 等,2017),历经 40 余年,其主要风险因素仍没有发生变化。

以下是 1977 年发现的一些风险因素(Jurkov 等,2014):

– 受污染的食物残渣。

– 受感染肉制品的销售。

– 被感染动物的交易。

– 未受影响地区的养殖场和企业与受 ASF影响地区的经济交通往来。

2.4 传播

2.4.1 丛林循环传播

从撒哈拉沙漠到南部的大多数非洲国家,ASF 均呈现地方性流行。欧洲撒丁岛是 ASF 迄今为止流行时间最长的地方(Mur 等,2018;Sánchez-Vizcaíno 等,2012)。ASFV 主要通过易感动物(家猪或野猪)与被感染动物、污染物或者被感染的猪肉产品接触,或者蜱叮咬进行传播。

历史上,疣猪和非洲本地猪(非洲野猪、假面野猪、森林猪等)被认为是 ASFV 在非洲自然环境中的主要储存宿主。后来的研究证明,ASFV 在非洲野猪中不能进行水平传播或垂直传播。为证明血清阳性疣猪能够直接将 ASFV 传播给家猪,研究者已经进行了多次尝试,但均未成功(Anderson 等,1998)。因此,软蜱被认为是 ASFV 丛林传播循环中的潜在参与者。疣猪生活在洞穴中,可经常接触软蜱(仔猪钝蜱仔猪亚种,非洲钝蜱仔猪亚种,非洲钝蜱)。

感染 ASFV 的蜱叮咬幼龄疣猪容易使其致病,血液中的 ASFV 滴度可能会达到 2~3 lg HAU/mL,这足以在软蜱中启动新一轮的丛林传播循环(Burrage,2013;Plowright 等,2002)。

在伊比利亚半岛,发现了另一种软蜱(游走钝缘蜱)可作为 ASFV 的传播媒介(Bastos 等,2006a)。毫无疑问,在西班牙暴发的游走钝缘蜱与 ASF 的流行有关。ASFV 在感染的软蜱中可存活 5 年,这会给 ASF 呈现地方性流行的国家带来严重问题(Boinas 等,2004)。最近的研究表明,目前在欧洲流行的 ASFV Georgia 2007/1 株可在游走钝缘蜱中有效复制,ASFV Georgia 2007/1 株是从葡萄牙南部采集的样本中分离到的(Diaz 等,2012)。

其他几种软蜱也具有复制 ASFV 的能力,在美国,波多钝缘蜱,土氏钝缘蜱,塔拉蜱,以及皮革钝蜱(Hess 等,1987)都是 ASFV 潜在的媒介(Hess 等,1987)。已经证明在皮革钝蜱中,ASFV 可持续存活 4 个月以上。然而,还没有证据证实病毒可经卵传播(Sánchez–Vizcaíno 等,2009)。

在非洲钝蜱中,ASFV 可经卵、气门及性接触进行传播(Hess 等,1989;Plowright 等,1970;Rennie 等,2001)。这些数据支持了在无猪参与的情况下,ASFV 可在蜱的种群中长期存在的假设。根据分类关系和丛林循环传播,ASFV 很可能在节肢动物中持续进化(Makarov 等,2016)。一些数据表明,ASFV 和软蜱具有一致的进化趋势,并且有多种宿主基因参与了这一进化过程。然而,由于家猪并不是 ASFV 的自然宿主,为此,ASFV 在家猪体内复制过程中可能会丢失某些基因(Afonso 等,2004;Burrage 等,2004;Dixon 和 Wilkinson,1988)。希望通过对软蜱基因组测序得出新的基因数据,这将有助于解析 ASFV 的进化和起源。

非洲野猪和软蜱共同构成了 ASFV 的丛林循环传播(Parker 等,1969)。软蜱也会将 ASFV 传播给家猪。蜱传播病毒的效率与蜱唾液腺和腰腺中的病毒滴度有关,病毒滴度要达到 4~6 lg HAU/mg (Bastos 等,2006b)。

约有 100 种软蜱为人所熟知,其中有 7 种在苏联境内被发现。其中,中亚的乳突钝缘蜱和高加索地区的疣皮钝蜱是回归热的传播媒介,此外还包括非洲的非洲钝蜱及伊比利亚半岛的游走钝蜱(Fillipova,1966),这些蜱对 ASFV 的流行具有重要意义。

在东欧国家,一些生物安全等级高的猪场也受到了 ASFV 的侵袭。在其他吸血昆虫的若干项研究中发现,在实验室环境中,厩舍苍蝇采食受感染动物的血液,会导致家猪被 ASFV 感染。这些研究表明，至少在猪群中，其他吸血昆虫可能在 ASFV 的传播中发挥作用(Olesen 等,2018)。在另一项研究中,对啮齿类动物和鸟类血液中的 ASFV 进行检测,结果均为阴性[欧洲食品安全局动物健康和福利(EFSA AHAW)小组 2014]。

2.4.2　污染物运输

影响病毒在自然环境中传播的一个最重要的因素是它在不同环境中的抵抗力。这是研究疾病分布、建模和风险评估的关键因素。根据 Kovalenko 和 Sidorov(1973)报道,ASFV Georgia 2007/1 株经口、鼻接种最小感染剂量是 10 HAU/mL。

带毒动物和患病动物通过唾液、鼻分泌物、粪便、尿液、生殖器分泌物和血液将 ASFV 排出体外。所有含有 ASFV 的排泄物都可能造成地面、饲料和水的污染。当粪便出现 ASFV 时,机体就会伴随着发热(Greig 和 Plowright,1970)。原发性发热 2~3 天后，鼻拭子内 ASFV 滴度可接近 4~5 lg HAU50/mL。在一些研究中,感染 70 天后可从粪便和口腔拭子中分离到 ASFV(de Carvalho Ferreira 等,2012)。在低温(4~6℃)环境中,患病动物粪便中的 ASFV 经过 159~253 天，尿液中的 ASFV 经过 60~87 天,仍能够分离到病毒(EFSA AHAW 小组,2014)。在炎热的气候环境中,ASFV 在饲养患病动物的圈舍中可存活 5~14 天,而西班牙的一项研究表明，在同样条件下病毒存活可长达 3 个月(Kovalenko 和 Sidorov,1973)。ASFV 在俄罗斯中部寒冷的气候环境中,pH 值为 4.5~4.6 的污染森林沙地中可存活 112 天(Smirnov 和 Butko,2011)。

根据之前确定的 ASFV 的最小感染剂量为 10 HAU/mL,粪便和尿液中的 ASFV,在 4℃时感染性可持续 8~15 天,在 37℃时感染性可持续 3~4 天(Davies 等,2017)。在冬季,患病的野猪排出的分泌物冷冻结冰及其污染的环境，在解冻后仍可能是潜在的感染源。目前已经建立一些用于野猪种群 ASFV 监测的方法。最佳的采样方法是用带有饲料引诱剂的绳子进行采样（Chichikin 等，2012)。野猪 ASFV 监测采样是选取粪便标本,因为 ASFV 在粪便中可长时间存活。粪便标本也可用于 ASFV 基因组的聚合酶链反应(polymerase chain reaction, PCR)检测(de Carvalho Ferreira 等,2012)。根据 de Carvalho Ferreira 等人(2014)的研究,粪便中的 ASFV DNA，在 4~12℃条件下持续到 98 天，在 37℃条件下持续到 35 天；尿液中的 ASFV DNA,在 4~37℃条件下,持续到 126 天;唾液中的 ASFV DNA，在 4℃条件下持续到 35 天，在 12~21℃条件下持续到 14 天。

2.5　感染免疫

ASF 的发病机制和免疫病理与大多数人类和动物出血热相似。ASFV 主要感染单核细胞(单核/巨噬细胞)——T 细胞介导免疫的最主要成分。被感染的单核细胞/巨噬细胞释放出大量的细胞因子,严重影响不同类型的细胞(淋巴细胞、内皮细胞)，导致细胞凋亡和损伤(Penrith,2009;Penrith 等,2004)。促炎性细胞因子(如 IL-1、IL-6、TNF-α)的产生使得机体出现类似于出血热的症状（如发热、血管损伤)(Salguero 等,2002;Sánchez-Cordón 等,2005)。ASFV 在单核/巨噬细胞中复制导致细胞的损伤和凋亡。受损伤和被破坏的单核细胞的成分也会激活内皮细胞，减慢凝血系统的凝血速度(Salguero 等,2008)。

ASFV 最初进入单核细胞，随后对多种类型细胞造成感染，特别是发病后期能够对多种类型细胞造成感染(如中性粒细胞、巨核细胞、扁桃体上皮细胞、肝细胞、肾细胞、粒细胞、可能的树突状细胞)(Greig 等,1967;Sierra 等,1990)。急性 ASF 中,淋巴器官(脾、淋巴结、胸腺)和肾脏出现严重的病理形态学改变和出血(Kleiboeker,2002;Ramiro-Ibáñez 等,1997)。

ASFV分离株不同,ASF的临床表现也各不相同,从不明显到慢性形式(EFSA AHAW小组,2014)。第一次出血性病变可能出现在ASFV感染后3天，与单核细胞/巨噬细胞被破坏的时间一致。ASFV感染平均潜伏期为2~7天。ASF的死亡率可能接近100%，但在不同的ASFV毒株之间有所不同(Mebus,1988)。急性ASF的典型临床症状包括高热、血性腹泻、呼吸困难、发绀和出血性病变。通常在感染后期可观察到中枢神经系统症状,如共济失调和惊厥。这与ASFV感染野猪表现的临床症状相同,与年龄和性别无关(Blome等,2013;Gabriel等,2011)。读者可以在欧洲食品安全署网站照片入口看到一些ASF的病理图片。

在俄罗斯和欧洲流行的ASF最初是由强毒ASFV(基因Ⅱ型)毒株引起的。在实验环境中,这些动物在感染后10天内死亡,没有幸存者。然而,自2012年以来,有许多报告表明,温和型ASFV的变异株在东欧和俄罗斯流行，引起的动物死亡数量越来越少，特别是在野猪中的死亡数量更少(Arias等,2018;Gallardo等,2018)。在俄罗斯和东欧发现了ASFV一些变异株,然而,尚未完全厘清基因变异和病程之间的相关性。

ASFV严重损害先天免疫系统［如干扰素(interferon, IFN)反应,Toll样受体(Toll-like receptor, TLR),主要组织相容性复合体(major histocompatibility complex, MHC)］。ASFV可调节宿主免疫反应的不同阶段,具有复杂的免疫逃避机制。ASFV的一些基因已被鉴定为毒力因子和IFN抑制剂(Afonso等,2004;Reis等,2017b)。最新的基因编辑方法可能有助于设计和生产安全、免疫原性好、缺乏毒力因子的重组ASFV，是未来极有希望的疫苗候选株。

2.6 诊断

由于尚无预防ASFV的疫苗,因此对ASF的防控要采取扑杀和预防的检疫隔离措施。为获得监测和根除方案所需的信息,必须对ASF进行实验室诊断。诊断结果阳性说明动物曾被ASFV感染过,诊断包括感染性病毒、DNA、特异性抗原和抗体的检测和鉴定(Agüero等,2004;Oura等,2013)。

ASF的诊断方法很多：用猪的生物试验和对猪巨噬细胞进行血球吸附试验测定传染性病毒(Orfei等,1968);采用初级(直接)免疫荧光法、免疫过氧化物酶法和酶联免疫吸附试验(enzyme-linked immunosorbent assay, ELISA)测定病毒特异性抗原；用PCR方法检测ASFV的DNA(Fernández-Pinero等,2013;James等,2010;King等,2003)。WOAH建议采用间接ELISA和间接免疫荧光或Western印迹法检测病毒特异性抗体(世界动物卫生组织,2012)。

自2015年年末以来,家畜流行病学、血清学和遗传学研究显示，血清阳性动物发病率显著增加,尤其是在欧洲联盟经济不发达的国家,阳性野猪的数量增加尤其明显。这些结果表明,一些动物可能患有慢性ASF(Olsevskis等,2016;Smietanka等,2016;Wozniakowski等,2016)。

Gallardo等(2015)通过对实验室和现场样本检测,评估了ASFV的DNA、抗原和抗体诊断方法(Gallardo等,2015)。研究者们同时调查了785份来自田间和实验室的猪样本,通过3种PCR方法检测ASFV基因组，结果表明这些猪被基因Ⅱ型ASFV感染(Agüero等,2003;Fernández-Pinero等,2013)。有研究表明,应用UPL-PCR(通用探针库）检测出的DNA阳性样本比用WOAH推荐的PCR检测出的DNA阳性样本多出3.3%。在临床感染的阶段,当血液和组织中病毒水平很高时,两种PCR方法都能够很容易地检测到病毒DNA。

Gallardo(2013)和Oura(2013)等人对ELISA检测方法和3种PCR检测方法的诊断效果进行了比较。结果表明，如果采样和储存条件不佳，

ELISA 方法对临床样品检测的敏感性会大大降低。这可能说明血清阳性动物组织中形成的抗原–抗体复合物,可以阻断 ASFV 抗原与特异性耦联物间的相互作用。在欧盟国家的大多数情况下,ASF 呈急性流行形式,并在所有组织中积累大量病毒,导致动物死亡(Gabriel 等,2011;Gimenez-Lirola 等,2016;Guinat 等,2016)。这说明 ELISA 方法和其他诊断方法同时应用是合理的。

利用免疫过氧化物酶法和间接 ELISA 法分别对 30 头试验感染 ASFV 的猪的基因型进行动态检测,结果表明,检测血清样本用 IPT 法比用间接 ELISA 法具有更高的敏感性。与间接 ELISA 相比,IPT 可检测到 ASFV 早期感染的抗体。与对比的检测方法不同,得到的间接 ELISA 的诊断敏感性也不同,与 IPT 相比,间接 ELISA 的诊断敏感性为 22%~50%。间接 ELISA 的敏感性之所以低,可能与从急性感染 ASFV 的动物中采集样本有关,因为这个时期抗体正在逐渐形成。然而,全面了解 ASF 流行情况,有必要对康复动物和病重动物的抗体进行检测。根据 WOAH 的建议(WOAH,2012),间接 ELISA 的阳性结果要通过免疫荧光替代方法(间接)和免疫印迹进行证实。作者认为,虽然 ASF 的诊断方法有很多,但是 UPL-PCR 联合 IPT 用于 ASFV 基因组和抗体检测是早期诊断最可靠的方法。

研究者通过间接 ELISA(WOAH-ELISA)和 Western 印迹法(WOAH-IB)对 ASFV 抗体血清标本检测,对血液和器官样本中的病毒基因组检测,并根据以上检测结果对坦桑尼亚塞伦盖蒂自然公园里疣猪的 ASF 的疫情进行评估。疣猪的血清间接 ELISA 检测阳性率为 100%(34/34),器官样本用 PCR 检测弱阳性,阳性率只有 8.8%(3/34)。然而,研究者没有从任何样本中分离到具有传染性的病毒。结果与其他报道的信息一致,大多数疣猪的血清呈阳性(Heuschele 和 Coggins,1969)。

应指出的是,2007—2012 年在俄罗斯流行 ASF 期间,同时采用 PCR 和直接免疫荧光技术,确保了诊断的 100%准确性。被感染猪的器官组织中的特异性抗体并不明显,只有 45%的动物被检测到抗体(33%的野猪和 49%的家猪)。家猪和野猪的特异性抗体水平为 $4.3\sim9\log_2$。我们注意到,即使在高效价的抗体中,也不能阻断细胞内抗原形成,也不影响使用直接免疫荧光和 ELISA 检测细胞内抗原(Strizhakova 等,2016)。

选择 ASF 诊断方法时,要重点考虑 ASFV 的感染阶段和流行所导致的疾病类型。对于急性型和亚急性型 ASF,在临床症状出现前可检测到传染性病毒和 DNA。在感染后 7~11 天,血清学转为阳性,可在患病的动物中检测到抗体。传染性病毒(或抗原)和 DNA 检测的阳性结果表明,在取样时,被检测的动物就已经被感染。另一方面,ASFV 抗体呈阳性,表明 ASFV 感染动物已超过 1 周和(或)ASFV 感染后存活下来。

2.7 防控

由于缺乏具体的预防措施,不易对 ASF 进行有效控制。只有采取快速诊断和严格扑杀受感染动物的策略才有助于根除 ASF 和阻止 ASFV 传播。目前还没有安全有效的 ASF 疫苗,但最近的疫苗研究取得了令人鼓舞的成果。本章将向读者介绍关于 ASF 疫苗研究的简短历史及其当前应用的研究方法。

2.7.1 灭活疫苗

为了生产 ASF 传统灭活疫苗,研究者已经做了诸多尝试,包括使用戊二醛固定被感染的巨噬细胞,用紫外线辐射、氟利昂,离子和非离子洗涤剂处理原代细胞和传代细胞的裂解物,用丙内酯灭活固定在牛红细胞、分枝杆菌和 γ 免疫球蛋白上的纯化病毒粒子(Blome 等,2014;Kovalenko 和 Sidorov,1973;Makarov 等,2016;Mebus,1988;

Petrov 等,2018)。目前尚未观察到各种灭活方法制备的灭活病毒疫苗能够提供有效保护作用。相反,在一些情况下,用强毒 ASFV 感染的免疫猪比用强毒 ASFV 感染的非免疫对照组猪发病更严重(Hess,1981;Mebus,1988;Stone 等,1968)。

2.7.2 亚单位疫苗

亚单位疫苗的研究促进了潜在保护蛋白的研究进程。纯化的被感染细胞或重组蛋白 p30、p54、p72 和 CD2v 对猪进行免疫,其免疫效果被认为对猪具有潜在的免疫保护作用 (Barderas 等,2001;Gómez-Puertas 等,1996;Gutiérrez-Castañeda 等,2008;Kollnberger 等,2002)。

用杆状病毒表达的 p30 和 p54 重组蛋白对猪进行免疫接种后,进行强毒攻毒,可延迟 ASF 的临床症状出现(Gómez-Puertas 等,1998)。用脂质体包裹的 ASFV 糖蛋白对猪进行免疫可以诱导抗体产生,但会加速被攻毒动物的死亡。从感染 ASFV 的巨噬细胞中纯化出的血清型特异性的 ASFV 主要糖蛋白 CD2v(gp 110~140)用脂质体包裹后,对猪进行免疫接种,67%被感染的猪受到免疫保护,没有死亡,但对再次感染不能产生免疫保护(Sereda 等,1994)。ASFV CD2v 直接参与敏感细胞吸附 ASFV 感染的过程(Rodríguez 等,1993)。ASFV CD2v 编码位点的基因分型与 ASFV 分离株的血清型的分布相对应(Malogolovkin 等,2015a)。用表达 ASFV 的 CD2v 基因的重组杆状病毒对猪进行免疫接种,可以对随后的强毒攻毒产生免疫保护(Argilaguet 等,2013)。推测该蛋白可能主要诱导 CTL 效应。

因此,大多数研究者认为 p30、p54 和 CD2v 蛋白是诱导 ASF 免疫防御的必需蛋白,但它们与 ASF 的免疫保护无关。

2.7.3 减毒活疫苗

减毒活疫苗是一种很有前途的工具,可用于全面分析宿主的防御机制和发现隐藏的免疫相关特征。最近发表的几篇综述重点介绍了减毒活疫苗的特点(Arias 等,2018;Rock 2017;Souto 等,2016)。

在 Pokrov 研究所,我们对 ASFV 的一些毒株和变异株进行了减毒、改造、筛选和分离,结果表明,用这些毒株和变异株感染家猪,并没有引起死亡,并且能够形成免疫保护,避免受到同源分离株的感染(Vishnjakov 等,1991)。

在抗 ASF 免疫保护的研究中,研究者发现了弱毒 ASFV 有 8 种血清型。减毒候选疫苗对同血清型的强毒 ASFV 攻毒后 10~14 天,75%~90%的动物受到免疫保护(Sereda 和 Balyshev,2011)。尽管所选的减毒毒株达到了规定的保护要求,且是无害的,但它们的病毒血症持续时间和水平,以及形成病毒特异性保护的时间等生物学特性上存在差异。因此,接种 104.0 HAU 50 剂量的减毒 FK-135 株(血清 4 型)后第 7~10 天,可对猪产生 92%~100%的保护,而 MK-200 株(血清 3 型)对动物产生 50%保护,且只有到第 14 天保护率可达 82%~92%。

减毒活疫苗的缺点是病毒的毒力存在差异,有可能在后续的使用过程中恢复部分毒力、亚临床感染有时会转变为慢性感染,以及对免疫功能低下的动物(如妊娠母猪)的免疫保护不足。

已经分离到的 ASFV 自然减毒株有 OURT88/3 或 NH/P68。免疫后的猪可对同源毒株的感染产生免疫保护(Boinas 等,2004;Leitão 等,2001;Mulumba-Mfumu 等,2017),保护率为 66%~100%,这主要取决于猪的种类、接种方式和剂量。自然减毒的 ASFV 作为疫苗具有明显的副作用,免疫接种的一些猪产生了不良反应,包括肺炎、运动障碍、坏死病灶、流产和死亡(Gallardo 等,2015)。

基因缺失的重组 ASFV 参与免疫逃避,如胸苷激酶(TK)、9GL(B119L)基因、NL(DP71L)基因,以及360和505多基因家族(MGF360/505)的某些成员的缺失,可导致原始毒株的毒力衰减,并诱导对同源毒株产生免疫保护(Afonso等,1998;Neilan等,2002;O'Donnell等,2017)。

MGF360和505的6名成员与9GL基因的同时缺失导致了ASFV Georgia07/01株毒力的丧失,而且对随后的同源病毒感染不能产生免疫保护(O'Donnell等,2016)。相反,与仅敲除毒力因子9GL的Georgia07/01分离株相比,同时敲除毒力因子9GL和UK的Georgia07/01分离株接种动物能够产生免疫保护。这些结果表明,连续敲除第二毒力因子的减毒ASFV的重组体更安全。到目前为止,基因编辑的重组ASFV毒株最有希望用作ASF候选疫苗。

2.7.4 DNA疫苗

从概念上讲DNA疫苗很有发展前景,可诱导抗原特异性的细胞免疫,但DNA疫苗在动物体内的安全性有待进一步分析。现今候选DNA疫苗存在的主要问题仍是体内细胞摄入DNA相对较低,特别是在大型哺乳动物中体内细胞摄入DNA更低。这个难题正在利用多种手段来解决(Leifert等,2004;van Drunen Littel-van den Hurk等,2004)。

ASFV病毒的HA(或CD2v)与白细胞CD2分子之间的相似性表明,它可以靶向表达CD2受体(CD48和CD58)的淋巴细胞提呈病毒抗原(Borca等,1994;Rodríguez等,1993)。构建的表达PQ嵌合蛋白(p54和p30)的DNA疫苗加入HA基因后,经3次肌肉免疫接种,可明显增强机体的体液免疫和细胞免疫反应。接种sHA能够增强免疫应答,可能是由于该分子中存在辅助T细胞表位。

另一种方法是利用特异性识别抗原提呈胞表面细胞抗原的单链抗体可变片段,将编码的病毒抗原靶向到抗原提呈细胞(Grossmann等,2009)。将APCH1保守的抗原表位基因融合到PQ嵌合体开放阅读框构建重组质粒pCMV-APCH1PQ,进行猪的免疫,结果证明,APCH1保守的抗原表位可用作基因佐剂。仅编码PQ的重组DNA质粒不能诱导猪体内产生抗体,而用pCMV-APCH1PQ免疫猪,不仅能产生抗PQ的特异性抗体,还能激活MHC-Ⅱ的辅助性T细胞,这表明APCH1分子具有免疫佐剂的功能。然而,这种候选DNA疫苗并不能对随后感染ASFV的猪产生免疫保护(Barderas等,2001)。

虽然用pCMV-sHAPQ免疫猪能诱导体液免疫,但对猪进行攻毒却不能够提供免疫保护。为了促进诱导抗体产生和增强CD8+T细胞特异性应答反应,编码抗原决定簇的p30、p54和sHA,融合有细胞泛素的重组质粒pCMV-UbsHAPQ被构建。与预期的一样,接种pCMV-UbsHAPQ并没有引起猪的体液反应,但对ASFV的攻毒提供了一定保护,证实了T细胞反应在防御ASFV中的重要性。增加DNA疫苗免疫接种次数并没有增强保护效果,这可能反映了第一次免疫后诱导的T细胞反应缺乏免疫促进作用。用pCMV-UbsHAPQ免疫2次,6头猪中有2头存活,只免疫1次4头存活。由此推测,加强免疫对免疫保护有负面影响。根据这些作者的研究,使用pCMVUbsHAPQ进行4次免疫的诱导抗体水平可能很低,造成疾病恶化,反过来可能抑制CD8+T细胞诱导的免疫保护作用(Argilaguet等,2012)。

利用重组DNA表达文库筛选后进行免疫,被认为是一种非常有希望的新发疫病预防用新型疫苗研发策略(Talaat和Stemke-Hale,2005)。对以ASFV基因组的短片段与pCMV-Ub质粒中的泛素基因相融合为代表的ASFVUblib文库中筛选获得的重组DNA进行的免疫保护作用研究,结果表明可特异性增强CTL的诱导(Lacasta等,2014)。获得4029个克隆(总130 000 bp),可覆盖约76%的病毒基因组。

BacMam病毒是杆状病毒表达载体,在哺乳动物启动子控制下编码病毒蛋白,以BacMam病毒为基础构建的疫苗能使目的基因在哺乳动物细胞中高效表达(Argilaguet 等,2013)。

用 10^7 pfu 的 BacMam-sHAPQ 对猪免疫 3次,每次间隔 15 天,3 次免疫后测定 BacMam-sHAPQ 免疫原性。随后,所有动物用 102 HAU50 剂量的同源分离株 E75 感染。和预期一致,对照组在感染前没有产生特异性免疫反应，而接种 BacMam-sHAPQ 的 6 头猪中有 4 头产生特异性 T 细胞免疫应答反应。因此,接种 BacMam-sHAPQ 疫苗后,在不诱导抗体产生的情况下,BacMam-sHAPQ 疫苗也可能对亚致死性的同源 ASFV 感染的猪提供免疫保护。此外,诱导防御与 T 细胞的激活也有直接关系。

2.7.5 抗病毒制剂

抗病毒药物和方法对研究 ASFV 的增殖很有意义,可能在体内抑制病毒复制。一些抗病毒药物的“特定”靶点是病毒 DNA 聚合酶。磷酰基乙酸(phosphonoacetic acid, PAA)是一种有效的抑制 DNA 聚合酶活性的抑制剂，其在细胞培养实验和动物实验中均有效地抑制 DNA 聚合酶活性。根据 ASF 的合成效率、重复性和治疗活性的数据分析,研究者选择了 3 种化合物 PAA,带有 7-氨基 1,3,5-三氮金刚烷(A-14)的 PAA 复合物和 PAA-230 的吡啶钾盐进行研究，在对照组中，防止了 80%以上的被感染动物 100%死亡。在病原微生物污染的条件下,PAA 与美替沙腙(一种特异性的抑制剂，可抑制病毒晚期蛋白)联合使用,防止了与患病动物接触的所有母猪死亡,反之,未使用该化合物的动物 100%死亡。研究表明，该方法可能降低了 ASF 病毒携带水平(药物的剂量、疗程和使用的多样性)。不用致弱强毒,就有可能制备型特异性血清。这使得获得血清的时间缩短了 2/3~11/12，这对病毒的血清分型很重要(Zubairov 等,2017)。

研究表明，氟喹诺酮类药物在体外能抑制 ASFV 的复制(Freitas 等,2016),特别是染料木黄酮可阻断 ASFV 感染 Vero 细胞和猪巨噬细胞(Arabyan 等,2018)。

靶向 ASFV 拓扑异构酶Ⅱ的小干扰 RNA 降低病毒的产量(高达 99.7%)和对多个细胞的感染(75.5%)(Freitas 等,2016)。

在未来非常有希望将 CRISPR-Cas9 技术用于 ASFV 防治。Cas-9 和靶向 p30 病毒蛋白(CP204L)的 71~78 密码子的引导 RNA,可使感染野猪肺细胞的病毒滴度降低 3/4。这种替代方法未来有可能用于设计和培育具有自然抗性的猪(Hubner 等,2018)。

2.7.6 预防与控制

在俄罗斯,ASF 的主要危险因素包括:动物和受污染的猪产品(生肉、火腿、脂肪和皮)流通不受控制;垃圾和交通工具;被感染的动物尸体(野猪或家猪)使用“一针”进行不正确的群体免疫及被感染的野猪。根据 WOAH 建议,相关部门在国家应急计划中列出了应对疫情的一系列卫生措施。在暴发 ASF 时，根据国家立法和我们的经验,建议采取 3 个步骤消毒:使用碱性溶液清洗、使用杀菌泡沫进行定期消毒和最后使用微粒气溶胶消毒剂进行终末消毒(Sereda 等,2015)。

为了监测对 ASF 的消毒效果,农场引入“哨兵”猪(占猪群总数的 10%)。持续监测 60 天后,用 PCR 和 ELISA 分别对“哨兵”猪的血液和血清进行检测。此外,还收集了与农场有关地区的地面土壤和污水样本,收集到的样品送到 ASF 参考实验室进行检测。如果 PCR 检测和 ELISA 检测结果均为阴性,说明消毒效果很好,为猪群补充做好了准备(Sereda 等,2016)。否则,如果“哨兵”猪和地面样本检测出现阳性,应重新消毒。

ASF 在野猪种群中的流行更为复杂。ASFV在猪群中多次传播，可导致病毒毒力发生改变。另外，ASFV 可以在秋冬季存活几个月，已证实感染野生动物的 ASFV 主要来源于野猪尸体。

尽管没有针对 ASF 的疫苗，但历史上在世界各地有许多消灭 ASF 的积极例子(Lyra，2006；Peritz，1981；Sánchez-Vizcaíno 等，2009；Wilkinson，1986)。全面了解 ASF 的传播、致病机制和有效的生物安全措施对降低引入 ASF 风险至关重要。

参考文献

Abrams CC, Goatley L, Fishbourne E, Chapman D, Cooke L, Oura CA et al (2013) Deletion of virulence associated genes from attenuated African swine fever virus isolate OUR T88/3 decreases its ability to protect against challenge with virulent virus. Virology 443:99–105. https://doi.org/10.1016/j.virol.2013.04.028

Achenbach JE, Gallardo C, Nieto-Pelegrin E, Rivera-Arroyo B, Degefa-Negi T, Arias M et al (2017) Identification of a new genotype of African swine fever virus in domestic pigs from Ethiopia. Transbound Emerg Dis 64(5):1393–1404. https://doi.org/10.1111/tbed.12511

Afonso CL, Zsak L, Carrillo C, Borca MV, Rock DL (1998) African swine fever virus NL gene is not required for virus virulence. J Gen Virol 79:2543–2547

Afonso CL, Piccone ME, Zaffuto KM, Neilan J, Kutish GF, Lu Z et al (2004) African swine fever virus multigene family 360 and 530 genes affect host interferon response. J Virol 78:1858–1864. https://doi.org/10.1128/JVI.78.4.1858-1864.2004

Agüero M, Fernández J, Romero L, Mascaraque CS, Arias M, Sánchez-Vizcaíno JM (2003) Highly sensitive PCR assay for routine diagnosis of African swine fever virus in clinical samples. J Clin Microbiol 41:4431–4434. https://doi.org/10.1128/JCM.41.9.4431-4434.2003

Agüero M, Fernández J, Romero LJ, Zamora MJ, Sánchez C, Belák S et al (2004) A highly sensitive and specific gel-based multiplex RT-PCR assay for the simultaneous and differential diagnosis of African swine fever and Classical swine fever in clinical samples. Vet Res 35(5):551–563. https://doi.org/10.1051/vetres:2004031

Alonso C, Borca M, Dixon L, Revilla Y, Rodriguez F, Escribano JM, Consortium IR (2018) ICTV virus taxonomy profile: Asfarviridae. J Gen Virol 99(5):613–614. https://doi.org/10.1099/jgv.0.001049

Anderson EC (1986) African swine fever: current concepts on its pathogenesis and immunology. Revue Scientifique et Technique, Office International Des Epizooties 5:477–486

Anderson EC, Hutchings GH, Mukarati N, Wilkinson PJ (1998) African swine fever virus infection of the bushpig (Potamochoerus porcus) and its significance in the epidemiology of the disease. Vet Microbiol 62:1–15. https://doi.org/10.1016/S0378-1135(98)00187-4

Arabyan E, Hakobyan A, Kotsinyan A, Karalyan Z, Arakelov V, Arakelov G et al (2018) Genistein inhibits African swine fever virus replication in vitro by disrupting viral DNA synthesis. Antivir Res 156:128–137. https://doi.org/10.1016/j.antiviral.2018.06.014

Argilaguet JM, Pérez-Martín E, Nofrarías M, Gallardo C, Accensi F, Lacasta A et al (2012) DNA vaccination partially protects against African swine fever virus lethal challenge in the absence of antibodies. PLoS One 7:e40942. https://doi.org/10.1371/journal.pone.0040942

Argilaguet JM, Pérez-Martín E, López S, Goethe M, Escribano JM, Giesow K et al (2013) BacMam immunization partially protects pigs against sublethal challenge with African swine fever virus. Antivir Res 98:61–65. https://doi.org/10.1016/j.antiviral.2013.02.005

Arias M, Romero L, Agüero M, Canals A, Zamora MJ, Sánchez-Vizcaíno JM (2001) Eradication strategies of infectious diseases: African swine fever and porcine reproductive and respiratory syndrome (PRRS). Magyar Allatorvosok Lapja 123:40–46

Arias M, Jurado C, Gallardo C, Fernandez-Pinero J, Sanchez-Vizcaino JM (2018) Gaps in African swine fever: analysis and priorities. Transbound Emerg Dis 65(Suppl 1):235–247. https://doi.org/10.1111/tbed.12695

Barderas MG, Rodríguez F, Gómez-Puertas P, Avilés M, Beitia F, Alonso C, Escribano JM (2001) Antigenic and immunogenic properties of a chimera of two immunodominant African swine fever virus proteins. Arch Virol 146:1681–1691. https://doi.org/10.1007/s007050170056

Bastos AP, Nix RJ, Boinas F, Mendes S, Silva MJ, Cartaxeiro C et al (2006a) Kinetics of African swine fever virus infection in Ornithodoros crraticus ticks. J Gen Virol 87:1863–1871. https://doi.org/10.1099/vir.0.81765-0

Bastos AP, Portugal RS, Nix RJ, Cartaxeiro C, Boinas F, Dixon LK et al (2006b) Development of a nested PCR and its internal control for the detection of African swine fever virus (ASFV) in Ornithodoros erraticus. Arch Virol 151:819–826. https://doi.org/10.1007/s00705-005-0654-2

Blome S, Gabriel C, Beer M (2013) Pathogenesis of African swine fever in domestic pigs and

European wild boar. Virus Res 173:122–130. pii: S0168-1702(12)00415-7. https://doi.org/10.1016/j.virusres.2012.10.026

Blome S, Gabriel C, Beer M (2014) Modern adjuvants do not enhance the efficacy of an inactivated African swine fever virus vaccine preparation. Vaccine 32(31):3879–3882. https://doi.org/10.1016/J.VACCINE.2014.05.051

Boinas FS, Hutchings GH, Dixon LK, Wilkinson PJ (2004) Characterization of pathogenic and non-pathogenic African swine fever virus isolates from Ornithodoros erraticus inhabiting pig premises in Portugal. J Gen Virol 85:2177–2187. https://doi.org/10.1099/vir.0.80058-0

Boinas FS, Wilson AJ, Hutchings GH, Martins C, Dixon LJ (2011) The persistence of African swine fever virus in field-infected Ornithodoros erraticus during the ASF endemic period in Portugal. PLoS One 6:e20383. https://doi.org/10.1371/journal.pone.0020383

Bool PH, OrdaS A, SaNchez Botija C (1970) Fluorescent antibody test for African swine fever. Revista Del Patronato de Biologia Animal 14:115–132

Borca MV, Kutish GF, Afonso CL, Irusta P, Carrillo C, Brun A et al (1994) An African swine fever virus gene with similarity to the T-lymphocyte surface antigen CD2 mediates hemadsorption. Virology 199:463–468. https://doi.org/10.1006/viro.1994.1146

Borca MV, O'Donnell V, Holinka LG, Ramirez-Medina E, Clark BA, Vuono EA et al (2018) The L83L ORF of African swine fever virus strain Georgia encodes for a non-essential gene that interacts with the host protein IL-1beta. Virus Res 249:116–123. https://doi.org/10.1016/j.virusres.2018.03.017

Bosch J, Rodriguez A, Iglesias I, Munoz MJ, Jurado C, Sanchez-Vizcaino JM, de la Torre A (2017) Update on the risk of introduction of African swine fever by wild boar into disease-free European union countries. Transbound Emerg Dis 64(5):1424–1432. https://doi.org/10.1111/tbed.12527

Brown A-A, Penrith ML, Fasina FO, Beltran-Alcrudo D (2018) The African swine fever epidemic in West Africa, 1996-2002. Transbound Emerg Dis 65(1):64–76. https://doi.org/10.1111/tbed.12673

Burmakina G, Malogolovkin A, Tulman ER, Zsak L, Delhon G, Diel DG et al (2016) African swine fever virus serotype-specific proteins are significant protective antigens for African swine fever. J Gen Virol 97(7):1670–1675. https://doi.org/10.1099/jgv.0.000490

Burrage TG (2013) African swine fever virus infection in Ornithodoros ticks. Virus Res 173:131–139. https://doi.org/10.1016/j.virusres.2012.10.010

Burrage TG, Lu Z, Neilan JG, Rock DL, Zsak L (2004) African swine fever virus multigene family 360 genes affect virus replication and generalization of infection in Ornithodoros porcinus ticks. J Virol 78:2445–2453. https://doi.org/10.1128/JVI.78.5.2445-2453.2004

Caporale V, Rutili D, Nannini D, di Francesco C, Ghinato C (1988) Epidemiology of classical swine fever in Italy from 1970 to 1985. Revue Scientifique et Technique, Office International Des Epizooties 7:599–617

Carlson J, O'Donnell V, Alfano M, Velazquez Salinas L, Holinka LG, Krug PW et al (2016) Association of the host immune response with protection using a live attenuated African swine fever virus model. Viruses 8(10):291. https://doi.org/10.3390/v8100291

Chichikin AY, Gazaev IK, Tsybanov SZ, Kolvasov D (2012) A non-contact method for selecting saliva from a wild boar in African swine fever. Veterinariya 6:26–28

Cisek AA, Dabrowska I, Gregorczyk KP, Wyzewski Z (2016) African swine fever virus: a new old enemy of Europe. Ann Parasitol 62(3):161–167

Costard S, Wieland B, de Glanville W, Jori F, Rowlands R, Vosloo W et al (2009) African swine fever: how can global spread be prevented? Philos Trans R Soc Lond B Biol Sci 364:2683–2696. Retrieved from 19687038%5CnPM

Costard S, Mur L, Lubroth J, Sanchez-Vizcaino JM, Pfeiffer DU (2013) Epidemiology of African swine fever virus. Virus Res 173:191–197. https://doi.org/10.1016/j.virusres.2012.10.030

Davies K, Goatley LC, Guinat C, Netherton CL, Gubbins S, Dixon LK, Reis AL (2017) Survival of African swine fever virus in excretions from pigs experimentally infected with the Georgia 2007/1 isolate. Transbound Emerg Dis 64(2):425–431. https://doi.org/10.1111/tbed.12381

de Carvalho Ferreira HC, Weesendorp E, Elbers ARW, Bouma A, Quak S, Stegeman JA, Loeffen WLA (2012) African swine fever virus excretion patterns in persistently infected animals: a quantitative approach. Vet Microbiol 160:327–340. https://doi.org/10.1016/j.vetmic.2012.06.025

de Carvalho Ferreira HC, Weesendorp E, Quak S, Stegeman JA, Loeffen WLA (2014) Suitability of faeces and tissue samples as a basis for non-invasive sampling for African swine fever in wild boar. Vet Microbiol 172:449–454. https://doi.org/10.1016/j.vetmic.2014.06.016

De Kock G, Robinson EM, Keppel JJG (1940) Swine fever in South Africa. Onderstepoort J Vet Sci 14:31–93

Diaz AV, Netherton CL, Dixon LK, Wilson AJ (2012) African swine fever virus strain Georgia 2007/1 in Ornithodoros erraticus ticks. Emerg Infect Dis 18:1026–1028. https://doi.org/10.3201/eid1806.111728

Dixon LK, Wilkinson PJ (1988) Genetic diversity of African swine fever virus isolates from soft ticks (Ornithodoros moubata) inhabiting warthog burrows in Zambia. J Gen Virol 69:2981–2993. https://doi.org/10.1099/0022-1317-69-12-2981

Dixon LK, Abrams CC, Chapman DDG, Goatley LC, Netherton CL, Taylor G, Takamatsu HH (2013) Prospects for development of African swine fever virus vaccines. Dev Biol 135:147–157. https://doi.org/10.1159/000170936

Edgar G, Hart L, Hayston JT (1952) Studies on the viability of the virus of swine fever. Report 14th International Veterinary Congress, vol 2, pp 387–391

EFSA AHAW Panel (EFSA Panel on Animal Health and Welfare) (2014) Scientific opinion on African swine fever. EFSA J 12(4):3628. https://doi.org/10.2903/j.efsa.2014.3628

Fernández-Pinero J, Gallardo C, Elizalde M, Robles A, Gómez C, Bishop R et al (2013) Molecular diagnosis of African swine fever by a new real-time PCR using universal probe library. Transbound Emerg Dis 60:1–11. https://doi.org/10.1111/j.1865-1682.2012.01317.x

Fillipova NA (1966) Arachnids. In: Fauna of the USSR

Freitas FB, Frouco G, Martins C, Leitao A, Ferreira F (2016) In vitro inhibition of African swine fever virus-topoisomerase II disrupts viral replication. Antivir Res 134:34–41. https://doi.org/10.1016/j.antiviral.2016.08.021

Gabriel C, Blome S, Malogolovkin A, Parilov S, Kolbasov D, Teifke JP, Beer M (2011) Characterization of African swine fever virus Caucasus isolate in European wild boars. Emerg Infect Dis 17(12):2342–2345. https://doi.org/10.3201/eid1712.110430

Galindo I, Almazán F, Bustos MJ, Viñuela E, Carrascosa AL (2000) African swine fever virus EP153R open reading frame encodes a glycoprotein involved in the hemadsorption of infected cells. Virology 266:340–351. https://doi.org/10.1006/viro.1999.0080

Gallardo C, Soler A, Nieto R, Carrascosa AL, De Mia GM, Bishop RP et al (2013) Comparative evaluation of novel African swine fever virus (ASF) antibody detection techniques derived from specific ASF viral genotypes with the OIE internationally prescribed serological tests. Vet Microbiol 162:32–43. https://doi.org/10.1016/j.vetmic.2012.08.011

Gallardo C, Fernandez-Pinero J, Pelayo V, Gazaev I, Markowska-Daniel I, Pridotkas G et al (2014) Genetic variation among African swine fever genotype II viruses, eastern and central Europe. Emerg Infect Dis 20(9):1544–1547. https://doi.org/10.3201/eid2009.140554

Gallardo C, Soler A, Nieto R, Sanchez MA, Martins C, Pelayo V et al (2015) Experimental transmission of African swine fever (ASF) low virulent isolate NH/P68 by surviving pigs. Transbound Emerg Dis 62(6):612–622. https://doi.org/10.1111/tbed.12431

Gallardo C, Nurmoja I, Soler A, Delicado V, Simón A, Martin E et al (2018) Evolution in Europe of African swine fever genotype II viruses from highly to moderately virulent. Vet Microbiol 219:70–79. https://doi.org/10.1016/j.vetmic.2018.04.001

Gimenez-Lirola LG, Mur L, Rivera B, Mogler M, Sun Y, Lizano S et al (2016) Detection of African swine fever virus antibodies in serum and oral fluid specimens using a recombinant protein 30 (p30) dual matrix indirect ELISA. PLoS One 11(9):e0161230. https://doi.org/10.1371/journal.pone.0161230

Gogin A, Gerasimov V, Malogolovkin A, Kolbasov D (2013) African swine fever in the North Caucasus region and the Russian Federation in years 2007-2012. Virus Res 173:198–203. https://doi.org/10.1016/j.virusres.2012.12.007

Goller KV, Malogolovkin AS, Katorkin S, Kolbasov D, Titov I, Höper D et al (2015) Tandem repeat insertion in African swine fever virus, Russia, 2012. Emerg Infect Dis 21:731–732. https://doi.org/10.3201/eid2104.141792

Gómez-Puertas P, Rodríguez F, Oviedo JM, Ramiro-Ibáñez F, Ruiz-Gonzalvo F, Alonso C, Escribano JM (1996) Neutralizing antibodies to different proteins of African swine fever virus inhibit both virus attachment and internalization. J Virol 70:5689–5694

Gómez-Puertas P, Rodríguez F, Oviedo JM, Brun A, Alonso C, Escribano JM (1998) The African swine fever virus proteins p54 and p30 are involved in two distinct steps of virus attachment and both contribute to the antibody-mediated protective immune response. Virology 243:461–471. https://doi.org/10.1006/viro.1998.9068

Greig A, Plowright W (1970) The excretion of two virulent strains of African swine fever virus by domestic pigs. J Hyg 68:673–682. https://doi.org/10.1017/S0022172400042613

Greig AS, Boulanger P, Bannister GL (1967) African swine fever. V. Cultivation of the virus in primary pig kidney cells. Can J Comp Med Vet Sci 31:24–31

Grossmann C, Tenbusch M, Nchinda G, Temchura V, Nabi G, Stone GW et al (2009) Enhancement of the priming efficacy of DNA vaccines encoding dendritic cell-targeted antigens by synergistic toll-like receptor ligands. BMC Immunol 10:43. https://doi.org/10.1186/1471-2172-10-43

Guinat C, Gubbins S, Vergne T, Gonzales JL, Dixon L, Pfeiffer DU (2016) Experimental pig-to-pig transmission dynamics for African swine fever virus, Georgia 2007/1 strain-CORRIGENDUM. Epidemiol Infect 144:3564–3566. https://doi.org/10.1017/S0950268816001667

Gutiérrez-Castañeda B, Reis AL, Corteyn A, Parkhouse RME, Kollnberger S (2008) Expression, cellular localization and antibody responses of the African swine fever virus genes B602L and K205R. Arch Virol 153:2303–2306. https://doi.org/10.1007/s00705-008-0246-z

Hammond RA, Detray DE (1955) A recent case of African swine fever in Kenya, East Africa. Am Vet Med Assoc 126:389–391
Hess WR (1981) African swine fever: a reassessment. Adv Vet Sci Comp Med 25:39–69
Hess WR, Endris RG, Haslett TM, Monahan MJ, McCoy JP (1987) Potential arthropod vectors of African swine fever virus in North America and the Caribbean basin. Vet Parasitol 26:145–155. https://doi.org/10.1016/0304-4017(87)90084-7
Hess WR, Endris RG, Lousa A, Caiado JM (1989) Clearance of African swine fever virus from infected tick (Acari) colonies. J Med Entomol 26:314–317
Heuschele WP, Coggins L (1969) Epizootiology of African swine fever virus in warthogs. Bull Epizoot Dis Afr 17:179–183
Hubner A, Petersen B, Keil GM, Niemann H, Mettenleiter TC, Fuchs W (2018) Efficient inhibition of African swine fever virus replication by CRISPR/Cas9 targeting of the viral p30 gene (CP204L). Sci Rep 8(1):1449. https://doi.org/10.1038/s41598-018-19626-1
James HE, Ebert K, McGonigle R, Reid SM, Boonham N, Tomlinson JA et al (2010) Detection of African swine fever virus by loop-mediated isothermal amplification. J Virol Methods 164:68–74. https://doi.org/10.1016/j.jviromet.2009.11.034
Jurado C, Fernandez-Carrion E, Mur L, Rolesu S, Laddomada A, Sanchez-Vizcaino JM (2018) Why is African swine fever still present in Sardinia? Transbound Emerg Dis 65(2):557–566. https://doi.org/10.1111/tbed.12740
Jurkov GG, Peskovatskov AP, Shpackov AK, Mackarevich VG, Cherevatenko BN, Balabanov VA (2014) Report on the results of studying the routes of African swine fever (ASF) entry and spreading over Odessa region in 1977
Karaulov AK, Shevtsov AA, Petrova ON, Korennoy FI, Vadopolas TV (2018) The forecast of African swine fever spread in Russia until 2025. Veterinaria i Kormlenie 3:12–14
King DP, Reid SM, Hutchings GH, Grierson SS, Wilkinson PJ, Dixon LK et al (2003) Development of a TaqMan PCR assay with internal amplification control for the detection of African swine fever virus. J Virol Methods 107:53–61. https://doi.org/10.1016/S0166-0934(02)00189-1
King K, Chapman D, Argilaguet JM, Fishbourne E, Hutet E, Cariolet R et al (2011) Protection of European domestic pigs from virulent African isolates of African swine fever virus by experimental immunisation. Vaccine 29:4593–4600. https://doi.org/10.1016/j.vaccine.2011.04.052
Kleiboeker SB (2002) Swine fever: classical swine fever and African swine fever. Vet Clin N Am Food Anim Pract 18:431–451
Kolbasov D, Titov I, Tsybanov S, Gogin A, Malogolovkin A (2018a) African swine fever virus, Siberia, Russia, 2017. Emerg Infect Dis 24(4):796–798. https://doi.org/10.3201/eid2404.171238
Kolbasov DV, Gogin A, Malogolovkin A (2018b) Ten years with African swine fever—lessons learned. In: 25th International Pig Veterinary Society congress, p 37
Kollnberger SD, Gutierrez-Castañeda B, Foster-Cuevas M, Corteyn A, Parkhouse RME (2002) Identification of the principal serological immunodeterminants of African swine fever virus by screening a virus cDNA library with antibody. J Gen Virol 83:1331–1342
Korennoy FI, Gulenkin VM, Gogin AE, Vergne T, Karaulov AK (2017) Estimating the basic reproductive number for African swine fever using the Ukrainian historical epidemic of 1977. Transbound Emerg Dis 64:1858–1866. https://doi.org/10.1111/tbed.12583
Kovalenko Y, Sidorov MA (1973) [Reservoirs and mode of circulation of African swine fever virus in nature]. Sel'skokhozyaistvennaya Biologiya 8:598–606
Kovalev YI (2017) Veterinariy v svinovodstve. In: Swine production in Russia 2015-2020: current challenges, risks and solutions. Novosibirsk, 18–19 May
Lacasta A, Ballester M, Monteagudo PL, Rodríguez JM, Salas ML, Accensi F et al (2014) Expression library immunization can confer protection against lethal challenge with African swine fever virus. J Virol 88(22):13322–13332. https://doi.org/10.1128/JVI.01893-14
Leifert JA, Rodriguez-Carreno MP, Rodriguez F, Whitton JL (2004) Targeting plasmid-encoded proteins to the antigen presentation pathways. Immunol Rev 199:40–53. https://doi.org/10.1111/j.0105-2896.2004.0135.x
Leitão A, Cartaxeiro C, Coelho R, Cruz B, Parkhouse RME, Portugal FC et al (2001) The non-haemadsorbing African swine fever virus isolate ASFV/NH/P68 provides a model for defining the protective anti-virus immune response. J Gen Virol 82:513–523
Lyra TMP (2006) The eradication of African swine fever in Brazil, 1978-1984. Revue Scientifique et Technique (International Office of Epizootics) 25:93–103
Makarov V, Nedosekov V, Sereda A, Matvienko N (2016) Immunological conception of African swine fever. Zool Ecol 26(3):236–243. https://doi.org/10.1080/21658005.2016.1182822
Malogolovkin A, Yelsukova A, Gallardo C, Tsybanov S, Kolbasov D (2012) Molecular characterization of African swine fever virus isolates originating from outbreaks in the Russian Federation between 2007 and 2011. Vet Microbiol 158(3–4):415–419. https://doi.org/10.1016/j.vetmic.2012.03.002
Malogolovkin A, Burmakina G, Titov I, Sereda A, Gogin A, Baryshnikova E, Kolbasov D (2015a) Comparative analysis of African swine fever virus genotypes and serogroups. Emerg Infect Dis

21(2):312–315. https://doi.org/10.3201/eid2102.140649
Malogolovkin A, Burmakina G, Tulman ER, Delhon G, Diel DG, Salnikov N et al (2015b) African swine fever virus CD2v and C-type lectin gene loci mediate serological specificity. J Gen Virol 96(4):866–873. https://doi.org/10.1099/jgv.0.000024
Mebus CA (1988) African swine fever. Adv Virus Res 35:251–269
Misinzo G (2012) African swine fever virus, Tanzania, 2010–2012. Emerg Infect Dis 193:319–328. https://doi.org/10.3201/eid1812.121083
Monteagudo PL, Lacasta A, Lopez E, Bosch L, Collado J, Pina-Pedrero S et al (2017) BA71DeltaCD2: a new recombinant live attenuated African swine fever virus with cross-protective capabilities. J Virol 91(21):e01058-17. https://doi.org/10.1128/JVI.01058-17
Mulumba-Mfumu LK, Achenbach JE, Mauldin MR, Dixon LK, Tshilenge CG, Thiry E et al (2017) Genetic assessment of African swine fever isolates involved in outbreaks in the democratic Republic of Congo between 2005 and 2012 reveals co-circulation of p72 genotypes I, IX and XIV, including 19 variants. Viruses 9(2):E31. https://doi.org/10.3390/v9020031
Mur L, Martínez-López B, Sánchez-Vizcaíno J (2012) Risk of African swine fever introduction into the European Union through transport-associated routes: returning trucks and waste from international ships and planes. BMC Vet Res 8:149. https://doi.org/10.1186/1746-6148-8-149
Mur L, Sanchez-Vizcaino JM, Fernandez-Carrion E, Jurado C, Rolesu S, Feliziani F et al (2018) Understanding African Swine Fever infection dynamics in Sardinia using a spatially explicit transmission model in domestic pig farms. Transbound Emerg Dis 65(1):123–134. https://doi.org/10.1111/tbed.12636
Neilan JG, Zsak L, Lu Z, Kutish GF, Afonso CL, Rock DL (2002) Novel swine virulence determinant in the left variable region of the African swine fever virus genome. J Virol 76:3095–3104. https://doi.org/10.1128/JVI.76.7.3095-3104.2002
Nurmoja I, Petrov A, Breidenstein C, Zani L, Forth JH, Beer M et al (2017a) Biological characterization of African swine fever virus genotype II strains from north-eastern Estonia in European wild boar. Transbound Emerg Dis 64(6):2034–2041. https://doi.org/10.1111/tbed.12614
Nurmoja I, Schulz K, Staubach C, Sauter-Louis C, Depner K, Conraths FJ, Viltrop A (2017b) Development of African swine fever epidemic among wild boar in Estonia—two different areas in the epidemiological focus. Sci Rep 7(1):12562. https://doi.org/10.1038/s41598-017-12952-w
O'Donnell V, Holinka LG, Sanford B, Krug PW, Carlson J, Pacheco JM et al (2016) African swine fever virus Georgia isolate harboring deletions of 9GL and MGF360/505 genes is highly attenuated in swine but does not confer protection against parental virus challenge. Virus Res 221:8–14. https://doi.org/10.1016/j.virusres.2016.05.014
O'Donnell V, Risatti GR, Holinka LG, Krug PW, Carlson J, Velazquez-Salinas L et al (2017) Simultaneous deletion of the 9GL and UK Genes from the African swine fever virus Georgia 2007 isolate offers increased safety and protection against homologous challenge. J Virol 91(1):e01760-16. https://doi.org/10.1128/JVI.01760-16
Oganesyan AS, Petrova ON, Korennoy FI, Bardina NS, Gogin AE, Dudnikov SA (2013) African swine fever in the Russian Federation: spatio-temporal analysis and epidemiological overview. Virus Res 173:204–211. https://doi.org/10.1016/j.virusres.2012.12.009
Okoth E, Gallardo C, Macharia JM, Omore A, Pelayo V, Bulimo DW et al (2013) Comparison of African swine fever virus prevalence and risk in two contrasting pig-farming systems in South-west and Central Kenya. Prev Vet Med 110:198–205. https://doi.org/10.1016/j.prevetmed.2012.11.012
Olesen AS, Lohse L, Hansen MF, Boklund A, Halasa T, Belsham GJ et al (2018) Infection of pigs with African swine fever virus via ingestion of stable flies (Stomoxys calcitrans). Transbound Emerg Dis 65:1152–1157. https://doi.org/10.1111/tbed.12918
Olsevskis E, Guberti V, Serzants M, Westergaard J, Gallardo C, Rodze I, Depner K (2016) African swine fever virus introduction into the EU in 2014: experience of Latvia. Res Vet Sci 105:28–30. https://doi.org/10.1016/j.rvsc.2016.01.006
Onashvili T, Donduashvili M, Borca M, Mamisashvili E, Goginashvili K, Tighilauri T et al (2012) Countermeasures for the control of African Swine Fever in Georgia. Int J Infect Dis 16:e268. https://doi.org/10.1016/j.ijid.2012.05.920
Orfei Z, Persechino A, Lupini PM, Cassone A (1968) Haemadsorption test in the diagnosis of African swine fever in Italy. Atti Soc Ital Sci Vet 21:850–854
Oura CAL, Edwards L, Batten CA (2013) Virological diagnosis of African swine fever—comparative study of available tests. Virus Res 173:150–158. https://doi.org/10.1016/j.virusres.2012.10.022
Owolodun OA, Yakubu B, Antiabong JF, Ogedengbe ME, Luka PD, John Audu B et al (2010) Temporal dynamics of African swine fever outbreaks in Nigeria, 2002-2007. Transbound Emerg Dis 57:330–339. https://doi.org/10.1111/j.1865-1682.2010.01153.x
Pan IC, Hess WR (1985) Diversity of African swine fever virus. Am J Vet Res 46:314–320
Pan IC, Shimizu M, Hess WR (1980) Replication of African swine fever virus in cell cultures. Am J Vet Res 41:1357–1367
Parker J, Plowright W, Pierce MA (1969) The epizootiology of African swine fever in Africa. Vet

Rec 85:668–674
Pastor MJ, Laviada MD, Sanchez-Vizcaino JM, Escribano JM (1989) Detection of African swine fever virus antibodies by immunoblotting assay. Can J Vet Res = Revue Canadienne de Recherche Veterinaire 53:105–107. https://doi.org/10.4314/nvj.v27i2.3518
Pejsak Z, Truszczyński M, Kozak E, Markowska-Daniel I (2014) [Epidemiological analysis of two first cases of African swine fever in wild boar in Poland] Analiza epidemiologiczna dwóch pierwszych przypadków afrykańskiego pomoru świń u dzików w Polsce. Medycyna Weterynaryjna 70:369–372. Retrieved from http://www.scopus.com/inward/record.url?eid=2-s2.0-84903171078&partnerID=tZOtx3y1
Penrith ML (2009) African swine fever. Onderstepoort J Vet Res 76:91–95
Penrith ML, Thomson GR, Bastos ADS, Phiri OC, Lubisi BA, du Plessis EC et al (2004) An investigation into natural resistance to African swine fever in domestic pigs from an endemic area in southern Africa. Revue Scientifique et Technique, Office International Des Épizooties 23:965–977. Retrieved from http://www.ncbi.nlm.nih.gov/pubmed/15861893%5Cnhttp://www.cabdirect.org/abstracts/20053080761.html?resultNumber=18&start=0&q=("'hog+cholera"+OR+"'classical+swine+fever")+AND+yr:[2000+TO+2012]+AND+(cattle+OR+sheep+OR+goats+OR+pigs+OR+poultry)+%5Cn%5Cn%5Cn&fq=gl_
Peritz FJ (1981) The evolution of African swine fever in Latin America and F.A.O.'s corresponding action programme. Bulletin de l'Office International Des Epizooties 93(469):499
Petrov A, Forth JH, Zani L, Beer M, Blome S (2018) No evidence for long-term carrier status of pigs after African swine fever virus infection. Transbound Emerg Dis 65:1318–1328. https://doi.org/10.1111/tbed.12881
Pietschmann J, Guinat C, Beer M, Pronin V, Tauscher K, Petrov A et al (2015) Course and transmission characteristics of oral low-dose infection of domestic pigs and European wild boar with a Caucasian African swine fever virus isolate. Arch Virol 160(7):1657–1667. https://doi.org/10.1007/s00705-015-2430-2
Plowright W, Perry CT, Peirce MA (1970) Transovarial infection with African swine fever virus in the argasid tick, Ornithodoros moubata porcinus, Walton. Res Vet Sci 11:582–584
Plowright W, Thomson GR, Neser JA (2002) African swine fever. In: Infectious diseases of livestock with special reference to Southern Africa, vol 2, pp 567–599. Retrieved from http://www.oie.int/eng/normes/mmanual/2008/pdf/2.08.01_ASF.pdf
Ramiro-Ibáñez F, Ortega A, Ruiz-Gonzalvo F, Escribano JM, Alonso C (1997) Modulation of immune cell populations and activation markers in the pathogenesis of African swine fever virus infection. Virus Res 47:31–40. https://doi.org/10.1016/S0168-1702(96)01403-7
Reis AL, Goatley LC, Jabbar T, Sanchez-Cordon PJ, Netherton CL, Chapman DAG, Dixon LK (2017a) Deletion of the African swine fever virus gene DP148R does not reduce virus replication in culture but reduces virus virulence in pigs and induces high levels of protection against challenge. J Virol 91(24):e01428-17. https://doi.org/10.1128/JVI.01428-17
Reis AL, Netherton C, Dixon LK (2017b) Unraveling the armor of a killer: evasion of host defenses by African swine fever virus. J Virol 91(6):e02338-16. https://doi.org/10.1128/JVI.02338-16
Rennie L, Wilkinson PJ, Mellor PS (2001) Transovarial transmission of African swine fever virus in the argasid tick Ornithodoros moubata. Med Vet Entomol 15:140–146. https://doi.org/10.1046/j.1365-2915.2001.00282.x
Rock DL (2017) Challenges for African swine fever vaccine development—"... perhaps the end of the beginning". Vet Microbiol 206:52–58. https://doi.org/10.1016/j.vetmic.2016.10.003
Rodríguez JM, Yáñez RJ, Almazán F, Viñuela E, Rodriguez JF (1993) African swine fever virus encodes a CD2 homolog responsible for the adhesion of erythrocytes to infected cells. J Virol 67:5312–5320
Salguero FJ, Ruiz-Villamor E, Bautista MJ, Sánchez-Cordón PJ, Carrasco L, Gómez-Villamandos JC (2002) Changes in macrophages in spleen and lymph nodes during acute African swine fever: expression of cytokines. Vet Immunol Immunopathol 90:11–22. https://doi.org/10.1016/S0165-2427(02)00225-8
Salguero FJ, Gil S, Revilla Y, Gallardo C, Arias M, Martins C (2008) Cytokine mRNA expression and pathological findings in pigs inoculated with African swine fever virus (E-70) deleted on A238L. Vet Immunol Immunopathol 124:107–119. https://doi.org/10.1016/j.vetimm.2008.02.012
Sánchez-Cordón PJ, Núñez A, Salguero FJ, Carrasco L, Gómez-Villamandos JC (2005) Evolution of T lymphocytes and cytokine expression in classical swine fever (CSF) virus infection. J Comp Pathol 132(4):249–260. https://doi.org/10.1016/j.jcpa.2004.10.002
Sanchez-Cordon PJ, Chapman D, Jabbar T, Reis AL, Goatley L, Netherton CL et al (2017) Different routes and doses influence protection in pigs immunised with the naturally attenuated African swine fever virus isolate OURT88/3. Antivir Res 138:1–8. https://doi.org/10.1016/j.antiviral.2016.11.021
Sanchez-Vizcaino JM, Slauson DO, Ruiz-Gonzalvo F, Valero F (1981) Lymphocyte function and cell-mediated immunity in pigs with experimentally induced African swine fever. Am J Vet Res 42:1335–1341

Sánchez-Vizcaíno JM, Martínez-Lópeza B, Martínez-Avilés M, Martins C, Boinas F, Vial L et al (2009) Scientific review on African Swine Fever. CFP/EFSA/AHAW/2007/2, pp 1–141. https://doi.org/10.2903/sp.efsa.2009.EN-5
Sánchez-Vizcaíno JM, Mur L, Martínez-López B (2012) African swine fever: an epidemiological update. Transbound Emerg Dis 59:27–35. https://doi.org/10.1111/j.1865-1682.2011.01293.x
Sereda AD, Balyshev VM (2011) [Antigenic diversity of African swine fever viruses]. Vopr Virusol 56(4):38–42. Retrieved from http://www.ncbi.nlm.nih.gov/pubmed/21899069
Sereda AD, Anokhina EG, Makarov VV (1994) [Glycoproteins from the African swine fever virus]. Vopr Virusol 39:278–281. Retrieved from http://www.ncbi.nlm.nih.gov/entrez/query.fcgi?cmd=Retrieve&db=PubMed&dopt=Citation&list_uids=7716925
Sereda A, Selyaninov Y, Egorova I, Balyshev V, Bureev I, Kushnir A et al (2015) Testing means and methods of disinfection for the African swine fever virus. Veterinariya 7:51–55
Sereda A, Zhivodeorov S, Baluishev V, Strizhackova O, Sindryackova I, Lunitsin A, Kolbasov D (2016) Experience of determining disinfection level in carrying out African swine fever eradication campaigns in pig farms using a sentinel method. Veterinariya 6:44–48
Sierra MA, Carrasco L, Gómez-Villamandos JC, Martin de las Mulas J, Méndez A, Jover A (1990) Pulmonary intravascular macrophages in lungs of pigs inoculated with African swine fever virus of differing virulence. J Comp Pathol 102:323–334. https://doi.org/10.1016/S0021-9975(08)80021-7
Smietanka K, Wozniakowski G, Kozak E, Niemczuk K, Fraczyk M, Bocian L et al (2016) African swine fever epidemic, Poland, 2014–2015. Emerg Infect Dis 22(7):1201–1207. https://doi.org/10.3201/eid2207.151708
Smirnov AM, Butko MP (2011) Stability of the pathogen and measures to combat African swine fever. Veterinarian 6:2–7
Souto R, Mutowembwa P, van Heerden J, Fosgate GT, Heath L, Vosloo W (2016) Vaccine potential of two previously uncharacterized African swine fever virus isolates from southern Africa and heterologous cross protection of an avirulent European isolate. Transbound Emerg Dis 63:224–231. https://doi.org/10.1111/tbed.12250
Stone SS, DeLay PD, Sharman EC (1968) The antibody response in pigs inoculated with attenuated African swine fever virus. Can J Comp Med (Gardenvale, Quebec) 32:455–460
Strizhakova OM, Lyska VM, Malogolovkin AS, Novikova MB, Sidlik MV, Nogina IV et al (2016) Validation of an ELISA kit for detection of antibodies against ASF virus in blood or spleen of domestic pigs and wild boars. Sel'skokhozyaistvennaya Biologiya 51(6):845–852. https://doi.org/10.15389/agrobiology.2016.6.845eng
Talaat AM, Stemke-Hale K (2005) Expression library immunization: a road map for discovery of vaccines against infectious diseases. Infect Immun 73(11):7089–7098. https://doi.org/10.1128/IAI.73.11.7089-7098.2005
Terpstra C, Wensvoort G (1986) African swine fever in the Netherlands. Tijdschr Diergeneeskd 111:389–392
van Drunen Littel-van den Hurk S, Babiuk SL, Babiuk LA (2004) Strategies for improved formulation and delivery of DNA vaccines to veterinary target species. Immunol Rev 199:113–125. https://doi.org/10.1111/j.0105-2896.2004.00140.x
Vishnjakov I, Mitin N, Karpov G, Kurinnov VJ (1991) Differentiation African and classical swine fever viruses. Veterinariya 4:28–31
Wilkinson PJ (1986) African swine fever in Belgium. State Vet J 40:123–129; 12 ref.
World Organisation for Animal Health (OIE) (2012) Manual of diagnostic tests and vaccines for terrestrial animal
Wozniakowski G, Kozak E, Kowalczyk A, Lyjak M, Pomorska-Mol M, Niemczuk K, Pejsak Z (2016) Current status of African swine fever virus in a population of wild boar in eastern Poland (2014-2015). Arch Virol 161(1):189–195. https://doi.org/10.1007/s00705-015-2650-5
Zani L, Forth JH, Forth L, Nurmoja I, Leidenberger S, Henke J et al (2018) Deletion at the 5′-end of Estonian ASFV strains associated with an attenuated phenotype. Sci Rep 8(1):6510. https://doi.org/10.1038/s41598-018-24740-1
Zubairov M, Selyaninov Y, Roshchin A, Khokhlov P (2017) Antiviral activity and therapeutic and preventive effect of phosphonoacetic acid and its derivatives. Electr J Chem Saf 1:146–157. https://doi.org/10.25514/CHS2017111440

第 3 章 猪瘟病毒

Dilip K. Sarma

3.1 引言

经典猪瘟病毒（classical swine fever virus, CSFV）是CSF或猪霍乱的病原体，CSF给养猪业带来了重大经济损失，并且可以发生跨界传播，在家猪和野猪中流行。CSF是WOAH名录中的传染病，由于其对养猪业具有重大经济影响，被WOAH列为法定报告传染病(Edwards 等，2000)。尽管许多国家已经成功地对CSF实施了控制和根除方案，但CSF仍是一个全球性问题。由于全球化的发展，人口频繁流动，生猪集约化贸易，CSFV易感宿主家猪增多，CSFV自然宿主野猪的数量也在增加，这可能继续成为影响全球猪的健康的重要问题(Moennig 和 Becher，2015)。患病动物的高发病率和高死亡率，造成了巨大的经济损失。

在世界各国，农村养猪对维持居民的生计和提高社会经济具有十分重要的意义。因此，对猪肉生产，特别是在猪肉生产质量和卫生方面进行重大改变都非常必要。养猪业面临的两大挑战是生物安全性差和病毒性疫病的泛滥。据报道，世界范围上大约有40%的猪为粗放饲养且生物安全性低(Postel 等，2017)。世界上有超过60%的猪在亚洲，大部分都是在住家庭院养殖。尽管每家庭院养殖数量较少，但生物安全标准低、兽医服务缺乏、管理不善和疫病预防能力差等因素，促进了疫病的不断传播。因此，具有良好的生物安全性设施、科学规范的管理和切实可行的防控方案的标准化养猪场，必然会减少疫病带来的经济负担和经济损失。

3.2 结构与分类

CSFV属于黄病毒科(flaviviridae)瘟病毒属(pestivirus)，与其他黄病毒属(flavivirus)、丙型肝炎病毒属(hepacivirus)、庚型肝炎病毒属(pegivirus)同属于一科(Beer 等，2015)。瘟病毒属也包括1型和2型牛病毒性腹泻病毒(bovine viral diarrhoea viruse, BVDV)，羊边界病病毒，几种来自长颈鹿、叉角羚的未分类的非典型瘟病毒，邦戈万那病毒和非典型猪瘟病毒，这些病毒的遗传性与抗原性都与CSFV存在一定关系。土耳其报道的边界病毒群中的两种新的瘟病毒，Aydin/04和Burdur/05，与CSFV在遗传性和抗原性上存在着密切的关系(Postel 等，2015)。

根据CSFV全基因组序列分析及其抗原差异性，瘟病毒属被分为11个不同的种(Smith 等，2017)，CSFV则属于瘟病毒属的C种。

CSFV衣壳呈二十面体对称结构，RNA被核蛋白包裹，表面有一层脂质囊膜，上有三个囊膜糖蛋白，直径为40~60nm(Moennig 等，2003)。CSFV病毒粒子结构示意图如图3.1所示。

3.3 基因组

CSFV基因组全长约12.3kb，为单股正链RNA病毒(Wengler 等，1995)。基因组含有一个大的开放阅读框(open reading frame, ORF)，编码一个约3898个氨基酸的多聚蛋白，并在基因组5'和3'端有2个非编码区(NTR)。5'端的NTR携带内部核糖体进入位点(Kolupaeva 等，2000)。病毒基因组编码的非结构自身蛋白酶(N^{pro})位于N端，接着编码4种结构蛋白，即核蛋白C，E^{rns}，E1和E2，以及8种非结构(non-structural, NS)蛋白，即N^{pro}，p7，NS2，NS3，NS4A，NS4B，NS5A和NS5B蛋白(Meyers 和 Thiel，1996)。NS5B蛋白是RNA依赖RNA聚合酶(Lackner 等，2006)，NS3是蛋白酶(Tautz 等，1997)。核蛋白C是小蛋白，在核衣壳的形成中起着重要的作用，并且对基因表达具有调控作用(Liu 等，1998)。E^{rns}(E0)糖蛋白被称为gp44，其大小约为44 kDa，由227个氨基酸组成，与病毒囊膜松散联结(van Gennip 等，2000)。E^{rns}糖蛋白缺乏跨膜结构域，与病毒粒子一起从被感染细胞中分泌出来(Lin 等，2000)。据报道，E^{rns}在病毒的吸附和进入宿主细胞中起一定作用(Hulst 等，2000)。E1是最小的囊膜糖蛋白，其分子量为

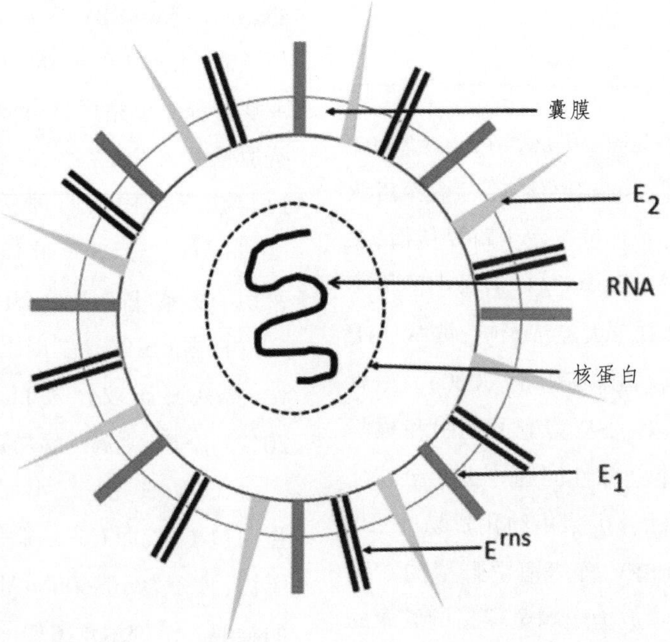

图 3.1 CSFV 示意图，显示了主要结构成分，例如，RNA、核蛋白、囊膜和囊膜糖蛋白。

33kDa，在病毒吸附宿主细胞过程中起作用(Fernandez-Sainz 等，2009)。E2 是主要的囊膜糖蛋白，含有 373 个氨基酸，分子量约为 55 kDa。E2 是主要的抗原蛋白，在病毒感染期间诱导免疫应答，产生中和抗体。CSFV 的 E2 包含 4 个抗原结构域(A，B，C 和 D)(Qi 等，2008)，这些抗原结构域位于 E2 蛋白近 N 端的 1/2 处。编码结构蛋白的基因位于基因组的 5'端，编码非结构蛋白的基因大部分位于基因组的 3'端(Moennig 等，2003)。CSFV 的非结构蛋白 p7 是一种分子量为 7 kDa 的小疏水性多肽，在孔隙形成过程中起作用，CSFV 的成孔蛋白参与病毒的致病过程（Gladue 等，2012)。其他非结构蛋白 NS2-3 至 NS5B 主要在病毒复制中起作用。图 3.2 为 CSFV 的基因组结构示意图。

3.4 吸附、侵入和复制

CSFV 的 2 个囊膜糖蛋白 E^{rns} 和 E2 在病毒感染中具有重要作用，负责病毒对宿主细胞的吸附和侵入。硫酸乙酰肝素可以在许多类型的细胞表面表达，常作为受体吸附病毒。此外，另一个细胞受体粘连蛋白受体已被证明是 CSFV 的宿主细胞受体(Chen 等，2015)。病毒吸附细胞后，通过网格蛋白介导的内吞作用进入细胞。病毒进入细胞依赖于动力蛋白（动力蛋白是一种 100kDa 的 GTP 酶，在与胞膜融合形成囊泡过程中起着至关重要的作用)，pH 值和胆固醇，以及 Rab5 和 Rab7(Shi 等，2016)。Rab 蛋白是外周膜蛋白，具有鸟苷三磷酸酶(GTPase)折叠子，可调节膜运输的许多过程，包括囊泡形成和膜融合。CSFV 的 2 个囊膜糖蛋白，即 E1 蛋白和 E2 蛋白通过病毒粒子中的 2 个半胱氨酸残基之间的二硫键形成异二聚体，异二聚体介导病毒进入细胞。病毒囊膜与细胞膜的融合依赖于 pH 值，并由核内体酸化启动。已经证明，E2 蛋白的 2 个多肽能够介导病毒囊膜和宿主细胞膜之间的融合。病毒进入细胞脱壳后，释放病毒基因组并翻译成蛋白。除结构蛋白外，非结构蛋白(如 NS3、NS4A、NS4B、NS5A 和 NS5B)对于 CSFV 复制也是必需的(Risager 等，2013)。非结构蛋白 NS2 具有自动蛋白酶活性，负责 NS2-3 蛋白的裂解(Lackner 等，2006)。先前研究表明，未裂解的 NS2-3 蛋白对形成具有传染性的 CSFV 病毒颗粒

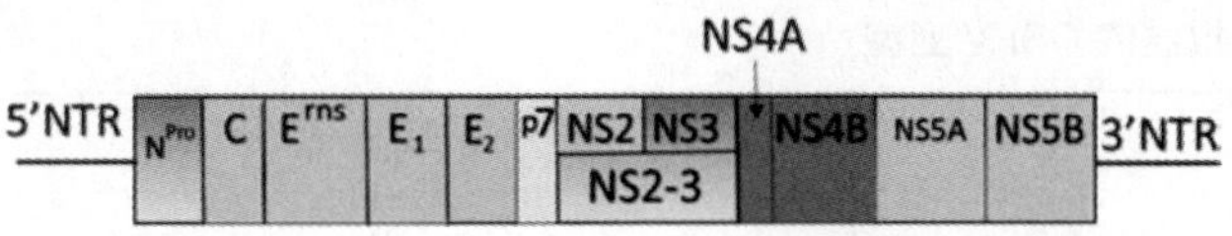

图 3.2　示意图(未按比例绘制)显示了 CSFV 的基因组结构。编码结构蛋白(C,E^{rns},E1 和 E2)的基因位于 5'NTR。编码非结构蛋白的 N^{pro} 基因也位于 5'NTR，而编码其他非结构蛋白（P7,NS2,NS3,NS4A,NS4B,NS5A 和 NS5B）的基因位于 3' NTR。致细胞病变 CSFV 毒株的 NS2–3 蛋白连续裂解为 NS2 和 NS3,仅能在致细胞病变的 CSFV 毒株感染的细胞内可检测到游离 NS3 蛋白(Meyer,2011)。

至关重要(Moulin 等,2007),致细胞病变 CSFV 毒株的 NS2–NS3 裂解是连续进行的。

3.5　遗传多样性

为了确定 CSFV 的遗传多样性，主要用 5' NTR,一段包含 150 个核苷酸(nt)的片段和 E2 蛋白中大小为 190nt 的片段进行 CSFV 的系统发育研究(Greiser–Wilke 等,2006)。除此之外,还用编码聚合酶的基因 NS5B 蛋白中大小为 409nt 的片段进行 CSFV 的系统发育研究(Paton 等,2000a)。但是，欧盟参考实验室认为使用编码 E2 蛋白的全长序列进行 CSFV 系统发育分析更可信(Postel 等,2012)。全长测序可用于准种分析、高分辨率分子流行病学研究及 CSFV 的毒力因子研究(Töpfer 等,2013)。

全球分离到的 CSFV 株可分 3 种基因型,即 1、2 和 3(Lowings 等,1996),并且每个基因型都有不同的基因亚型，例如,1.1、1.2、1.3、2.1、2.2、2.3、3.1、3.2、3.3 和 3.4(Paton 等,2000a)。古巴报道了 CSFV 的另一种亚型 1.4(Postel 等,2013)。最近，在巴西报道了两种新的亚型流行(1.5 和 1.6)(Silva 等,2017)。此外,在厄瓜多尔也报道了 CSFV 的一种新基因亚型 1.7 流行(Rios 等,2018)。

最近,基于 GenBank 中完整的 E2 基因序列，提出了一种新的 CSFV 基因分型标准。依据该基因分型标准,CSFV 被分为 5 个基因型(1~5)和 14 个基因亚型（每个基因型 1 和 2 为 7 个亚型）(Rios 等,2018)。

CSFV 毒株系统发育分析表明，基因型和地理位置之间存在相关性(Bartak 和 Greiser–Wilke，2000)。例如,亚型 1.1、1.2 和 1.3 在亚洲、南美和俄罗斯被发现,亚型 2.1、2.2 和 2.3 在欧洲和亚洲部分地区被发现，亚型 3.1、3.2、3.3 和 3.4 仅在亚洲被发现。历史上流行的毒株,包括属于基因 1 型的 CSFV 疫苗株，目前流行的大部分基因 2 型毒株,大部分基因 3 型毒株都分布在亚洲(Paton 等，2000a)。20 世纪 90 年代在西欧暴发 CSFV 时,分离到的毒株大多数属于基因 2 型(原英文为 Group 2，为了全文命名一致,翻译为基因 2 型)。近年来,在欧洲和亚洲被分离到的基因 1 型和 3 型野毒株转化为基因 2 型(Cha 等,2007)。表 3.1 为在全球不同大陆 CSFV 基因亚型的分布。

据报道，在印度流行的 CSFV 主要基因亚型是 1.1(Sarma 等,2011),但后来,一些研究者通过对来自印度不同地区的 CSFV 分离株进行系统发育分析,也发现了其他亚型,如 1.2、2.1 和 2.2 的存在(Patil 等,2010;Roychoudhury 等,2014)。研究者已经对来自印度的 CSFV 亚型 1.1、2.2 和 2.1 的完整基因组序列进行了报道（Gupta 等,2011；Kumar 等,2014;Ahuja 等,2015)。最近的一项研究中发现了基因亚型 2.1 分离株的遗传多样性,基因亚型 2.1 分离株可进一步分为 10 个亚基因亚型(2.1a~j),只有亚基因亚型 2.1d 分离株在印度流行,其余 9 个亚基因亚型在中国流行(Gong 等，2016)。

据报道,CSFV 基因 1 型随着时间推移,保持相对稳定没有发生变化,而 CSFV 基因 2 型自1980

表 3.1 一些大洲不同国家报告的 CSFV 亚型

	CSFV 亚型/基因型										
分布	1.1	1.2	1.3	1.4	1.5	1.6	1.7	2.1	2.2	2.3	3
亚洲	+	+	+	−	−	−	−	+	+	+	+
北美洲(古巴)	−	+	−	+	−	−	−	−	−	−	+
南美洲	+	−	+	−	+	+	+	−	−	−	−
欧洲	−	−	−	−	−	−	−	+	+	+	−
非洲(马达加斯加)	−	−	−	−	−	−	−	+	−	−	−
澳大利亚	−	−	−	−	−	−	−	−	−	−	−
南极洲	−	−	−	−	−	−	−	−	−	−	−

+:报道;−:未报道。

年以来逐渐蔓延，在 2000—2010 年间出现波动(Kwon 等,2015)。与基因 1 型相比,基因 2 型的遗传多样性水平增加了许多倍。CSFV 基因 1 型和 2 型的 E2 基因的进化速率分别为 5.76×10^{-4} 和 17.29×10^{-4} 替换/(位点·年)。与基因 1 型相比,基因 2 型在低免疫压力的猪群中短时间内广泛流行,并且有着比较快的进化速度(Kwon 等,2015),这可能是免疫压力低所致。但是,CSFV 的遗传多样性并没有引起抗原的多样性。

3.6 流行病学

3.6.1 地理分布

一种类似 CSF 的疫病于 1822 年在法国(Hanson,1957)首次被报道,随后于 1833 年在美国俄亥俄州也有该病的相关记载，很快就在欧洲和美国广泛流行,并在 20 世纪末传到全球的多个地方。许多国家在没有接种疫苗的情况下也成功地控制了 CSF(Edwards 等,2000),北美(古巴除外)、大洋洲和北欧部分地区已成功消灭了 CSF。加拿大自 1963 年以来就没有出现过 CSF,美国于 1961 年开始实施正式猪瘟根除计划,有关 CSF 病例最后记载的时间是 1976 年(Wise,1986)。巴西部分地区消灭了 CSF。自 1991 年以来,乌拉圭就没有关于 CSF 的报道,已全面消灭了 CSF。智利的最后一例 CSF 病例记录时间是 1996 年 8 月，从 1997 年 10 月开始在全国范围内禁止使用 CSF 疫苗,并且从 1998 年 4 月起,智利也宣布消灭了 CSF(没有接种疫苗)(Edwards 等,2000)。

表 3.2 列出了截至 2018 年 5 月,根据世界动物卫生组织(2018)发布的《陆生动物卫生法典》的规定被认为无 CSF 的成员国。尽管日本宣布为无 CSF 国家，但是在宣布后 26 年，于 2018 年 9 月 CSFV 再次出现,感染了家猪和野猪。再次出现的 CSFV 属于 2.1 基因亚型内的一个新的分枝(Postel 等,2019)。在亚洲和东南亚,经常暴发 CSF,据报道，这些地区的病毒种类最丰富 (Paton 和 Greiser-Wilke,2003)。在非洲大陆除马达加斯加以外，其他地区还没有 CSF 确诊的报道(Penrith 等,2011),CSF 在马达加斯加呈地方性流行。这是由于在 1965 年马达加斯加从欧洲引进猪时,引入了 CSF，自此 CSF 呈现地方流行性 (Penrith 等，2011)。在印度所有的养猪州都有关于 CSF 的报道(Sarma 等,2008a)。

3.6.2 宿主易感性

猪科成员易感染 CSFV,CSFV 的自然宿主是家猪和野猪,卷毛野猪被认为天生易感 CSFV(WOAH,2007),但人类不易感。印度有报道称在

表3.2 根据《陆生动物卫生法典》规定被认为无CSF的成员国

澳大利亚	丹麦	卢森堡	斯洛伐克
奥地利	芬兰	墨西哥	斯洛文尼亚
比利时	法国	新喀里多尼亚	西班牙
保加利亚	德国	新西兰	瑞典
加拿大	匈牙利	挪威	瑞士
智利	爱尔兰	巴拉圭	荷兰
哥斯达黎加	意大利	波兰	英国
捷克共和国	日本	葡萄牙	美国
	列支敦士登	罗马尼亚	

家猪和野猪中分离和检测到CSFV (Sarma等，2008b;Barman等,2014)。还证实了疣猪(非洲野猪属普通疣猪)和丛林猪(非洲野猪属假面野猪)对CSFV具有易感性(Everett等,2011)。据报道，在印度阿萨姆邦濒临灭绝的侏儒猪中检测到CSFV(Barman等,2012)。虽然CSFV不能在反刍动物中复制,但在猪中发现了反刍动物瘟病毒。实验动物通常不易被感染，但CSFV已成功地适应了兔体生长,并使病毒对猪的毒力减弱。

3.6.3 病毒抵抗力

在物理和化学条件下检测了CSFV的抵抗力。由于CSFV的囊膜属于脂质,因此去污剂和脂质溶剂很容易将病毒灭活(McKissick 和Gustafson，1967),使用1%福尔马林,2%次氯酸盐,6%甲酚，5%苯酚和2%氢氧化钠可灭活病毒(Moennig和Plagemann,1992)。在5℃时,CSFV的半衰期平均为2~4天,但在30℃时,仅为1~3小时,而在100℃以上时,则不到1分钟(Downing等,1977)。病毒在温度不同的粪便中存活率存在明显差异(Weesendorp等,2008)。在不同的时间和温度条件下研究病毒灭活,CSFV在90℃和70℃分别在1分钟内和5分钟内被灭活(Rehman,1987)。CSFV在低温下相对稳定，因此实验室样品的处理和诊断样品的运输应在低温条件下进行。尽管将样品在室温下(20~25℃)短期放置,病毒可能不会被破坏,但建议将诊断样品保存在4℃(Edwards等，2000)。CSFV在冷藏肉中可以存活数月,而在冻肉中可以存活数年。病毒在pH值为5~10的环境中仍可存活，但在pH值为3以下及pH值为10以上时则迅速失活(Terpstra,1991)。在冬季,CSFV在被污染的围栏中最多可以存活4周。

3.6.4 传播

CSFV可以通过不同的方式和途径传播。健康的猪可以通过与被感染的猪直接接触而被感染,CSFV可通过被感染动物的唾液、尿液和粪便排出(Laeven等,1999)。家猪可通过直接接触或者间接接触被感染的野猪肉传播。CSFV可经精液传播,母猪在配种时被感染。有研究者已经报道了由于用被CSFV污染的精液进行人工授精,发生病毒传播的情况(Floegel等,2000)。由于该病毒可以穿过胎盘屏障，因此该病毒可以从被感染的母猪经子宫垂直传播给仔猪。患病猪的新鲜猪肉已被证明是传播CSFV的危险因素。还有证据表明鸟类可以传播CSFV (FAO,2010)。车辆、设备、衣物和污染的针头等传播媒介可以传播病毒。叮咬昆虫可以机械传播病毒(Van Oirschot,2004)。

除上述因素外，其他因素也可能增加CSFV的传入和传播风险。CSF的临床症状多变且缺乏典型的临床特征,早期难以发现,而且容易与其他疫病的症状混淆(例如,猪皮炎肾病综合征、猪繁殖与呼吸综合征和先天性震颤),现场未能及时准确进行诊断都会促进 CSFV 的传播。迟报、控制措施缺乏、卫生条件差和管理不善也会促进该病的传播。在某些区域,家猪与野猪数量的增加及家猪与野猪的紧密接触,增加了 CSFV 传播的风险。由于印度等许多国家/地区缺少有组织的生猪屠宰场,为此屠宰前不对生猪进行检疫。患病猪直接运到菜市场进行屠宰,增加了病毒的传播机会。

3.6.5 发病机制

在自然条件下,CSFV 大多经口腔或鼻腔途径感染。进入宿主后,CSFV 首先感染扁桃体上皮细胞,然后扩散到附近的淋巴结。在 24 小时内发展成病毒血症,病毒从最初复制的部位传播到其他淋巴组织(脾脏、淋巴集结、淋巴结和胸腺)、内皮细胞、骨髓和循环系统中的白细胞。CSFV 对网状内皮细胞系统中的巨噬细胞和树突状细胞具有特殊的亲和力。在胰腺、大脑、心脏、胆囊和膀胱、下颌唾液腺、肾上腺、甲状腺、肝脏和肾脏也能观察到病毒(Trautwein 和 Leiss,1988)。CSFV 在感染后 3~4 天,影响到许多上皮细胞,并且出现在被感染猪的排泄物和分泌物中。在感染后期,病毒可以在角质细胞、毛囊上皮细胞和真皮间充质细胞间传播。CSFV 可引起淋巴组织耗竭,导致免疫抑制,使得猪更易被其他病原体感染。骨髓受损造成白细胞减少症和血小板减少症。CSFV 破坏内皮细胞,引起许多部位出血。在慢性 CSF 病例中,沉积在肾脏的抗原-抗体复合物导致肾小球肾炎。CSFV 通过感染妊娠母猪,穿过胎盘,引起部分或全部胎儿被感染。宫内感染造成的影响取决于妊娠的阶段,可能产生木乃伊胎、流产、死胎或活仔被持续感染。

淋巴细胞减少症是 CSFV 感染的最典型症状之一。被强毒 CSFV 感染后,引起血小板、NK 细胞、Th 细胞、Tc 细胞、CDT 细胞、CD3-、CD4-、CD8-,以及 CD3+、CD4+、CD8+淋巴细胞的数量显著减少(Zhou 等,2009)。许多研究证明,中等毒力 CSFV 或强毒 CSFV 感染可以引起淋巴细胞凋亡,从而导致淋巴细胞减少症和血小板减少症(Sanchez-Cordon 等,2005)。CSFV 感染巨噬细胞释放的 TNF,可诱导被感染细胞及其周边未被感染细胞凋亡(Choi 等,2004)。IL-1 是一种重要的抗感染的炎性细胞因子,可引起发热、痛觉过敏、血管舒张和低血压(Contassot 等,2012)。多项研究表明,强毒 CSFV 感染后 2 天,在血液中就可检测到 IFN-α,在感染后 3~5 天就达到最高水平。CSFV 感染引起的严重淋巴细胞减少症与 IFN-α 反应有关,被感染动物可能表现出 B 和 T 淋巴细胞衰竭。事实上,淋巴细胞的凋亡与 IFN-α 的提前过表达密切相关(Summerfield 等,2006),这在强毒 CSFV 感染中很常见。但是低毒力的 CSFV 感染不会诱导或降低 IFN-α 和促炎细胞因子的表达水平(Summerfield 等,2006)。CSFV 对网状内皮细胞系统的吞噬细胞具有高度亲和力,这些细胞被感染会导致血管通透性增加、淋巴细胞减少、血小板减少、凝血障碍,以及胸腺和骨髓萎缩(Gomez-Villamandos 等,2003)。进一步研究表明,CSFV 能够诱导骨髓淋巴细胞和中性粒细胞凋亡(Sato 等,2000)。

3.6.6 临床症状和病理变化

根据 CSFV 分离株的毒力和宿主反应,CSF 的临床表现可分为不同形式,如最急性、急性、慢性和持续性。自然感染 CSFV 通常会引起不同的临床表现。CSFV 强毒株感染,通常潜伏期短,引起最急性型感染,但随着最急性型感染消失,这种情况已经发生了改变(Meyers 和 Thiel,1996)。因此,在临床上 CSF 感染分为急性感染(一过性的或致死性)、慢性感染、先天性感染和轻度感染。CSF 的潜伏期通常为 3~10 天(Moennig 和 Greiser-Wilke,2008)。自然感染条件下,有时在病毒感染后 4 周或更晚之后才表现出临床症状(Laeven 等,1999)。急性感染的临床表现更常见于仔猪到 12 周龄的猪。急性型 CSF 的特征性临床表现是高热，通常高于41℃，但成年猪的体温不会超过40℃。被感染的动物表现出厌食、蜷缩、嗜睡、结膜炎、呼吸道症状和短暂便秘，随后出现腹泻。可能有神经症状，如步态蹒跚、后肢无力、运动不协调和抽搐。腹部、耳、尾巴和四肢内侧也有充血或出血现象（Moennig 等，2003）。

剖检病死猪，可观察到皮下瘀血肿胀、水肿、浅表淋巴结出血。脾脏局部充血、出血、典型梗死。肠系膜血管严重充血和心脏心外膜表面斑点状出血，肺的非萎缩性出血及肝大和充血。肾脏的被膜下出血，类似于“火鸡卵”肾是急性感染的特征。会厌软骨、胆囊黏膜和回肠连接处可见点状出血和斑块状出血。

尽管大多数淋巴组织有典型的肉眼可见的病理变化，但没有特征性的病理组织学变化。淋巴结和扁桃体通常可见局部出血到弥漫性出血、淋巴滤泡耗竭和坏死的细胞碎片、脾髓出血和脾小体中淋巴细胞耗竭。肾脏显示皮质髓质区出血，肾间质内有单核细胞浸润，表现出特征性的间质性肾炎。肺部表现为间质性肺炎，肺泡广泛出血，暴露在空气中有纤维蛋白渗出。肺泡间隔增厚，单核细胞浸润，肺泡毛细血管充血。细支气管周围区域出血和淋巴组织增生，淋巴聚集。肝脏的变化是中央静脉和肝窦腔充血，肝小叶中心变性，局部坏死。派尔集合淋巴结的淋巴滤泡内可见淋巴细胞坏死和耗竭。大脑皮质血管形成袖套，伴有局灶性或弥漫性胶质增生。

患急性 CSF 的动物通常在感染后 2~4 周死亡。患病动物死亡率取决于动物的年龄和病毒毒力，死亡率最高可以达 100%（Moennig 等，2003）。然而并不是每个毒株的毒力都得到测定（Mittelholzer 等，2000）。CSFV 感染后引起的免疫抑制，使得二次感染的进程和症状变得复杂多样。

慢性型 CSF 的初期症状类似于急性型感染，但随病程发展，出现非特异特征性临诊症状，如间歇性发热、腹泻和发育迟缓。慢性感染的猪可能会在 2~3 个月死亡。患有慢性型 CSF 的猪从临床症状出现到死亡期间，CSFV 通常一直持续存在。免疫系统开始产生抗体，就可在血清样品中检测到，但是这些抗体还不能清除宿主体内的病毒（Depner 等，1996）。慢性型 CSF 缺乏典型的病理变化，尤其是器官和浆膜无出血变化。患有慢性型 CSF，且出现腹泻的病猪，在回肠、回盲瓣和直肠上出现坏死和溃疡性病变。在慢性病例中，患病猪的盲肠和结肠中经常出现边缘凸起溃疡，称为纽扣状溃疡。

先天性 CSF 患病猪是在母体妊娠期间被感染，病毒可以通过胎盘屏障，垂直感染子宫内的胎儿。子宫内发生感染的程度取决于妊娠阶段。妊娠前 3 个月被感染导致反复配种和胎儿木乃伊化，而妊娠后 3 个月被感染主要导致流产、畸形、弱仔或新生仔猪死亡。在这种条件下出生的仔猪通常具有免疫耐受性，不仅携带病毒，而且还可以传播病毒。有时仔猪表现出虚弱和震颤。先天性感染的过程被称为迟发性 CSF（Kaden 等，2005；Moennig等，2003）。妊娠后 50~70 天被非致病性 CSFV 感染，出生的仔猪会被持续感染，免疫耐受的仔猪一般能成活，但是有的也会出现死亡，成活下来的仔猪在存活期间能够传播病毒。

轻症 CSF 通常发生在老龄动物中，其特征是一过性发热、厌食和康复。康复的动物获得持久免疫力。

3.6.7 感染免疫

CSFV 与宿主免疫系统的相互作用很复杂，包括诱导先天性和适应性免疫反应。囊膜糖蛋白 E^{rns} 和 E2 都可以诱导产生中和抗体，并且可以各自提供免疫保护（Konig 等，1995）。但中和抗体主要还是由 E2 蛋白产生，且 E2 蛋白单独也能够诱导动物产生保护性抗体。被 CSFV 感染的动物能产生抗非结构蛋白 NS3 的抗体，但这些抗体没有中和能力。CSFV 的其他蛋白质免疫原性较低，虽然在病毒结构蛋白和非结构蛋白中能够检测到 T 细胞表位（Armengol 等，2002），但对细胞免疫在

CSFV 感染中的作用知之甚少。据报道,CSFV 感染可引起 IFN 的激活基因和其他免疫应答基因的表达(Summerfield 和 Ruggli,2015)。在 CSFV 感染后，中和抗体通常在 2 周内被检测到，但慢性型 CSF 病猪被感染后 1 个月才能检测到中和抗体。被动免疫可以对出生后前 5 周的仔猪提供保护，但这并不能阻止病毒的复制，而且也不能彻底清除病毒。

3.7 实验室诊断

3.7.1 样本选择

剖检动物的扁桃体组织、淋巴结、脾脏和回肠末端都可用作 CSFV 实验室诊断样本。眼瞬膜(半透明的第三眼睑)也可用作 CSFV 检测,因为眼瞬膜自我分解速度比内脏器官缓慢(Teifke 等,2005)。在病毒血症期间，也可以采集活猪的全血样本用于病毒检测，但是如果有可能的话应该多采几头猪的全血样本(Van Oirschot,2004)。血清样本可以从疑似发病 3 周以上的猪和恢复期的猪进行采集,用于实验室病毒特异性抗体的检测。

3.7.2 病毒分离

通过细胞培养分离 CSFV 仍被认为是诊断 CSF 的金标准。细胞可用于分离组织、全血或血浆中的 CSFV。在感染的早期选择全血或血浆检测 CSFV 比用白细胞检测 CSFV 更敏感 (Gisler 等,1999)。猪的肾原代细胞或 PK-15,SK6,PS 和 STE(猪睾丸上皮样细胞)等细胞系都可用于 CSFV 分离培养。由于 CSFV 感染细胞不会产生细胞病变,因此需要用免疫学和分子生物学技术检测细胞培养中的病毒。

3.7.3 抗原检测

免疫学技术已被广泛用于疫病的实验室诊断,且在过去的几十年中得到了迅速发展。已在野外条件下对荧光抗体试验检测 CSFV 抗原的敏感性进行了评估(Bouma 等,2001)。

分别采集疑似感染 CSFV 猪和实验感染 CSFV 猪的扁桃体样本用于 CSFV 检测,其中,从疑似感染 CSFV 猪的扁桃体组织中检测病毒抗原的灵敏度预估为 75%，从实验感染 CSFV 猪样本中检测病毒抗原的灵敏度预估为 99%。使用直接荧光抗体试验,在 2 小时内就可在冰冻的组织切片中检测到 CSFV 抗原(Van Oirschot,2004)。CSFV 感染后第 2 天，就可以从被感染猪的扁桃体中检测到 CSFV 抗原。除被感染猪的扁桃体外,在淋巴结、脾脏和回肠中也能检测到病毒抗原。但是荧光抗体试验不能将 CSFV 和其他瘟病毒区分，除非在测试中使用特异性单克隆抗体 (Van Oirschot,2004)。双抗体夹心 ELISA 已经被标准化,并用于临床样品中的 CSFV 抗原的检测 (Sarma 和 Sarma,1995)。对自然感染猪和供人食用被屠宰猪的组织样品分别利用双抗夹心 ELISA 法和斑点 ELISA 法进行检测，并对两种检测方法进行比较分析,统计分析表明,这两种方法检测结果基本一致(Sarma 和 Meshram,2008)。据报道,抗原 ELISA 有望实现对早期被感染，且没有出现临床症状的猪进行检测 (Raut 等,2015)。ELISA 是一种快速的、大样本筛选方法，但其敏感性不如 RT-PCR (Clavijo 等,1998;Van Oirschot,2004)。

已有报道称，现已建立了用于 CSFV 抗原快速检测的多克隆/单克隆胶体金试纸条(Zhang 等,2007),以及用于 CSFV 抗原检测的快速免疫磁珠法(Conlan 等,2009)。

3.7.4 核酸检测

分子诊断技术特别是基于 PCR 的诊断技术被广泛应用于病毒核酸的检测，且分子生物学检测技术与病毒分离技术相当，甚至更好 (Paton 等,2000b;Handel 等,2004)。此外,基于 PCR 的诊断技术比传统的病毒分离技术能更早地检测出 CSFV,为了检测被感染组织中的病毒 RNA,有学者已报道了一种荧光原位杂交的检测方法(Nagarajan 和 Saikumar,2012)。PCR 检测方法也可以在福尔马林固定的组织中检测到病毒 RNA（Singh 等，2005)。也有报道利用 RT-PCR 方法检测冷冻组织中 CSFV RNA（Chopade 等,2010)。除了常规的 RT-PCR,用于 CSFV 核酸检测的实时 RT-PCR 方法(Zhao 等,2008;Wen 等,2010)也被建立,具有更高的特异性和敏感性。

反转录环介导等温扩增(reverse transcription loop-mediated isothermal amplification, RT-LAMP)技术也被应用于 CSFV 的快速检测（Chen 等，2010),结果表明,该方法的灵敏度比标准的 RT-PCR 方法高 100 倍,RT-PCR 检测阴性的样品，通过 RT-LAMP 检测呈阳性。此外,除了 CSFV,其他病毒，例如,BVDV 和 PRRSV 用 CSFV 的 RT-LAMP 未能检测到。由于 RT-LAMP 成本低、检测快速,因此可以用作现场 CSFV 监测的工具,特别是在发展中国家（Chen 等,2010）应用。在 RT-LAMP 检测方法中也可使用羟色酚蓝染料进行 CSFV 的 RNA 的检测(Wongsawat 等,2011)。除应用于诊断外,分子生物学技术还可用于每次暴发的 CSFV 中分离株间的相互关系分析，以此了解流行病学和病毒进化。

3.7.5 抗体检测

一些用于检测被感染动物血清中的 CSFV 特异性抗体的方法已经被建立。病毒中和法是检测 CSFV 抗体最常用的方法。除此之外,用于 CSFV 抗体检测的封闭 ELISA,间接 ELISA 和过氧化物酶联中和试验也被建立(Terpstra 等,1984;Clavijo 等,2001;Langedijk 等,2001;Lin 等,2005)。免疫层析试纸条也用于 CSFV 抗体的快速检测(Li 等,2012)。但是这些检测方法不适用于免疫接种的动物与野毒感染动物的鉴别，并且由于 CSFV 与 BVDV 存在交叉免疫反应，在检测病毒抗体时可能会出现假阳性结果。过氧化物酶联中和试验、荧光抗体试验、病毒中和法和 ELISA 可用于血清学的诊断或监测，这些方法也是 WOAH 规定的,在进行国际贸易时用于猪的血清样品筛查。

虽然在 CSFV 抗体检测方面取得了一些进展，但 CSFV 与反刍动物瘟病毒的交叉免疫反应给 CSF 的诊断带来了挑战。在土耳其的绵羊和山羊中检测到两个反刍动物瘟病毒 Aydin/04 和 Burdur/05 与 CSFV 存在着免疫交叉反应。两种病毒全基因组测序表明，它们在 CSFV 和边界病病毒之间单独形成一个群，但在遗传和抗原上与 CSFV 关系较近(Becher 等,2012;Postel 等,2015)。相比之下，最近在猪中发现的非典型猪瘟病毒不会干扰 CSF 的诊断(Postel 等,2017)。

3.7.6 预防与控制

在全球范围内用于预防和控制 CSF 的两个主要策略是系统性疫苗接种和根除政策。CSF 流行国家的预防和控制措施主要为实施疫苗接种计划。修饰的改良式活毒疫苗或减毒活疫苗常用于 CSF 的预防，一些国家也使用 E2 的亚单位疫苗(Huang 等,2014)。尽管改良式活毒疫苗比 E2 亚单位疫苗能提供更好的免疫保护，但接种改良式活毒疫苗后,无法区分自然感染与疫苗免疫动物。尽管如此,由于 CSF 在大多数国家呈现地方性流行,而市场上更先进的 CSF 标记疫苗(DIVA)和辅助的鉴别诊断技术有限，因此改良式活毒疫苗仍被广泛应用(Greiser-Wilke 和 Moennig,2004)。中国于 20 世纪 50 年代研制的 C 株弱毒活疫苗已成为中国控制 CSF 的主要措施，该疫苗是根据 C-株 CSFV 在兔体内连续传代毒力致弱(兔化弱毒疫苗)研制而成,随后在细胞培养中开发了减毒疫苗。普遍认为 CSFV 的 C 株疫苗是控制 CSF 的金标准疫苗(Dewulf 等,2004)。据报道,由于 CSFV 特异性 IFN-γ 的产生,C 株疫苗接后的 6 天就能提供早期的免疫保护作用，接种后 2~3 周就能检测到中和抗体，可以激发针对所有基因型的 CSFV 保护性免疫应答(Suradhat 和 Damrongwatanapokin,2003)。CSFV 不同毒株已用于减毒活疫苗的生产。除 C 株外，还包括 GPE、Thiverval、Brescia、

PAV-250等有效疫苗株。这些疫苗株用于预防接种也是安全的,并且可产生较长的免疫力(de Smit等,2001)。由于被CSFV感染动物和疫苗接种动物无法区分,所以用于进行国际贸易交易的动物及其产品在前18个月禁止使用全病毒的减毒活疫苗(Greiser-Wilke和Moennig,2004;Moennig等,2003)。

为了防止来自野猪的CSFV感染扩散,已在野猪中用C株活疫苗进行了一些试验(Kaden等,2000)。由于难以对野猪进行肌肉接种,因此已经尝试使用以诱饵形式口服减毒活疫苗。这种口服疫苗的缺点之一是这种免疫接种方式并不适合于所有动物。诱饵主要被成年动物采食,可产生良好的免疫力,但年幼动物没有机会采食到诱饵。该疫苗的另一个缺点是难以区分疫苗接种动物和被CSFV感染的动物。已经探索了开发标记疫苗或被称之为DIVA疫苗的几种可能性。对于DIVA疫苗,可以使用病毒的小部分,例如,表达单个的蛋白(Moormann等,2000)或仅使用蛋白的一部分,如单肽(Dong和Chen,2005),或删除编码囊膜蛋白E^{rns}基因。E2亚单位疫苗(DIVA)在某些国家/地区已有销售。此外,2014年欧洲药品管理局批准了一种候选的嵌合标记疫苗"CP7_E2alf"(Renson等,2013)。尽管这个疫苗还处于研究阶段,但可以预料这可能是对家猪和野猪进行紧急疫苗接种的有力工具。重组DNA疫苗也有助于诱导免疫(Hammond等,2001)。也可通过缺失全病毒突变体(van Gennip等,2002)或嵌合病毒(van Gennip等,2000;de Smit等,2001)来实现CSFV感染的免疫保护。嵌合病毒是基于CSFV的C株的感染性克隆,其中主要抗原蛋白E^{rns}或E2或其中一部分被BVDV的相应部分取代(de Smit等,2001),反之亦然(Reimann等,2004)。

据报道,使用基因1型CSFV疫苗可以为基因2型CSFV存活提供很好的免疫环境,使之不断进化,逃避免疫监测。研究表明,对流行地区CSF改良式活毒疫苗使用免疫剂量不足会导致CSFV发生进化,这也突出了制定缓解策略的必要性,以最大限度地减少疫苗逃逸突变体出现的相关风险(Yoo等,2018)。为了减缓CSFV的进化,除了对流行的CSFV进行连续监测和分子生物学特性分析外,还需要制定正确有效的疫苗接种策略。此外,需要寻找更有效和稳定的疫苗株,以此根除CSF并切断CSFV的进化。

考虑到传统的CSF减毒疫苗反复使用,难以与被CSFV感染动物区分,已经开展抗病毒药物的研究,许多药物,如前列腺素和衣霉素已经证明在PK-15和SK-6细胞中能抑制CSFV的增殖和扩散(Freitas等,1998;Tyborowska等,2007)。最近,检测了新设计的糖基转移酶底物(硫糖基的类似物)抗CSFV的活性,其中两种糖复合物不仅抑制了CSFV的增殖,而且还有效抑制了SK-6细胞中病毒蛋白的产生(Gawolek等,2017)。据报道,使用干扰小RNA抑制CSFV基因组复制和病毒颗粒产生,结果表明,干扰小RNA可使病毒基因组拷贝数减少4~12倍,并且在72~84小时对传染性病毒的产生抑制多达467倍(Xu等,2008)。因此,应用干扰小RNA策略控制CSFV有望替代传统的CSFV预防方法。

虽然疫苗已经用于CSF的控制,但在野猪中应用疫苗控制CSF仍具有挑战性。C株CSFV减毒疫苗制成诱饵用于野猪口服免疫。但是,要在野猪中成功地进行大规模口服疫苗接种,关键是要研制出易于检测、适口和有效摄入的口服诱饵,这是相当困难的(Rossi等,2015)。

3.8 展望

考虑到CSF对养猪生产的影响,控制和根除CSF对保障粮食安全和农村生活、促进经济增长具有重要意义。尽管许多国家取得了重大成就,但全球根除这一疫病进展缓慢。许多亚洲国家的经济和社会因素对CSF的控制产生了不利影响。在一些亚洲国家,一个突出的问题是,要保证优质疫苗的供应及其大规模使用，才能为大量的猪提供充分的免疫保护。世界上约有65%的猪在亚洲，并且随着对猪肉和猪肉产品需求的增加，应把应对此类疫病(尤其是CSF)作为维持猪的产量的首要任务。在许多发达国家，面临的挑战主要是野猪疫病的控制。利用现有的诊断方法将感染动物与疫苗接种动物区分开是另一个需要解决的重要问题。由于新瘟病毒的出现及其与CSFV存在交叉反应，在野外条件下快速、特异性诊断CSF，对发展中国家而言更为迫切。在全球范围内根除CSF应该是未来的主要目标，理想的疫苗毒株和新型抗病毒药物及持续的监测对于完成这一艰巨的任务至关重要。

参考文献

Ahuja A, Bhattacharjee U, Chakraborty AK et al (2015) Complete genome sequence of classical swine fever virus subgenogroup 2.1 from Assam, India. Genome Announc 3:e01437-14

Armengol E, Wiesmuller KH, Wienhold D et al (2002) Identification of T-cell epitopes in the structural and non-structural proteins of classical swine fever virus. J Gen Virol 83:551–560

Barman NN, Bora DP, Tiwari AK, Kataria RS et al (2012) Classical swine fever in the pygmy hog. Rev Sci Tech Off Int Epiz 31:919–930

Barman NN, Bora DP, Khatoon E et al (2014) Classical swine fever in wild hog: report of its prevalence in northeast India. Transbound Emerg Dis 63:540–547

Bartak P, Greiser-Wilke I (2000) Genetic typing of classical swine fever virus isolates from the territory of the Czech Republic. Vet Microbiol 77:59–70

Becher P, Schmeiser S, Oguzoglu TC et al (2012) Complete genome sequence of a novel pestivirus from sheep. J Virol 86:11412. https://doi.org/10.1128/JVI.01994-12

Beer M, Goller KV, Staubach C et al (2015) Genetic variability and distribution of classical swine fever virus. Anim Health Res Rev 16:33–39

Bouma A, Stegeman JA, Engel B et al (2001) Evaluation of diagnostic tests for the detection of classical swine fever in the field without a gold standard. J Vet Diagn Investig 13:383–388

Cha SH, Choi EJ, Park JH et al (2007) Phylogenetic characterization of classical swine fever viruses isolated in Korea between 1988 and 2003. Virus Res 126:256–261

Chen L, Fan X-Z, Wang Q et al (2010) A novel RT-LAMP assay for rapid and simple detection of classical swine fever virus. Virol Sin 25:59–64

Chen J, He WR, Shen L et al (2015) The laminin receptor is a cellular attachment receptor for classical swine fever virus. J Virol 89:4894–4906

Choi C, Hwang KK, Chae C (2004) Classical swine fever virus induces tumor necrosis factor-alpha and lymphocyte apoptosis. Arch Virol 149:875–889

Chopade NA, Deshmukh VV, Rautmare SS et al (2010) Detection of classical swine fever virus from frozen tissue by RT-PCR. Anim Sci Rep 4:56–59

Clavijo A, Zhou EM, Vydelingum S et al (1998) Development and evaluation of a novel antigen capture assay for the detection of classical swine fever virus antigen. Vet Microbiol 60:155–168

Clavijo A, Lin M, Riva J et al (2001) Application of competitive enzyme-linked immunosorbent assay for the serologic diagnosis of classical swine fever virus infection. J Vet Diagn Investig 13:357–360

Conlan JV, Khounsy S, Blacksell SD et al (2009) Development and evaluation of a rapid immunomagnetic bead assay for the detection of classical swine fever virus antigen. Trop Anim Health Prod 41:913–920

Contassot E, Beer HD, French LE (2012) Interleukin-1, inflammasomes, autoinflammation and the skin. Swiss Med Wkly 142:w1359

de Smit AJ, Bouma A, Deekluijver EP et al (2001) Duration of the protection of an E2 subunit marker vaccine against classical swine fever after a single vaccination. Vet Microbiol 78:307–317

Depner EKR, Rodriguez OA, Pohlenz O et al (1996) Persistent classical swine fever virus infection in pigs infected after weaning with a virus isolated during the 1995 epidemic in Germany: clinical, virological, serological and pathological findings. Eur J Vet Pathol 2:61–66

Dewulf J, Laevens H, Koenen F et al (2004) Efficacy of E2-sub-unit marker and C-strain vaccines in reducing horizontal transmission of classical swine fever virus in weaner pigs. Prev Vet Med 65:121–133

Dong XN, Chen YH (2005) Candidate peptide-vaccines induced immunity against CSFV and identified sequential neutralizing determinants in antigenic domain A of glycoprotein E2. Vaccine 24:1906–1913

Downing DR, Carbrey EA, Stewart WC (1977) Preliminary findings on a thermal inactivation curve for Hog Cholera virus. In: Agri. Res. Seminar on Hog Cholera/Classical Swine Fever and African Swine Fever. USDA Veterinary Laboratory, Ames, Iowa
Edwards S, Fukusho A, Lefevre PC et al (2000) Classical swine fever: the global situation. Vet Microbiol 73:103–119
Everett H, Crooke H, Gurrala R et al (2011) Experimental infection of common warthogs (Phacochoerus africanus) and bushpigs (Potamochoerus larvatus) with classical swine fever virus. I: susceptibility and transmission. Transbound Emerg Dis 58:128–134
FAO (2010) Good practices for biosecurity in the pig sector—issues and options in developing and transition countries. FAO animal production and health paper 169, Rome
Fernandez-Sainz I, Holinka LG, Gavrilov BK et al (2009) Alteration of the N-inked glycosylation condition in E1 glycoprotein of classical swine fever virus strain Brescia alters virulence in swine. Virology 386:210–216
Floegel G, Wehrend A, Depner KR et al (2000) Detection of classical swine fever virus in semen of infected boars. Vet Microbiol 77:109–116
Freitas TR, Caldas LA, Rebello MA (1998) Prostaglandin A1 inhibits replication of classical swine fever virus. Mem Inst Oswaldo Cruz 93:815–818
Gawolek GP, Chaubey B, Szewczyk B et al (2017) Novel thioglycosyl analogs of glycosyltransferase substrates as antiviral compounds against classical swine fever virus and Hepatitis C virus. Eur J Med Chem 137:247–262
Gisler ACF, Nardi NB, Nonnig RB et al (1999) Classical swine fever virus in plasma and peripheral blood mononuclear cells of acutely infected swine. J Vet Med B 46:585–593
Gladue DP, Holinka LG, Largo E et al (2012) Classical swine fever virus p7 protein is a viroporin involved in virulence in swine. J Virol 86:6778–6791. https://doi.org/10.1128/JVI.00560-12
Gomez-Villamandos JC, Salguero FJ, Ruiz-Villamor E et al (2003) Classical swine fever: pathology of bone marrow. Vet Pathol 40:157–163. https://doi.org/10.1354/vp.40-2-157
Gong W, Wu J, Lu Z, Zhang L et al (2016) Genetic diversity of subgenotype 2.1 isolates of classical swine fever virus. Infect Genet Evol 41:218–226
Greiser-Wilke I, Moennig V (2004) Vaccination against classical swine fever virus: limitations and new strategies. Anim Health Res Rev 5:223–226
Greiser-Wilke I, Dreier S, Haas L et al (2006) Genetic typing of classical swine fever viruses—a review. Dtsch Tierarztl Wochenschr 113:134–138
Gupta PK, Saini M, Dahiya SS et al (2011) Molecular characterization of lapinized classical Swine Fever vaccine strain by full-length genome sequencing and analysis. Anim Biotechnol 22:111–117
Hammond JM, Jansen ES, Morrissy CJ et al (2001) A prime-boost vaccination strategy using naked DNA followed by recombinant porcine adenovirus protects pigs from classical swine fever. Vet Microbiol 80:101–119
Handel K, Kehler H, Hils K et al (2004) Comparison of reverse transcriptase–polymerase chain reaction, virus isolation, and immunoperoxidase assays for detecting pigs infected with low, moderate, and high virulent strain of classical swine fever virus. J Vet Diagn Investig 16:132–138
Hanson RP (1957) Origin of hog cholera. J Am Vet Med Assoc 131:211–218
Huang YL, Deng MC, Wang FI et al (2014) The challenges of classical swine fever control; modified live and E2 subunit vaccines. Virus Res 179:1–11
Hulst MM, van Gennip HGP, Moormann RJM (2000) Passage of classical swine fever virus in cultured swine kidney cells selects virus variants that bind to heparin sulfate due to a single amino acid change in envelope protein E-rns. J Virol 74:9553–9561
Kaden V, Lange E, Fischer U et al (2000) Oral immunisation of wild boar against classical swine fever: evaluation of the first field study in Germany. Vet Microbiol 73:239–252
Kaden V, Steyer H, Schnabel J et al (2005) Classical swine fever (CSF) in wild boar: the role of the transplacental infection in the perpetuation of CSF. J Vet Med B Infect Dis Vet Public Health 52:161–164
Kolupaeva VG, Pestova TV, Hellen CU (2000) Ribosomal binding to the internal ribosomal entry site of classical swine fever virus. RNA 6:1791–1807
Konig M, Lengsfeld T, Pauly T et al (1995) Classical swine fever virus: Independent induction of protective immunity by two structural glycoproteins. J Virol 69:6479–6486
Kumar R, Rajak KK, Chandra T et al (2014) Whole-genome sequence of a classical swine fever virus isolated from the Uttarakhand State of India. Genome A 2:e00371-14
Kwon T, Yoon SH, Kim KW et al (2015) Time-calibrated phylogenomics of the classical swine fever viruses: genome-wide Bayesian coalescent. PLoS 10:e0121578. https://doi.org/10.1371/journal.pone.0121578
Lackner T, Thiel HJ, Tautz N (2006) Dissection of a viral autoprotease elucidates a function of a cellular chaperone in proteolysis. Proc Natl Acad Sci USA 103:1510–1515
Laeven AH, Koenen OF, Deluyker EH et al (1999) Experimental infection of slaughter pigs with classical swine fever virus: transmission of the virus, course of the disease and antibody

response. Vet Rec 145:243–248
Langedijk JPM, Middel WG, Meloen RH et al (2001) Enzyme-linked immunosorbent assay using a virus type specific peptide based on a subdomain of envelope protein E(rns) for serologic diagnosis of pestivirus infections in swine. J Clin Microbiol 39:906–912
Li X, Li W, Shi X et al (2012) Development of an immunochromatographic strip for rapid detection of antibodies against classical swine fever virus. J Virol Meth 180:32–37
Lin M, Lin F, Mallory M, Clavijo A (2000) Deletions of structural glycoprotein E2 of classical swine fever virus strain Alfort/187 resolve a linear epitope of monoclonal antibody WH303 and the minimal N-terminal domain essential for binding immunoglobulin G antibodies of a pig hyperimmune serum. J Virol 74:11619–11625
Lin M, Trottier E, Mallory M (2005) Enzyme-linked immunosorbent assay based on a chimeric antigen bearing antigenic regions of structural proteins Erns and E2 for serodiagnosis of classical swine fever virus infection. Clin Diagn Lab Immunol 12:877–881
Liu JJ, Wong ML, Chang TJ (1998) The recombinant nucleocapsid protein of classical swine fever virus can act as a transcriptional regulator. Virus Res 53:75–80
Lowings P, Ibata G, Needham J et al (1996) Classical swine fever diversity and evolution. J Gen Virol 77:1311–1321
McKissick GE, Gustafson DP (1967) In vivo demonstration of lability of hog cholera virus to lipolytic agents. Am J Vet Res 28:909–914
Meyer D (2011) Epitope mapping of the structural protein Erns of classical swine fever virus. Thesis submitted in partial fulfilment of the requirement for the degree Doctor Rerurm Naturatium. Institute of Virology, Hannover
Meyers G, Thiel HJ (1996) Molecular characterization of pestiviruses. Adv Virus Res 47:53–118
Mittelholzer C, Moser C, Tratschin J et al (2000) Analysis of classical swine fever virus replication kinetics allows differentiation of highly virulent from avirulent strains. Vet Microbiol 74:293–308
Moennig V, Becher P (2015) Pestivirus control programs: how far have we come and where are we going? Anim Health Res Rev 16:83–87
Moennig V, Greiser-Wilke I (2008) Classical swine fever virus. In: Mahy BWJ, van Regenmortel MHV (eds) Encyclopedia of virology. Elsevier, Amsterdam, pp 525–532
Moennig V, Plagemann PGW (1992) The pestiviruses. Adv Virus Res 41:53–98
Moennig V, Floegel NG, Greiser-Wilke I (2003) Clinical signs and epidemiology of classical swine fever: a review of new knowledge. Vet J 165:11–20
Moormann RJ, Bouma A, Kramps JA et al (2000) Development of a classical swine fever subunit marker vaccine and companion diagnostic test. Vet Microbiol 73:209–219
Moulin HR, Seuberlich T, Bauhofer O et al (2007) Nonstructural proteins NS2-3 and NS4A of classical swine fever virus: Essential features for infectious particle formation. Virology 365:376–389
Nagarajan K, Saikumar G (2012) Fluorescent in-situ hybridization technique for the detection and localization of classical swine fever virus in infected tissues. Veterinarski Arhiv 82:495–504
OIE (2007) Manual of diagnostic tests and vaccines for terrestrial animals. http://www.oie.int/eng/norms/manual/asummary.htm
OIE (2018) Terrestrial animal health code, 27th edn. World Organization for animal health, Paris, France
Patil SS, Hemadri D, Shankar BP et al (2010) Genetic typing of recent classical swine fever isolates from India. Vet Microbiol 141:367–373
Paton DJ, Greiser-Wilke I (2003) Classical swine fever—an update. Res Vet Sci 75:169–178. https://doi.org/10.1016/S0034-5288(03)00076-6
Paton DJ, McGoldrick A, Greiser-Wilke I et al (2000a) Genetic typing of classical swine fever virus. Vet Microbiol 73:137–157
Paton DJ, McGoldrick A, Belak S (2000b) Classical swine fever virus: a ring test to evaluate RT-PCR detection methods. Vet Microbiol 73:159–174
Penrith ML, Vosloo W, Mather C (2011) Classical swine fever (hog cholera): Review of aspects relevant to control. Transbound Emerg Dis 58(3):187–196
Postel A, Schmeiser S, Bernau J et al (2012) Improved strategy for phylogenetic analysis of classical swine fever virus based on full-length E2 encoding sequences. Vet Res 43:50
Postel A, Schmeiser S, Perera CL et al (2013) Classical swine fever virus isolates from Cuba a new subgenotype 1.4. Vet. Microbiol 161:334–338
Postel A, Schmeiser S, Oguzoglu TC et al (2015) Close relationship of ruminant pestiviruses and classical swine fever virus. Emerg Infect Dis 21(4):668–672
Postel A, Busch SA, Petrov A et al (2017) Epidemiology, diagnosis and control of classical swine fever: recent developments and future challenges. Transbound Emerg Dis 64:1–14
Postel A, Nishi T, Kameyama K et al (2019) Re-emergence of classical swine fever, Japan, 2018. Emerg Infect Dis 25(6):1228–1231. https://doi.org/10.3201/eid2506.181578
Qi Y, Liu LC, Zhang BQ et al (2008) Characterization of antibody responses against a neutralizing epitope on the glycoprotein E2 of classical swine fever virus. Arch Virol 153:1593–1598

Raut S, Dattatraya RK, Kishore KR et al (2015) Detection of classical swine fever virus antigen and nucleic acid on blood of experimentally infected piglets. Adv Anim Vet Sci 3:1–6
Rehman S (1987) Virucidal effect of the heat treatment of waste food for swine. Tierarztliche Umschau 42(11):892–896
Reimann I, Depner K, Trapp S et al (2004) An avirulent chimeric Pestivirus with altered cell tropism protects pigs against lethal infection with classical swine fever virus. Virology 322:143
Renson P, Dimna ML, Keranflech A et al (2013) CP7-E2alf oral vaccination confers partial protection against early classical swine fever virus challenge and interferes with pathogeny-related cytokine responses. Vet Res 44:9. https://doi.org/10.1186/1297-9716-44-9
Rios L, Nunez JI, de Arce HD et al (2018) Revisiting the genetic diversity of classical swine fever virus: a proposal for new genotyping and sub genotyping schemes of classification. Transbound Emerg Dis 65(4):963–971
Risager PC, Fahnøe U, Gullberg M et al (2013) Analysis of classical swine fever virus RNA replication determinants using replicons. J Gen Virol 94:1739–1748
Rossi S, Staubach C, Blome S et al (2015) Controlling of CSFV in European wild boar using oral vaccination: a review. Front Microbiol 6:1–11. https://doi.org/10.3389/fmicb.2015.01141
Roychoudhury P, Sarma DK, Rajkhowa S et al (2014) Predominance of genotype 1.1 and emergence of genotype 2.2 classical swine fever viruses in north-eastern region of India. Transbound Emerg Dis 61(Suppl. 1):69–77
Sanchez-Cordon PJ, Nunez A, Salguero FJ (2005) Lymphocyte apoptosis and thrombocytopenia in spleen during classical swine fever: role of macrophages and cytokines. Vet Pathol 42:477–488
Sarma DK, Meshram DJ (2008) Comparison of sandwich and dot ELISA for detection of CSF virus antigen in pigs. Ind Vet J 85:915–918
Sarma DK, Sarma PC (1995) ELISA for detection of hog cholera virus antigen. Ind J Anim Sci 65:650–651
Sarma DK, Krishna L, Misri J (2008a) Classical swine fever in pigs and its status in India: a review. Ind J Anim Sci 78:1311–1317
Sarma DK, Mishra N, Rajukumar K et al (2008b) Isolation and characterization of classical swine fever virus from pigs in Assam. Ind J Anim Sci 78:37–39
Sarma DK, Mishra N, Vilcek S et al (2011) Phylogenetic analysis of recent classical swine fever virus (CSFV) isolates from Assam, India. Comp Immunol Microbiol Infect Dis 34:11–15
Sato M, Mikami O, Kobayashi M et al (2000) Apoptosis in the lymphatic organs of piglets inoculated with classical swine fever virus. Vet Microbiol 75:1–9
Shi BJ, Liu CC, Zhou J et al (2016) Entry of classical swine fever virus into PK-15 cells via a pH-, dynamin-, and cholesterol-dependent, clathrin-mediated endocytic pathway that requires Rab5 and Rab7. J Virol 90:9194–9208
Silva MN, Silva DM, Leite AS et al (2017) Identification and genetic characterization of classical swine fever virus isolates in Brazil: a new subgenotype. Archiv Virol 162:817–822
Singh VK, Kumar GS, Paliwal OP (2005) Detection of classical swine fever virus in archival formalin-fixed tissues by reverse transcription-polymerase chain reaction. Res Vet Sci 79:81–84
Singh B, Bardhan D, Verma MR et al (2016) Incidence of classical swine fever in pigs in India and its economic evaluation with a simple mathematical model. Anim Sci Rep 10:3–9
Smith DB, Meyers G, Bukh J et al (2017) Proposed revision to the taxonomy of the genus Pestivirus, family *Flaviviridae*. J Gen Virol 98:2106–2112
Summerfield A, Ruggli N (2015) Immune responses against classical swine fever virus: between ignorance and lunacy. Front Vet Sci 2:1–9. https://doi.org/10.3389/fvets.2015.00010
Summerfield A, Alves M, Ruggli N et al (2006) High IFN-alpha responses associated with depletion of lymphocytes and natural IFN-producing cells during classical swine fever. J Interf Cytokine Res 26:248–255
Suradhat S, Damrongwatanapokin S (2003) The influence of maternal immunity on efficacy of a classical swine fever vaccine against classical swine fever virus, genogroup 2.2, infection. Vet Microbiol 92:187–194
Tautz N, Elbers K, Stoll D, Meyers G, Thiel HJ (1997) Serine protease of pestiviruses: determination of cleavage sites. J Virol 71:5415–5422
Teifke JP, Lange E, Klopfleisch R et al (2005) Nictitating membrane as a potentially useful postmortem diagnostic specimen for classical swine fever. J Vet Diagn Investig 17:341–345
Terpstra C, Bloemraad M, Gielkens AL (1984) The neutralizing peroxidase- linked assay for detection of antibody against swine fever virus. Vet Microbiol 9:113–120
Terpstra C (1991) Hog cholera: an update of present knowledge. Brit Vet J 147:397–406
Töpfer A, Höper D, Blome S et al (2013) Sequencing approach to analyze the role of quasispecies for classical swine fever. Virology 438:14–19
Trautwein G, Leiss B (1988) Pathology and pathogenesis of the disease, classical swine fever and related infections. Martinus Nijhoff Publishing, Boston, MA, pp 27–54
Tyborowska J, Zdunek E, Szewczyk B (2007) Effect of N-glycosylation inhibition on the synthesis and processing of classical swine fever virus glycoproteins. Acta Biochim Pol 54:813–819
van Gennip HGP, van Rijn PA, Widjojoatmodjo MN et al (2000) Chimeric classical swine fever

viruses containing envelope protein E-RNS or E2 of bovine viral diarrhoea virus protect pigs against challenge with CSFV and induce a distinguishable antibody response. Vaccine 19:447–459

van Gennip HGP, Bouma A, van Rijn PA et al (2002) Experimental non-transmissible marker vaccines for classical swine fever (CSF) by trans-complementation of E(rns) or E2 of CSFV. Vaccine 20:1544–1556

Van Oirschot VJT (2004) Hog cholera. In: JAW C, Tustin RC (eds) Infectious diseases of livestock, 2nd edn. Oxford University Press, Oxford, pp 975–986

Weesendorp E, Stegeman A, Willie LA et al (2008) Survival of classical swine fever virus at various temperatures in faeces and urine derived from experimentally infected pigs. Vet Microbiol 132:249–259

Wen G, Yang J, Luo Q et al (2010) A one-step real-time reverse transcription-polymerase chain reaction detection of classical swine fever virus using a minor groove binding probe. Vet Res Commun 34:359–369

Wengler G, Bradley DW, Collett MS et al (1995) Family *Flaviviridae*. Virus taxonomy. In: Murphy FA, Fauquet CM et al (eds) Sixth report of the international committee on taxonomy of viruses. Springer, New York, pp 415–427

Wise GH (1986) Eradication of hog cholera from the United States. In: Woods GT (ed) Practices in veterinary public health and preventive medicine in the United States. Iowa State University Press, Ames, IA, pp 199–223

Wongsawat K, Dharakul T, Narat P et al (2011) Detection of nucleic acid of classical swine fever virus by reverse transcription loop-mediated isothermal amplification (RT-LAMP). Sci Res 3:447–452

Xu X, Guo H, Xiao C et al (2008) In vitro inhibition of classical swine fever virus replication by siRNAs targeting Npro and NS5B genes. Antivir Res 78:188–193

Yoo SJ, Kwon T, Kang K et al (2018) Genetic evolution of classical swine fever virus under immune environments conditioned by genotype 1-based modified live virus vaccine. Transbound Emerg Dis 65:735–745. https://doi.org/10.1111/tbed.12798

Zhang C, Huang Y, Zheng H et al (2007) Study on colloidal gold strip in detecting classical swine fever virus. Fujian J Anim Hus Vet Med 6:13

Zhao JJ, Cheng D, Li N et al (2008) Evaluation of a multiplex real-time RT-PCR for quantitative and differential detection of wild-type viruses and C-strain vaccine of classical swine fever virus. Vet Microbiol 126:1–10

Zhou YC, Wang Q, Fan XZ et al (2009) The changes of peripheral blood leucocytes subpopulation after challenge with CSFV virulent strain Shimen. Chin J Virol 25:303–308

第 4 章 冠状病毒

A. N. Vlasova, Q. Wang, K. Jung, S. N. Langel,
Yashpal Singh Malik, L. J. Saif

4.1　引言

所有已知的猪冠状病毒(coronavirus,CoV)属于尼多病毒目、冠状病毒科;冠状病毒亚科的 α 属冠状病毒、β 属冠状病毒和 δ 属冠状病毒(图 4.1)(de Groot 等,2008)。通常会对猪的胃肠道、呼吸道、周围神经系统和中枢神经系统造成影响。已经鉴定出 5 种猪 CoV:①1946 年首次定义的传染性胃肠炎病毒 (transmissible gastroenteritis virus, TGEV);②1984 年分离到的 TGEV 的突变体——猪呼吸道冠状病毒(porcine respiratory coronavirus, PRCV); ③1977 年分离到的猪流行性腹泻病毒(porcine epidemic diarrhoea virus, PEDV);④1962 年分离到的猪血凝性脑脊髓炎病毒(porcine haemagglutinating encephalomyelitis virus,PHEV);⑤2012 年报道的猪 δ 冠状病毒 (porcine deltacoronavirus, PDCoV)。TGEV 和 PRCV,同猫和狗的 CoV 关系较近,属于同种 α 属冠状病毒,PEDV 和人源 CoV(229E 和 NL63)在 α 属冠状病毒中单独形成一个种。PHEV 和 PDCoV 分别属于 β 属冠状病毒(β 属冠状病毒的一个种)和 δ 属冠状病毒。PDCoV 与亚洲豹猫和白鼬猫群德尔塔冠状病毒关系密切(Ma 等,2005)。PRCV 可引起猪的原发性亚临床感染,而猪肠道致病性病原体 α 属冠状病毒[(TGEV,PEDV,猪肠道冠状病毒(swine enteric coronavirus, SeCoV), 猪肠道 α 属冠状病毒(porcine enteric alphacoronavirus, PEAV)]和 PDCoV 引起的肠道疫病的严重程度取决于动物年龄和免疫状态。其中,猪冠状病毒只有一种血清型。

据报道,TGEV 和 PEDV 已经在欧亚大陆和美国传播。最近,在欧洲鉴定并报道了一株致病性的 TGEV/PEDV 重组变异的 SeCoV (Akimkin 等,2016;Belsham 等,2016;Boniotti 等,2016)。SeCoV

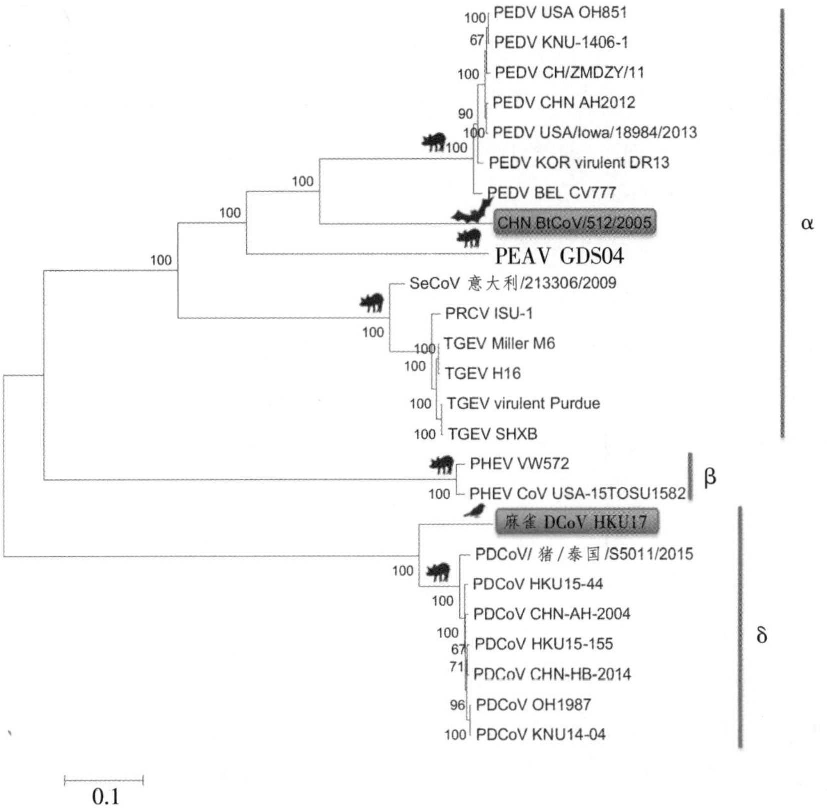

图 4.1　猪冠状病毒系统发育树:α、β、δ。框内为潜在的原始非猪冠状病毒。通过 1000 次重复的自举检验来验证每个节点的可靠性。

以 TGEV 作为主体骨架，并含有 PEDV S 蛋白，其感染引起的临床表现与 TGEV 和 PEDV 感染引起的没有区别（表 4.1）。此外，2017 年发现了一种与哺乳仔猪严重腹泻有关的新型蝙蝠-HKU2 样猪 CoV（PEAV，GDS04 株）（Gong 等，2017）。然而，此病毒的流行程度和对宿主猪的适应状况尚不清楚。

目前，PHEV 是唯一在世界范围内流行的猪 β 属冠状病毒（Li 等，2016）。在缺乏母源抗体的新生仔猪（通常是从无感染的猪群中购买的）中，PHEV 是引起脑脊髓炎或者以呕吐和消瘦为特征的疫病。实际上，仔猪通过初乳获得母源抗体的免疫保护作用可持续 15 周，而 4 周龄以上的猪和成年猪被感染时，主要表现为亚临床症状，因此，研究者认为很少产生严重的经济影响。然而，最近报道在美国出现一种罕见的猪呼吸道病（类似于流感）（Lorbach 等，2017）并发现 PHE 在成年猪中的流行增加，这表明潜在的定向变异可能会引起流行病学的重大变化。为了清楚定位这种复杂的流行病学，需要进行持续的监测和开发先进的、快速可靠的检测工具和技术，以提供确切的鉴别诊断（Kim 等，2001；Masuda 等，2016）。

CoV 有囊膜且病毒粒子呈多形性，直径为 60~160nm。猪 CoV 是单股正链、多腺苷酸、大基因组（25~30kb）的传染性 RNA 病毒。它的基因组谱、复制机制，以及蛋白质表达与其他人和动物 CoV 相同（Enjuanes 和 Van der Zeijst，1995；Gonzalez 等，2003；Laude 等，1993）。大多数猪 CoV 具有 4 种基本结构蛋白：大的表面糖（S）蛋白，形成一层长杆形突起的纤突蛋白，定义为冠状；囊膜（E）蛋白；完整的膜糖（M）蛋白和核衣壳（N）蛋白。然而，PHEV 包含一种血凝素酯酶（HE）蛋白，形成第二层更短的表面刺突（de Groot 等，2008）。TGEV、PEDV 和 PDCoV 也能够通过 ORF3（TGEV 和 PEDV），ORF6（PDCoV）和 ORF7（TGEV 和 PDCoV）转录 1~2 个辅助蛋白。完整的基因组顺序是 5'UTR-ORF1ab、S、ORF3、E、M、ORF6、N 和ORF7-3'UTR。

TGEV 和 PRCV 整个基因组核苷酸和氨基酸序列的同源性为 96%~98%，这表明 PRCV 是由 TGEV 进化而来的。PRCV 基因组具有两个典型的特征，即 S 基因 N 端碱基的大量缺失（621~681nt），导致 S 蛋白变小及 ORF3 序列变异，这两个特征可能解释了 PRCV 对组织偏嗜性的改变（Ballesteros 等，1997；Sanchez 等，1999）。

但没有证据表明 PEDV 存在不同的血清型（Lin 等，2015a），从遗传学上讲，PEDV 分离株可分为两个群：①经典株（与来自欧亚大陆的原始分离株 CV777 的基因相似）；②新出现的 PEDV 株。所有经典 PEDV 株都会发生纤突基因（S-INDEL）的插入和缺失（Lin 等，2016；Valsova 等，2014），但新出现的大部分强毒 PEDV 分离株不存在 S 基因的插入和缺失（Valsova 等，2014）。因此，最初于 2010 年出现并传播到美国、欧洲和其他亚洲地区的强毒 PEDV 分离株被称为新出现的非 S-INDEL PEDV 株（Diep 等，2017；Oka 等，2014；Suzuki 等，2015）。PEDV 经典毒株和新出现毒株重组体基因组中的 S 基因与经典毒株相同，存在基因的缺失和插入，被称为 S-INDEL 株，在亚洲、欧洲和美国传播。此外，报道了 PEDV 其他不常见的变异株，其 S 蛋白的 N-末端存在片段缺失（194~216aa），被称为是 S1 NTD-del 型 PEDV。已报道的 PEDV 基因缺失株（S-INDEL 和 S1 NTF）与肠道型 TGEV 进化为呼吸型 PRCV 的组织嗜性改变不同，PEDV 基因缺失株仍然保持嗜肠道性，但毒性较低（Suzuki 等，2016；Hou 等，2017）。

猪肠道 CoV（TGEV、PEDV 和 PDCoV）具有高度传染性，与腹泻、呕吐等严重疫病的发生相关，并且增加幼龄猪的死亡率（通常为 100%）。猪肠道 CoV 在养殖猪的国家呈现散发流行（地方流行性）或大规模流行（图 4.2）。到目前为止，还没有针对猪肠道 CoV 的任何特异治疗方法，有的地区至今还在抵制根除猪肠道 CoV 政策。本章回顾了家养猪和野生猪中 CoV 的持续进化引起的疫病，以及其他可能的宿主（禽类或蝙蝠物种）或包括食肉

表 4.1　猪肠道冠状病毒发病机制比较

	症状	临床发病	损伤	发病率	死亡率	受影响最大的年龄组
TGEV	腹泻、呕吐、脱水	24 小时	空肠、回肠	100%	高达 100%	新生仔猪到 3 周龄
PEDV	腹泻、呕吐、脱水	24~36 小时	空肠、回肠	100%	高达 100%	新生仔猪到 3 周龄
SeCoV	腹泻、呕吐、脱水	24~36 小时	空肠、回肠	100%	高达 100%	新生仔猪到 3 周龄
PEAV	腹泻、脱水	3~4 天	?	100%	?	新生仔猪到 3 周龄
PDCoV	腹泻、呕吐、脱水	1~3 天	空肠、回肠、结肠	高达 100%	40%~80%	新生仔猪到 3 周龄

动物在内的二级宿主，或通过跨界传播、重组和缺失逃逸变异产生的疾病。此外，还回顾了虽然 PRCV 失去肠道偏嗜性但仍能够诱导针对 TGEV 的免疫保护性反应。

4.2　发病机制与临床症状

4.2.1　TGEV

感染 TGEV 后 24 小时内，空肠和回肠的成熟肠上皮细胞大量坏死，导致酶活性（碱性磷酸酶、乳分解酶等）降低，消化紊乱，细胞电解质（包括钠）平衡紊乱。这些变化主要导致肠内液体滞留，导致急性吸收不良性腹泻（Moon，1978）。仔猪血管外蛋白丢失和大量脱水可能导致死亡（Butler 等，1974）。后者还可导致代谢性酸中毒和高钾血症，造成心功能异常。

TGE 的病变主要局限于胃肠道，可以看到胃和小肠胀满，充满凝乳块，有时还可以观察到皮下出血（Hooper 和 Haelterman 等，1966a）。小肠壁薄而透明。空肠绒毛显著萎缩和回肠轻度萎缩是 TGE 的主要病变，且新生仔猪的肠绒毛萎缩与 3 周龄以上仔猪的肠绒毛萎缩更为明显（Moon，1978；Hooper 和 Haelterman 等，1966b）。严重感染 TGEV 导致小于 2 周龄的仔猪死亡率增加（通常为 100%），而年龄较大的猪死亡率降低。尽管任何年龄的猪对 TGEV 均易感，但血清反应呈阳性的猪群和大于 5 周龄猪的死亡率通常较低。在临床上，猪对 TGEV 易感性与年龄有关的机制是由于被感染新生猪的肠绒毛上皮细胞和从隐窝迁移的新分化的肠上皮细胞代谢缓慢（Moon，1978）。

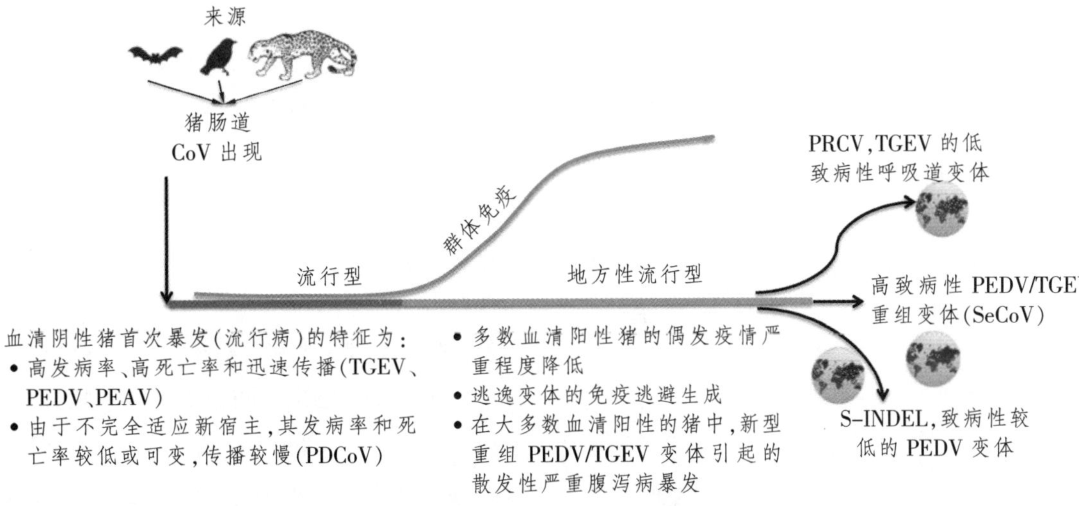

图 4.2　猪肠道 CoV 进化的不同阶段。

这些病变与PEDV/PDCoV的病理变化相似(Debouck等,1981;Jung等,2015a),但比轮状病毒(rotavirus,RV)引起的病理损伤严重(Bohl等,1978)。地方流行性TGEV引起病猪的病理变化和绒毛萎缩程度异常多变(Pritchard,1978)。

除肠道外,肺(肺泡巨噬细胞)和乳腺组织也被认为是TGEV复制的部位(Kemeny等,1975)。迄今为止的报告显示,TGEV通过口腔、鼻腔感染可导致猪患肺炎,但乳腺被感染的临床意义尚不明确(Underdahl等,1975)。然而,在被TGEV感染的母猪中常见无乳症,且TGEV在畜群中传播迅速。

4.2.2 PRCV

PRCV在猪的1型和2型肺泡细胞中高效复制,多见于鼻腔、气管、支气管、细支气管的上皮细胞和肺泡,有时也见于肺泡巨噬细胞(Atanasova等,2008;Pensaert等,1986;O'Toole等,1989)。在血液和支气管淋巴结中也可以观察到PRCV。实验感染后,鼻腔中PRCV通常会持续存在4~6天。随着病毒中和抗体效价增加(Atanasova等,2008),肺部病变和临床症状随之消退。尽管PRCV有时在肠上皮细胞中存在,但它并不能有效地扩散到邻近上皮细胞(Cox等,1990),并且排泄物中病毒含量很低,可能无法检测到。

PRCV感染主要引起呼吸道症状。临床症状表现为间质性肺炎和细支气管肺炎,支气管周围和血管周围形成整齐的淋巴组织细胞套(Atanasova等,2008;Cox等,1990;Halbur等,1993;Jung等,2007)。PRCV引起的支气管性间质肺炎主要表现为:①巨噬细胞和淋巴细胞浸润使肺泡间隔增厚;②2型肺细胞肥大和增生;③由气管上皮坏死及细支气管周或血管周围淋巴组织细胞炎症导致的肺泡和细支气管管腔内有细胞碎片和炎性白细胞聚集。

4.2.3 PEDV

感染PEDV后22小时,可有明显临床症状,与病毒复制出现的峰值相一致。临床表现为吸收不良性水样腹泻、呕吐、抑郁和厌食,引起的病理损伤与TGEV引起的病理损伤在临床上没有区别(Debouck等,1981;Coussement等,1982)。

仔猪发病率接近100%,母猪发病率不一。1周龄以下的新生仔猪常因严重脱水而死亡,死亡率高达50%~100%,其中年龄较大的猪死亡率较低且1周内即可恢复正常。在母猪中,腹泻程度一致,通常只表现为抑郁和厌食。同样,育肥猪可能会产生水样粪便,并可能在1周内出现厌食和抑郁。与TGEV一样,与断奶仔猪相比,感染PEDV的仔猪肠道细胞代谢速度慢和未发育成熟的先天免疫系统可能会加重临床症状、引起更高的死亡率和更慢的康复速度(Jung等,2015a;Moon等,1975;Annamalai等,2015)。

PEDV的暴发,一般持续3~4周,但是,如果大型养殖场有多个独立的单元,且母猪的乳源免疫水平不一,则暴发时间可能更长。妊娠母猪被PEDV感染能够为仔猪提供足够的母源性免疫保护,阻断猪流行性腹泻暴发。急性暴发后,腹泻可能持续存在,并在断奶仔猪中反复出现,类似于地方流行性TGE(Martelli等,2008)。

自然感染和实验感染的哺乳仔猪损伤严重程度和病毒复制水平会因为典型的PEDV、新出现的非S-INDEL和S-INDEL PEDV毒株的不同而异(Jung等,2015a;Coussement等,1982;Kim和Chae,2003;Pospischil等,1981;Sueyoshi等,1995;Lin等,2015b;Madson等,2014)。病变仍然局限于小肠,表现为小肠肿胀,充满了水状的淡黄色液体。镜下观察可见合胞体,小肠上皮细胞,尤其是绒毛近端上皮细胞发生空泡化和脱落。与TGEV相似,PEDV感染导致肠上皮细胞变性,绒毛长度与肠腺隐窝深度比值减少和酶活性降低。尽管在结肠上皮细胞中检测到PEDV抗原,但没有观察到相关的组织病理学变化(Debouck等,1981)。

在PEDV感染期间被安乐死的猪,在其血清和不同组织(包括肺、脾、肝和肌肉)中均检测到了病毒(Suzuki等,2016;Jung等,2014,2015a;Lohse等,2017;Chen等,2016a;Park和Shin,2014)。血清中病毒滴度最高值为 7~8 log_{10} GE/mL，与粪便中病毒滴度最高峰值一致（11~12 log_{10} GE/mL）(Jung 等,2015a)。此外,新出现的 PEDV 毒株流行期间,在 40.8%(20/49)的母猪乳汁样本中可检测到 PEDV RNA(Sun 等,2012)。

4.2.4 PDCoV

在哺乳仔猪和老龄猪被 PDCoV 感染后的 3 天内,可观察到临床症状。尽管临床症状相似,但与 PEDV 和 TGEV 感染引起的临床症状相比并不明显(Chen 等,2015;Hu 等,2016;Jung 等,2015b;Ma 等,2015)。主要表现为由肠上皮细胞大量损失而引起的吸收不良性急性水样腹泻，其他症状还包括呕吐、脱水、体重减轻、嗜睡和死亡。被感染的结肠上皮细胞空泡化,抑制水和电解质再吸收。血清阴性猪在任何年龄都易感染 PDCoV,仔猪发病率高达 100%。通过对 2014 年美国、中国和泰国记载病例的评估，乳猪死亡率高达 40%~80%与 PDCoV 感染相关(Anson,2014)。繁殖场的 PDCoV 感染具有自我限制性，并且当妊娠母猪产生足以保护其后代的母源性免疫时,感染就会终止。

严重的病变为肠壁变薄和透明(空肠至结肠),其中含有大量黄色液体和气体,胃里经常充满了凝固的牛奶。

PDCoV 在大肠和小肠的上皮细胞中复制,出现类似于 TGEV 和 PEDV 感染的病变,但较轻微。组织病理学变化为组织病理损失严重，出现多病灶性至弥漫性、轻度至极度萎缩性空肠肠炎和回肠肠炎，在某些部位伴有盲肠和结肠上皮细胞轻度空泡化（Chen 等,2015;Hu 等,2016;Jung 等,2015b;Ma 等,2015)。急性感染中,在空肠中段至回肠段的绒毛上皮细胞中可检测到 PDCoV 抗原(Jung 等,2015b),但在十二指肠和盲肠/结肠中很少检测到 PDCoV(Jung 等,2016a)。PDCoV 抗原也可能存在于肠固有层免疫细胞、派尔集合淋巴结和肠系膜淋巴结(Hu 等,2016)。固有层中可见炎性细胞(巨噬细胞、淋巴细胞和中性粒细胞)浸润。被 PDCoV 感染的空肠和结肠上皮细胞出现急性坏死(Jung 等,2016a),导致肠绒毛明显萎缩,但十二指肠和大肠病理变化不明显,这与十二指肠、盲肠或结肠上皮细胞中的 PDCoV 阳性抗原较少一致(Chen 等,2015;Jung 等,2015b)。在急性期病毒血症的血清中检测到的 PDCoV-RNA 浓度较低(Chen 等,2015;Ma 等,2015;Jung 等,2016b)。猪在临床上恢复正常后，在肠道淋巴组织中可发现大量的PDCoV 抗原(Hu 等,2016)。此外,在多个器官中检测到少量或中等量的 PDCoV RNA,但未检测到抗原，提示可能与病毒血症有关。粪便中 PDCoV 含量降低(与 PEDV 和 TGEV 相比)可能表明其不完全适应于猪，并导致其在猪群中的传播速度缓慢,以及哺乳猪的死亡率低(Jung 等,2015b)。

4.3 流行病学

4.3.1 TGEV

1946 年首次从暴发急性腹泻高死亡率的仔猪中检测到 TGEV(Doyle 和 Hutchings,1946)。自此,采取集约化养猪国家和地区,包括欧洲、亚洲(日本、韩国、马来西亚和中国台湾地区)、美洲(北、中、南)和非洲(加纳、扎伊尔)都报道了该病。尽管疫苗被广泛应用，但是在 20 世纪 60 年代至 80 年代引起美国和全球仔猪肠道疫病和死亡率的主要原因仍然是 TGEV 感染（Laude 等,1993;Pensaert 等,1986,1993;Brown 和 Cartwright,1986;Pensaert,1989;Yaeger 等,2002)。TGEV 基因缺失突变株 PRCV 的存在和广泛流行,降低了 TGE 的临床影响。目前,在北美、欧洲和亚洲 TGEV/PRCV 血清阴性猪群中，仔猪因被 TGEV 感染而出现大量腹泻的零星暴发尚待证实。然而,这可能需要仔细鉴别 TGEV 和新出现的 TGEV/PEDV 重组体。

TGE 存在两种明显的流行形式:流行性和地方流行性。在血清学阴性的畜群中,TGE 流行性表现尤为突出。TGEV 传入畜群后,迅速传播给任何年龄的猪,特别是在冬季,被感染的猪表现为食欲缺乏、呕吐或腹泻。乳猪临床症状为明显的迅速脱水。泌乳母猪通常表现为厌食、无乳和产奶量减少,由此进一步增加了仔猪死亡率。

出现地方流行性 TGE 是由于猪群中不断引入易感猪而导致病毒和疫病在猪群中持续存在。这是一种典型的原发性暴发 TGE 的后遗症,常发生在规律性的生产仔猪、扩群的猪中,或猪血清阳性和易感猪混群饲养时也会发生 (Stepanek 等,1979)。在地方流行性畜群中,TGEV 在成年猪中传播缓慢(Pritchard,1987)。大部分母猪都具有抵抗力且无症状,并将不同水平的母源性免疫力传递给后代。在断奶后 6 天至 2 周的猪群中,可观察到轻度 TGEV 腹泻,死亡率低于 10%。

4.3.2 PRCV

当 PRCV 感染呼吸道时,粪便中的 PRCV 含量有限或没有(Pensaert,1989)。PRCV 于 1984 年在比利时首次被分离出来(Pensaert,1986)。在美国,从没有 TGEV 感染或疫苗接种史的畜群中检测到 PRCV(Hill 等,1990;Wesley 等,1990)。PRCV 感染的猪可产生中和 TGEV 的抗体。

自首次报道以来,该病毒在欧洲迅速传播(Laude 等,1993;Brown 和 Cartwright,1986;Have,1990;van Nieuwstadt 等,1989)并呈现世界范围内流行,包括进入无 TGEV 的国家(Laude 等,1993;Pensaert,1989;Pensaert,1993)。1995 年,美国血清学调查表明,艾奥瓦州不同畜群中的临床健康猪的血清 PRCV 呈阳性(Wesley 等,1997)。

4.3.3 PEDV

从 20 世纪 70 年代初至 80 年代末,经典的 PEDV 毒株在欧洲引起了几次高死亡率的流行,但 2000 年以后的报道非常少。在意大利,2005 年和 2006 年发生了一场波及 63 个猪群的流行病,所有年龄段的猪均发病,但死亡率主要局限于哺乳仔猪(Martelli 等,2008)。由于该病的临床重要性较低,因此在 2014 年欧洲出现新的 PEDV 变异体之前,研究者一直没有对其进行任何监督研究。值得注意的是,在欧洲猪群中,经典 PEDV 流行的历史情况是未知的。2014 年,一种新型非 S-INDEL 毒株在乌克兰暴发流行;同时在欧洲国家(德国、比利时、法国、荷兰和斯洛文尼亚)也有 S-INDEL 株的暴发(Lin 等,2016)。

20 世纪 70 年代末,中国报道了经典 PEDV 毒株的感染。自那时起,尽管在猪群中使用了疫苗(靶向 PEDV 原始毒株 CV777),但猪流行性腹泻仍在猪场中传播,并成为病毒性腹泻的主要原因(Wang 等,2016a;Xuan 等,1984)。日本于1982 年首次检测到 PEDV(Takahashi 等,1983),20 世纪 90 年代仍在哺乳仔猪中持续流行,死亡率为 30%~100%(Sueyoshi 等,1995;Kuwahara 等,1988)。PEDV 于 1993 年在韩国首次报道,并在 20 世纪 90 年代确诊的哺乳仔猪肠道病毒感染中占 50% 以上(Chae 等,2000;Hwang 等,1994)。在印度,528 份猪(2~6 个月大)的血清样本中,有 21.2%的 PEDV 抗体呈现阳性 (Barman 等,2003)。泰国在 2007—2008 年间,8 个省份中各个年龄段的猪被 PEDV 感染(Puranaveja 等,2009),而且新生仔猪的死亡率高达 100%。

2013 年,其他国家也出现了新型 PEDV 毒株。在同一时期,美国猪场也暴发了由高致病性非 S-INDEL PEDV 毒株引起的疫情 (Stevenson 等,2013),随后在 2014 年 1 月暴发了由较温和的 S-INDEL PEDV 毒株引起的疫情(Wang 等,2014a)。然而,早在 2013 年 6 月采集的猪样本中就发现了 S-INDEL 毒株 (Vlasova 等,2014)。2013—2014 年,PEDV 造成了美国猪存栏量减少近 10%(700 万头猪),并造成相关利润的损失。据报道,到 2018 年 PEDV 已经传播到美国 39 个州和波多黎各。在美国,前一次大规模暴发 PEDV 是在 2014 年春季,并且已经报道了新出现的非 S-INDEL PEDV 毒株向包括加拿大和墨西哥在内的邻国传播(Lin 等,2016)。

2013 年,新型非 S-INDEL PEDV 株在日本出现(Masuda 等,2015),随后传播到韩国(Kim 等,2015)、越南(Vui 等,2014)、泰国(Cheun-Aron 等,2015)、中国台湾地区(Lin 等,2014)和菲律宾(Kim 等,2016)。2013 和 2014 年,日本(Suzuki 等,2015)和韩国(Lee 等,2014)的猪场分别检测到了 S-INDEL PEDV 毒株。到目前为止,在日本的猪体中只检测到了在 S 蛋白 N 端有大片段缺失的其他 PEDV 变异株。除了无致病力的 TTR-2 毒株(Suzuki 等,2016),从 2013 年 12 月至 2015 年 6 月收集的腹泻猪粪便或肠道样本中检测到 15 个 S 基因 N 端有大片段缺失的新的 PEDV 野毒突变株(Diep 等,2017),缺失片段大小为 582~648nt。有趣的是,这些样本中至少含有 2 株截然不同的大片段基因缺失的 PEDV 毒株,并且大多数 PEDV 毒株被证实含有完整的 S 基因。在原发性猪流行性腹泻和复发性猪流行性腹泻的暴发中均发现 S 基因大片段缺失的变异。综上所述,在亚洲和欧洲存在经典的、新型非 S-INDEL 和 S-INDEL PEDV 毒株,但目前只有新出现的非 S-INDEL 和 S-INDEL PEDV 毒株被证实在美洲传播。到目前为止,在非洲和澳大利亚大陆还未有报告提及 PEDV 感染。

4.3.4 PDCoV

2007—2011 年,中国的初步调查,确定了猪和野生鸟类中存在德尔塔冠状病毒(delta coronavirus,DCoV)(Woo 等,2012)。事实上,在 2005—2006 年,就已经在小型哺乳动物(包括鼬獾和亚洲豹猫)中分离出 DCoV。由于 DCoV 的螺旋酶、S基因与 PDCoV 关系密切,提示 DCoV 在亚洲食肉动物、猪和鸟类之间存在跨界传播(Dong 等,2007),但 PDCoV 的起源尚不清楚。

2014 年初,俄亥俄州首次暴发了与 PDCoV 相关的猪腹泻。通过 RT-PCR 对俄亥俄州 5 个猪场腹泻猪的肠道内容物或粪便样本进行检测,PDCoV 阳性率为 92.9%(Wang 等,2014b)。PDCoV 的序列与 2012 年出现的两株 PDCoV 毒株,即 HKU15-44 和 HKU15-155 的核苷酸具有高度的相似性。同一时期,在美国各州发现了另外 2 个基因相似的 PDCoV 株,USA/IA/2014/8734 和 SDCV/USA/Illinois121/2014(Li 等,2014;Marthaler等,2014a)。在此之后,证实 PDCoV 在美国 19 个州存在,但仍没有 PEDV 传播范围广。在美国,猪中的 PDCoV 来源尚不清楚;然而,血清学证据表明,在 2014 年之前,PDCoV 曾在美国的猪群中传播(Sinha 等,2015;Thachil 等,2015)。

加拿大(2014 年 3 月)、韩国(2014 年 4 月)、中国(2015 年)、泰国(2015 年)及越南和老挝证实存在 PDCoV。韩国的一项研究表明,2014—2015 年在 59 个养殖场采集腹泻猪的 691 个粪便样本中,仅在 1 个养殖场中检测到 2 份 PDCoV RNA 阳性样本(Lee 等,2016a)。有 2 株韩国 PDCoV 株——SL2 和 SL5 与美国 PDCoV 株在基因上关系密切,但与韩国 KNU14-04 株不同。2012 年,在中国香港地区发现 PDCoV(Woo 等,2012),随后在中国内地腹泻猪中也检测到 PDCoV。已经报道 PDCoV 的高患病率(>30%)并且与 PEDV 频繁发生混合感染(51%)。所有中国 PDCoV 毒株与全球 PDCoV 毒株的核苷酸具有较高同源(>98.9%)。泰国 PDCoV 毒株与中国 CHN-AH-2004 毒株的核苷酸具有最高的同源性(≥98.4%),但不同于中国和美国毒株,该毒株已单独形成一个分支。在老挝发现了与泰国 PDCoV 谱系遗传关系密切的毒株,而在越南发现了与美国 PDCoV 谱系关系密切的毒株(Saeng-Chuto 等,2017)。

4.4 感染免疫

4.4.1 TGEV/PRCV

成年猪被感染可产生持续 6 个月至数年的较高血清抗体水平(Stepanek 等,1979)。血清抗体的存在提供了 TGEV 或 PRCV 感染的生物学证据,它们与 TGEV 的免疫保护作用关系尚不确定。被 TGEV 感染的猪恢复正常后，会对攻毒产生短期的免疫保护，如肠黏膜免疫(Brim 等,1995;Saif等,1994;VanCott 等,1993,1994)。

免疫保护作用依赖于肠道浆细胞产生的分泌型 IgA(sIgA)抗体(Saif 等,1994;VanCott 等,1993,1994)。猪口服 TGEV 后,在肠道和血清中检测到 IgA-TGEV 抗体和抗体分泌细胞,但非口服TGEV,则检测不到(Saif 等,1994;VanCott 等,1993, 1994;Kodama 等,1980)。血清中存在 IgA 抗体(可能来自肠道)被认为是 TGE 主动免疫的一个指标。除了局部抗体反应外,细胞介导免疫(cell-mediated immunity, CMI) 在防止 TGEV 感染方面也很重要(Brim 等,1995;Frederick 等,1976;Shimizu 和 Shimizu,1979)。维 A 酸可能通过增加 CD8+T 细胞向淋巴结和小肠的转运，增强 TGEV 灭活疫苗免疫后仔猪的 CMI(Chen 等,2016)。人们认为,自然杀伤细胞活性低可能与仔猪和临产母猪对 TGEV 感染的高度易感有关(Cepica 和 Derbyshire,1984)。一株强毒 (SHXB) 但未减毒 (STC3)的 TGEV 毒株阻碍了猪肠道树突状细胞或单核细胞衍生的树突状细胞在体内外摄入抗原、迁移和刺激 T 细胞扩增的能力(Zhao 等,2014),提示 TGEV 具有免疫抑制的潜能。

接种 PRCV 的猪产生了有效的全身性和支气管相关但非肠道相关的抗体、抗体分泌细胞和 T 细胞反应(VanCott 等,1993;Brim 等,1994)。除了 TGEV 和 PRCV 引起的母猪乳汁中 IgA 抗体水平差异外,还建议根据乳汁中 IgA 抗体,研究病毒表位的可能差异(De Diego 等,1992)。

随着 PRCV 的广泛分布,TGE 在欧洲的流行大幅度减少,研究表明,PRCV 感染可诱导机体对 TGEV 产生局部免疫，这一点已经根据大部分被检测猪的排毒和腹泻持续时间缩短被证实(Brim 等,1995;VanCott 等,1994;Cox 等,1993;Wesley 和 Woods,1996)。这种局部免疫与抗 TGEV 中和抗体的快速增加，以及肠内 IgG 和 IgA 抗体分泌细胞的含量相关(Saif 等,1994;VanCott 等,1994)。经 TGEV 攻毒后,PRCV-IgG 和 IgA ASC 从支气管相关淋巴组织向暴露在 PRCV 环境中的猪肠道迁移，这可以说明其快速的记忆免疫应答反应诱导机体产生了局部免疫保护作用(VanCott 等,1994)。但是新生猪需要在 PRCV 感染后一周才能对随后的 TGEV 攻击产生部分免疫力(Wesley 和 Woods,1996)。

通过初乳获得的循环被动抗体(主要是 IgG)可保护新生仔猪免受全身性感染,但并不能对肠道感染产生免疫保护(Hooper 和 Healterman,1996a;Saif 和 Sestak,2006)。在哺乳期前 7 天,乳汁中 IgA 占优势,IgG 减少。研究人员介绍了猪对 TGEV 感染的被动免疫机制(Saif 和 Sestak,2006;Chattha 等,2015;Saif 和 Jackwood,1990;Saif 和 Bohl,1979),TGE 康复猪可以通过初乳或乳汁(乳源性免疫)将 TGEV 中和抗体传送给乳猪(Hooper 和 Healterman,1996a),使乳猪获得被动免疫,清除肠腔中摄入的 TGEV。仔猪定期从免疫母猪获得母源抗体或通过持续饲喂血清抗体试验，也能够清除肠腔中的 TGEV。

乳汁中的 IgA-TGEV 抗体可在肠道中稳定存在,并提供最好的免疫保护,但是,如果通过非口服免疫或全身性免疫(Bohl 和 Saif,1975)或通过初乳摄入 IgG 抗体使得乳汁中 IgG 抗体效价很高,IgG 抗体同样具有保护作用(Stone 等,1977)。在 TGEV 感染和肠道内抗原刺激后,IgA 免疫细胞移向乳腺，在乳腺内产生 IgA 抗体并进入初乳和乳汁，在乳猪被动免疫保护中发挥重要作用(Saif 和 Sestak,2006;Saif 和 Jackwood,1990;Saif 和 Bohl,1979;Bohl 和 Saif,1975)。“肠-乳”免疫轴在 TGEV 感染的猪中被首次提出，第一次支持了共同黏膜免疫系统的概念(Bohl 等,1972;Saif 等,1972)。

4.4.2 PEDV

尽管所有年龄段的猪都易感染 PEDV，但死亡率最高的是 1 周龄和更小的仔猪，它们的存活取决于获得的母源抗体，尤其是病毒中和和 sIgA（通过免疫或以前感染的母猪的初乳和乳汁）。关于 TGEV 感染中有关母源成分安全性的阐述也适用于 PEDV（在 TGEV 章节中描述）(Chattha 等，2015；Langel 等，2016)。猪在断奶时失去了母源性的免疫保护，容易被 PEDV 感染。PEDV 感染产生的体液免疫反应与 TGEV 完全相同（在 TGEV 章节中描述）(Saif 和 Sestak，2006；Chattha 等，2015)。在血清中出现病毒中和抗体，但在临床上的免疫保护作用并不明显，因为免疫保护作用主要依赖于肠黏膜中的 sIgA 抗体(Chattha 等，2015；Langel 等，2016)。免疫保护可能不会持续很长时间，但是在再次接触 PEDV 时，会快速产生免疫记忆反应，可能会减低再次发病的严重程度，甚至阻止它的发生。

至少有 11 种蛋白质，ORF1ab 编码的非结构蛋白质(nsp1、nsp3、nsp5、nsp7、nsp14–16)、结构蛋白质(E、M、N)，辅助蛋白 ORF3 被认为是 IFN 拮抗剂，可以使 PEDV 逃避宿主 IFN 作用(Ding 等，2014；Wang 等，2015；Zhang 等，2016a)。从 TGEV 感染中观察到，先天性免疫应答的下降(具体是自然杀伤细胞的频率和功能)(Annamalai 等，2015)可能会增加哺乳猪和日龄偏大的(断奶，育肥，成年)猪被 PEDV 感染的严重性(Derbyshire 等，1969)。

4.4.3　PDCoV

猪被 PDCoV 感染的免疫反应尚不清楚，但推测与 TGEV 和 PEDV 相似。无菌猪通过口服接种 PDCoV 原始毒株或组织培养的 PDCoV 毒株(OH-FD22)，在免疫接种后 14 天，血清出现 IgG、IgA 和病毒中和抗体，24 天抗体水平达到最高峰值，此时猪的临床表现恢复正常且粪便中检测不到病毒(Hu 等，2016)。与 TGEV 和 PEDV 相似，来自免疫母猪的初乳和乳汁的母源抗体，特别是 IgA 和病毒中和抗体，具有中和 PDCoV 的能力，保护仔猪肠道免受感染(Bohl 等，1972；Saif 等，1972)。

4.5　诊断

4.5.1　TGEV/PRCV

TGEV 感染与其他肠道病原体(RV、PEDV、PDCoV 和球虫)感染引起的临床症状和萎缩性肠炎相同，为此在实验室要对 TGE 确诊，必须对粪便中病毒抗原或核酸进行检测，分离到病毒或检测到 TGEV 抗体。

PRCV 的确诊也采用类似的方法，但 PRCV 检测主要采用呼吸道样本(鼻拭子或肺匀浆)。根据临床症状评估、组织病理学损伤和病毒抗原在组织中的分布可做出疑似诊断。

通过免疫荧光(immunofluorescence, IF)(Pensaert 等，1970)或免疫组化技术(Shoup 等，1996)，TGEV 高毒保守的 N 蛋白的单克隆抗体可用于早期被 TGEV 感染的福尔马林固定组织或冰冻组织中小肠上皮细胞 TGEV 抗原的检测。

基于单抗或多克隆抗体的 ELISA 可用于检测细胞培养物、肠内容物和粪便中的 TGEV 抗原(Lanza 等，1995；Sestak 等，1996，1999a，van Nieuwstadt 等，1988）或鼻拭子或肺匀浆中的 PRCV 抗原(Lanza 等，1995)；但 ELISA 检测方法敏感性普遍低于 RT–PCR 检测方法。目前，检测 TGEV 和鉴别 TGEV、PRCV、PDCoV 和PEDV 比较常用的方法是 RT–PCR 或实时 qPCR(Kim 等，2000，2001；Masuda 等，2016；Costantini 等，2004；Ogaw 等，2009)。PRCV/TGEV 的鉴别诊断通常是利用 PCR 对 PRCV 株 S 基因的缺失进行检测。多重 RT–PCR 和实时 qPCR 已被应用于与猪腹泻相关的主要病毒(RV、TGEV、PDCoV 和PEDV)的同时鉴定(Masuda 等，2016；Ogawa 等，2009)。此外，多重微阵列杂交分析也用于包括 TGEV 在内的 8 种 CoV 的快速鉴别检测(Chen 等，2005)。

透射电子显微镜用于观察被感染猪粪便和肠内容物中的 TGEV。此外，与透射电子显微镜相比，免疫电子显微镜用于 TGEV 和 PEDV、PDCoV 的鉴别检测具有更敏感、更准确的优点。

主要用猪肾原代细胞（Bohl 和 Kumagai，1965）和传代细胞或细胞系（Laude 等，1981）、猪甲状腺细胞（Witte，1971）和 McClurkin 猪睾丸（swine testicle, ST）细胞系（McClurkin 和 Norman，1966）从被感染猪粪便或肠道内容物中分离 TGEV。野毒株被首次培养，观察不到典型的细胞病变效应（cytopathic effect, CPE），需要盲传几代培养后才可观察到 CPE（Bohl 和 Kumagai，1965）。CPE 包括细胞肿胀变圆，呈气球状。为了观察到病毒性的 CPE 或斑块，在 ST 细胞培养基中加入胰酶或胰蛋白酶，以及使用较老的细胞可以提高病毒性 CPE 和斑块的检测（Bohl，1979）。

选择 ST 细胞和猪肾细胞最适合用于从鼻拭子或肺组织匀浆中分离 PRCV。PRCV 和 TGEV 产生的 CPE 与 PEDV 在 Vero 细胞中生长形成的合胞体相似（Hofmann 和 Wyler，1988；Ksiazek 等，2003）。细胞培养物中的病毒可以通过病毒中和、IF 或免疫电子显微镜使用特异性 TGEV 抗血清或特异性单克隆抗体（Garwes 等，1988），以及病毒特异性 RT-PCR 进行确定（Enjuanes 和 Van der Zeijst，1995；Laude 等，1993；Kim 等，2000）。

TGEV 抗体可以通过一些血清学检测来鉴定。然而，TGEV 血清学是复杂的，因为 TGEV 和 PRCV 产生的病毒中和抗体在数量和质量上相似（Pensaert，1989）。用单克隆抗体阻断 ELISA 能容易鉴别 TGEV 与 PRCV（Garwes 等 1988；Bernard 等，1989；Callebaut 等，1989），当然，阻断 ELISA 用于群体抗体检测效果会更好，因为某些 TGEV 或 PRCV 抗体效价较低的猪的个体可能会出现假阴性（Callebaut 等，1989；Sestak 等，1999b；Simkins 等，1993）。此外，商品化的 ELISA 鉴别 PRCV 和 TGEV 的准确性较低（Sestak 等，1999a）。

4.5.2 PEDV

为了与 TGEV 鉴别，确诊 PEDV 必须进行临床症状观察和病毒 RNA、病毒抗原或 PEDV 特异性抗体的实验室检测。对于 PEDV RNA 的检测，目前应用最广泛的实验室技术是常规 PCR（Kim 等，2001；Ishikawa 等，1997）或实时荧光定量 RT-PCR（Kim 等，2007），对不同临床样本中的病毒 RNA 检测具有灵敏、特异、快速的特点。尽管环介导等温扩增（loop-mediated isothermal amplification，LAMP）（Ren 和 Li，2011；Yu 等，2015）方法在检测 PEDV RNA 方面具有高度的敏感性，但目前仍未应用于实验室诊断。原位杂交可用于检测固定组织中的 PEDV RNA（Stadler 等，2015）。

利用 IF 或免疫组化对腹泻初期和肠细胞脱落前被安乐致死的仔猪小肠组织进行检测，可实现对 PEDV 及其抗原的直接检测。利用透射电子显微镜或免疫电子显微镜可以直接在腹泻猪的粪便中观察到病毒颗粒。然而，要区分 PEDV 和其他 CoV（即 TGEV 和 PDCoV），必须使用免疫电子显微镜，因为通过病毒粒子形态观察难以区分所有的 CoV。

与粪便相比，使用肠内容物/匀浆可以提高用 Vero 细胞分离 PEDV 的成功率（Oka 等，2014；Chen 等，2014）。基于多克隆抗体和单克隆抗体的 ELISA 试验可用于检测粪便中 PEDV 抗原（Callebaut 等，1982；Carvajal 等，1995）。

成对的血清样本是对地方流行性猪流行性腹泻血清学诊断的必要条件。最近，在唾液中观察到抗 PEDV 的 IgG 和 IgA 抗体，这表明它们可能适合于监测先前接触 PEDV 的畜群（Bjustrom-Kraft 等，2016）。利用适应细胞培养的全病毒抗原，从感染的 Vero 细胞或哺乳动物表达系统（Wang 等，2015；Gerber 等，2014；Gerber 和 Opriessnig，2015；Okda 等，2015；Paudel 等，2014））中提取的 S/N 病毒蛋白（Knuchel 等，1992；Oh 等 2005）建立的间接 ELISA 法，被用于 PEDV 抗体的检测（Carvajal 等，1995；Hofmann 和 Wyler，1990；Kweon 等，1994；Thomas 等，2015）。此外，基于单克隆抗体或多克隆抗体建立的阻断和竞争性 ELISA 也被应用于

PEDV 抗体检测(Carvajal 等,1995;Okda 等,2015;van Nieuwstadt 和 Zetstra,1991)。血清中抗 PEDV 蛋白的 IgG 抗体在感染后 9~14 天被检测到,在感染后 21 天左右达到高峰，然后逐渐下降(Okda 等,2015)。抗 PEDV 的病毒中和抗体评估可通过在 Vero 细胞进行病毒中和试验完成(Thomas 等,2015;Okda 等,2015; Paudel 等,2014),这些血清学分析通常用于病毒接触前的筛查和评估疫苗的可行性。

总的来说,建议通过实验室检测,鉴别 PEDV 感染与 TGEV、SeCoV 和 PDCoV 感染。特别是对于 TGEV(主体)和 PEDV(S 蛋白)重组体 SeCoV,应优先选择检测 TGEV(S 基因以外的任何基因)和 PEDV(S 基因)。

4.5.3 PDCoV

PDCoV 感染与相关的 PEDV、TGEV 和 RV 感染的鉴别诊断必须进行实验室检测。所讨论的 TGEV 和 PEDV 诊断方法都适用于 PDCoV 诊断。PDCoV 感染的确诊是通过在粪便或肠道物质/组织中检测 PDCoV RNA 或抗原实现的。基于靶向 PDCoV 的 M 或 N 保守基因的 RT-PCR(Marthaler 等,2014b;Wang 等,2014c),基于病毒特异性单克隆抗体或多克隆抗体的 IF 或免疫组化(Chen 等,2015;Jung 等,2015b;Ma 等,2015) 和利用原位杂交(Jung 等,2015b),这些方法可实现对 PDCoV 感染的确诊。目前已经建立了用于肠道和粪便中的 PDCoV 检测和(或)PEDV 鉴别检测的双重实时 RT-PCR 检测方法(Zhang 等,2016b)。

直接电子显微镜可以在猪腹泻的粪便中观察到 PDCoV 病毒颗粒。但是,PDCoV 与 PEDV 或 TGEV(Jung 等,2015a)必须使用超免血清或恢复期血清的免疫电子显微镜进行鉴别。然而,除了少数毒株外(OHFD22)(Hu 等,2015),利用 LLC-PK 或 ST 细胞中分离 PDCoV 的成功率较低。其他用于诊断 PDCoV 的血清学方法还有 IF、病毒中和和 ELISA。以细胞培养的病毒抗原或 S1 和 N 病毒蛋白为基础的 ELISA 法可用于测定不同类型血清和乳汁中的 PDCoV 抗体含量 (Thachil 等,2015;Okda 等,2016;Su 等,2016)。

4.6 传播

4.6.1 TGEV

TGEV 在冷冻条件下更稳定，在室温或高温下易失活。在一项实验研究中,液体肥料中病毒的传染性在 5℃下保持 8 周以上，在 20℃下保持 2 周,在 35℃下保持 24 小时(Haas 等,1995)。此外,该病毒对光具有高度的敏感性，在阳光或紫外线照射下不到 6 小时就被灭活(Cartwright 等,1965;Haelterman,1962)。TGEV 在 1% Lysovet(苯酚和醛)、0.03%福尔马林、0.01% β-丙内酯、1mmol 二乙烯亚胺、次氯酸钠、氢氧化钠、碘、季铵化合物、乙醚和氯仿中也易被灭活 (VanCott 等,1993;Brown,1981)。TGEV 野毒株对胰蛋白酶具有抗性,在猪胆汁中及 pH 值为 3 时相对稳定(Laude 等,1981),病毒可以在胃和小肠中存活。

在温带气候中,TGE 具有一定的季节性,大多在冬季暴发。据推测,这可能是由于病毒在冷冻条件下更稳定，而对热或阳光较敏感(Haelterman,1962),在冬季,病毒很容易在畜群中通过污染物或被感染动物传播。TEGV 季节性流行的 3 个可能宿主是:①地方流行性 TGEV 的畜群;②除猪以外的宿主;③带毒的猪。犬、猫和狐狸也可能是病毒携带者，推动了病毒在农场中的传播(Haelterman,1962;McClurkin 等,1970)，由于病毒的传播在任何时期都可发生，因此被排出的病毒(由犬排出)对猪随时具有传染性(Haelterman,1962;Reynolds 等,1980)。

饲养场越冬椋鸟(紫翅椋鸟)的数量增加有助于在寒冷季节 TGEV 的机械传播。在一项研究中,喂食椋鸟 TGEV 后的 32 小时,在椋鸟的粪便中检测到 TGEV (Pilchard,1965)。同样，家蝇也是

TGEV 的机械传播媒介(Pilchard,1965)。在猪群中的果蝇中可检测到 TGEV 抗原，而通过实验饲喂果蝇 TGEV,TGEV 在果蝇内可持续存在 3 天(Gough 和 Jorgenson,1983)。值得注意的是,在中欧进行的调查证实，近 30%的野猪体内存在 TGEV 抗体(Sedlak 等,2008)。在感染后 104 天仍然能检测到 TGEV 的排出(Underdahl 等,1975),但目前尚不清楚感染性病毒颗粒是否在那个时候还能被排出或传播。有研究曾经在猪 TGE 暴发后的 3 个月、4 个月和 5 个月,在猪群中增加“哨兵”猪，结果显示在引进的猪中没有出现新的病例(Derbyshire 等,1969)。

4.6.2 PRCV

猪群密度、农场间距离,以及季节都会对PRCV的流行病学产生影响(Pensaert,1989; Have,1990)。任何年龄猪都可通过接触或空气传播感染PRCV。养猪密度高会增加本地区 PRCV 传播的风险,病毒可以传播几公里。

4.6.3 PEDV

类似于其他肠道病毒的传播,PEDV 也主要通过直接或间接粪口传播。非免疫农场通常在新购进猪后的 4 天暴发急性 PEDV。病毒主要通过被感染的猪进入农场,也可通过被污染的饲料、卡车、靴子或其他污染物进入农场。农场工人也可作为病毒传播载体(Dee 等,2014,2016;Schumacher 等,2016)。一些研究证明 PEDV 能够通过气溶胶传播(Alonso 等,2014),但未在其他研究中发现这一结果(Niederwerder 等,2016)。4 周龄猪被新型非 S-INDEL PEDV 株感染,病毒随排泄物排出可持续 14~16 天(Crawford 等,2015)。尽管如此,一些猪在通过口腔初次接触病毒 42 天后,可在其粪便中检测到 PEDV RNA,但不能检测到具有感染性的病毒颗粒。

与 TGEV 类似,在养殖场初次暴发 PEDV 后,如果新生仔猪和断奶仔猪的数量足够多，完全可以维持病毒的持续存在，则 PEDV 就有可能成为地方流行性病毒。值得注意的是,来自韩国的报告显示，野猪 PEDV 的感染率是 9.75%(Lee 等,2016b)，但它们在维持和传播 PEDV 方面的作用尚不清楚。

4.6.4 PDCoV

PDCoV 主要通过粪口途径传播。粪便、呕吐物和其他受污染的污物是病毒的主要来源。实验感染 PDCoV 引起的腹泻可持续 5~10 天，病毒 RNA 随粪便排出可持续 19 天(Hu 等,2016;Ma 等,2015)。病猪恢复正常后，仍会随粪便排出 PDCoV RNA。因此,PDCoV 的另一个宿主可能是亚临床感染或恢复期带毒的猪。

4.7 预防和控制

4.7.1 TGEV

在生产中对新生仔猪进行临床治疗通常没有意义,但是给仔猪补充 1 周或更长时间的电解质/葡萄糖溶液可降低其死亡率(Bohl,1981)。通过对仔猪采取保温措施和抗生素溶液饮水（治疗继发感染)可以改善仔猪的健康状况。加强生物安全措施，减少被感染动物及受 TGEV 影响的农场中被污染车辆引入易感畜群的机会。TGEV 不仅可以通过被感染的动物活体传播，而且还可以通过屠宰感染 TGEV 动物的未加工的组织传播(Forman,1991)。

研究者已经尝试了多种免疫母猪方法以诱导母源性免疫，进而保护新生仔猪（Chattha 等,2015;Saif 和 Jackwood,1990;Bohl 和 Saif,1975)。过去已经评估了几种病毒疫苗(强毒、弱毒、灭活和重组亚单位)的不同给药途径(口服、鼻内、皮下、肌内和乳房内)(Saif 和 Sestak,2006;Moxley 和 Olson,1989),其中通过肌内注射,非口服或乳房注射弱毒活疫苗、灭活疫苗或亚单位疫苗免疫妊娠母猪并没有提供完全免疫保护，但可有效降低仔猪死亡率(Brim 等,1994)。与强毒肠道自然感染不同,弱毒病毒不能完全激活肠道-乳腺-分泌型轴 IgA,并且观察不到后续诱导的免疫反应。由 Merck Animal Health 生产的两种修饰的 TGEV 商业疫苗,即 PROSYSTEM® TGE/Rota 和PROSYSTEM® TREC,可经口服和肌内注射联合给药。这些疫苗可以有效诱导以前接触过 TGEV 猪的免疫反应,但不能对猪的全群提供免疫保护。

研究者通过所有母猪接触 TGEV 强毒(被感染猪的肠内容物或肠组织)来增强群体免疫的方式,实现提高母源性(母乳)免疫力的目的(Bohl 和 Saif,1975;Bohl 等,1972)。在实际生产中,这种做法被称为返饲,可快速提高妊娠母猪的免疫力(特别是 TGEV 暴发后 2 周或更长时间内分娩的母猪),减少新生仔猪的损失。但这也可能导致其他病原体 (可能存在于含有 TEGV 粪便/肠内容物中)传播到邻近的畜群。在小规模农场,可以实现群体免疫,且 TGEV 的感染具有自限性。相比之下,在具有连续分娩体系和持续引入易感动物的大型农场 (≤200 头母猪) 中,TGEV 在初次暴发后,经常成为地方流行性病毒 (Saif 和 Sestak,2006),可以尝试通过返饲的方法,消灭畜群中的 TGEV。返饲后 3 周,断奶仔猪中就不再出现 TGE,这样即使 TGEV 再次传入农场,畜群中也不会出现易感宿主。

4.7.2 PEDV

由于缺乏特异性抗 PEDV 药物,治疗 PEDV 的重点是如何减轻腹泻。被 PEDV 感染的猪必须给予充足的饮水,以此来减少脱水,因为脱水会加剧疫病的严重程度。对急性期育肥猪暂时停止饲喂可能有益于康复。

PEDV 与 TGEV 防控措施相同,即采取适当的生物安全措施,避免 PEDV 传入饲养场。现有流行病学研究表明,PEDV 在饲养场之间传播主要通过动物和人员流动及污染的饲料。建议加强处理废料。

与亚洲的情况相反,在欧洲认为 PEDV 感染(流行的毒株主要是温和型毒株 S-INDEL) 对经济影响很小,不需要研发疫苗(Lee,2015)。然而,在亚洲,经典 PEDV 暴发对经济影响非常严重,需要研发疫苗,用于 PEDV 的预防和控制。在中国,基于 CV777 的 PEDV 灭活疫苗和弱毒疫苗分别于 1995 年和 1998 年获得批准(Wang 等,2016b)。不久之后,基于 PEDV 经典株的 KPEDV-9 和 DR13 的弱毒疫苗分别在 1999 年和 2004 年在韩国商品化(Kweon 等,1999;Song 等,2007)。自 1997 年以来,以适应于细胞培养的 PEDV 经典株 P-5V (Nisseiken Co.Ltd,日本)为基础而研发的 PEDV 商品化弱毒疫苗在日本被用于母猪的免疫(Sato 等,2011)。在强毒非 S-INDEL PEDV 毒株出现前,在亚洲应用的这些基于 PEDV 经典株的疫苗,有效地控制了猪流行性腹泻。本领域研究证明,经典的 PEDV 疫苗对新出现的强毒非 S-INDEL PEDV 毒株造成猪的严重腹泻不产生免疫保护作用(Lee,2015)。

妊娠母猪(返饲方法)故意接触 PEDV,以此促进母源性免疫的快速产生,从而缩短了农场猪流行性腹泻的进程和严重程度 (Chattha 等,2015)。然而,如 TGEV 部分所述,这种方法可能促进了其他病原体在农场的传播。作者最近证明,与低剂量 PEDV 接种母猪和模拟感染母猪相比,高剂量强毒 PEDV 接种母猪可以显著提高仔猪存活率(图 4.3)(Langel 等,2016)。这一新发现表明,目前应用的返饲方法控制 PEDV 策略,可以通过给妊娠母猪的全群接种高剂量 PEDV 进行改进。

自 2013 年 PEDV 暴发以来,美国针对新出现的非 S-INDEL PEDV 毒株有条件地批准了 2 种 PEDV 疫苗:基于甲病毒的疫苗(Harris vaccines™,现在是默克动物保健公司) 和灭活疫苗(Zoetis) (2014)。第一种疫苗是在 2014 年 6 月研发的,使用复制缺陷型委内瑞拉马脑脊髓炎病毒包装系统表达一种新型非 S-INDEL PEDV 株的 S 蛋白。第二种疫苗是 2015 年 9 月开发的加佐剂灭活全病

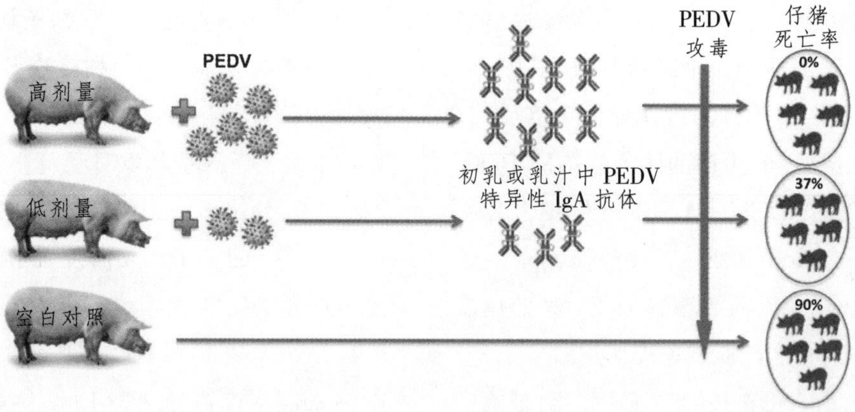

图 4.3 妊娠猪的 PEDV 剂量可能影响黏膜免疫反应和母源性免疫。在 3~4 周的准备期，母猪接受高剂量、低剂量 PEDV 或空白对照。所有仔猪在产后 3~5 天接受 PEDV 攻毒。(Langel et al. 2016)

毒(非 S-INDEL PEDV)疫苗(Crawford 等,2016)。2016 年 10 月,中国批准了基于非 S-INDEL PEDV 毒株 AJ1102 的灭活疫苗(Wang 等,2016b)。在韩国,证明基于非 S-INDEL 毒株 KNU-141112 的灭活疫苗能够对后备母猪及其哺乳仔猪产生保护作用(Baek 等,2016)。然而,尚未评估这些疫苗/候选疫苗在该领域的功效。迄今为止,研究者已经使用不同的方法建立了 PEDV 经典毒株和新型非 S-INDEL PEDV 毒株的反向遗传学平台 (Beall 等,2016;Jengarn 等,2015;Li 等,2013), 并且在将来可用于对安全有效的 PEDV 进行合理的设计。

4.7.3 PDCoV

对 TGEV 和 PEDV 采取的控制和预防措施同样适用于 PDCoV。在没有任何用来控制 PDCoV 的疫苗或抗病毒药物的情况下, 口服碳酸氢盐溶液和自由饮水可以减轻哺乳猪的酸中毒和脱水。在并发/继发细菌感染的情况下,给予抗生素可能是有益的。在死亡率很高的情况下,必须选择返饲技术来激活母源性免疫,以此降低仔猪死亡率。此外,在 PDCoV 流行期间,必须实施严格的生物安全计划,来减少通过污染物传播 PDCoV。全进/全出系统和彻底消毒(使用酚类消毒剂,漂白剂,过氧化物,醛或碘制剂)可以阻断 PDCoV 的流行周期。

参考文献

Akimkin V, Beer M, Blome S et al (2016) New chimeric porcine coronavirus in swine feces, Germany, 2012. Emerg Infect Dis 22:1314–1315

Alonso C, Goede DP, Morrison RB et al (2014) Evidence of infectivity of airborne porcine epidemic diarrhea virus and detection of airborne viral RNA at long distances from infected herds. Vet Res 45:73

Annamalai T, Saif LJ, Lu Z et al (2015) Age-dependent variation in innate immune responses to porcine epidemic diarrhea virus infection in suckling versus weaned pigs. Vet Immunol Immunopathol 168:193–202

Anon (2014) USDA to require reports of PED. J Am Vet Med Assoc 244:1234

Atanasova K, Van Gucht S, Barbé F, Lefebvre DJ, Chiers K, Van Reeth K (2008) Lung cell tropism and inflammatory cytokine-profile of porcine respiratory coronavirus infection. Open Vet Sci J 2:117–126

Baek PS, Choi HW, Lee S et al (2016) Efficacy of an inactivated genotype 2b porcine epidemic diarrhea virus vaccine in neonatal piglets. Vet Immunol Immunopathol 174:45–49

Ballesteros ML, Sanchez CM, Enjuanes L (1997) Two amino acid changes at the N-terminus of

transmissible gastroenteritis coronavirus spike protein result in the loss of enteric tropism. Virology 227:378–388
Barman N, Barman B, Sarma K et al (2003) Indian J Anim Sci 73:576–578
Beall A, Yount B, Lin CM et al (2016) Characterization of a pathogenic full-length cDNA clone and transmission model for porcine epidemic diarrhea virus strain PC22A. MBio 7:e01451-15
Belsham GJ, Rasmussen TB, Normann P et al (2016) Characterization of a novel chimeric swine enteric coronavirus from diseased pigs in Central Eastern Europe in 2016. Transbound Emerg Dis 63:595–601
Bernard S, Bottreau E, Aynaud JM et al (1989) Natural infection with the porcine respiratory coronavirus induces protective lactogenic immunity against transmissible gastroenteritis. Vet Microbiol 21:1–8
Bjustrom-Kraft J, Woodard K, Gimenez-Lirola L et al (2016) Porcine epidemic diarrhea virus (PEDV) detection and antibody response in commercial growing pigs. BMC Vet Res 12:99
Bohl E (1979) Diagnosis of diarrhea in pigs due to transmissible gastroenteritis virus or rotavirus. In: Bricout F, Scherrer R (eds) Viral enteritis in humans and animals. INSERM, Paris, pp 341–343
Bohl EH (1981) Transmissible gastroenteritis. In: Leman AD, Glock RD, Mengeling WL, Penny RHC, Scholl E, Straw B (eds) Diseases of swine, 5th edn. University Press, Ames, IA, pp 195–208
Bohl EH, Kumagai T (1965) The use of cell cultures for the study of swine transmissible gastroenteritis virus. In: United States Livestock Sanitary Association meeting
Bohl EH, Saif LJ (1975) Passive immunity in transmissible gastroenteritis of swine: immunoglobulin characteristics of antibodies in milk after inoculating virus by different routes. Infect Immun 11:23–32
Bohl EH, Gupta RK, Olquin MV et al (1972) Antibody responses in serum, colostrum, and milk of swine after infection or vaccination with transmissible gastroenteritis virus. Infect Immun 6:289–301
Bohl EH, Kohler EM, Saif LJ et al (1978) Rotavirus as a cause of diarrhea in pigs. J Am Vet Med Assoc 172:458–463
Boniotti MB, Papetti A, Lavazza A et al (2016) Porcine epidemic diarrhea virus and discovery of a recombinant swine enteric coronavirus, Italy. Emerg Infect Dis 22:83–87
Brim TA, VanCott JL, Lunney JK et al (1994) Lymphocyte proliferation responses of pigs inoculated with transmissible gastroenteritis virus or porcine respiratory coronavirus. Am J Vet Res 55:494–501
Brim TA, VanCott JL, Lunney JK et al (1995) Cellular immune responses of pigs after primary inoculation with porcine respiratory coronavirus or transmissible gastroenteritis virus and challenge with transmissible gastroenteritis virus. Vet Immunol Immunopathol 48:35–54
Brown I, Cartwright S (1986) New porcine coronavirus? Vet Rec 119:282–283
Brown TT Jr (1981) Laboratory evaluation of selected disinfectants as virucidal agents against porcine parvovirus, pseudorabies virus, and transmissible gastroenteritis virus. Am J Vet Res 42:1033–1036
Butler DG, Gall DG, Kelly MH et al (1974) Transmissible gastroenteritis. Mechanisms responsible for diarrhea in an acute viral enteritis in piglets. J Clin Invest 53:1335–1342
Callebaut P, Debouck P, Pensaert M (1982) Enzyme-linked immunosorbent assay for the detection of the coronavirus-like agent and its antibodies in pigs with porcine epidemic diarrhea. Vet Microbiol 7:295–306
Callebaut P, Pensaert MB, Hooyberghs J (1989) A competitive inhibition ELISA for the differentiation of serum antibodies from pigs infected with transmissible gastroenteritis virus (TGEV) or with the TGEV-related porcine respiratory coronavirus. Vet Microbiol 20:9–19
Cartwright SF, Harris HM, Blandford TB et al (1965) A cytopathic virus causing a transmissible gastroenteritis in swine. I. Isolation and properties. J Comp Pathol 75:387–396
Carvajal A, Lanza I, Diego R et al (1995) Evaluation of a blocking ELISA using monoclonal antibodies for the detection of porcine epidemic diarrhea virus and its antibodies. J Vet Diagn Investig 7:60–64
Cepica A, Derbyshire JB (1984) Antibody-dependent and spontaneous cell-mediated cytotoxicity against transmissible gastroenteritis virus infected cells by lymphocytes from sows, fetuses and neonatal piglets. Can J Comp Med 48:258–261
Chae C, Kim O, Choi C et al (2000) Prevalence of porcine epidemic diarrhoea virus and transmissible gastroenteritis virus infection in Korean pigs. Vet Rec 147:606–608
Chattha KS, Roth JA, Saif LJ (2015) Strategies for design and application of enteric viral vaccines. Annu Rev Anim Biosci 3:375–395
Chen W, Yan M, Yang L et al (2005) SARS-associated coronavirus transmitted from human to pig. Emerg Infect Dis 11:446–448
Chen Q, Li G, Stasko J et al (2014) Isolation and characterization of porcine epidemic diarrhea viruses associated with the 2013 disease outbreak among swine in the United States. J Clin

Microbiol 52:234–243
Chen Q, Gauger P, Stafne M et al (2015) Pathogenicity and pathogenesis of a United States porcine deltacoronavirus cell culture isolate in 5-day-old neonatal piglets. Virology 482:51–59
Chen Q, Gauger PC, Stafne MR et al (2016a) Pathogenesis comparison between the United States porcine epidemic diarrhea virus prototype and S-INDEL-variant strains in conventional neonatal piglets. J Gen Virol 97:1107–1121
Chen X, Tu C, Qin T et al (2016b) Retinoic acid facilitates inactivated transmissible gastroenteritis virus induction of CD8(+) T-cell migration to the porcine gut. Sci Rep 6:24152
Cheun-Arom T, Temeeyasen G, Srijangwad A et al (2015) Complete genome sequences of two genetically distinct variants of porcine epidemic diarrhea virus in the eastern region of Thailand. Genome Announc 3:e00634-15
Costantini V, Lewis P, Alsop J et al (2004) Respiratory and fecal shedding of porcine respiratory coronavirus (PRCV) in sentinel weaned pigs and sequence of the partial S-gene of the PRCV isolates. Arch Virol 149:957–974
Coussement W, Ducatelle R, Debouck P et al (1982) Pathology of experimental CV777 coronavirus enteritis in piglets. I. Histological and histochemical study. Vet Pathol 19:46–56
Cox E, Hooyberghs J, Pensaert MB (1990) Sites of replication of a porcine respiratory coronavirus related to transmissible gastroenteritis virus. Res Vet Sci 48:165–169
Cox E, Pensaert MB, Callebaut P (1993) Intestinal protection against challenge with transmissible gastroenteritis virus of pigs immune after infection with the porcine respiratory coronavirus. Vaccine 11:267–272
Crawford K, Lager K, Miller L et al (2015) Evaluation of porcine epidemic diarrhea virus transmission and the immune response in growing pigs. Vet Res 46:49
Crawford K, Lager KM, Kulshreshtha V et al (2016) Status of vaccines for porcine epidemic diarrhea virus in the United States and Canada. Virus Res 226:108–116
De Diego M, Laviada MD, Enjuanes L et al (1992) Epitope specificity of protective lactogenic immunity against swine transmissible gastroenteritis virus. J Virol 66:6502–6508
de Groot R, Ziebuhr J, Poon L et al (2008) Revision of the family *Coronaviridae*. Taxonomic proposal of the Coronavirus Study Group to the ICTV Executive Committee. http://talk.ictvonline.org/media/p/1230.aspx
Debouck P, Pensaert M, Coussement W (1981) The pathogenesis of an enteric infection in pigs, experimentally induced by the coronavirus-like agent, CV 777. Vet Microbiol 6:157–165
Dee S, Clement T, Schelkopf A et al (2014) An evaluation of contaminated complete feed as a vehicle for porcine epidemic diarrhea virus infection of naive pigs following consumption via natural feeding behavior: proof of concept. BMC Vet Res 10:176
Dee S, Neill C, Singrey A et al (2016) Modeling the transboundary risk of feed ingredients contaminated with porcine epidemic diarrhea virus. BMC Vet Res 12:51
Derbyshire JB, Jessett DM, Newman G (1969) An experimental epidemiological study of porcine transmissible gastroenteritis. J Comp Pathol 79:445–452
Diep NV, Norimine J, Sueyoshi M et al (2017) Novel porcine epidemic diarrhea virus (PEDV) variants with large deletions in the spike (S) gene coexist with PEDV strains possessing an intact S gene in domestic pigs in Japan: a new disease situation. PLoS One 12:e0170126
Ding Z, Fang L, Jing H et al (2014) Porcine epidemic diarrhea virus nucleocapsid protein antagonizes beta interferon production by sequestering the interaction between IRF3 and TBK1. J Virol 88:8936–8945
Dong BQ, Liu W, Fan XH et al (2007) Detection of a novel and highly divergent coronavirus from Asian leopard cats and Chinese ferret badgers in southern China. J Virol 81:6920–6926
Dong N, Fang L, Zeng S et al (2015) Porcine deltacoronavirus in mainland China. Emerg Infect Dis 21:2254–2255
Doyle LP, Hutchings LM (1946) A transmissible gastroenteritis in pigs. J Am Vet Med Assoc 108:257–259
Enjuanes L, Van der Zeijst B (1995) Molecular basis of transmissible gastroenteritis virus epidemiology. In: Siddell SG (ed) The Coronaviridae. Plenum Press, New York, pp 337–376
Forman AJ (1991) Infection of pigs with transmissible gastroenteritis virus from contaminated carcases. Aust Vet J 68:25–27
Frederick GT, Bohl EH, Cross RF (1976) Pathogenicity of an attenuated strain of transmissible gastroenteritis virus for newborn pigs. Am J Vet Res 37:165–169
Garwes DJ, Stewart F, Cartwright SF et al (1988) Differentiation of porcine coronavirus from transmissible gastroenteritis virus. Vet Rec 122:86–87
Gerber PF, Opriessnig T (2015) Detection of immunoglobulin (Ig) A antibodies against porcine epidemic diarrhea virus (PEDV) in fecal and serum samples. MethodsX 2:368–373
Gerber PF, Gong Q, Huang YW et al (2014) Detection of antibodies against porcine epidemic diarrhea virus in serum and colostrum by indirect ELISA. Vet J 202:33–36
Gong L, Li J, Zhou Q et al (2017) A new bat-HKU2-like coronavirus in swine, China, 2017. Emerg Infect Dis 23. https://doi.org/10.3201/eid2309.170915
Gonzalez JM, Gomez-Puertas P, Cavanagh D et al (2003) A comparative sequence analysis to

revise the current taxonomy of the family Coronaviridae. Arch Virol 148:2207–2235
Gough PM, Jorgenson RD (1983) Identification of porcine transmissible gastroenteritis virus in house flies (Musca domestica Linneaus). Am J Vet Res 44:2078–2082
Haas B, Ahl R, Bohm R et al (1995) Inactivation of viruses in liquid manure. Rev Sci Tech 14:435–445
Haelterman EO (1962) Epidemiological studies of transmissible gastroenteritis of swine. In: United States Livestock Sanitary Association meeting
Halbur PG, Paul PS, Vaughn EM et al (1993) Experimental reproduction of pneumonia in gnotobiotic pigs with porcine respiratory coronavirus isolate AR310. J Vet Diagn Investig 5:184–188
Have P (1990) Infection with a new porcine respiratory coronavirus in Denmark: serologic differentiation from transmissible gastroenteritis virus using monoclonal antibodies. Adv Exp Med Biol 276:435–439
Hill H, Biwer J, Woods R, Wesley R (1990) Porcine respiratory coronavirus isolated from two U.S. swine herds. In: American Association of Swine Practitioners meeting
Hofmann M, Wyler R (1988) Propagation of the virus of porcine epidemic diarrhea in cell culture. J Clin Microbiol 26:2235–2239
Hofmann M, Wyler R (1990) Enzyme-linked immunosorbent assay for the detection of porcine epidemic diarrhea coronavirus antibodies in swine sera. Vet Microbiol 21:263–273
Hooper BE, Haelterman EO (1966a) Concepts of pathogenesis and passive immunity in transmissible gastroenteritis of swine. J Am Vet Med Assoc 149:1580–1586
Hooper BE, Haelterman EO (1966b) Growth of transmissible gastroenteritis virus in young pigs. Am J Vet Res 27:286–291
Hou Y, Lin CM, Yokoyama M et al (2017) Deletion of a 197-amino-acid region in the N-terminal domain of spike protein attenuates porcine epidemic diarrhea virus in piglets. J Virol 91:e00227-17
Hu H, Jung K, Vlasova AN et al (2015) Isolation and characterization of porcine deltacoronavirus from pigs with diarrhea in the United States. J Clin Microbiol 53:1537–1548
Hu H, Jung K, Vlasova AN et al (2016) Experimental infection of gnotobiotic pigs with the cell-culture-adapted porcine deltacoronavirus strain OH-FD22. Arch Virol 161:3421–3434
Hwang EK, Kim JH, Jean YH et al (1994) Current occurrence of porcine epidemic diarrhea in Korea. RDA J Agric Sci 36:587–596
Ishikawa K, Sekiguchi H, Ogino T et al (1997) Direct and rapid detection of porcine epidemic diarrhea virus by RT-PCR. J Virol Methods 69:191–195
Janetanakit T, Lumyai M, Bunpapong N et al (2016) Porcine deltacoronavirus, Thailand, 2015. Emerg Infect Dis 22:757–759
Jengarn J, Wongthida P, Wanasen N et al (2015) Genetic manipulation of porcine epidemic diarrhea virus (PEDV) recovered from a full length infectious cDNA clone. J Gen Virol 96:2206–2218
Jung K, Alekseev KP, Zhang X et al (2007) Altered pathogenesis of porcine respiratory coronavirus in pigs due to immunosuppressive effects of dexamethasone: implications for corticosteroid use in treatment of severe acute respiratory syndrome coronavirus. J Virol 81:13681–13693
Jung K, Wang Q, Scheuer KA et al (2014) Pathology of US porcine epidemic diarrhea virus strain PC21A in gnotobiotic pigs. Emerg Infect Dis 20:662–665
Jung K, Annamalai T, Lu Z et al (2015a) Comparative pathogenesis of US porcine epidemic diarrhea virus (PEDV) strain PC21A in conventional 9-day-old nursing piglets vs. 26-day-old weaned pigs. Vet Microbiol 178:31–40
Jung K, Hu H, Eyerly B et al (2015b) Pathogenicity of 2 porcine deltacoronavirus strains in gnotobiotic pigs. Emerg Infect Dis 21:650–654
Jung K, Hu H, Saif LJ (2016a) Porcine deltacoronavirus induces apoptosis in swine testicular and LLC porcine kidney cell lines in vitro but not in infected intestinal enterocytes in vivo. Vet Microbiol 182:57–63
Jung K, Hu H, Saif LJ (2016b) Porcine deltacoronavirus infection: etiology, cell culture for virus isolation and propagation, molecular epidemiology and pathogenesis. Virus Res 226:50–59
Kemeny LJ, Wiltsey VL, Riley JL (1975) Upper respiratory infection of lactating sows with transmissible gastroenteritis virus following contact exposure to infected piglets. Cornell Vet 65:352–362
Kim O, Chae C (2003) Experimental infection of piglets with a Korean strain of porcine epidemic diarrhoea virus. J Comp Pathol 129:55–60
Kim L, Chang KO, Sestak K et al (2000) Development of a reverse transcription-nested polymerase chain reaction assay for differential diagnosis of transmissible gastroenteritis virus and porcine respiratory coronavirus from feces and nasal swabs of infected pigs. J Vet Diagn Investig 12:385–388
Kim SY, Song DS, Park BK (2001) Differential detection of transmissible gastroenteritis virus and porcine epidemic diarrhea virus by duplex RT-PCR. J Vet Diagn Investig 13:516–520
Kim SH, Kim IJ, Pyo HM et al (2007) Multiplex real-time RT-PCR for the simultaneous detection and quantification of transmissible gastroenteritis virus and porcine epidemic diarrhea virus. J Virol Methods 146:172–177

Kim SH, Lee JM, Jung J et al (2015) Genetic characterization of porcine epidemic diarrhea virus in Korea from 1998 to 2013. Arch Virol 160:1055–1064
Kim YK, Cho YY, An BH et al (2016) Molecular characterization of the spike and ORF3 genes of porcine epidemic diarrhea virus in the Philippines. Arch Virol 161:1323–1328
Knuchel M, Ackermann M, Muller HK et al (1992) An ELISA for detection of antibodies against porcine epidemic diarrhoea virus (PEDV) based on the specific solubility of the viral surface glycoprotein. Vet Microbiol 32:117–134
Kodama Y, Ogata M, Simizu Y (1980) Characterization of immunoglobulin A antibody in serum of swine inoculated with transmissible gastroenteritis virus. Am J Vet Res 41:740–745
Ksiazek TG, Erdman D, Goldsmith CS et al (2003) A novel coronavirus associated with severe acute respiratory syndrome. N Engl J Med 348:1953–1966
Kuwahara H, Nunoya T, Samejima T et al (1988) Passage in piglets of a coronavirus associated with porcine epidemic diarrhea. J Jpn Vet Med Assoc 41:169–173
Kweon C, Kwon BJ, Jung TS et al (1993) Isolation of porcine epidemic diarrhea virus (PEDV) infection in Korea. Korean J Vet Res 33:249–254
Kweon CH, Kwon BJ, Kang YB et al (1994) Cell adaptation of KPEDV-9 and serological survey on porcine epidemic diarrhea virus (PEDV) infection in Korea. Korean J Vet Res 34:321–326
Kweon CH, Kwon BJ, Lee JG et al (1999) Derivation of attenuated porcine epidemic diarrhea virus (PEDV) as vaccine candidate. Vaccine 17:2546–2553
Langel SN, Paim FC, Lager KM et al (2016) Lactogenic immunity and vaccines for porcine epidemic diarrhea virus (PEDV): historical and current concepts. Virus Res 226:93–107
Lanza I, Shoup DI, Saif LJ (1995) Lactogenic immunity and milk antibody isotypes to transmissible gastroenteritis virus in sows exposed to porcine respiratory coronavirus during pregnancy. Am J Vet Res 56:739–748
Laude H, Gelfi J, Aynaud JM (1981) In vitro properties of low- and high-passaged strains of transmissible gastroenteritis coronavirus of swine. Am J Vet Res 42:447–449
Laude H, Van Reeth K, Pensaert M (1993) Porcine respiratory coronavirus: molecular features and virus-host interactions. Vet Res 24:125–150
Lee C (2015) Porcine epidemic diarrhea virus: an emerging and re-emerging epizootic swine virus. Virol J 12:193
Lee S, Park GS, Shin JH et al (2014) Full-genome sequence analysis of a variant strain of porcine epidemic diarrhea virus in South Korea. Genome Announc 2:e00753-14
Lee JH, Chung HC, Nguyen VG et al (2016a) Detection and phylogenetic analysis of porcine deltacoronavirus in Korean swine farms, 2015. Transbound Emerg Dis 63:248–252
Lee DU, Kwon T, Je SH et al (2016b) Wild boars harboring porcine epidemic diarrhea virus (PEDV) may play an important role as a PEDV reservoir. Vet Microbiol 192:90–94
Li C, Li Z, Zou Y et al (2013) Manipulation of the porcine epidemic diarrhea virus genome using targeted RNA recombination. PLoS One 8:e69997
Li G, Chen Q, Harmon KM et al (2014) Full-length genome sequence of porcine deltacoronavirus strain USA/IA/2014/8734. Genome Announc 2:e00278-14
Li Z, He W, Lan Y et al (2016) The evidence of porcine hemagglutinating encephalomyelitis virus induced nonsuppurative encephalitis as the cause of death in piglets. PeerJ 4:e2443
Lin CN, Chung WB, Chang SW et al (2014) US-like strain of porcine epidemic diarrhea virus outbreaks in Taiwan, 2013-2014. J Vet Med Sci 76:1297–1299
Lin CM, Gao X, Oka T et al (2015a) Antigenic relationships among porcine epidemic diarrhea virus and transmissible gastroenteritis virus strains. J Virol 89:3332–3342
Lin CM, Annamalai T, Liu X et al (2015b) Experimental infection of a US spike-insertion deletion porcine epidemic diarrhea virus in conventional nursing piglets and cross-protection to the original US PEDV infection. Vet Res 46:134
Lin CM, Saif LJ, Marthaler D et al (2016) Evolution, antigenicity and pathogenicity of global porcine epidemic diarrhea virus strains. Virus Res 226:20–39
Lohse L, Krog JS, Strandbygaard B et al (2017) Experimental infection of young pigs with an early European strain of porcine epidemic diarrhoea virus and a recent US strain. Transbound Emerg Dis 64:1380–1386
Lorbach JN, Wang L, Nolting JM et al (2017) Porcine hemagglutinating encephalomyelitis virus and respiratory disease in exhibition swine, Michigan, USA, 2015. Emerg Infect Dis 23:1168–1171
Ma G, Feng Y, Gao F et al (2005) Biochemical and biophysical characterization of the transmissible gastroenteritis coronavirus fusion core. Biochem Biophys Res Commun 337:1301–1307
Ma Y, Zhang Y, Liang X et al (2015) Origin, evolution, and virulence of porcine deltacoronaviruses in the United States. MBio 6:e00064
Ma Y, Zhang Y, Liang X et al (2016) Two-way antigenic cross-reactivity between porcine epidemic diarrhea virus and porcine deltacoronavirus. Vet Microbiol 186:90–96
Madson DM, Magstadt DR, Arruda PH et al (2014) Pathogenesis of porcine epidemic diarrhea virus isolate (US/Iowa/18984/2013) in 3-week-old weaned pigs. Vet Microbiol 174:60–68
Martelli P, Lavazza A, Nigrelli AD et al (2008) Epidemic of diarrhoea caused by porcine epidemic

diarrhoea virus in Italy. Vet Rec 162:307–310
Marthaler D, Jiang Y, Collins J et al (2014a) Complete genome sequence of strain SDCV/USA/Illinois121/2014, a porcine deltacoronavirus from the United States. Genome Announc 2:e00218-14
Marthaler D, Raymond L, Jiang Y et al (2014b) Rapid detection, complete genome sequencing, and phylogenetic analysis of porcine deltacoronavirus. Emerg Infect Dis 20:1347–1350
Masuda T, Murakami S, Takahashi O et al (2015) New porcine epidemic diarrhoea virus variant with a large deletion in the spike gene identified in domestic pigs. Arch Virol 160:2565–2568
Masuda T, Tsuchiaka S, Ashiba T et al (2016) Development of one-step real-time reverse transcriptase-PCR-based assays for the rapid and simultaneous detection of four viruses causing porcine diarrhea. Jpn J Vet Res 64:5–14
McClurkin AW, Norman JO (1966) Studies on transmissible gastroenteritis of swine. II. Selected characteristics of a cytopathogenic virus common to five isolates from transmissible gastroenteritis. Can J Comp Med Vet Sci 30:190–198
McClurkin AW, Stark SL, Norman JO (1970) Transmissible gastroenteritis (TGE) of swine: the possible role of dogs in the epizootiology of TGE. Can J Comp Med 34:347–349
Moon HW (1978) Mechanisms in the pathogenesis of diarrhea: a review. J Am Vet Med Assoc 172:443–448
Moon HW, Kemeny LJ, Lambert G et al (1975) Age-dependent resistance to transmissible gastroenteritis of swine. III. Effects of epithelial cell kinetics on coronavirus production and on atrophy of intestinal villi. Vet Pathol 12:434–445
Moxley RA, Olson LD (1989) Clinical evaluation of transmissible gastroenteritis virus vaccines and vaccination procedures for inducing lactogenic immunity in sows. Am J Vet Res 50:111–118
Niederwerder MC, Nietfeld JC, Bai J et al (2016) Tissue localization, shedding, virus carriage, antibody response, and aerosol transmission of porcine epidemic diarrhea virus following inoculation of 4-week-old feeder pigs. J Vet Diagn Investig 28:671–678
Ogawa H, Taira O, Hirai T et al (2009) Multiplex PCR and multiplex RT-PCR for inclusive detection of major swine DNA and RNA viruses in pigs with multiple infections. J Virol Methods 160:210–214
Oh JS, Song DS, Yang JS et al (2005) Comparison of an enzyme-linked immunosorbent assay with serum neutralization test for serodiagnosis of porcine epidemic diarrhea virus infection. J Vet Sci 6:349–352
Oka T, Saif LJ, Marthaler D et al (2014) Cell culture isolation and sequence analysis of genetically diverse US porcine epidemic diarrhea virus strains including a novel strain with a large deletion in the spike gene. Vet Microbiol 173:258–269
Okda F, Liu X, Singrey A et al (2015) Development of an indirect ELISA, blocking ELISA, fluorescent microsphere immunoassay and fluorescent focus neutralization assay for serologic evaluation of exposure to North American strains of porcine epidemic diarrhea virus. BMC Vet Res 11:180
Okda F, Lawson S, Liu X et al (2016) Development of monoclonal antibodies and serological assays including indirect ELISA and fluorescent microsphere immunoassays for diagnosis of porcine deltacoronavirus. BMC Vet Res 12:95
O'Toole D, Brown I, Bridges A et al (1989) Pathogenicity of experimental infection with 'pneumotropic' porcine coronavirus. Res Vet Sci 47:23–29
Park JE, Shin HJ (2014) Porcine epidemic diarrhea virus infects and replicates in porcine alveolar macrophages. Virus Res 191:143–152
Paudel S, Park JE, Jang H et al (2014) Comparison of serum neutralization and enzyme-linked immunosorbent assay on sera from porcine epidemic diarrhea virus vaccinated pigs. Vet Q 34:218–223
PED vaccine gains conditional approval. J Am Vet Med Assoc 2014;245:267
Pensaert M (1989) Transmissible gastroenteritis virus (respiratory variant). In: Pensaert M (ed) Virus infections of porcines, vol 2. Elsevier Science Publishers, Amsterdam, pp 154–165
Pensaert M, Haelterman EO, Burnstein T (1970) Transmissible gastroenteritis of swine: virus-intestinal cell interactions. I. Immunofluorescence, histopathology and virus production in the small intestine through the course of infection. Arch Gesamte Virusforsch 31:321–334
Pensaert M, Callebaut P, Vergote J (1986) Isolation of a porcine respiratory, non-enteric coronavirus related to transmissible gastroenteritis. Vet Q 8:257–261
Pensaert M, Cox E, van Deun K et al (1993) A sero-epizootiological study of porcine respiratory coronavirus in Belgian swine. Vet Q 15:16–20
Pilchard EI (1965) Experimental transmission of transmissible gastroenteritis virus by starlings. Am J Vet Res 26:1177–1179
Pospischil A, Hess RG, Bachmann PA (1981) Light microscopy and ultrahistology of intestinal changes in pigs infected with epizootic diarrhoea virus (EVD): comparison with transmissible gastroenteritis (TGE) virus and porcine rotavirus infections. Zentralbl Veterinarmed B 28:564–577

Pritchard GC (1987) Transmissible gastroenteritis in endemically infected breeding herds of pigs in East Anglia, 1981-85. Vet Rec 120:226–230

Puranaveja S, Poolperm P, Lertwatcharasarakul P et al (2009) Chinese-like strain of porcine epidemic diarrhea virus, Thailand. Emerg Infect Dis 15:1112–1115

Ren X, Li P (2011) Development of reverse transcription loop-mediated isothermal amplification for rapid detection of porcine epidemic diarrhea virus. Virus Genes 42:229–235

Reynolds DJ, Garwes DJ, Lucey S (1980) Differentiation of canine coronavirus and porcine transmissible gastroenteritis virus by neutralization with canine, porcine, and feline sera. Vet Microbiol 5:283–290

Saeng-Chuto K, Lorsirigool A, Temeeyasen G et al (2017) Different lineage of porcine deltacoronavirus in Thailand, Vietnam and Lao PDR in 2015. Transbound Emerg Dis 64:3–10

Saif LJ (1999) Enteric viral infections of pigs and strategies for induction of mucosal immunity. Adv Vet Med 41:429–446

Saif LJ, Bohl EH (1979) Role of SIgA in passive immunity of swine to enteric viral infections. In: Ogra P, Dayton D (eds) Immunology of breast milk. Raven Press, New York, NY, pp 237–248

Saif L, Jackwood DJ (1990) Enteric virus vaccines: theoretical considerations, current status, and future approaches. In: Saif L, Theil KW (eds) Viral diarrheas of man and animals. CRC Press, Inc., Boca Raton, FL, pp 313–329

Saif LJ, Sestak K (2006) Transmissible gastroenteritis virus and porcine respiratory coronavirus. In: Straw BE, Zimmerman JJ, D'Allaire D, Taylor DJ (eds) Diseases of swine, 9th edn. Blackwell Publishing Company, Ames, IA, pp 489–516

Saif LJ, Bohl EH, Gupta RK (1972) Isolation of porcine immunoglobulins and determination of the immunoglobulin classes of transmissible gastroenteritis viral antibodies. Infect Immun 6:600–609

Saif LJ, Bohl EH, Kohler EM et al (1977) Immune electron microscopy of transmissible gastroenteritis virus and rotavirus (reovirus-like agent) of swine. Am J Vet Res 38:13–20

Saif LJ, van Cott JL, Brim TA (1994) Immunity to transmissible gastroenteritis virus and porcine respiratory coronavirus infections in swine. Vet Immunol Immunopathol 43:89–97

Sanchez CM, Izeta A, Sanchez-Morgado JM et al (1999) Targeted recombination demonstrates that the spike gene of transmissible gastroenteritis coronavirus is a determinant of its enteric tropism and virulence. J Virol 73:7607–7618

Sato T, Takeyama N, Katsumata A et al (2011) Mutations in the spike gene of porcine epidemic diarrhea virus associated with growth adaptation in vitro and attenuation of virulence in vivo. Virus Genes 43:72–78

Schumacher LL, Woodworth JC, Jones CK et al (2016) Evaluation of the minimum infectious dose of porcine epidemic diarrhea virus in virus-inoculated feed. Am J Vet Res 77:1108–1113

Sedlak K, Bartova E, Machova J (2008) Antibodies to selected viral disease agents in wild boars from the Czech Republic. J Wildl Dis 44:777–780

Sestak K, Lanza I, Park SK et al (1996) Contribution of passive immunity to porcine respiratory coronavirus to protection against transmissible gastroenteritis virus challenge exposure in suckling pigs. Am J Vet Res 57:664–671

Sestak K, Zhou Z, Shoup DI et al (1999a) Evaluation of the baculovirus-expressed S glycoprotein of transmissible gastroenteritis virus (TGEV) as antigen in a competition ELISA to differentiate porcine respiratory coronavirus from TGEV antibodies in pigs. J Vet Diagn Investig 11:205–214

Sestak K, Meister RK, Hayes JR et al (1999b) Active immunity and T-cell populations in pigs intraperitoneally inoculated with baculovirus-expressed transmissible gastroenteritis virus structural proteins. Vet Immunol Immunopathol 70:203–221

Shimizu M, Shimizu Y (1979) Lymphocyte proliferative response to viral antigen in pigs infected with transmissible gastroenteritis virus. Infect Immun 23:239–243

Shoup DI, Swayne DE, Jackwood DJ et al (1996) Immunohistochemistry of transmissible gastroenteritis virus antigens in fixed paraffin-embedded tissues. J Vet Diagn Investig 8:161–167

Simkins RA, Weilnau PA, Van Cott J et al (1993) Competition ELISA, using monoclonal antibodies to the transmissible gastroenteritis virus (TGEV) S protein, for serologic differentiation of pigs infected with TGEV or porcine respiratory coronavirus. Am J Vet Res 54:254–259

Sinha A, Gauger P, Zhang J et al (2015) PCR-based retrospective evaluation of diagnostic samples for emergence of porcine deltacoronavirus in US swine. Vet Microbiol 179:296–298

Song DS, Oh JS, Kang BK et al (2007) Oral efficacy of Vero cell attenuated porcine epidemic diarrhea virus DR13 strain. Res Vet Sci 82:134–140

Song D, Zhou X, Peng Q et al (2015) Newly emerged porcine deltacoronavirus associated with diarrhoea in swine in China: identification, prevalence and full-length genome sequence analysis. Transbound Emerg Dis 62:575–580

Stadler J, Zoels S, Fux R et al (2015) Emergence of porcine epidemic diarrhea virus in southern Germany. BMC Vet Res 11:142

Stepanek J, Mensik J, Franz J, Hornich M (1979) Epizootiology, diagnosis and prevention of viral diarrhea in piglets under intensive husbandry conditions. In: 21st World Veterinary congress,

Moscow
Stevenson GW, Hoang H, Schwartz KJ et al (2013) Emergence of porcine epidemic diarrhea virus in the United States: clinical signs, lesions, and viral genomic sequences. J Vet Diagn Investig 25:649–654
Stone SS, Kemeny LJ, Woods RD et al (1977) Efficacy of isolated colostral IgA, IgG, and IgM(A) to protect neonatal pigs against the coronavirus of transmissible gastroenteritis. Am J Vet Res 38:1285–1288
Su M, Li C, Guo D et al (2016) A recombinant nucleocapsid protein-based indirect enzyme-linked immunosorbent assay to detect antibodies against porcine deltacoronavirus. J Vet Med Sci 78:601–606
Sueyoshi M, Tsuda T, Yamazaki K et al (1995) An immunohistochemical investigation of porcine epidemic diarrhoea. J Comp Pathol 113:59–67
Sun RQ, Cai RJ, Chen YQ et al (2012) Outbreak of porcine epidemic diarrhea in suckling piglets, China. Emerg Infect Dis 18:161–163
Sun D, Wang X, Wei S et al (2016) Epidemiology and vaccine of porcine epidemic diarrhea virus in China: a mini-review. J Vet Med Sci 78:355–363
Suzuki T, Murakami S, Takahashi O et al (2015) Molecular characterization of pig epidemic diarrhoea viruses isolated in Japan from 2013 to 2014. Infect Genet Evol 36:363–368
Suzuki T, Shibahara T, Yamaguchi R et al (2016) Pig epidemic diarrhoea virus S gene variant with a large deletion non-lethal to colostrum-deprived newborn piglets. J Gen Virol 97:1823–1828
Takahashi K, Okada K, Ohshima K (1983) An outbreak of swine diarrhea of a new-type associated with coronavirus-like particles in Japan. Nihon Juigaku Zasshi 45:829–832
Thachil A, Gerber PF, Xiao CT et al (2015) Development and application of an ELISA for the detection of porcine deltacoronavirus IgG antibodies. PLoS One 10:e0124363
Thomas JT, Chen Q, Gauger PC et al (2015) Effect of porcine epidemic diarrhea virus infectious doses on infection outcomes in naive conventional neonatal and weaned pigs. PLoS One 10:e0139266
Underdahl NR, Mebus CA, Torres-Medina A (1975) Recovery of transmissible gastroenteritis virus from chronically infected experimental pigs. Am J Vet Res 36:1473–1476
van Nieuwstadt AP, Zetstra T (1991) Use of two enzyme-linked immunosorbent assays to monitor antibody responses in swine with experimentally induced infection with porcine epidemic diarrhea virus. Am J Vet Res 52:1044–1050
van Nieuwstadt AP, Cornelissen JB, Zetstra T (1988) Comparison of two methods for detection of transmissible gastroenteritis virus in feces of pigs with experimentally induced infection. Am J Vet Res 49:1836–1843
van Nieuwstadt AP, Zetstra T, Boonstra J (1989) Infection with porcine respiratory coronavirus does not fully protect pigs against intestinal transmissible gastroenteritis virus. Vet Rec 125:58–60
VanCott JL, Brim TA, Simkins RA et al (1993) Isotype-specific antibody-secreting cells to transmissible gastroenteritis virus and porcine respiratory coronavirus in gut- and bronchus-associated lymphoid tissues of suckling pigs. J Immunol 150:3990–4000
VanCott JL, Brim TA, Lunney JK et al (1994) Contribution of antibody-secreting cells induced in mucosal lymphoid tissues of pigs inoculated with respiratory or enteric strains of coronavirus to immunity against enteric coronavirus challenge. J Immunol 152:3980–3990
Vlasova AN, Marthaler D, Wang Q et al (2014) Distinct characteristics and complex evolution of PEDV strains, North America, May 2013–February 2014. Emerg Infect Dis 20:1620–1628
Vui DT, Tung N, Inui K et al (2014) Complete genome sequence of porcine epidemic diarrhea virus in Vietnam. Genome Announc 2
Wang L, Byrum B, Zhang Y (2014a) New variant of porcine epidemic diarrhea virus, United States, 2014. Emerg Infect Dis 20:917–919
Wang L, Byrum B, Zhang Y (2014b) Detection and genetic characterization of deltacoronavirus in pigs, Ohio, USA, 2014. Emerg Infect Dis 20:1227–1230
Wang L, Byrum B, Zhang Y (2014c) Porcine coronavirus HKU15 detected in 9 US states, 2014. Emerg Infect Dis 20:1594–1595
Wang D, Fang L, Shi Y et al (2015) Porcine epidemic diarrhea virus 3C-like protease regulates its interferon antagonism by cleaving NEMO. J Virol 90:2090–2101
Wang L, Hayes J, Sarver C et al (2016a) Porcine deltacoronavirus: histological lesions and genetic characterization. Arch Virol 161:171–175
Wang D, Fang L, Xiao S (2016b) Porcine epidemic diarrhea in China. Virus Res 226:7–13
Wesley RD, Woods RD (1996) Induction of protective immunity against transmissible gastroenteritis virus after exposure of neonatal pigs to porcine respiratory coronavirus. Am J Vet Res 57:157–162
Wesley RD, Woods RD, Hill HT et al (1990) Evidence for a porcine respiratory coronavirus, antigenically similar to transmissible gastroenteritis virus, in the United States. J Vet Diagn Investig 2:312–317
Wesley RD, Woods RD, McKean JD et al (1997) Prevalence of coronavirus antibodies in Iowa

swine. Can J Vet Res 61:305–308
Witte KH (1971) Isolation of the virus of transmissible gastroenteritis (TGE) from naturally infected piglets in cell culture. Zentralbl Veterinarmed B 18:770–778
Woo PC, Lau SK, Lam CS et al (2012) Discovery of seven novel mammalian and avian coronaviruses in the genus deltacoronavirus supports bat coronaviruses as the gene source of alphacoronavirus and betacoronavirus and avian coronaviruses as the gene source of gammacoronavirus and deltacoronavirus. J Virol 86:3995–4008
Xuan H, Xing D, Wang D et al (1984) Culture of porcine epidemic diarrhea virus by using pig intestinal monolayer cell. J People's Liber Army Vet Univ 4:202–208
Yaeger M, Funk N, Hoffman L (2002) A survey of agents associated with neonatal diarrhea in Iowa swine including Clostridium difficile and porcine reproductive and respiratory syndrome virus. J Vet Diagn Investig 14:281–287
Yu X, Shi L, Lv X et al (2015) Development of a real-time reverse transcription loop-mediated isothermal amplification method for the rapid detection of porcine epidemic diarrhea virus. Virol J 12:76
Zhang J (2016) Porcine deltacoronavirus: overview of infection dynamics, diagnostic methods, prevalence and genetic evolution. Virus Res 226:71–84
Zhang Q, Shi K, Yoo D (2016a) Suppression of type I interferon production by porcine epidemic diarrhea virus and degradation of CREB-binding protein by nsp1. Virology 489:252–268
Zhang J, Tsai YL, Lee PY et al (2016b) Evaluation of two singleplex reverse transcription-insulated isothermal PCR tests and a duplex real-time RT-PCR test for the detection of porcine epidemic diarrhea virus and porcine deltacoronavirus. J Virol Methods 234:34–42
Zhao S, Gao Q, Qin T et al (2014) Effects of virulent and attenuated transmissible gastroenteritis virus on the ability of porcine dendritic cells to sample and present antigen. Vet Microbiol 171:74–86

第 5 章 细环病毒

Souvik Ghosh, Alyssa Kleymann, Yashpal Singh Malik,
Nobumichi Kobayashi

5.1 引言

细环病毒(torque teno virus,TTV)首次被发现于一例输血后感染非甲-非庚型肝炎的老年患者血清中(Nishizawa 等,1997)。由于早期源于人的 TTV 分离株与圆环病毒的结构和分子相似,因此最初将 TTV 归属于圆环病毒科(Miyata 等,1999)。后来,根据 TTV 与圆环病毒基因组的特征和核苷酸序列同源性的差异,ICTV 于 2005 年将 TTV 归入圆环病毒科的一个新属“指环病毒属(Anelloviridae)”,此后于 2009 年又将其单独列为一个科“指环病毒科”(Biagini 等,2005,2012)。尽管首次发现的病毒以患者(T.T.)姓名首字母缩写暂时命名为 TTV,但来源于拉丁语词“torque”(项链)和“tenuis”(小或薄)的首字母缩写 TTV(torque teno virus),实际上是指病毒的基因组组成 (Biagini 等,2012;Nishizawa 等,1997)。

TTV 已在其他多种动物宿主中检出 [(獾、蝙蝠、骆驼、猫、牛、犬、马、非人灵长类动物、负鼠、猪、鸽子、松貂、家禽、啮齿类动物、海狮、海龟、树鼩和野猪)(de Souza 等,2018;Li 等,2015;Manzin 等,2015;Zhang 等,2017)](表 5.1)。虽然 TTV 与多种疫病状况和混合感染有关，特别是在人和猪中，但 TTV 在健康宿主中也普遍存在，因此对 TTV 的致病机制仍未可知(Kekarainen 和 Segalés,2012;Manzin 等,2015;Meng,2012)。

由于 TTV 与影响猪经济效益的重要疫病相关 (Ghosh 等,2018;Kekarainen 和 Segalés,2012;Meng,2012),本章主要关注各种动物 TTV 中猪的细环病毒。

5.2 分类

在 2009 年,ICTV 将 TTV 归为一个新的单链 DNA 病毒科,即“指环病毒科”(Biagini 等,2012)。指环病毒科具有广泛的宿主范围，包括 14 个属 [甲型细环病毒属(alphatorquevirus)、乙型细环病毒属(betatorquevirus)、丙型细环病毒属(deltatorquevirus)、丁型细环病毒属(epsilontorquevirus)、戊型细环病毒属(etatorquevirus)、庚型细环病毒属(gammatorquevirus)、环转病毒属(gyrovirus)、壬型细环病毒属(iotatorquevirus)、Kappa 细环病毒属(Kappatorquevirus)、Lambda 细环病毒属(Lambdatorquevirus)、Mu-细环病毒属(Mutorquevirus)、Nu-细环病毒属(Nutorquevirus)、辛型细环病毒属(thetatorquevirus)和己型细环病毒属(zetatorquevirus)]和两个新拟定的属[Omega-细环病毒属(Omegatorquevirus)和 Sigma-细环病毒属(Sigmatorquevirus)](表 5.1)(de Souza 等,2018;ICTV,2018)。

猪细环病毒(Torque teno sus virus,TTSuV)被分为两个属,Iota-细环病毒属(种类:猪细环病毒 1a,TTSuV1a 和猪细环病毒 1b,TTSuV1b)和 Kappa-细环病毒属 [(种类：猪细环病毒 k2a,TTSuVk2a 和猪细环病毒 k2b,TTSuVk2b)(Cornelissen-Keijsers 等,2012;ICTV,2018)](表 5.1)。

5.3 结构、基因组与复制

TTV 的直径通常为 30~32nm,无囊膜,二十面体结构 (Biagini 等,2012;Kekarainen 和 Segalés,2012)。病毒基因组为环状单股负链 DNA 病毒,病毒基因组全长为 2000~3900bp(取决于宿主种类)(TTSuV 中约为 2800bp)(Biagini 等,2012;Cornelissen-Keijsers 等,2012;Cortey 等,2011;Okamoto 等,2002)。所有 TTV,包括 TTSuV,在病毒基因组的非翻译区(untranslated region,UTR)保留了一段短的高度保守序列(Biagini 等,2012;Kekarainen 和 Segalés,2012)。

根据 TTSuV 的全基因组(Cornelissen-Keijsers 等,2012;Cortey 等,2011;Kekarainen 和 Segalés,2012) 可以推断出其含有 3 个 ORF，推测认为,ORF1 可能编码病毒衣壳蛋白,ORF2 可能编码一种具有酪氨酸磷酸酶特性的蛋白或与抑制 NF-κb

表5.1　指环病毒科的分类

属[a]	病毒种类[a,b]数量[a,b],名称(未分类的病毒数量[c])	宿主[a,b,c]
甲型细环病毒属(alphatorquevirus)	30;细环病毒1-29,负鼠细环病毒(5)	非洲绿猴,黑猩猩,人类,负鼠
乙型细环病毒属(betatorquevirus)	12;小细环病毒1-12(4)	非洲绿猴,黑猩猩,人类
丙型细环病毒属(deltatorquevirus)	2;树鼩细环病毒;柔暮鼠细环病毒	普通树鼩,啮齿类动物
丁型细环病毒属(epsilontorquevirus)	1;绒猴细环病毒	绒猴
戊型细环病毒属(etatorquevirus)	2;猫细环病毒,猫细环病毒2	猫
庚型细环病毒属(gammatorquevirus)	15;中细环病毒1-15	非洲绿猴,黑猩猩,人类
环转病毒属(gyrovirus)	1;鸡贫血病毒(8)	灰色暴风雨海燕,鸡,人类
壬型细环病毒属(iotatorquevirus)	2;猪细环病毒1a,猪细环病毒1b	猪
Kappa细环病毒属(Kappatorquevirus)	2;猪细环病毒k2a,猪细环病毒k2b	猪
Lambda细环病毒属(Lambdatorquevirus)	6;海豹细环病毒1-3,8和9,海狮细环病毒1	海豹,海狮
Mu-细环病毒属(Mutorquevirus)	1;马细环病毒1	马
Nu-细环病毒属(Nutorquevirus)	2;海豹细环病毒4和5	海豹
辛型细环病毒属(thetatorquevirus)	1;犬细环病毒	犬
已型细环病毒属(zetatorquevirus)	1;夜猴细环病毒	夜猴
拟定属[b]		
Omega-细环病毒属(Omegatorquevirus)	6;啮齿动物细环病毒3-8	啮齿类动物
Sigma-细环病毒属(Sigmatorquevirus)	3;昭短尾叶鼻蝠细环病毒,吸血蝠细环病毒,白耳负鼠细环病毒	蝙蝠,负鼠

[a] 基于ICTV,https://talk.ictvonline.org/ictv-reports/ictv_9th_report/ssdna-viruses-2011/w/ssdna_viruses/139/anelloviridae(accessed 16 April 2019)。

[b] 基于de Souza等(2018)。

[c] 基于https://www.ncbi.nlm.nih.gov/genomes/GenomesGroup.cgi?taxid=687329(accessed16 April 2019)。

通路有关(Biagini等,2012;Kekarainen和Segalés,2012;Zheng等,2007),ORF3可能是由剪接产生的,并编码一个未知功能蛋白(Cortey等,2011;Kekarainen和Segalés,2012)。已经发现TTSuV具有选择性剪接模式,且能够产生不同亚型蛋白(Martínez-Guinó等,2011)。

由于缺乏适当的细胞培养体系,限制了对TTV复制周期中的主要过程研究(Meng,2012),然而,在各组织、分泌物和排泄物中对TTV的检测及转染的研究提示,TTV可在不同类型细胞中进行复制(Biagini等,2012;Manzin等,2015;Meng,2012;Nieto等,2013)。根据观察,由TTV ORF1推导的氨基酸序列含有与其他动植物环状单链DNA病毒基因组的Rep蛋白相似的保守基序(Biagini等,2012)。

5.4　猪细环病毒

尽管在多种动物中都发现了TTV,但是猪TTV与影响猪经济效益的其他重要疫病,特别是猪圆环病毒感染相关(Kekarainen和Segalés 2012;Manzin等,2015),所以与其他非人类宿主相比,对猪TTV或TTSuV的研究要更多。

5.4.1　病毒多样性

TTSuV中的核苷酸突变率高于典型DNA病毒突变率,且与RNA病毒相当(Cortey等,2011)。鉴于明显缺乏选择压力(疫苗接种和剔除患病动物),自然选择和漂移被认为是TTSuV进化的主要驱动力(Cortey 等,2012)。基于核苷酸序列同源性和遗传进化分析,ICTV 已将 TTSuV 至少分为 2 个属 4 个种 (Cornelissen-Keijsers 等,2012;ICTV,2018)。TTSuV1a 和 TTSuV1b 属于 Iota-细小病毒属而 TTSuVk2a 和 TTSuVk2 属于 Kappa-细小病毒属。在种系发育方面,最近的一项研究提出了另外一种 TTSuV1(TTSuV1c)(Ramos 等,2018)。

TTSuV1 与 TTSuV2 之间的核苷酸序列的相似性约为 50%(Kekarainen 和 Segalés,2012)。TTSuV1 毒株(核苷酸序列差异>30%)比 TTSuV2 毒株(<15%)具有更高的种内变异(Cortey 等,2011)。全球动物贸易直接影响了 TTSuV 种群的基因组成,对种群的基因组成造成改变和破坏,该种群可能自猪群驯化以来就在其中传播(Cortey 等,2012)。

5.4.2 流行病学

TTSuV1 和 TTSuVk2a 在健康猪和病猪中均高度流行并广泛分布,在世界不同国家的感染率为 16.8%~100%(Manzin 等,2015)。另一方面,虽然关于 TTSuVk2b 的信息有限,但对来自 17 个国家猪的血清抗体监测显示,TTSuVk2b 的感染率为 0~100%,随后另两项研究(一项来自意大利,另一项来自美国)表明,TTSuVk2b 的感染率分别为 11.5% 和 24.7% (Blois 等,2014;Cornelissen-Keijsers 等,2012;Rogers 等,2017)。在世界范围内,经常发现不同种属的 TTSuV 在猪群中发生混合感染(Cornelissen-Keijsers 等,2012;Ghosh 等,2018;Kekarainen 和 Segalés,2012;Ramos 等,2018)。

尽管指环病毒的主要传播途径为粪-口传播,但研究认为 TTSuV 存在水平传播和垂直传播两种传播途径,这是因为已经在猪胎儿的血液、组织、初乳和精液中检测到了病毒 DNA(Aramouni 等,2010;Kekarainen 等,2007;Martínez-Guinó 等,2009,2010;Sibila 等,2009b;Tshering 等,2012)。TTSuV1 的 DNA 已经在人类、家畜和野生动物的血清中被广泛检测出,同时报道称在牛、马和羊的血清中检测 TTSuV1 也呈现阳性 (Singh 和 Ramamoorthy,2018;Ssemadaali 等,2016)。此外,已有研究表明 TTSuV 可污染某些特定的细胞系、人用药物、零售猪肉产品和兽用疫苗 (Leblanc 等,2014;Meng,2012;Monini 等,2016)。这些研究结果表明,TTSuV 在猪和异源宿主中广泛传播的风险,以及成为人兽共患病的风险正在增加。

关于不同年龄猪中 TTSuV 的感染频率和病毒载量的报告有所矛盾。虽然一些研究提出检测频率或病毒载量与动物年龄无关,但是相关证据表明患病率随年龄增加而增加,并且保育猪和幼龄猪中 TTSuV 至少有一个属的感染率较高(Blois 等,2014;de Castro 等,2015;de Menezes Cruz 等,2016;Ramos 等,2018;Teixeira 等,2015;Xiao 等,2012)。在猪中常见 TTSuV 的持续性感染和病毒血症 (Aramouni 等,2010;Kekarainen 等,2007;Kekarainen 和 Segalés,2012;Martínez-Guinó 等,2009,2010;Sibila 等,2009a,b)。

5.4.3 发病机制与临床

目前,关于 TTSuV 的致病性仍不明确(Meng,2012;Ramos 等,2018),由于在健康猪中 TTSuV 广泛存在且持续传播,因此认为 TTSuV 本身不能诱发严重疫病(Kekarainen 和 Segalés,2012;Manzin 等,2015)。此外,一些研究人员甚至提出了 TTV 可能是组成宿主共生菌群的一部分(Griffiths,1999;Manzin 等,2015)。TTSuV 因与影响经济效益的重要猪疫病,尤其是因圆环病毒而引起的疫病相关而备受关注(Kekarainen 和 Segalés,2012)。

猪圆环病毒-2 型(porcine circovirus-2, PCV2)是 PCV-2 系统性疫病 (PCV-2 systemic disease,PCV2-SD,以前称为断奶后仔猪多系统衰竭综合征)及其他一些猪的疫病主要病原体,统称为猪圆环病毒相关疫病(porcine circovirus-associated disease, PCVAD)(Segalés 等,2013)。PCV2-SD 被认为是影响养猪业发展的最重要的疫病之一(Meng,2012;Segalés 等,2013)。

与健康的猪相比,PCV2-SD 猪中的 TTSuVk2a 的感染率和病毒载量明显偏高,但抗

TTSuV2 抗体效价较低(Aramouni 等,2011;Huang 等,2011;Nieto 等,2013;Kekarainen 等,2006)。另一方面，所有研究团队中发现的 TTSuV1 病毒载量均很高(Kekarainen 等,2006),然而,无菌猪实验接种 TTSuV1 和 PCV2 后，可诱导出 PCV2-SD (Ellis 等,2008)。在另一项研究中，在患有PCV2-SD 猪的血清中,TTSuVk2b 的病毒载量比健康猪高(Cornelissen-Keijsers 等,2012)。TTSuV1 和 TTSuV2 均可通过改变宿主的免疫系统从而促进 PCV-2 相关淋巴病变的发展(Lee 等,2015)。

除 PCV2 外，推测 TTSuV 与其他猪病原体(猪瘟病毒、猪戊型肝炎病毒、猪肺炎支原体、猪繁殖与呼吸综合征病毒和猪细小病毒）间也存在相关性（Opriessnig 和 Halbur,2012;Pérez 等,2011;Savic 等,2010;Zhang 等,2014b)。

基于以上观察结果,有人提出 TTSuV 可能作为诱发或加重某些猪疫病的复杂因子/辅助因子(Kekarainen 和 Segalés,2012;Meng,2012;Zhang 等,2012)。另一方面，在相互矛盾的研究中未发现 TTSuV 与患 PCV2-SD 猪之间有任何的相关性,这些发现对 TTSuV 在 PCVAD 中的作用提出了质疑(Lee 等,2010;Ramos 等,2018;Rogers 等,2017;Teixeira 等,2015;Vargas-Ruiz 等,2017)。

5.4.4 诊断

TTSuV 已经证明在健康和患病动物中广泛存在,因此,如果动物感染此病,无法根据临床症状进行初步诊断。PCR 和 qPCR 检测方法已成功用于检测和区分不同种属的 TTSuV 筛查中,并且后一种方法在评估各种样品(包括零售猪肉产品)中病毒 DNA 载量具有明显优势 (Ghosh 等,2018;Cornelissen -Keijsers 等,2012;Huang 等,2010;Leblanc 等,2014;Monini 等,2016;Ramos 等,2018;Rogers 等,2017;Segalés 等,2009;Teixeira 等,2015;Vargas-Ruiz 等,2018)。PCR/qPCR 筛选检测主要基于 UTR,因为该区域似乎在 TTSuV 基因组中是保守的，且又有足够的差异来区分病毒种属(Huang 等,2010;Segalés 等,2009)。由于在与其他病原体混合感染中经常发现 TTSuV，因此研发了一种可以同时检测出不同种属 TTSuV 和其他猪病毒的多重 qPCR 检测方法(Pérez 等,2012)。

其他用于检测和(或)区分不同种属 TTSuV 的检测方法包括:宏基因组学、ELISA、免疫印迹法、基于荧光微珠的 4 重链免疫测定法(用于同时检测出抗 TTSuV1、TTSuV2、PRRSV-1 和 PRRSV-2 的 IgG 抗体）和使用多毒株合并探针的原位杂交法(Huang 等,2011;Giménez-Lirola 等,2014;Lee 等,2014;Nieto 等,2015;Zhang 等,2014a)。

5.4.5 免疫

目前,尚无针对 TTSuV 的疫苗获得批准。在一项研究中,用 DNA 和蛋白质联合疫苗免疫猪产生了对 TTSuVk2a 的特异性抗体,并且可以显著降低自然感染期间 TTSuVk2a 病毒血症(Jiménez-Melsió 等,2015),然而,其他种属的 TTSuV DNA 载量却并不受这种免疫方法的影响。

参考文献

Aramouni M, Segalés J, Cortey M, Kekarainen T (2010) Age-related tissue distribution of swine Torque teno sus virus 1 and 2. Vet Microbiol 146(3–4):350–353. https://doi.org/10.1016/j.vetmic.2010.05.036

Aramouni M, Segalés J, Sibila M, Martin-Valls GE, Nieto D, Kekarainen T (2011) Torque teno sus virus 1 and 2 viral loads in postweaning multisystemic wasting syndrome (PMWS) and porcine dermatitis and nephropathy syndrome (PDNS) affected pigs. Vet Microbiol 153(3–4):377–381. https://doi.org/10.1016/j.vetmic.2011.05.046

Biagini P, Todd D, Bendinelli M, Hino S, Mankertz A, Mishiro S, Niel C, Okamoto H, Raidal S, Ritchie BW, Teo GC (2005) Anellovirus. In: Fauquet CM, Mayo MA, Maniloff J, Desselberger U, Ball LA (eds) Virus taxonomy: eighth report of the international committee on taxonomy of viruses. Elsevier/Academic Press, London, pp 335–341

Biagini P, Bendinelli M, Hino S, Kakkola L, Mankertz A, Niel C, Okamoto H, Raidal S, Teo CG,

Todd D (2012) Family Anelloviridae. In: King AMQ, Adams MJ, Carstens EB, Lefkowitz EJ (eds) Virus taxonomy: classification and nomenclature of viruses: ninth report of the international committee on taxonomy of viruses. Academic Press, London, pp 331–341

Blois S, Mallus F, Liciardi M, Pilo C, Camboni T, Macera L, Maggi F, Manzin A (2014) High prevalence of co-infection with multiple Torque teno sus virus species in Italian pig herds. PLoS One 9(11):e113720. https://doi.org/10.1371/journal.pone.0113720

Cornelissen-Keijsers V, Jiménez-Melsió A, Sonnemans D, Cortey M, Segalés J, van den Born E, Kekarainen T (2012) Discovery of a novel Torque teno sus virus species: genetic characterization, epidemiological assessment and disease association. J Gen Virol 93(Pt 12):2682–2691. https://doi.org/10.1099/vir.0.045518-0

Cortey M, Macera L, Segalés J, Kekarainen T (2011) Genetic variability and phylogeny of Torque teno sus virus 1 (TTSuV1) and 2 (TTSuV2) based on complete genomes. Vet Microbiol 148(2–4):125–131. https://doi.org/10.1016/j.vetmic.2010.08.013

Cortey M, Pileri E, Segalés J, Kekarainen T (2012) Globalisation and global trade influence molecular viral population genetics of Torque teno sus viruses 1 and 2 in pigs. Vet Microbiol 156(1–2):81–87. https://doi.org/10.1016/j.vetmic.2011.10.026

de Castro AMMG, Baldin CM, Favero CM, Gerber PF, Cipullo RI, Richtzenhain LJ (2015) Torque teno sus virus 1 and 2 viral loads in faeces of porcine circovirus 2-positive pigs. Acta Vet Brno 84:91–95

de Menezes Cruz AC, Silveira RL, Baez CF, Varella RB, de Castro TX (2016) Clinical aspects and weight gain reduction in swine infected with porcine circovirus type 2 and Torque teno sus virus in Brazil. Vet Microbiol 195:154–157. https://doi.org/10.1016/j.vetmic.2016.09.012

de Souza WM, Fumagalli MJ, de Araujo J, Sabino-Santos G Jr, Maia FGM, Romeiro MF, Modha S, Nardi MS, Queiroz LH, Durigon EL, Nunes MRT, Murcia PR, Figueiredo LTM (2018) Discovery of novel anelloviruses in small mammals expands the host range and diversity of the Anelloviridae. Virology 514:9–17. https://doi.org/10.1016/j.virol.2017.11.001

Ellis JA, Allan G, Krakowka S (2008) Effect of coinfection with genogroup 1 porcine Torque teno virus on porcine circovirus type 2-associated postweaning multisystemic wasting syndrome in gnotobiotic pigs. Am J Vet Res 69(12):1608–1614. https://doi.org/10.2460/ajvr.69.12.1608

Ghosh S, Soto E, Illanes O, Navarro R, Aung MS, Malik YS, Kobayashi N, Fuentealba C (2018) High detection rates of Torque teno sus virus in co-infection with important viral pathogens in porcine kidneys on St. Kitts Island, Lesser Antilles. Transbound Emerg Dis 65(5):1175–1181. https://doi.org/10.1111/tbed.12960

Giménez-Lirola LG, Gerber PF, Rowland RR, Halbur PG, Huang YW, Meng XJ, Opriessnig T (2014) Development and validation of a 4-plex antibody assay for simultaneous detection of IgG antibodies against Torque teno sus virus 1 (TTSuV1), TTSuV2, and porcine reproductive and respiratory syndrome virus types 1 and 2. Res Vet Sci 96(3):543–550. https://doi.org/10.1016/j.rvsc.2014.02.014

Griffiths P (1999) Time to consider the concept of a commensal virus? Rev Med Virol 9(2):73–74

Huang YW, Dryman BA, Harrall KK, Vaughn EM, Roof MB, Meng XJ (2010) Development of SYBR green-based real-time PCR and duplex nested PCR assays for quantitation and differential detection of species- or type-specific porcine Torque teno viruses. J Virol Methods 170(1–2):140–146. https://doi.org/10.1016/j.jviromet.2010.09.018

Huang YW, Harrall KK, Dryman BA, Beach NM, Kenney SP, Opriessnig T, Vaughn EM, Roof MB, Meng XJ (2011) Expression of the putative ORF1 capsid protein of Torque teno sus virus 2 (TTSuV2) and development of Western blot and ELISA serodiagnostic assays: correlation between TTSuV2 viral load and IgG antibody level in pigs. Virus Res 158(1–2):79–88. https://doi.org/10.1016/j.virusres.2011.03.013

ICTV (2018) Family Anelloviridae, Virus Taxonomy: 2018b Release. https://talk.ictvonline.org/ictv-reports/ictv_9th_report/ssdna-viruses-2011/w/ssdna_viruses/139/anelloviridae. Accessed 12 Apr 2019

Jiménez-Melsió A, Rodriguez F, Darji A, Segalés J, Cornelissen-Keijsers V, van den Born E, Kekarainen T (2015) Vaccination of pigs reduces Torque teno sus virus viremia during natural infection. Vaccine 33(30):3497–3503. https://doi.org/10.1016/j.vaccine.2015.05.064

Kekarainen T, Segalés J (2012) Torque teno sus virus in pigs: an emerging pathogen? Transbound Emerg Dis 59(Suppl 1):103–108. https://doi.org/10.1111/j.1865-1682.2011.01289.x.

Kekarainen T, Sibila M, Segalés J (2006) Prevalence of swine Torque teno virus in post-weaning multisystemic wasting syndrome (PMWS)-affected and non-PMWS-affected pigs in Spain. J Gen Virol 87(Pt 4):833–837

Kekarainen T, López-Soria S, Segalés J (2007) Detection of swine Torque teno virus genogroups 1 and 2 in boar sera and semen. Theriogenology 68(7):966–971

Leblanc D, Houde A, Gagné MJ, Plante D, Bellon-Gagnon P, Jones TH, Muehlhauser V, Wilhelm B, Avery B, Janecko N, Brassard J (2014) Presence, viral load and characterization of Torque teno sus viruses in liver and pork chop samples at retail. Int J Food Microbiol 178:60–64. https://doi.org/10.1016/j.ijfoodmicro.2014.03.005

Lee SS, Sunyoung S, Jung H, Shin J, Lyoo YS (2010) Quantitative detection of porcine Torque teno virus in porcine circovirus-2-negative and porcine circovirus-associated disease-affected pigs. J Vet Diagn Investig 22(2):261–264

Lee Y, Lin CM, Jeng CR, Pang VF (2014) Detection of Torque teno sus virus 1 and 2 in porcine tissues by in situ hybridization using multi-strained pooled probes. Vet Microbiol 172(3–4):390–399. https://doi.org/10.1016/j.vetmic.2014.06.002

Lee Y, Lin CM, Jeng CR, Chang HW, Chang CC, Pang VF (2015) The pathogenic role of Torque teno sus virus 1 and 2 and their correlations with various viral pathogens and host immunocytes in wasting pigs. Vet Microbiol 180(3–4):186–195. https://doi.org/10.1016/j.vetmic.2015.08.027

Li L, Giannitti F, Low J, Keyes C, Ullmann LS, Deng X, Aleman M, Pesavento PA, Pusterla N, Delwart E (2015) Exploring the virome of diseased horses. J Gen Virol 96(9):2721–2733. https://doi.org/10.1099/vir.0.000199

Manzin A, Mallus F, Macera L, Maggi F, Blois S (2015) Global impact of Torque teno virus infection in wild and domesticated animals. J Infect Dev Ctries 9(6):562–570. https://doi.org/10.3855/jidc.6912

Martínez-Guinó L, Kekarainen T, Segalés J (2009) Evidence of Torque teno virus (TTV) vertical transmission in swine. Theriogenology 71(9):1390–1395. https://doi.org/10.1016/j.theriogenology.2009.01.010

Martínez-Guinó L, Kekarainen T, Maldonado J, Aramouni M, Llorens A, Segalés J (2010) Torque teno sus virus (TTV) detection in aborted and slaughterhouse collected foetuses. Theriogenology 74(2):277–281. https://doi.org/10.1016/j.theriogenology.2010.02.011

Martínez-Guinó L, Ballester M, Segalés J, Kekarainen T (2011) Expression profile and subcellular localization of Torque teno sus virus proteins. J Gen Virol 92(Pt 10):2446–2457. https://doi.org/10.1099/vir.0.033134-0

Meng XJ (2012) Emerging and re-emerging swine viruses. Transbound Emerg Dis 59(Suppl 1):85–102. https://doi.org/10.1111/j.1865-1682.2011.01291.x

Miyata H, Tsunoda H, Kazi A, Yamada A, Khan MA, Murakami J, Kamahora T, Shiraki K, Hino S (1999) Identification of a novel GC-rich 113-nucleotide region to complete the circular, single-stranded DNA genome of TT virus, the first human circovirus. J Virol 73(5):3582–3586

Monini M, Vignolo E, Ianiro G, Ostanello F, Ruggeri FM, Di Bartolo I (2016) Detection of Torque teno sus virus in pork bile and liver sausages. Food Environ Virol 8(4):283–288

Nieto D, Kekarainen T, Aramouni M, Segalés J (2013) Torque teno sus virus 1 and 2 distribution in tissues of porcine circovirus type 2-systemic disease affected and age-matched healthy pigs. Vet Microbiol 163(3–4):364–367. https://doi.org/10.1016/j.vetmic.2013.01.005

Nieto D, Martínez-Guinó L, Jiménez-Melsió A, Segalés J, Kekarainen T (2015) Development of an indirect ELISA assay for the detection of IgG antibodies against the ORF1 of Torque teno sus viruses 1 and 2 in conventional pigs. Vet Microbiol 180(1–2):22–27. https://doi.org/10.1016/j.vetmic.2015.08.023

Nishizawa T, Okamoto H, Konishi K, Yoshizawa H, Miyakawa Y, Mayumi M (1997) A novel DNA virus (TTV) associated with elevated transaminase levels in posttransfusion hepatitis of unknown etiology. Biochem Biophys Res Commun 241(1):92–97

Okamoto H, Takahashi M, Nishizawa T, Tawara A, Fukai K, Muramatsu U, Naito Y, Yoshikawa A (2002) Genomic characterization of TT viruses (TTVs) in pigs, cats and dogs and their relatedness with species-specific TTVs in primates and tupaias. J Gen Virol 83(Pt 6):1291–1297

Opriessnig T, Halbur PG (2012) Concurrent infections are important for expression of porcine circovirus associated disease. Virus Res 164(1–2):20–32. https://doi.org/10.1016/j.virusres.2011.09.014

Pérez LJ, de Arce HD, Frias MT, Perera CL, Ganges L, Núñez JI (2011) Molecular detection of Torque teno sus virus in lymphoid tissues in concomitant infections with other porcine viral pathogens. Res Vet Sci 91(3):e154–e157. https://doi.org/10.1016/j.rvsc.2011.02.012

Pérez LJ, Perera CL, Frías MT, Núñez JI, Ganges L, de Arce HD (2012) A multiple SYBR Green I-based real-time PCR system for the simultaneous detection of porcine circovirus type 2, porcine parvovirus, pseudorabies virus and Torque teno sus virus 1 and 2 in pigs. J Virol Methods 179(1):233–241. https://doi.org/10.1016/j.jviromet.2011.11.009

Ramos N, Mirazo S, Botto G, Teixeira TF, Cibulski SP, Castro G, Cabrera K, Roehe PM, Arbiza J (2018) High frequency and extensive genetic heterogeneity of TTSuV1 and TTSuVk2a in PCV2-infected and non-infected domestic pigs and wild boars from Uruguay. Vet Microbiol 224:78–87. https://doi.org/10.1016/j.vetmic.2018.08.029

Rogers AJ, Huang YW, Heffron CL, Opriessnig T, Patterson AR, Meng XJ (2017) Prevalence of the novel Torque teno sus virus species k2b from pigs in the United States and lack of association with post-weaning multisystemic wasting syndrome or mulberry heart disease. Transbound Emerg Dis 64(6):1877–1883. https://doi.org/10.1111/tbed.12586

Savic B, Milicevic V, Bojkovski J, Kureljusic B, Ivetic V, Pavlovic I (2010) Detection rates of the swine Torque teno viruses (TTVs), porcine circovirus type 2 (PCV2) and hepatitis E virus (HEV) in the livers of pigs with hepatitis. Vet Res Commun 34(7):641–648. https://doi.org/10.1007/s11259-010-9432-z

Segalés J, Martínez-Guinó L, Cortey M, Navarro N, Huerta E, Sibila M, Pujols J, Kekarainen T (2009) Retrospective study on swine Torque teno virus genogroups 1 and 2 infection from 1985 to 2005 in Spain. Vet Microbiol 134(3–4):199–207. https://doi.org/10.1016/j.vetmic.2008.08.002

Segalés J, Kekarainen T, Cortey M (2013) The natural history of porcine circovirus type 2: from an inoffensive virus to a devastating swine disease? Vet Microbiol 165(1–2):13–20. https://doi.org/10.1016/j.vetmic.2012.12.033

Sibila M, Martínez-Guinó L, Huerta E, Llorens A, Mora M, Grau-Roma L, Kekarainen T, Segalés J (2009a) Swine Torque teno virus (TTV) infection and excretion dynamics in conventional pig farms. Vet Microbiol 139(3–4):213–218. https://doi.org/10.1016/j.vetmic.2009.05.017

Sibila M, Martínez-Guinó L, Huerta E, Mora M, Grau-Roma L, Kekarainen T, Segalés J (2009b) Torque teno virus (TTV) infection in sows and suckling piglets. Vet Microbiol 137(3–4):354–358. https://doi.org/10.1016/j.vetmic.2009.01.008

Singh G, Ramamoorthy S (2018) Potential for the cross-species transmission of swine Torque teno viruses. Vet Microbiol 215:66–70. https://doi.org/10.1016/j.vetmic.2017.12.017

Ssemadaali MA, Effertz K, Singh P, Kolyvushko O, Ramamoorthy S (2016) Identification of heterologous Torque teno viruses in humans and swine. Sci Rep 6:26655. https://doi.org/10.1038/srep26655

Teixeira TF, Cibulski SP, dos Santos HF, Wendlant A, de Sales Lima FE, Schmidt C, Franco AC, Roehe PM (2015) Torque teno sus virus 1 (TTSuV1) and 2 (TTSuV2) viral loads in serum of postweaning multisystemic wasting syndrome (PMWS)-affected and healthy pigs in Brazil. Res Vet Sci 101:38–41. https://doi.org/10.1016/j.rvsc.2015.05.016

Tshering C, Takagi M, Deguchi E (2012) Detection of Torque teno sus virus types 1 and 2 by nested polymerase chain reaction in sera of sows at parturition and of their newborn piglets immediately after birth without suckling colostrum and at 24 hr after suckling colostrum. J Vet Med Sci 74(3):315–319

Vargas-Ruiz A, Ramírez-Álvarez H, Sánchez-Betancourt JI, Quintero-Ramírez V, Rangel-Rodríguez IC, Vázquez-Perez JA, García-Camacho LA (2017) Retrospective study of the relationship of Torque teno sus virus 1a and Torque teno sus virus 1b with porcine circovirus associated disease. Can J Vet Res 81(3):178–185

Vargas-Ruiz A, García-Camacho LA, Ramírez-Alvarez H, Rangel-Rodriguez IC, Alonso-Morales RA, Sánchez-Betancourt JI (2018) Molecular characterization of the ORF2 of Torque teno sus virus 1a and Torque teno sus virus 1b detected in cases of postweaning multisystemic wasting syndrome in Mexico. Transbound Emerg Dis 65(6):1806–1815. https://doi.org/10.1111/tbed.12956

Xiao CT, Giménez-Lirola L, Huang YW, Meng XJ, Halbur PG, Opriessnig T (2012) The prevalence of Torque teno sus virus (TTSuV) is common and increases with the age of growing pigs in the United States. J Virol Methods 183(1):40–44. https://doi.org/10.1016/j.jviromet.2012.03.026

Zhang Z, Wang Y, Fan H, Lu C (2012) Natural infection with Torque teno sus virus 1 (TTSuV1) suppresses the immune response to porcine reproductive and respiratory syndrome virus (PRRSV) vaccination. Arch Virol 157(5):927–933. https://doi.org/10.1007/s00705-012-1249-3

Zhang B, Tang C, Yue H, Ren Y, Song Z (2014a) Viral metagenomics analysis demonstrates the diversity of viral flora in piglet diarrhoeic faeces in China. J Gen Virol 95(Pt 7):1603–1611. https://doi.org/10.1099/vir.0.063743-0.

Zhang Z, Wang Y, Dai W, Dai D (2014b) Detection and distribution of Torque teno sus virus 1 in porcine reproductive and respiratory syndrome virus positive/negative pigs. Vet Microbiol 172(3–4):367–370. https://doi.org/10.1016/j.vetmic.2014.05.008

Zhang Z, Dai W, Dai D (2017) Molecular characterization of pigeon Torque teno virus (PTTV) in Jiangsu province. Comput Biol Chem 69:10–18. https://doi.org/10.1016/j.compbiolchem.2017.04.012

Zheng H, Ye L, Fang X, Li B, Wang Y, Xiang X, Kong L, Wang W, Zeng Y, Ye L, Wu Z, She Y, Zhou X (2007) Torque teno virus (SANBAN isolate) ORF2 protein suppresses NF-kappaB pathways via interaction with IkappaB kinases. J Virol 81(21):11917–11924

第 6 章 捷申病毒

Yashpal Singh Malik, Sudipta Bhat, Anastasia N. Vlasova, Fun-In Wang, Nadia Touil, Souvik Ghosh, Kuldeep Dhama, Mahendra Pal Yadav, Raj Kumar Singh

6.1 引言

捷申病毒也被称为捷申病、塔尔凡病、猪脊髓灰质炎、地方性良性轻瘫、Klobauk 病、猪传染性瘫痪。捷申病毒引发的猪脑炎最初被记述为捷申病，它是一种死亡率很高的强致死性脑脊髓灰质炎，90 多年前在捷克共和国的捷申小镇被发现(1929 年)(Trefny,1930)。猪捷申病毒(porcine teschovirus, PTV)在全世界范围内的猪群中普遍存在(Knowles,2015)。虽然大多数感染为亚临床感染，但有些临床感染，会影响到机体的多个系统，包括神经(脊髓灰质炎)、生殖、肠道、呼吸系统和皮肤。20 世纪 20 年代末至 50 年代，欧洲及其相邻的国家都报道过 PTV 疫情的暴发(Knowles, 2015)。在这之后，这种疫病就变得非常罕见并在西欧消失了。但近来，2009 年，海地发生了两次新的 PTV 疫情，随后 2011 年在加拿大又再次发生(Deng 等,2012;Salles 等,2011)。迄今为止，已经发现了 13 种不同类型的 PTV，并且还发现了其他的 PTV 14~22 型，这些类型在全世界范围内传播(Knowles,2015)。不同类型 PTV 感染自然宿主(家猪和野猪)会引起不同的临床症状。PTV 1 型被认为是导致所有年龄段猪暴发重大疫情的毒力最强的病毒之一。如今，在大多数猪群中普遍发现致病性较低病毒株，引起幼龄动物的无症状感染。

6.2 流行病学

6.2.1 病毒

6.2.1.1 分类

1999 年，PTV 被归类于肠道病毒属，称作猪肠道病毒(porcine enterovirus, PEV)(科，小 RNA 病毒科;目，小 RNA 病毒目)。PEV 被分为 3 个类群:①PEV 1~7 型和 11~13 型;②PEV 8 型;③PEV 9 和 10 型。在重新分类前，根据理化性质、血清学试验、猪肾(porcine kidney, PK)细胞产生的 CPE 类型和不同的细胞培养宿主范围等因素，将 PEV 原来的 11 个血清型分为 3 个亚群(Ⅰ、Ⅱ和Ⅲ)(Knowles 等,1979)。1999 年，对当时现有的 PEV Ⅰ群(以前称为 PEV 1~7 和 PEV 11~13)序列进行了全基因组分析，发现它们不同于其他肠道病毒。根据其遗传差异将其归类为一个新属，捷申病毒(该名称来源于捷申病)是包含多个血清型的单一种类 PTV(Kaku 等,2001;Zell 等,2001)。该种类的病毒也被重新命名为捷申病毒 A，以此来消除其对天然宿主的含混指称，但病毒名称仍保留为 PTV。近来，中国提出了另外 9 个 PTV(PTV 14~22)基因型(Yang 等,2018)。大多数严重的临床疫病(捷申病毒脑脊髓炎)都与 PTV 1 型有关，而其他类型疫病临床症状较温和(塔尔凡病)。

6.2.1.2 病毒特征和基因组

PTV 病毒体直径较小(25~30nm)，无囊膜。二十面体衣壳由 60 个壳粒组成，每个壳粒由 3 个表面蛋白和 1 个内蛋白组成。表面蛋白序列的变异使壳粒呈现出多样性，这些多样性与病毒的某些特性相关，如抗原性、受体识别、浮密度和 pH 值稳定性等。

病毒基因组含有一个 7.1kb 的单股正链 RNA，有一个大的 ORF，编码一个长的多聚合蛋白，它进一步裂解可形成 12 个单独的病毒蛋白。基因组在 5’末端与基因连接病毒蛋白蛋白连接，接下来是 5’非编码区(noncoding region, NCR)、一个前导(L)蛋白、4 种结构蛋白(VP 1~4)和 7 种非结构蛋白(2A~C、3A~D)、3’NCR 和多聚腺苷 A 尾(图 6.1)。在 4 种结构蛋白中，VP1 对分子流行病学和基因分型具有十分重要的意义，虽然在 VP2 上也有中和位点，但中和位点主要位于 VP1上。

6.2.1.3 病毒理化特性

成熟的 PTV 病毒体非常稳定，能够抵抗各种环境条件，包括:pH 值范围为 2~9、在粪便能够保持长期存活。加热、脂溶剂和几种消毒剂都不能灭活 PTV(Derbyshire 和 Arkell,1971)。使用卤离子、亚氯酸钠、高温和 70%乙醇可有效灭活 PTV。

6.2.1.4　血清型与毒株变异

根据 VP1 基因序列的多样性或通过交叉中和试验,可知 PTV 具有 13 种已知的基因型。PTV 血清型 1~10 以前被称为猪肠道病毒Ⅰ群，之前的 PEV 1~7 已被重新命名为 PTV 1~7,PEV 11~13 被命名为 PTV 8~10(Kaku 等,2001)。2011 年,在西班牙猪群中首次鉴定出 PTV 12，推测其是 VP1 基因突变的结果(Cano-Gomez 等,2011)。同样,在匈牙利的野猪粪便中也首次检测到了 PTV 13 型(Boros 等,2012)。2018 年,中国的一项研究提出了 9 个 PTV 基因型(PTV 14~22)。然而,这些病毒株仍未通过血清学方法进行验证(Yang 等,2018)。负压选择和同源重组是结构蛋白基因和非结构蛋白基因遗传多样性的两个重要原因,是 PTV 进化的主要驱动机制(Lin 等,2012)。首先形成了 3 个群，这些群进一步进化产生 13 个不同的血清型(图 6.2)。另一个关于 PTV 进化演变的特征是其毒力的逐渐变化,产生较少的致病性变异株。

6.2.2　宿主

猪是 PTV 已知的唯一宿主,但是也发现野猪感染 PTV。然而,与野猪易感性相关的文献报道较少(Cano-Gómez 等,2013)。处于生长期的猪(哺乳期或断奶)更容易感染 PTV。

6.2.3　地理分布

最近来自马达加斯加、中欧和东欧等地区的疫情暴发报告,显示与 PTV 相关的轻症脑脊髓炎在全世界范围内都有发生,而 PTV 引起致死性脑脊髓炎较为罕见。WOAH 报告了捷申病毒引起的脑脊髓炎流行情况,1996 年、1999 年和 2005 年在白俄罗斯;2002—2004 年在摩尔达维亚;2002 年在罗马尼亚;2004 年在俄罗斯;1996—2005 年在乌克兰;1997 年和 2000—2002 年在拉脱维亚;1996—2000 年、2004—2005 年在马达加斯加;2001 年在乌干达;2002 年在日本;2000 年和 2004 年在中国台湾地区;2009 年和 2011 年分别在海地和加拿大。

6.2.4　死亡率和发病率

各年龄段的商品猪群都会受到 PTV 感染,有时会发生地方性流行。轻症类型,塔尔凡病,有较低的发病率和死亡率，通常临床疫病仅见于幼龄猪和断奶后的仔猪中。新发感染猪群中,由于无抗体,以及入侵的毒株至少具有中等毒力,可能会引起较高的死亡率,如最近 2009 年在海地暴发的疫情,据报告,死亡率为 40%,发病率为 60%(Deng 等,2012)。而 2001 年,加拿大报告的死亡率为 100%(Salles 等,2011)。

6.2.5　传播

采食是 PTV 最常见的感染途径,肠道和相关淋巴结是病毒增殖的主要部位。处于恢复期的猪通过粪便和口腔分泌物排毒可长达 7 周。PTV 性质稳定，在 15℃环境中可以存活 5 个月以上,并且在污染物上易于扩散(Horak 等,2016)。因为已在颅脑(包括嗅球)中检测出 PTV,提示 PTV 极有可能导致鼻内感染(Chiu 等,2013,2014)。在流行病区，在尿液中检测到了具有传染性的 PTV 毒株,证明通过尿液排毒也是一个重要的传播途径,

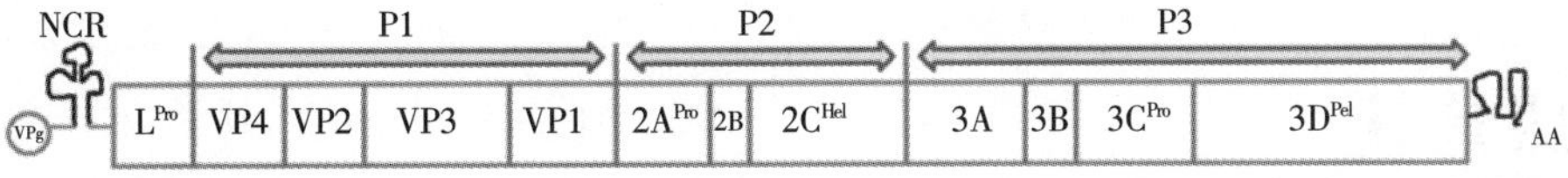

图 6.1　捷申病毒基因组结构示意图。如图所示:基因连接病毒蛋白、非编码 RNA、前导(L)蛋白、病毒蛋白、结构蛋白(P1)、非结构蛋白(P2~P3)、蛋白酶、解旋酶和聚合酶排列。

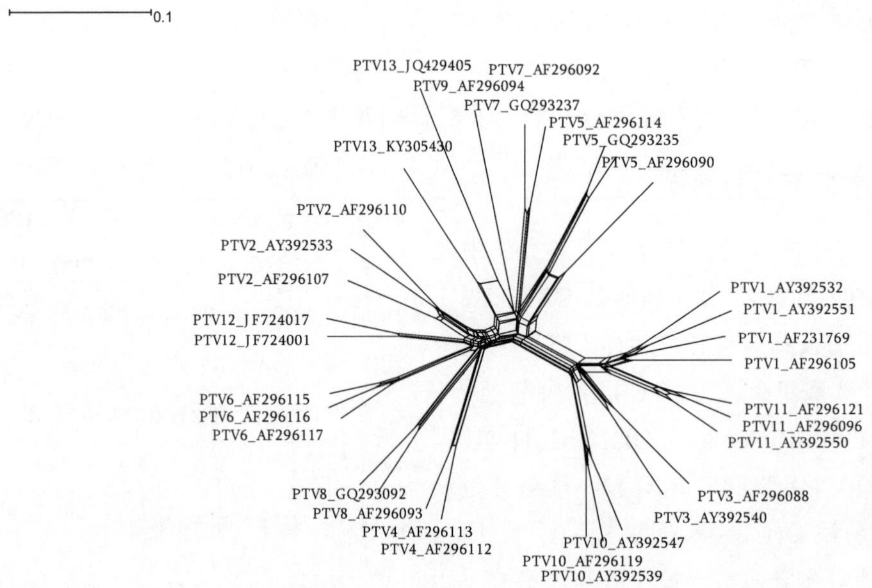

图 6.2 基于 VP1 氨基酸序列，应用 Splits Tree4 软件对 PTV 的 13 种类型代表毒株建立遗传进化树（Huson 和 Bryant，2006）。

而粪便(粪便和尿液混合物)则使得病毒的存活和传播更加容易(Tsai 等,2016)。截至目前,尚未报告其是人兽共患传染病。

6.3 发病机制

PTV 通过采食进入体内后，在扁桃体和肠道中进行增殖。扁桃体在病毒入侵、存活和抵御感染中具有重要作用。病毒在大肠和回肠中的滴度高于其他部分肠道(Long,1985;Chiu 等,2013)。

强毒株通过血液循环进入中枢神经系统 1~2 天后,表现为体温升高,随后发生腹泻。10~11 天出现弛缓性和痉挛性瘫痪(Long,1985)。呼吸系统麻痹(窒息)是导致病畜死亡的主要原因(Knowles,2015)。人工经鼻腔接种病毒可出现神经症状。病毒通过血液循环到达胎盘,引起繁殖障碍。经妊娠母猪的鼻腔和口腔进行实验室感染，可导致胎儿感染(Chiu 等,2014)。

6.4 临床症状

PTV 感染通常引起亚临床症状。不同血清型 PTV 毒力不同,其感染后引起的临床症状也不同,在临床上常见的是几种血清型混合感染，临床症状主要是由强毒株 PTV-1 感染引起。捷申病毒脑脊髓炎的潜伏期为 14 天(Knowles,2015),但 PTV-1 强毒株“Zabreh” 在实验条件下感染仔猪,5~7 天就出现临床症状。

患捷申病毒脑脊髓炎的病畜，其初期临床症状表现为发热、厌食、倦怠和行动紊乱,随后出现瘫痪或麻痹。发病早期 2~3 天出现共济失调,进一步发展为麻痹或瘫痪,但很少发展成完全瘫痪,一般出现临床症状 3~4 天后，发生死亡(Yamada 等,2014)。繁殖障碍表现为死产-木乃伊胎-死胎-不育综合征(Dunne 等,1965)。细小病毒感染也会引起类似的繁殖综合征,其感染率比 PTV 更加普遍。在妊娠早期至中期(40~70 天)感染引起死胎和木乃伊胎，但在后期感染可能导致死产(Lin 等,2012)。实验室人工感染与野外自然感染均可引起流产(Kirkbride 和 McAdaragh,1978)。虽然从雄性生殖道中分离到了病毒，但向子宫内人工实验接种含病毒的精子，并不会引起胚胎感染或阻止受孕。然而,在健康猪和腹泻猪中分离到的病毒,人工感染 PTV 可引起宿主腹泻(排除其他病原体)。

关于 PTV 引起呼吸道感染的报告并不多。最近捷克共和国报告了1例患有典型非化脓性脊髓炎的先天性小眼综合征病猪(Andrysikova 等,2018)。表6.1显示了一些血清型与某些特定疫病的相关性,但更多血清型不引起任何临床症状。

6.5 诊断

6.5.1 临床诊断

临床上出现发热,随后出现行动失调和瘫痪/麻痹等症状的病猪,疑似感染 PTV。如果有母猪出现死产或木乃伊胎等症状,疑似 PTV 感染诱发的繁殖障碍。

6.5.2 样品采集

由于病变主要发生在大脑、小脑、间脑、延髓、颈髓和腰髓,所以除血清学检测外,其他所有检测诊断采样的优先部位是脑和脊髓(Alexandersen,2012)。用于病毒分离的理想样本是从最近死亡或剖检的病猪上采集的样品。通过病毒中和实验或 ELISA 对血清样本对比检测可评估血清是否阳性(Hübschle 等,1983)。

6.5.3 病理变化

尸体剖检未发现与 PTV 诱发的脊髓灰质炎相关的肉眼可见的病变。对患有非化脓性脊髓灰质炎病猪进行中枢神经系统病理组织学检查,常在血管周围观察到淋巴套等病变。发病晚期常出现神经元变性(肿胀、染色质溶解和坏死)和轴突变性(Yamada 等,2007)。肉眼可见病变是浆液性纤维素性心包炎,伴有混浊的心包积液,易形成凝血块。偶尔会出现局灶心肌坏死伴有细胞浸润。

6.5.4 病毒分离

PTV 对猪原代细胞和传代肾细胞非常易感,所以猪原代细胞和传代肾细胞常用于病毒分离。PTV 在其他的细胞系中也能很好地增殖,如猪肾细胞(IBRS-2)。培养分离到的 PTV 可通过病毒中和(VN)试验和免疫荧光抗体试验进行鉴定(Knowles,2015),但 VN 和免疫荧光抗体检测耗时较长,分别需要72小时和12小时。

6.5.5 核酸检测

PTV 诊断的金标准是核酸检测。“Palmquist RT-PCR”用1对引物就能够对猪萨佩罗病毒(porcine sapeloviru, PSV)和 PTV 进行鉴别检测,而多重“Zell RT-PCR”用3对引物就能对 PTV、PSV 和 PEV 进行鉴别检测(Palmquist 等,2002;Zell 等,2000)。这些 RT-PCR 检测方法特异性强,敏感度高且快速,与伪狂犬病毒、猪繁殖呼吸综合征病毒、猪细小病毒、猪冠状病毒、猪呼肠孤病毒、小 RNA 样病毒等均无交叉反应。

中国建立了 RT-LAMP 快速检测系统(Wang 等,2011)。为便于病毒性疫病的检测,开发了基于核酸的多重 PCR 检测方法用于猪病毒的检测。基于探针的实时定量,RT-PCR 显示出了高敏感性和特异性,并且实现了一次能够同时对10份样本进行检测(Zhang 等,2013)。

多种基因型的 PTV 已传播到世界各地,所以通过对 PTV 的基因分型研究,有助于对 PTV 流行病学的了解。在 VP1 和 VP2 中鉴定出来的中和抗

表6.1 PTV 感染相关的自然或实验室条件下相关综合征

综合征	相关 PTV 血清型
脊髓灰质炎	PTV-1,PTV-2,PTV-3,PTV-5,PTV-12,PTV-13
繁殖障碍	PTV-1,PTV-3,PTV-6
腹泻	PTV-1,PTV-2,PTV-3,PTV-5
肺炎	PTV-1,PTV-2,PTV-3
心包炎和心肌炎	PTV-2,PTV-3

原表位被用于 PTV 的分型和分子流行病学研究。引物和靶基因的一些细节如表 6.2 所示。

6.5.6 抗体检测

PTV 是一种普遍存在的病毒;因此,单一阳性血清学试验并不能证明 PTV 感染。然而,与对照血清样本相比,抗体效价升高 4 倍,且表现出相关的临床症状,可被认为是 PTV 阳性。ELISA 也可被用于进行 PTV 感染的检测与分型。

6.5.7 鉴别诊断

PTV 疾病需要与下列疾病进行区分:

• 病毒性疾病:伪狂犬病(Aujeszky 病)、经典猪瘟(急性猪霍乱)、猪乙型脑炎、猪血凝性脑脊髓炎、狂犬病、猪繁殖与呼吸综合征(强毒株)。

• 细菌性疾病:猪链球菌性脑膜脑炎、大肠杆菌肠毒血症诱发的水肿病。

• 中毒:盐(脱水)、铅、杀虫剂。

6.6 预防与控制

与大多数病毒感染一样,PTV 的控制主要取决于预防而非治疗。如果 PTV 感染是轻症型的宿主,一过性瘫痪后,如果能够恢复正常食欲,它们可能会存活。但发生捷申病毒脑脊髓炎需要向监管部门报告(Knowles,2015)。

6.6.1 免疫

抗 PTV 免疫反应主要是通过 IgG 和 IgM 抗体介导的体液免疫应答。当病毒经口腔侵入时,肠道内产生具有保护性的 IgA 抗体,但由于抗体生成受到破坏,肠道可能会出现持续性病毒感染。母源抗体可有效预防病毒血症和病毒经胎盘传播。PTV 缓慢扩散至子宫内,导致不同发育阶段的胎儿死亡(Wang 和 Pensaert,1989)。胎儿的 PTV 抗体主要为 IgM 型,其次是 IgG,在 70 天时开始形成,在

表 6.2 用于 PTV 检测和基因分型的 RT-PCR 试验

检测法	引物序列	靶基因	参考文献
RT-PCR	5'AGTTTTGGATTATCTTGTGCCC-3'	5'NTR	Zell 等(2000)
	5'-CCAGCCGCGACCCTGTCAGGCAGCAC-3'		
nRT-PCR	5'-TGAAAGACCTGCTCTGGCGCGAG-3'		
	5'-GCTGGTGGGCCCCAGAGAAATCTC-3'		
双重 PCR(PTV 和 PSV)	5'-GTGGCGACAGGGTACAGAAGAG-3	5'UTR	Palmquist 等(2002)
	5'-GGCCAGCCGCGACCCTGTCAG-3'		
RT-PCR(PTV-1 特异性)	ATGCCTTTGAGACCTGTTAATGA	PTV-1	Zell 等(2000)
	CAACATTAGTCATCTTTGTAATTGT	VP3-VP1	
RT-PCR(基因分型)	GCATCHAAYGARAAYCC	VP1	La Rosa 等(2006)
	CCAAAYCCAAARTCYTG		
RT-PCR(基因分型)	CACCARYTGCTTAARTGYKGTTGG	VP2	Kaku 等(2007)
	CACAGGGTTGCTGAAGARTTTGT		
RT-LAMP	CACATCAATGACACGGGTTTTCCC	3DPoL	Wang 等(2011)
	TCGCCTTCTTTCACAAGAATCCCG		
多重 RT-PCR(PTV、PRRSV、CSFV)	GTGGCGACAGGGTACAGAAGAG	5'UTR	Liu 等(2011)
	GGCCAGCCGCGACCCTGTCAG		
实时 RT-PCR	5'-CTCCTGACTGGGCAATGGG-3'	5'UTR	Zhang 等(2013)
	5'-TGTCAGGCAGCACAAGTCCA-3'		

90 天时成熟(Wang 等,1973)。这些抗体可以保护这个阶段的胎儿免受感染。从初乳中获得的母源抗体可在断奶后产生保护作用(Wang 和 Pensaert,1989)。

6.6.2 交叉保护

抗体介导的免疫在预防 PTV 感染中具有更重要的作用(Alexandersen 等,2012)。但由于 PTV 血清型/基因型多样性,因此不太可能发生交叉免疫保护。

6.6.3 预防

成功控制 PTV 诱发脑脊髓炎的措施,包括限制行动、检疫、屠宰和定期疫苗接种。

(1)在中欧和马达加斯加 PTV 发病高峰期,应用减毒疫苗和灭活疫苗可有效预防 PTV,但后来随着疫病减少,就停止了疫苗的使用。由于该病是由几种血清型 PTV 引起的,因此研发对多种血清型 PTV 都具有保护作用的疫苗具有很大挑战性。通过疫苗接种控制高致病性的捷申病,保护珍贵的种畜、母猪或猪群最经济有效。

(2)限制从受 PTV 影响的国家进口猪肉可能有助于抑制致高病性 PTV-1 毒株传播。有效的控制措施就是加强检疫和扑杀。

(3)对于新引进的种猪,在混群前 1 个月,要让其充分暴露在有 PTV 流行的地方,使其产生免疫力。

(4)封闭式猪群可降低外源性病毒侵入的风险,但是由于相对耐药的 PTV 可通过多种污染物传播,仍可带来传播风险。

参考文献

Alexandersen S, Knowles NJ, Dekker A, Belsham G, Zhang Z, Koenen F (2012) Picornaviruses. In: Zimmerman J, Karriker LA, Ramirez A, Schwartz KJ, Stevenson GW (eds) Diseases of swine, 10th edn. Blackwell Publishing Press, Ames, IA, pp 587–620

Andrysikova R, Sydler T, Kümmerlen D, Pendl W, Graage R, Moutelikova R, Prodelalova J, Voelter K (2018) Congenital microphthalmic syndrome in a swine. Case Rep Vet Med 2018:1–7, 2051350, 6 p

Boros A, Nemes C, Pankovics P, Kapusinszky B, Delwart E, Reuter G (2012) Porcine teschovirus in wild boars in Hungary. Arch Virol 157(8):1573–1578

Cano-Gomez C, Palero F, Buitrago MD, García-Casado MA, Fernández-Pinero J, Fernández-Pacheco P, Agüero M, Gómez-Tejedor C, Jiménez-Clavero MÁ (2011) Analyzing the genetic diversity of teschoviruses in Spanish pig populations using complete VP1 sequences. Infect Genet Evol 11(8):2144–2150

Cano-Gómez C, García-Casado MA, Soriguer R, Palero F, Jiménez-Clavero MA (2013) Teschoviruses and sapeloviruses in faecal samples from wild boar in Spain. Vet Microbiol 165(1–2):115–122

Chiu S, Hu S, Chang C, Chang CY, Huang CC, Pang VF, Wang FI (2013) The role of porcine teschovirus in causing diseases in endemically infected pigs. Vet Microbiol 161(1/2):88–95

Chiu S, Yang CL, Chen YM, Hu SC, Chiu KC, Lin YC, Chang CY, Wang FI (2014) Multiple models of porcine teschovirus pathogenesis in endemically infected pigs. Vet Microbiol 168(1):69–77

Deng MY, Millien M, Jacques-Simon R, Flanagan JK, Bracht AJ, Carrillo C, Barrette RW, Fabian A, Mohamed F, Moran K, Rowland J, Swenson SL, Jenkins-Moore M, Koster L, Thomsen BV, Mayr G, Pyburn D, Morales P, Shaw J, Burrage T, White W, McIntosh MT, Metwally S (2012) Diagnosis of porcine teschovirus encephalomyelitis in the Republic of Haiti. J Vet Diagn Investig 24(4):671–678

Derbyshire JB, Arkell S (1971) Activity of some chemical disinfectants against Talfan virus and porcine adenovirus type-2. Br Vet J 127(3):137–142

Dunne HW, Gobble JL, Hokanson JF, Kradel DC, Bubash GR (1965) Porcine reproductive failure associated with a newly identified "SMEDI" group of picorna viruses. Am J Vet Res 26(115):1285–1297

Horak S, Killoran K, Leedom Larson KR (2016) Porcine teschovirus. Swine Health Information Center and Center for Food Security and Public Health, Ames, IA

Hübschle OJ, Rajanarison I, Koko M, Rakotondramary E, Rasiofomanana P (1983) ELISA for testing swine sera for antibodies against Teschen virus. Dtsch Tierarztl Wochenschr 90(3):86–88

Huson DH, Bryant D (2006) Application of phylogenetic networks in evolutionary studies. Mol Biol Evol 23(2):254–267

Kaku Y, Sarai A, Murakami Y (2001) Genetic reclassification of porcine enteroviruses. J Gen Virol

82(2):417–424

Kaku Y, Murakami Y, Sarai A, Wang Y, Ohashi S, Sakamoto K (2007) Antigenic properties of porcine teschovirus 1 (PTV-1) Talfan strain and molecular strategy for serotyping of PTVs. Arch Virol 152(5):929–940

Kirkbride CA, McAdaragh JP (1978) Infectious agents associated with fetal and early neonatal death and abortion in swine. J Am Vet Med Assoc 172(4):480–483

Knowles NJ (2015) Teschovirus encephalomyelitis (previously enterovirus encephalomyelitis or Teschen/Talfan disease). In: Manual of diagnostic tests and vaccines for terrestrial animals. OIE, Paris

Knowles NJ, Buckley LS, Pereira HG (1979) Classification of porcine enteroviruses by antigenic analysis and cytopathic effects in tissue-culture—description of 3 new serotypes. Arch Virol 62(3):201–208

La Rosa G, Muscillo M, Di Grazia A, Fontana S, Iaconelli M, Tollis M (2006) Validation of RT-PCR assays for molecular characterization of porcine teschoviruses and enteroviruses. J Vet Med B Infect Dis Vet Public Health 53(6):257–265

Lin W, Cui S, Zell R (2012) Phylogeny and evolution of porcine teschovirus 8 isolated from pigs in China with reproductive failure. Arch Virol 157(7):1387–1391

Liu S, Zhao Y, Hu Q, Lv C, Zhang C, Zhao R, Hu F, Lin W, Cui S (2011) A multiplex RT-PCR for rapid and simultaneous detection of porcine teschovirus, classical swine fever virus, and porcine reproductive and respiratory syndrome virus in clinical specimens. J Virol Methods 172(1–2):88–92

Long JF (1985) Pathogenesis of porcine polioencephalomyelitis. In: Comparative pathobiology of viral diseases, vol I. CRC Press, Boca Raton, FL, pp 179–197

Palmquist JM, Munir S, Taku A, Kapur V, Goyal SM (2002) Detection of porcine teschovirus and enterovirus type II by reverse transcription-polymerase chain reaction. J Vet Diagn Investig 14(6):476–480

Salles MWS, Scholes SFE, Dauber M, Strebelow G, Wojnarowicz C, Hassard L, Acton AC, Bollinger TK (2011) Porcine teschovirus polioencephalomyelitis in western Canada. J Vet Diagn Investig 23(2):367–373

Trefny L (1930) Hromadna onemocnemi vepru na Tesinsku. Zverolek Obz 23:235–236

Tsai ATH, Kuo CC, Kuo YC, Yang JL, Chang CY, Wang FI (2016) The urinary shedding of porcine teschovirus in endemic field situation. Vet Microbiol 182:150–155

Wang JT, Pensaert MB (1989) Porcine enterovirus (reproductive disorders). In: Virus infections of porcines. Elsevier, Amsterdam, pp 235–239

Wang JT, Dunne HW, Griel LC, Hokanson JF, Murphy DM (1973) Mortality, antibody development, and viral persistence in porcine fetuses incubated in utero with SMEDI (entero-) virus. Am J Vet Res 34(6):785–791

Wang B, Wang Y, Tian ZJ, An TQ, Peng JM, Tong GZ (2011) Development of a reverse transcription loop-mediated isothermal amplification assay for detection of porcine teschovirus. J Vet Diagn Investig 23(3):516–518

Yamada M, Kaku Y, Nakamura K, Yoshii M, Yamamoto Y, Miyazaki A, Tsunemitsu H, Narita M (2007) Immunohistochemical detection of porcine teschovirus antigen in the formalin-fixed paraffin-embedded specimens from pigs experimentally infected with porcine teschovirus. J Vet Med Physiol Pathol Clin Med 54(10):571–574

Yamada M, Miyazaki A, Yamamoto Y, Nakamura K, Ito M, Tsunemitsu H, Narita M (2014) Experimental teschovirus encephalomyelitis in gnotobiotic pigs. J Comp Pathol 150(2–3):276–286

Yang T, Li R, Yao Q, Zhou X, Liao H, Ge M, Yu X (2018) Prevalence of porcine teschovirus genotypes in Hunan, China: identification of novel viral species and genotypes. J Gen Virol 99(9):1261–1267

Zell R, Krumbholz A, Henke A, Birch-Hirschfeld E, Stelzner A, Doherty M, Hoey E, Dauber M, Prager D, Wurm R (2000) Detection of porcine enteroviruses by nRT-PCR: differentiation of CPE groups I-III with specific primer sets. J Virol Methods 88(2):205–218

Zell R, Dauber M, Krumbholz A, Henke A, Birch-Hirschfeld E, Stelzner A, Prager D, Wurm R (2001) Porcine teschoviruses comprise at least eleven distinct serotypes: molecular and evolutionary aspects. J Virol 75(4):1620–1631

Zhang C, Wang Z, Hu F, Liu Y, Qiu Z, Zhou S, Cui S, Wang M (2013) The survey of porcine teschoviruses in field samples in China with a universal rapid probe real-time RT-PCR assay. Trop Anim Health Prod 45(4):1057–1061

第 7 章　黄病毒

Shailendra K. Saxena, Swatantra Kumar, Amrita Haikerwal

7.1 引言

黄病毒科的虫媒病毒是造成人和动物死亡的主要原因。黄病毒分为 3 个属，即黄热病毒属、瘟病毒属、丙型肝炎病毒属(Simmonds 等，2017)。“黄病毒”这个术语来自拉丁语 flavus，意思为黄色的，定义为由黄热病毒引起的黄疸。已有 70 多种病毒被归类为黄病毒，其中 16 种通过蜱传播，40 种通过蚊虫传播，其余病毒暂未发现传播媒介(Mukhopadhyay 等，2005)。最主要的蚊媒黄病毒是流行性乙型脑炎病毒，也称日本脑炎病毒(Japanese encephalitis virus, JEV)、寨卡病毒(Zika virus, ZIKV)、登革热病毒(dengue virus, DENV)、西尼罗病毒(West Nile virus, WNV)和黄热病毒(yellow fever virus, YFV)。黄病毒可分为不同的血清学群，如 DENV 血清群、JEV 血清群和血凝活性较差的 YFV 群。根据分子系统发育分析，黄病毒可进一步分为几个进化分枝、簇和种系 (Kuno 等，1998)。根据临床表现，黄病毒也被归类为：①嗜神经病毒、JEV (Solomon，2004) 和 DENV (Garg 等，2017)，感染中枢神经系统造成神经功能损伤，临床表现为脑炎；②ZIKV，感染表现为吉兰-巴雷综合征和小头畸形(Broutet 等，2016)。动物在黄病毒的生命周期中扮演着至关重要的角色，例如，猪是 JEV(Mansfield 等，2017)的增殖宿主，而候鸟是 WNV 的贮存宿主(Pérez-Ramírez 等，2014)。

尽管在发病机制、偏嗜性和翻译机制等方面存在差异，但黄病毒全基因组结构相似(Neufeldt 等，2018)，属于单股正链 RNA 病毒，大小为 40~60nm，基因组约 11kb，被衣壳蛋白包裹，表面有 180 个拷贝的囊膜糖蛋白突起。

黄病毒基因组仅有一个 ORF，编码一个 3400 个氨基酸组成的多聚蛋白，5’ 端带有一个 1 型的帽状结构。多聚蛋白的 N-末端是结构蛋白：衣壳(C)、前膜(prM)和构成病毒颗粒的囊膜(E)。其余部分是 7 种非结构蛋白：NS1、NS2A、NS2B、NS3、NS4A、NS4B 和 NS5(图 7.1)。NS1 蛋白是黄病毒的分泌蛋白 (Muller 和 Young，2013)，NS2A 在病毒的早期复制过程中和感染性病毒粒子的增殖过程中具有重要作用。NS2B 作为辅助因子，协助 NS3 募集到 ER 膜上(Li 等，2016)。NS3 具有多种活性，如蛋白酶活性、核苷酸 5’-三磷酸酶(NTPase)活性和解旋酶活性(Li 等，2014)。NS4A 具有膜曲率诱导活性，是一种完整的膜蛋白，虽然 NS4B 无酶活性，但通过与 NS3 相互作用则对病毒复制起重要作用(Apte-Sengupta 等，2014)。NS5 的 N 末端具有鸟嘌呤-N7-甲基转移酶活性、鸟嘌呤转移酶活性和核苷-2-O-甲基转移酶活性，负责病毒基因组的甲基化和 5’RNA 的帽化。NS5 的 C 末端区域有 RNA 依赖 RNA 聚合酶(RNA polymerase, RdRp)，并引发病毒 RNA 的合成(Zou 等，2011)。

7.2 病毒复制

JEV/WNV 通过被感染蚊虫的叮咬进入宿主细胞，如皮肤树突状细胞(Lannes 等，2017)。附着因子或受体与二级受体发生内化作用后，参与到病毒与靶细胞的结合。CD209，C 型凝聚素受体(DC-SIGN)，对 JEV 和 WNV 等病毒的吸附具有至关重要的作用(Wang 等，2016)。黄病毒通过网

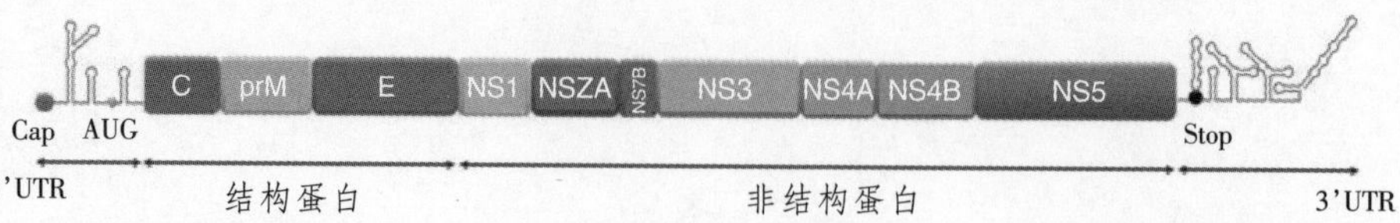

图 7.1 黄病毒基因组。黄病毒基因组全长约 10.5kb，编码 3 种结构蛋白：衣壳、前膜、囊膜及 7 种非结构蛋白：NS1、NS2A、NS2B、NS3、NS4A、NS4B 和 NS5。基因组在 3’和 5’端包含非翻译区。黄病毒 RNA 通过粗面内质网进行 cap 依赖性翻译。

格蛋白介导的内吞作用进入宿主细胞，使病毒暴露在细胞核内酸性区间，结果引起病毒和核膜融合，导致囊膜糖蛋白的构象重排(Nour 等，2013)。通过膜融合诱导病毒 RNA 释放到细胞质中，在粗面内质网膜利用核糖体进行帽依赖性翻译，从而合成多肽。病毒蛋白酶和宿主信号肽酶将多聚蛋白加工成 3 种结构蛋白和 7 种非结构蛋白。蛋白质的裂解与细胞内质网膜有关，其中非结构蛋白调节内质网膜的内陷，并且通过负链 RNA [(-) RNA] 的合成完成 RNA 的扩增 (Paul 和 Bartenschlager，2015)。病毒复制酶复合物产生的 (+) RNA 链通过出芽方式进入内质网腔，随后在病毒衣壳装配化过程中将产生的核酸包裹在病毒颗粒中，这一过程在与复制位点相反的区域里完成(Shi 和 Suzuki，2018)。成熟的病毒通过弗林蛋白酶将 prM 裂解为 M (一种高尔基体驻留蛋白酶)(图 7.2)。具有传染性的成熟病毒粒子通过一种被称为外溢的方式离开细胞(Murray 等，2008)。

7.3　流行性乙型脑炎病毒

7.3.1　地理分布

JEV 是单股正链 RNA 病毒，属于黄病毒科黄病毒属。遗传进化分析表明马来群岛的 JEV 起源于一种古老的病毒(Solomon 等，2003)。JEV 很可能在 1000 年前就已演变出了不同的基因型（Ⅰ～Ⅴ)，随后在亚洲地区广泛流行。1871 年，日本报道了首例具有临床表现的日本脑炎(Japanese encephalitis, JE)病例，此后约半个世纪，临床发现的病例超过 6000 例 (WHO，2019)。因此，在 1927 年、1934 年和 1935 年报道了几次疫情的暴发。1935 年，从人脑中分离出病原体。10 年后，通过将

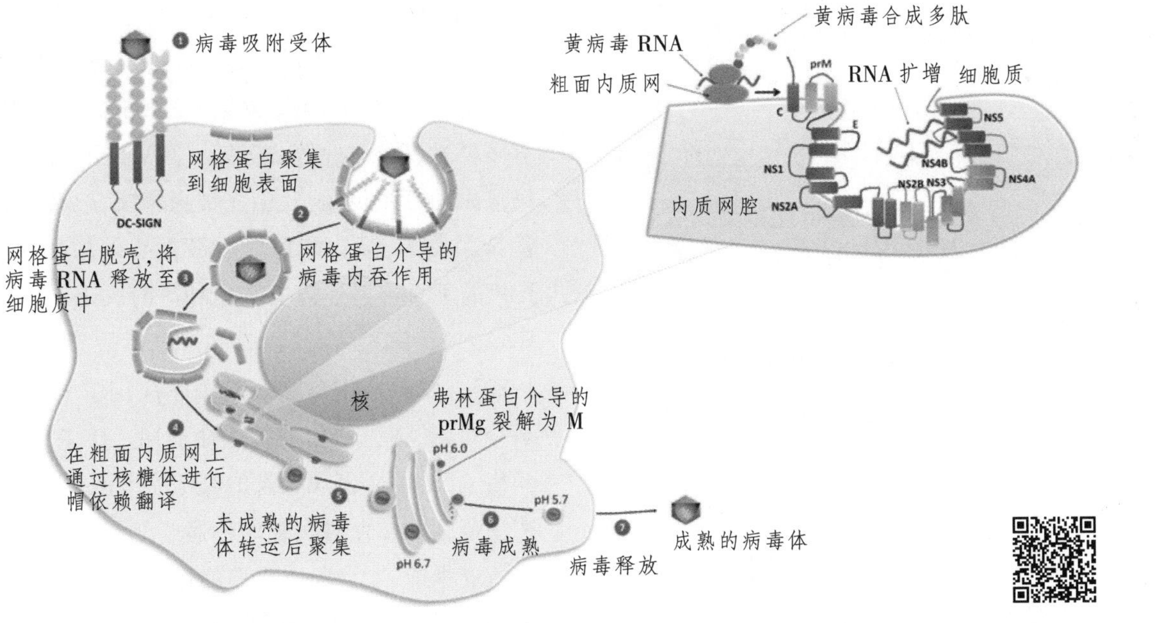

图 7.2　黄病毒复制。①DC-SIGN 对 JEV 和 WNV 的吸附具有重要的作用；②网格蛋白聚集到细胞表面；③黄病毒通过网格蛋白介导的内吞作用进入宿主细胞；④与网格蛋白联结的病毒与核膜融合、脱壳，将病毒 RNA 释放至细胞质中；⑤在粗面内质网膜上通过核糖体进行帽依赖性翻译合成多肽；⑥病毒在内质网(ER)表面聚集；⑦未成熟的病毒体转运到高尔基体进一步发育成熟，成熟病毒通过弗林蛋白酶使得 prM 裂解为 M(一种高尔基体驻留蛋白酶)；⑧成熟的具有感染性的病毒粒子从被感染细胞内释放出来。图中的插画表示黄病毒 RNA 在粗面内质网膜通过核糖体的帽依赖性翻译，合成多肽。多聚蛋白通过病毒蛋白酶和宿主信号肽酶的作用裂解为 3 种结构蛋白和 7 种非结构蛋白。蛋白质的裂解与细胞内质网膜有关，其中非结构蛋白调节内质网膜的内陷，并且通过负链 RNA 的合成来完成 RNA 的扩增。(扫码看彩图)

分离到的病原体接种到猴体内，JEV 才被证实(Erlanger 等，2009)。一些国家报告 JE 首个的病例时间如下，韩国在 1933 年、中国在 1940 年、菲律宾在 1950 年、马来西亚在 1952 年、印度在 1955 年。1938 年，证明在 JEV 生命周期中，涉水鸟类、三带喙库蚊为传播媒介，猪为增殖宿主(van den Hurk 等，2009)。

7.3.2 基因型

JEV 分为 5 个基因型，即 GⅠ~GⅤ(Schuh等，2014)。最主要的基因型是 GⅢ，其在日本、中国、韩国、菲律宾、印度、尼泊尔和斯里兰卡等亚洲国家广泛分布。然而，在过去的几十年中，在几个国家中，包括中国、韩国和泰国，GⅠ已取代了 GⅢ(Fulmali 等，2011)。此外，大部分在热带地区流行的 GⅠ是 GⅠ-a 而在温带地区流行的是 GⅠ-b。在过去的 60 年一直没有监测到 GⅤ流行，但在 60 年后又再次出现(Schuh 等，2013a)。GⅣ仅限于印度尼西亚地区，其原因尚不明确(Schuh 等，2013b)。

7.3.3 流行

JE 是一种人兽共患的病毒性传染病，在东南亚、东亚和澳大利亚等地主要表现为病毒性脑炎，每年约有 6.8 万例临床病例和 2.4 万死亡病例(Campbell 等，2011)。根据 WHO 的报告显示，西太平洋和东南亚等地区有 24 个国家存在 JEV 的地方性传播，约有 30 亿人处于感染 JEV 的风险中(WHO，2019)。据报告，地方流行性的 JEV 具有广泛的宿主范围。JEV 感染人的全球流行病学的系统性回顾分析表明，JE 的实际发病率比 WHO 报告的高出约 10 倍(Wang 和 Liang，2015)。利用监测系统监测家畜的病毒血症，对了解 JEV 的宿主范围、传播媒介和生态分布至关重要。在马来西亚的家畜中，JEV 血清阳性率最高的是犬和猪(Kumar 等，2018)。在中国，对 JE 发病率的流行病学调查显示，血清 JEV 阳性率最高的是猪(Zhang 等，2017)。

不同国家都记载了马的周期性 JE 病例，其死亡率为 5%~30%。韩国的流行病学数据显示，家畜中的发病率存在不确定性，其中，野生鸟类为 86.7%，牛为 51.7%，马为 49.7%，山羊为 12.1%(Wang 和 Liang，2015)。类似的 JE 监测结果显示，在印度不同地区的马的血清阳性率为 10%(Gulati 等，2011)。

7.3.4 传播

JEV 流行地区频繁出现人被感染的病例可能与密集的养猪业和农业水稻种植有关，这表明与猪有关的 JEV 感染在农村循环流行（Su 等，2014)。猪作为 JEV 增殖宿主，在 JEV 的生命周期中扮演着关键角色，且被感染猪的病毒血症水平很高(Cappelle 等，2016)。JEV 感染宿主的临床表现多种多样，被感染猪表现为轻度症状，而终末宿主，人和马被感染则表现出严重的症状（图 7.3）(Impoinvil 等，2013)。

此外，其他脊椎动物，如犬、鸭、爬行动物和鸡的血清阳性率也较高，这说明 JEV 的宿主范围具有多样性(Oliveira 等，2017)。JEV 主要通过被感染蚊虫的叮咬传播，其中三带喙库蚊、杂鳞库蚊和伪杂鳞库蚊是传播 JEV 的最重要的蚊虫种类。JEV 的垂直传播还需要其他种类的蚊子，包括环库蚊、尖音库蚊、白纹伊蚊和海滨伊蚊(Rosen 等，1989)。蚊虫与水鸟之间的关系决定了 JEV 的生态分布，被称为鸟相关的生命周期。JE 在低密度养猪地区，也会出现大量的其他家畜流行，这表明宿主范围多样性对病毒的传播非常重要。在病毒传播过程中，家禽起着重要的作用，因为病毒血症水平高的鸭子和鸡易感染蚊虫（Lord 等，2015)。此外，在实验室条件下，被感染的三带喙库蚊更易选择叮咬牛而不是猪。在印度，JEV 的自然感染也证实了这一点(Philip Samuel 等，2008)。已经证实不通过传播媒介也可传播 JEV，其中，猪口鼻分泌物中的 JEV 更容易感染其他易感的猪(Ricklin 等，2016a)。

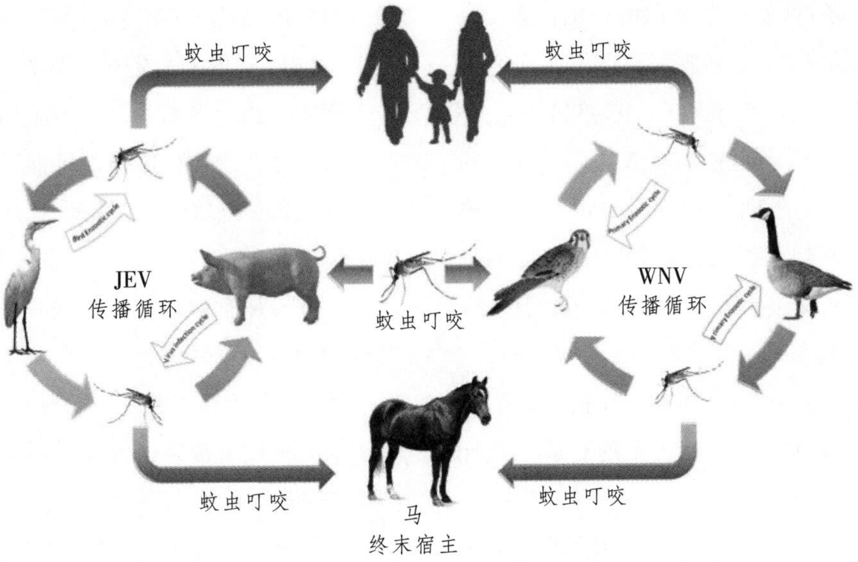

图 7.3 黄病毒的传播途径。传播循环途径开始于被感染的蚊虫叮咬的增殖宿主。猪和白鹭是 JEV 的增殖宿主和贮存宿主。而候鸟是 WNV 的增殖宿主和贮存宿主。在两个传播循环中人和马是终末宿主。

7.3.5 病毒感染与生态

JE 广泛的地理分布表明,约 50%的世界人口生活在有 JE 流行的疫区。JE 的出现可用一些生物学、生态学和社会经济学因素来解释。生物学因素包括生物类传播媒介的种群动力、与其他黄病毒的血清交叉反应及无传播媒介传播等(Kilpatrick 和 Randolph,2012)。生态学因素包括多样化的农业生产实践、夏季平均气温升高、水鸟迁徙规模增大及猪的养殖数量增加等(Tian 等,2015)。JEV 的季节性在中国台湾尤为突出,1991—2005 年的 5~8 月,发生在中国台湾的 JE 病例愈演愈烈(Lin 等,2017)。

同样,多雨季节增加了蚊虫的数量,而蚊虫数量的增加也与各种气象因素变化有关(Ramesh 等,2015)。社会经济学因素包括国际旅行、公共卫生系统不健全及城市周边发展等(Luo 等,1995)。除上述风险因素外,在流行疫区缺乏有效的大规模疫苗接种计划也可能会导致 JE 病例的出现。

7.3.6 感染免疫

宿主清除 JEV 感染的免疫反应包括先天性免疫应答反应和适应性免疫应答反应两种。体液免疫反应已经被充分的研究证实,但细胞介导免疫反应目前仍是世界范围内关注的研究领域。T 细胞在有效维持抗体介导免疫反应中的重要作用表明,体液免疫和细胞介导免疫的协同作用可预防免疫性病理损失,并且有助于了解 JEV 在中枢神经系统中的持续存在(Larena 等,2011)。同样条件下,宿主介导的抗病毒免疫反应也是临床上造成中枢神经系统病理损失的关键因素,其被称为脑炎。多项研究已证实中和抗体可预防 JEV 感染。然而,抗 NS1 的 IgG1 抗体介导的免疫应答也具有保护性(Li 等,2012)。在 JEV 急性感染期,中和抗体能够阻断病毒复制和病毒扩散,而中和抗体还可减少慢性感染中的细胞病理学效应,从而引起较重的组织损伤。黄病毒之间抗原相似性较大,导致感染的抗体依赖性增强(Pierson 等,2008)。

已经确认在 JEV 感染人的过程中有淋巴细胞的参与,然而,其确切作用机制尚不清楚,目前已成为全球新的研究领域。用致死剂量病毒预处理过的 T 细胞进行小鼠颅内接种,结果表明 T 细胞具有免疫保护作用(Murali-Krishna 等,1996)。临床研究表明病毒 NS3 蛋白在诱导 CD8+和

CD4+T 细胞反应方面发挥主要作用。JEV 感染个体引起的 CD8+和 CD4+T 细胞反应的机制研究显示,具有抗原性的氨基酸序列是 193~324 位氨基酸(Kumar 等,2004)。最近使用合成的肽库进行的一项研究表明,JEV 感染人的过程中 T 记忆细胞的反应具有不同靶向性。感染 JEV 的健康人中 CD8+T 细胞应答占据主导地位,而在已痊愈的 JE 患者中 CD4+T 细胞应答更具优势(Turtle 等,2016)。

此外,在病毒感染期间,IFN-γ 在宿主防御机制中的功能是清除中枢神经系统中的病毒,并且 T 细胞也参与 IFN-γ 的产生 (Larena 等,2013)。同样,在 JEV 感染早期,细胞因子的微环境对宿主免疫应答起着至关重要的作用。树突状细胞(DC)是产生促炎症细胞因子的主要来源,在 JEV 感染后,DC 分泌较高水平的 IL-8、IL-6 和 TNF-α,并更加有效地促进致敏的 Treg 细胞增殖(Gupta 等,2014)。为了确保感染成功,JEV 同时调节适应性免疫反应和先天性免疫反应。病毒基因组编码长度为 3400 个氨基酸的多聚蛋白,通过病毒蛋白酶和宿主信号肽酶裂解,产生 3 种结构蛋白和 7 种非结构蛋白。病毒蛋白参与调节宿主防御机制,其特征属于病毒性免疫病理(Kumar 等,2016)。JEV 在动物中的免疫发病理不明确;但是,对猪的研究表明,病毒的偏嗜性不只限于中枢神经系统,而且还涉及次级淋巴器官(Ricklin 等,2016b)。

7.3.7 诊断

在大多数动物中,JEV 感染都是无症状感染,这对家畜群体中 JEV 的鉴别和流行监测提出了挑战。多种临床诊断方法都可用于 JEV 的诊断,包括临床诊断、血清学诊断和病理学诊断等 (表 7.1)。用于马的最理想的诊断方法是从被感染马的中枢神经系统组织中分离病毒(WHA,2016)。病毒的体外分离培养主要用原代细胞及其传代细胞系进行。空斑试验可用于测定病毒滴度、感染率和分离株的致病性。传染性样本应在生物安全等级 3 级(BL-3)的实验室进行处理。血清学的诊断并不稳定,由于黄病毒间存在血清学交叉反应或动物曾被 JEV 感染,这些都可能导致出现假阳性(Hobson-Peters,2012)。因此,必须在进行病毒滴度测定的基础上,进行血清学的诊断。检测动物被 JEV 感染的最有效血清学方法是空斑减少中和试验(WOAH,2012)。

此外,对于急性感染 JEV 病例,可以使用 ELISA 检测 IgM 特异性抗体(Litzba 等,2010)。在动物中,通过检测 NS1 抗体,区分自然感染动物和疫苗接种动物(Konishi 等,2006)。核酸检测是用于 JEV 检测最常用的诊断平台之一。现有的 3 种核酸检测方法分别是反转录-聚合酶链反应、实时荧光 PCR 和 RT-LAMP。3 种检测血液样品中的 JEV 均具有高度特异性(Dhanze 等,2015)。JEV 不同基因型间至少有 12%的变异,一些基于核苷酸序列检测的方法已被用于基因分型,特别是基于 GⅠ和 GⅢ的 RT-PCR 方法(Chen 等,2014)。

7.3.8 预防与控制

有几种候选疫苗(表 7.2)可供人类接种,然而,家畜用疫苗还有待开发。通过给动物接种减毒活疫苗或灭活病毒疫苗可以延长其预期寿命。在韩国,已经给猪接种 Anyang 300(减毒活疫苗)近 30 年,患 JE 猪的数量显著减少(Nah 等,2015)。另一种方法是使用编码病毒结构蛋白的 DNA 疫苗或 DNA 疫苗与灭活病毒疫苗联合使用(Konishi 等,2000)。疫苗接种的主要作用是预防胎儿畸形、产生免疫记忆及中和抗体。

此外,通过给马接种疫苗可有效预防 JEV。另一种有效防治方法是通过控制流行 JEV 疫区的虫媒种群数量,防止病毒传播(Benelli 等,2016)。在动物养殖区域内通过喷除虫菊酯类杀虫剂,以及使用防蚊网可阻止传播媒介接触动物(Karunara-

表7.1　JEV和WNV诊断

序号	方法	检测对象
JEV诊断		
1	酶联免疫吸附试验	JEV抗原
2	空斑减少中和试验	中和抗体效价
3	间接免疫荧光试验	抗体与病毒抗原的特异性反应
4	血凝抑制试验	通过滴定法检测和测量病毒浓度
5	补体结合试验	抑制病毒抗原的抗体滴度
6	反转录-聚合酶链反应	病毒RNA
7	反转录环介导等温扩增试验	病毒
WNV诊断		
1	抗原捕获	病毒抗原
2	生物快速检测系统	病毒抗原
3	乳鼠病毒分离	传染性病毒
4	绿猴肾细胞培养物病毒分离	传染性病毒
5	常规反转录-聚合酶链反应	病毒RNA
6	核苷酸序列扩增	病毒RNA
7	TaqMan技术(实时RT-PCR)	病毒RNA
8	反转录环介导等温扩增试验	病毒RNA

tne和Hemingway,2000)。

7.4　西尼罗病毒

7.4.1　地理分布

WNV属于黄病毒科、黄病毒属,与JEV和圣路易斯脑炎病毒具有相似的免疫学特性。该病毒1937年在乌干达西尼罗地区一位发热女性的血液中首次分离到，被认为是一种特殊的病原体(Smithburn等,1940)。WNV在初级宿主和贮存宿主的蚊媒-候鸟之间传播,人类和其他几种哺乳动物则是偶然宿主和终末宿主。人和马中约有20%的有症状病例，人中有1%的病例表现出神经症状,死亡率为1%~10%(Danis等,2011),而在马中有90%的病例患有神经系统疾病,死亡率为30%~40%(Ward等,2006)。WNV分布遍及非洲各地、中东国家、亚欧大陆,随后蔓延至北美和南美,影响美国公众、家畜和野生动物的健康(Karabatsos,1978)。1999—2001年,美国报告了149例与WNV感染人有关的脑炎和脑膜炎临床病例，出现马脑炎病例814例,以及死亡鸟11 932只(其中乌鸦的死亡数量最多)(Komar等,2003)。美国国家虫媒病毒监测系统(national arboviral surveillance system, ArboNET)是由疾病预防与控制中心管理的虫媒病毒监测系统。除蚊虫叮咬外,WNV在鸟中可能是通过接触或经口传播。

7.4.2　流行

WNV在被发现之初并没被视为公共卫生的一个潜在威胁;然而,在20世纪50年代,中东地区流行的WNV引起的发热和脑炎，确定了其流行病学和生态学地位。在1951年以色列暴发的疫情中明确了WNV的临床症状，大多数被感染的患儿,主要表现为头痛、腹痛、发热和呕吐等(Hurlbut等,1956)。20世纪50年代在埃及曾多次暴发WNV疫情,人群中的血清阳性率很高,随后启动了包括动物血清学调查、WNV传播媒介鉴定,以

表 7.2 JEV 和 WNV 疫苗

序号	疫苗名称	状态	人类/动物	备注
JEV 疫苗				
1	Anyang300 疫苗	可用	动物	使用 Anyang300 毒株的 JEV 活疫苗
2	JEV G1 疫苗	临床试验:Ⅰ期	动物	含重组猪 GM-CSF 的灭活 JEV G1(KV1899 毒株)
3	JENCEVAC	已完成Ⅲ期试验	人类	鼠脑提纯乙脑(JE)灭活疫苗
4	IXIARO	基于 SA-14-14-2,在一些国家获批	人类	非洲绿猴肾细胞灭活疫苗
5	ChimeriVax	基于Ⅰ~Ⅲ期临床试验,在一些国家获批	人类	由黄热病(YF)17D 构建的减毒重组活病毒
6	JENVAC	已完成Ⅲ期试验	人类	使用 JEV 印度毒株开发的非洲绿猴肾细胞衍生疫苗
WNV 疫苗				
1	West Nile innovator	商品化生产	动物	WNV 全病毒灭活疫苗
2	Vetera 疫苗	商品化生产	动物	WNV 全病毒灭活疫苗
3	West Nile-innovator DNA	商品化生产	动物	质粒 DNA PrM/E
4	Recombitek	商品化生产	动物	表达 PrM/E 的重组金丝雀痘病毒
5	Prevenile	召回	动物	表达 WNV PrM/E 的重组 YF17D
6	ChimeriVax-WN02	临床试验,Ⅱ期	人类	嵌合表达 WNV PrM/E 的重组 YF17D
7	Chimeric WN/DEN4-3'delta30	临床试验,Ⅰ期	人类	嵌合表达 WNV PrM/E 的重组 DV4
8	Clinical trial VRC303	临床试验,Ⅰ期	人类	表达 PrM/E 的 DNA 质粒
9	WN-80E	临床试验,Ⅰ期	人类	缺乏跨膜区域的可溶性 E

及节肢动物、鸟类、人类和马类动物等有关 WNV 感染性实验的深入研究。这些研究的发现极大地加深了研究者对 WNV 的流行病学和临床特征的理解。在动物中的血清学调查表明,病毒的宿主范围很广,包括鸟类和其他哺乳动物(包括马类)。

有趣的是,一项研究表明,只能在蚊虫分离到病毒,而在其他节肢动物中不能分离到病毒,这确定了蚊虫是其主要传播媒介。1957 年,在以色列疫情暴发期间, 有研究者首次报告了老年人出现严重的神经系统病变 (Hayes 和 Gubler,2006)。1974 年,南非暴发 WNV,报告了患有脑膜炎或脑炎的患者。然而,有神经系统症状的患者呈现散发特点,大多数患者临床表现为轻度发热。WNV 在俄罗斯、印度、西班牙和南非等地区表现为间歇性暴发,在 20 世纪 70 年代末期和 20 世纪 80 年代,大规模暴发的次数并不多。1999 年,WNV 在北美被检测到, 尤其是在皇后区和纽约市发现的 150 多种鸟类死亡,WNV 检测呈阳性。研究者从死亡的鸟中分离到了病毒, 并通过 WNV 特异性 RNA 序列确定了死于感染 WNV 的鸟的数量。1999 年在美国纽约市从乌鸦上分离到高致病性的毒株。2003 年,在得克萨斯州和墨西哥的鸟类中分离到弱毒 WNV,表明西方国家出现的 WNV 毒株表型发生变异(Davis 等,2004)。2006 年,在监测急性脑炎综合征期间,印度阿萨姆邦报告了多例 WNV 病例,阳性率为 11.65%。此外,在印度西孟加拉国的临床样品检测中也报告出现了 WNV(Khatun 和 Chatterjee,2017)。

7.4.3 传播

WNV 通过蚊-鸟-蚊的途径循环传播，其中蚊虫主要是库蚊属(Turell 等，2001)(图 7.3)。然而，美国已从蚊虫的 10 个种属 29 种不同的蚊虫中分离到 WNV。在美国北部地区的鸟类中发现大量可作为动物流行病传播媒介的尖音库蚊。同样，五带淡色库蚊、黑须库蚊、环喙库蚊是北美其他地区流行的 WNV 传播媒介(Sardelis 等，2001)。在非洲，单纹家蚊是将 WNV 传播给人类的最重要传播媒介。已在东半球某些地区的软蜱和硬蜱中分离到了 WNV，但并不认为蜱是动物中流行 WNV 的重要传播媒介。鸟类是 WNV 的增殖宿主，被认为是 WNV 的天然病毒库；仅在北美就有 111 种鸟类可被 WNV 感染(CDC，2002)。这些被感染的鸟类初期携带病毒量较高，增加了将 WNV 传播给采食蚊虫的概率；然而，这些被感染鸟类虽然会患上病毒血症，但能存活下来，并产生抗病毒免疫力，然而也有少数种类的鸟类出现死亡。被感染的候鸟将 WNV 病毒传入地中海和欧洲国家。各种哺乳动物都可以通过自然或实验室被 WNV 感染。WNV 自然感染的哺乳动物种类广泛，如人类、少数种类的蝙蝠、花栗鼠、猫、臭鼬和兔子等。

7.4.4 感染免疫

蚊虫叮咬后，WNV 可在短时间内于被叮咬对象表皮的朗格汉斯细胞中进行复制(Garcia-Tapia 等，2007)，进而感染淋巴组织中的树突状细胞，分泌细胞因子如 INF-α 和 INF-β(Liu 等，2006)。可以从短暂性病毒血症期或急性感染期的血液中分离到 WNV。这些树突状细胞也分泌 TNF-α(Diamond 等，2003)和巨噬细胞移动抑制因子(Arjona 等，2007)，诱导血脑屏障的完整性改变，导致脑内血浆和细胞蛋白渗漏，进而造成 WNV 侵入神经。这一过程诱导了脑内细胞因子/趋化因子的产生和免疫细胞的迁移，导致神经元的严重炎症反应和坏死，进而造成血脑屏障的破坏，并抑制神经前体细胞的分化和增殖。WNV 可在神经元中长期存在，从而引起先天性免疫应答和适应性免疫应答。

7.4.4.1 先天性免疫

伴随 WNV 感染，朗格汉斯细胞被激活，树突状细胞过表达 MHC Ⅱ分子、CD80、上皮钙黏蛋白和 CD54 (Byrne 等，2001)。这些细胞在 24~48 小时归巢至引流淋巴结，开始激活先天性免疫应答，包括分泌各种细胞因子和趋化因子、激活补体系统、促进淋巴结中的白细胞增殖、加工和向 T 细胞递呈抗原。病原体识别受体，如 Toll 样受体(TLR-3、TLR-7)、视黄酸诱导基因 I 和黑色素瘤分化相关基因 5，通过与 RNA 病毒结合，在识别和应答中发挥重要作用。因此，它们的结合将刺激转录因子，如 IFN 调节因子(regulatory factor，IRF)3 和 7，以及导致 IFN 刺激基因的表达，激发抗病毒通路以限制 WNV 感染。IFN 启动 Janus 激酶/信号传感器和转录激活因子通路，激活具有抗 WNV 活性的各种成分(Platanias，2005)。IFN 刺激基因促进 IFN-β 的生成以抑制病毒的翻译。然而，病毒的一些非结构蛋白，如 NS2A、NS2B、NS3、NS4A、NS4B 和 NS5 等，在病毒细胞感染过程中的多个阶段都可阻断 IFN 的信号传递。IFN 的表达可激活 T 细胞，进而抑制 WNV 感染。γδ T 细胞是先天性免疫应答和适应性免疫应答之间的桥梁，产生 Th1 或 Th2 型细胞因子，但与 αβT 细胞相比，这些细胞能够识别的抗原范围有限 (Poccia 等，2005)。γδ T 细胞产生 IFN-γ，IFN-γ 在限制病毒传播方面发挥关键作用，其可通过多种方式限制病毒传播，如通过其抗病毒活性激活 Th 细胞和 MHCI 的过表达，以及激活吞噬细胞等。

7.4.4.2 适应性免疫

被 WNV 感染的树突状细胞激活 B 细胞，产生抗 WNV 的特异性免疫球蛋白 IgM 和 IgG，清除血液中的病毒。WNV 粒子主要由两个抗体表位组成，如前膜蛋白或膜蛋白(prM 或 M)和囊膜蛋白(E)等。中和抗体可阻断 WNV 靶向结合细胞膜，而抗体亚群是抗 prM 蛋白或 M 蛋白的抗体(Nybakken 等,2005)。抗体通过阻止病毒与细胞吸附、融合,进而抑制病毒进入细胞。NS1 是一种非结构糖蛋白,可与宿主细胞表面受体结合。一项研究显示，预防性或以剂量依赖性方式给小鼠接种 NS1 抗体，可以对 WNV 感染小鼠引起的脑炎产生免疫保护。在黄病毒感染引起脑炎的过程中,杀伤性 CTL 在清除周围系统中的病毒发挥着重要的免疫作用。当 WNV 感染细胞与 CTL 接触,就会促进 CTL 增殖,释放细胞因子并杀死被感染细胞(Camenga 等,1974)。研究表明,细胞免疫反应和体液免疫反应协同作用,可以减少血液中的病毒,并防止中枢神经系统感染。

7.4.4.3 免疫损伤

WNV 进入中枢神经系统的机制尚不清楚;可能涉及逆向轴突运输(Samuel 等,2007)从神经元至脊髓或穿过血脑屏障(Samuel 和 Diamond,2006)。WNV 可能通过感染巨噬细胞或内皮细胞破坏血脑屏障或通过细胞因子表达上调(TNF-α、巨噬细胞移动抑制因子)破坏血脑屏障(Wang 等,2004)。在 WNV 攻击中枢神经系统的过程中,引起神经元死亡或损伤,特别是包括脊髓神经元、皮质神经元、中脑神经元和小脑神经元等在内的靶向区域。在小脑内,WNV 介导 CXCL10 区域性表达上调,吸引具有 CXCR3 受体的 T 细胞,即 CD8+T 细胞,以降低病毒载量(Zhang 等,2008)。小鼠的各项研究显示,在脑中趋化因子 CCL3-5 可与受体 CCR5 高效结合;在一项实验中靶向敲除 CCR5,可导致白细胞的运输数量减少,病毒载量增加及神经元死亡。然而,对 WNV 感染引起脑炎时导致的血管周围 T 细胞浸润了解有限(Bouffard 等,2004)。

7.4.5 诊断

WNV 已被认为对公共卫生构成了威胁。因此，对一些人和动物病例进行密切监测是十分必要的。已经开发出了多种血清学诊断方法用于检测人、蚊虫和非人类脊椎动物宿主中的 WNV。WNV 和其他虫媒病毒感染引起脑炎的临床症状或临床表现类似，大多为无症状或非特异性。为此，对于不同的病原体鉴别应考虑进行实验室诊断。实验室诊断可用于检测人类及其他哺乳动物、鸟类和爬行动物血液、血清、脑脊液和组织中的 WNV(CDC,2013)。通过对 IgM 检测可对 WNV 进行鉴定，目前被美国食品药品监督管理局批准的在市面上销售的四种试检测剂盒可实现对 IgM 进行检测。许多检测方法可用于检测人类样本中的 WNV、WNV 蛋白或 WNV RNA，但这些检测方法很少用于其他脊椎动物和蚊虫的检测(见表 7.1)。

7.4.6 病毒感染与风险

WNV 的传播规模取决于气候条件，包括降水、降水量和最适的环境温度,以及蚊媒的地理分布。若干研究已经证实了环境温度、降水量与尖音库蚊种群数量之间的相关性(Paz 和 Albersheim,2008)。暴雨和洪水会破坏蚊虫在死水中的栖息繁殖地,而干旱条件会促进 WNV 的传播,因为鸟类(增殖宿主）在水资源缺乏期间会与蚊虫共享水源。在另一方面,与 WNV 感染引起的临床症状和临床诊断相关的风险因素也非常关键,因为 WNV 感染大多为无症状感染。人类中仅有 1%的 WNV 病例出现神经后遗症,如脑膜炎、急性弛缓性麻痹和脑炎(Gray 和 Webb,2014)。针对 WNV 感染的诊断可能具有挑战性，因为其临床症状与其他多种病毒感染引起的脑炎临床症状类似,如 JEV 脑炎、库京病毒脑炎和圣路易斯脑炎。基于 IgM 和 IgG 抗体的特异性检测方法可以鉴定病毒。一些风险因素具有宿主特异性,例如,与丙型肝炎病毒混合感染、有心血管疾病史、免疫抑制和慢性肾病等的潜在宿主,都容易被 WNV 感染。感染者的迁移也可能是 WNV 在新地区扩散的原因之一。

7.4.7 WNV 的预防与控制

通过实施综合病媒管理计划，可有效防控人兽共患虫媒病毒性疾病。通过一个健全的监测系统对被感染成年蚊子和幼虫的数量监测，进而通过控制传播媒介来降低感染率。通过改变蚊虫的来源或栖息地,可减少其数量。使用驱蚊剂、浸渍蚊帐、食幼虫鱼、油、空间喷雾器和杀虫剂及防止积水、固体废物管理及坑洼地的消毒等措施来减少蚊虫繁殖点的数量（Haikerwal 等,2017）。预防 WNV 感染最有效的方法是给人类和动物都接种疫苗。然而,目前市场还没有获批的疫苗(Brandler 和 Tangy,2013)。商业化的兽用 WNV 疫苗可用于马和其他哺乳动物。人用疫苗分为临床实验疫苗(见表 7.2)和临床前实验疫苗。各种候选疫苗正在进行临床前的动物模型试验,例如,亚单位疫苗、DNA 疫苗、灭活病毒疫苗、重组病毒载体疫苗、嵌合疫苗、减毒活疫苗等。

7.5 结论

黄病毒所带来的疾病负担使其在全世界范围内都易发生传播流行。因此，在这里对 JEV 和 WNV 进行了细致的讨论;它们最显著的特点是在东南亚国家和西方国家都有流行。通过对病毒重要特征,如病毒背景、发病率和流行率、传播循环路径、免疫生物学、诊断、风险因素及预防和控制等进行调查，可有助于了解其复杂特性。JEV 是 JE 的病原体,JE 是一种儿童疾病（不局限于儿童),可导致儿童发病和死亡。监测系统已报告了 JEV 感染动物的病例,如猪、马、野鸟、牛、山羊等。JEV 的重新出现主要发生在稻田中有适合于蚊虫繁殖的环境、温度和降水量的时候。参与 JEV 传播循环的猪是增殖宿主,水鸟和候鸟是贮存宿主,蚊虫是传播媒介。与 JEV 相似,WNV 也可引起神经系统疾病,如人和马可患脑炎,但更多见于马。WNV 主要在初级宿主和储存宿主的蚊虫和候鸟之间进行传播，人和其他几种哺乳动物则作为偶然宿主和终末宿主。由此,监测系统已检测到大量的 WNV 禽类宿主。这些被感染的候鸟已经将病毒从乌干达传播到中东国家，然后传播到欧洲和北美等地区。由于结构的相似性,两种病毒可以共存。所以,要通过对所有因素的综合考虑和国际合作的方法，实现对这些有着广泛传播媒介病毒的发生及新流行进行有效控制和预防。

7.6 展望

全球化、森林采伐、气候变化、全球贸易和旅行、人口增长及微生物进化等因素是黄病毒出现和重新出现的主要原因。为有效控制黄病毒,需要根除导致病毒传播媒介产生的自然因素和人为因素。目前,蚊虫的滋生地已不局限于温带、热带或亚热带地区;全球化带来的发达的交通运输,使得全世界都将面临黄病毒感染的风险。人兽共患虫媒病毒疾病的预防和控制可以通过建立综合病媒管理计划来完成。健全监测被感染成年蚊子和幼虫数量的系统,可控制病毒感染率的上升。然而,为了减少这些疾病的损失，必须全面了解黄病毒在各种宿主(包括自然宿主/贮存动物)中的免疫生物学特性和发病机制。

参考文献

Apte-Sengupta S, Sirohi D, Kuhn RJ (2014) Coupling of replication and assembly in flaviviruses. Curr Opin Virol 9:134–142

Arjona A, Foellmer HG, Town T, Leng L, McDonald C, Wang T, Wong SJ, Montgomery RR, Fikrig E, Bucala R (2007) Abrogation of macrophage migration inhibitory factor decreases West Nile virus lethality by limiting viral neuroinvasion. J Clin Invest 117(10):3059–3066

Benelli G, Jeffries CL, Walker T (2016) Biological control of mosquito vectors: past, present, and future. Insects 7(4):E52

Bouffard JP, Riudavets MA, Holman R, Rushing EJ (2004) Neuropathology of the brain and spinal cord in human West Nile virus infection. Clin Neuropathol 23(2):59–61
Brandler S, Tangy F (2013) Vaccines in development against West Nile virus. Viruses 5(10):2384–2409
Broutet N, Krauer F, Riesen M, Khalakdina A, Almiron M, Aldighieri S, Espinal M, Low N, Dye C (2016) Zika virus as a cause of neurologic disorders. N Engl J Med 374(16):1506–1509
Byrne SN, Halliday GM, Johnston LJ, King NJ (2001) Interleukin-1β but not tumor necrosis factor is involved in West Nile virus-induced Langerhans cell migration from the skin in C57BL/6 mice. J Investig Dermatol 117(3):702–709
Camenga DL, Nathanson N, Cole GA (1974) Cyclophosphamide-potentiated West Nile viral encephalitis: relative influence of cellular and humoral factors. J Infect Dis 130(6):634–641
Campbell GL, Hills SL, Fischer M, Jacobson JA, Hoke CH, Hombach JM, Marfin AA, Solomon T, Tsai TF, Tsu VD, Ginsburg AS (2011) Estimated global incidence of Japanese encephalitis: a systematic review. Bull World Health Organ 89(10):766–774, 774A–774E
Cappelle J, Duong V, Pring L, Kong L, Yakovleff M, Prasetyo DB, Peng B, Choeung R, Duboz R, Ong S, Sorn S, Dussart P, Tarantola A, Buchy P, Chevalier V (2016) Intensive circulation of Japanese encephalitis virus in peri-urban sentinel pigs near Phnom Penh, Cambodia. PLoS Negl Trop Dis 10(12):e0005149
Centers for Disease Control and Prevention (2013) West Nile virus in the United States: guidelines for surveillance, prevention, and control. https://www.cdc.gov/westnile/resources/pdfs/wnvguidelines.pdf. Accessed 28-2-2018
Centers for Disease Control and Prevention (CDC) (2002) West Nile virus activity—United States, 2001. MMWR Morb Mortal Wkly Rep 51(23):497–501
Chen YY, Lin JW, Fan YC, Chiou SS (2014) Detection and differentiation of genotype I and III Japanese encephalitis virus in mosquitoes by multiplex reverse transcriptase-polymerase chain reaction. Transbound Emerg Dis 61(1):37–43
Danis K, Papa A, Theocharopoulos G, Dougas G, Athanasiou M, Detsis M, Baka A, Lytras T, Mellou K, Bonovas S, Panagiotopoulos T (2011) Outbreak of West Nile virus infection in Greece, 2010. Emerg Infect Dis 17(10):1868–1872
Davis CT, Beasley DW, Guzman H, Siirin M, Parsons RE, Tesh RB, Barrett AD (2004) Emergence of attenuated West Nile virus variants in Texas, 2003. Virology 330(1):342–350
Dhanze H, Bhilegaonkar KN, Ravi Kumar GV, Thomas P, Chethan Kumar HB, Suman Kumar M, Rawat S, Kerketta P, Rawool DB, Kumar A (2015) Comparative evaluation of nucleic acid-based assays for detection of Japanese encephalitis virus in swine blood samples. Arch Virol 160(5):1259–1266
Diamond MS, Sitati EM, Friend LD, Higgs S, Shrestha B, Engle M (2003) A critical role for induced IgM in the protection against West Nile virus infection. J Exp Med 198(12):1853–1862
Erlanger TE, Weiss S, Keiser J, Utzinger J, Wiedenmayer K (2009) Past, present, and future of Japanese encephalitis. Emerg Infect Dis 15(1):1–7
Fulmali PV, Sapkal GN, Athawale S, Gore MM, Mishra AC, Bondre VP (2011) Introduction of Japanese encephalitis virus genotype I, India. Emerg Infect Dis 17(2):319–321
Garcia-Tapia D, Hassett DE, Mitchell WJ Jr, Johnson GC, Kleiboeker SB (2007) West Nile virus encephalitis: sequential histopathological and immunological events in a murine model of infection. J Neurovirol 13(2):130–138
Garg RK, Malhotra HS, Jain A, Kumar N (2017) Dengue encephalopathy: very unusual neuroimaging findings. J Neurovirol 23(5):779–782
Gray TJ, Webb CE (2014) A review of the epidemiological and clinical aspects of West Nile virus. Int J Gen Med 7:193–203
Gulati BR, Singha H, Singh BK, Virmani N, Khurana SK, Singh RK (2011) Serosurveillance for Japanese encephalitis virus infection among equines in India. J Vet Sci 12(4):341–345
Gupta N, Hegde P, Lecerf M, Nain M, Kaur M, Kalia M, Vrati S, Bayry J, Lacroix-Desmazes S, Kaveri SV (2014) Japanese encephalitis virus expands regulatory T cells by increasing the expression of PD-L1 on dendritic cells. Eur J Immunol 44(5):1363–1374
Haikerwal A, Bhatt ML, Saxena SK (2017) Reducing the global burden of dengue: steps toward preventive methods. Arch Prev Med 2(1):28–33
Hayes EB, Gubler DJ (2006) West Nile virus: epidemiology and clinical features of an emerging epidemic in the United States. Annu Rev Med 57:181–194
Hobson-Peters J (2012) Approaches for the development of rapid serological assays for surveillance and diagnosis of infections caused by zoonotic flaviviruses of the Japanese encephalitis virus serocomplex. J Biomed Biotechnol 2012:379738
Hurlbut HS, Rizk F, Taylor RM, Work TH (1956) A study of the ecology of West Nile virus in Egypt. Am J Trop Med Hyg 5(4):579–620
Impoinvil DE, Baylis M, Solomon T (2013) Japanese encephalitis: on the One Health agenda. Curr Top Microbiol Immunol 365:205–247
Karabatsos N (1978) Supplement to international catalogue of arboviruses including certain other viruses of vertebrates. Am J Trop Med Hyg 27(2 Pt 2 Suppl):372–440

Karunaratne SH, Hemingway J (2000) Insecticide resistance spectra and resistance mechanisms in populations of Japanese encephalitis vector mosquitoes, *Culex tritaeniorhynchus* and Cx. gelidus, in Sri Lanka. Med Vet Entomol 14(4):430–436
Khatun T, Chatterjee S (2017) Emergence of West Nile virus in West Bengal, India: a new report. Trans R Soc Trop Med Hyg 111(4):178–184
Kilpatrick AM, Randolph SE (2012) Drivers, dynamics, and control of emerging vector-borne zoonotic diseases. Lancet 380(9857):1946–1955
Komar N, Langevin S, Hinten S, Nemeth N, Edwards E, Hettler D, Davis B, Bowen R, Bunning M (2003) Experimental infection of North American birds with the New York 1999 strain of West Nile virus. Emerg Infect Dis 9(3):311–322
Konishi E, Yamaoka M, Kurane I, Mason PW (2000) Japanese encephalitis DNA vaccine candidates expressing premembrane and envelope genes induce virus-specific memory B cells and long-lasting antibodies in swine. Virology 268(1):49–55
Konishi E, Shoda M, Kondo T (2006) Analysis of yearly changes in levels of antibodies to Japanese encephalitis virus nonstructural 1 protein in racehorses in central Japan shows high levels of natural virus activity still exist. Vaccine 24(4):516–524
Kumar P, Sulochana P, Nirmala G, Haridattatreya M, Satchidanandam V (2004) Conserved amino acids 193-324 of non-structural protein 3 are a dominant source of peptide determinants for CD4+ and CD8+ T cells in a healthy Japanese encephalitis virus-endemic cohort. J Gen Virol 85(Pt 5):1131–1143
Kumar S, Chitti SV, Kant R, Saxena SK (2016) Insights into the immunopathogenesis during Japanese encephalitis virus infection. J Immune Serum Biol 3(1):1–4
Kumar K, Arshad SS, Selvarajah GT, Abu J, Toung OP, Abba Y, Bande F, Yasmin AR, Sharma R, Ong BL, Rasid AA, Hashim N, Peli A, Heshini EP, Shah AKMK (2018) Prevalence and risk factors of Japanese encephalitis virus (JEV) in livestock and companion animal in high-risk areas in Malaysia. Trop Anim Health Prod 50(4):741–752
Kuno G, Chang GJ, Tsuchiya KR, Karabatsos N, Cropp CB (1998) Phylogeny of the genus Flavivirus. J Virol 72(1):73–83
Lannes N, Summerfield A, Filgueira L (2017) Regulation of inflammation in Japanese encephalitis. J Neuroinflammation 14(1):158
Larena M, Regner M, Lee E, Lobigs M (2011) Pivotal role of antibody and subsidiary contribution of CD8+ T cells to recovery from infection in a murine model of Japanese encephalitis. J Virol 85(11):5446–5455
Larena M, Regner M, Lobigs M (2013) Cytolytic effector pathways and IFN-γ help protect against Japanese encephalitis. Eur J Immunol 43(7):1789–1798
Li Y, Counor D, Lu P, Duong V, Yu Y, Deubel V (2012) Protective immunity to Japanese encephalitis virus associated with anti-NS1 antibodies in a mouse model. Virol J 9:135
Li K, Phoo WW, Luo D (2014) Functional interplay among the flavivirus NS3 protease, helicase, and cofactors. Virol Sin 29(2):74–85
Li XD, Deng CL, Ye HQ, Zhang HL, Zhang QY, Chen DD, Zhang PT, Shi PY, Yuan ZM, Zhang B (2016) Transmembrane domains of NS2B contribute to both viral RNA replication and particle formation in Japanese encephalitis virus. J Virol 90(12):5735–5749
Lin CL, Chang HL, Lin CY, Chen KT (2017) Seasonal patterns of Japanese encephalitis and associated meteorological factors in Taiwan. Int J Environ Res Public Health 14(11):E1317
Litzba N, Klade CS, Lederer S, Niedrig M (2010) Evaluation of serological diagnostic test systems assessing the immune response to Japanese encephalitis vaccination. PLoS Negl Trop Dis 4(11):e883
Liu WJ, Wang XJ, Clark DC, Lobigs M, Hall RA, Khromykh AA (2006) A single amino acid substitution in the West Nile virus nonstructural protein NS2A disables its ability to inhibit alpha/beta interferon induction and attenuates virus virulence in mice. J Virol 80(5):2396–2404
Lord JS, Gurley ES, Pulliam JR (2015) Rethinking Japanese encephalitis virus transmission: a framework for implicating host and vector species. PLoS Negl Trop Dis 9(12):e0004074
Luo D, Ying H, Yao R, Song J, Wang Z (1995) Socio-economic status and micro-environmental factors in relation to the risk of Japanese encephalitis: a case-control study. Southeast Asian J Trop Med Public Health 26(2):276–279
Mansfield KL, Hernández-Triana LM, Banyard AC, Fooks AR, Johnson N (2017) Japanese encephalitis virus infection, diagnosis and control in domestic animals. Vet Microbiol 201:85–92
Miller S, Krijnse-Locker J (2008) Modification of intracellular membrane structures for virus replication. Nat Rev Microbiol 6(5):363–374
Mukhopadhyay S, Kuhn RJ, Rossmann MG (2005) A structural perspective of the flavivirus life cycle. Nat Rev Microbiol 3(1):13–22
Muller DA, Young PR (2013) The flavivirus NS1 protein: molecular and structural biology, immunology, role in pathogenesis and application as a diagnostic biomarker. Antiviral Res 98(2):192–208
Murali-Krishna K, Ravi V, Manjunath R (1996) Protection of adult but not newborn mice against lethal intracerebral challenge with Japanese encephalitis virus by adoptively transferred virus-

specific cytotoxic T lymphocytes: requirement for L3T4+ T cells. J Gen Virol 77(Pt 4):705–714
Murray CL, Jones CT, Rice CM (2008) Architects of assembly: roles of Flaviviridae non-structural proteins in virion morphogenesis. Nat Rev Microbiol 6(9):699–708
Nah JJ, Yang DK, Kim HH, Song JY (2015) The present and future of veterinary vaccines for Japanese encephalitis in Korea. Clin Exp Vaccine Res 4(2):130–136
Neufeldt CJ, Cortese M, Acosta EG, Bartenschlager R (2018) Rewiring cellular networks by members of the Flaviviridae family. Nat Rev Microbiol 16(3):125–142
Nour AM, Li Y, Wolenski J, Modis Y (2013) Viral membrane fusion and nucleocapsid delivery into the cytoplasm are distinct events in some flaviviruses. PLoS Pathog 9(9):e1003585
Nybakken GE, Oliphant T, Johnson S, Burke S, Diamond MS, Fremont DH (2005) Structural basis of West Nile virus neutralization by a therapeutic antibody. Nature 437(7059):764–769
OIE (2012) Manual of diagnostic tests and vaccines for terrestrial animals, 7th edn. World Health Organization for Animal Health (OIE), Paris
Oliveira ARS, Cohnstaedt LW, Strathe E, Hernández LE, McVey DS, Piaggio J, Cernicchiaro N (2017) Meta-analyses of the proportion of Japanese encephalitis virus infection in vectors and vertebrate hosts. Parasit Vectors 10(1):418
Paul D, Bartenschlager R (2015) Flaviviridae replication organelles: Oh, what a tangled web we weave. Annu Rev Virol 2(1):289–310
Paz S, Albersheim I (2008) Influence of warming tendency on Culex pipiens population abundance and on the probability of West Nile fever outbreaks (Israeli Case Study: 2001–2005). Ecohealth 5(1):40–48
Pérez-Ramírez E, Llorente F, Jiménez-Clavero MÁ (2014) Experimental infections of wild birds with West Nile virus. Viruses 6(2):752–781
Philip Samuel P, Arunachalam N, Hiriyan J, Tyagi BK (2008) Host feeding pattern of Japanese encephalitis virus vector mosquitoes (Diptera: Culicidae) from Kuttanadu, Kerala, India. J Med Entomol 45(5):927–932
Pierson TC, Fremont DH, Kuhn RJ, Diamond MS (2008) Structural insights into the mechanisms of antibody-mediated neutralization of flavivirus infection: implications for vaccine development. Cell Host Microbe 4(3):229–238
Platanias LC (2005) Mechanisms of type-I- and type-II-interferon-mediated signalling. Nat Rev Immunol 5(5):375–386
Poccia F, Agrati C, Martini F, Capobianchi MR, Wallace M, Malkovsky M (2005) Antiviral reactivities of gammadelta T cells. Microbes Infect 7(3):518–528
Ramesh D, Muniaraj M, Samuel PP, Thenmozhi V, Venkatesh A, Nagaraj J, Tyagi BK (2015) Seasonal abundance & role of predominant Japanese encephalitis vectors Culex tritaeniorhynchus & Cx. gelidus Theobald in Cuddalore district, Tamil Nadu. Indian J Med Res 142(Suppl):S23–S29
Ricklin ME, García-Nicolás O, Brechbühl D, Python S, Zumkehr B, Nougairede A, Charrel RN, Posthaus H, Oevermann A, Summerfield A (2016a) Vector-free transmission and persistence of Japanese encephalitis virus in pigs. Nat Commun 7:10832
Ricklin ME, Garcìa-Nicolàs O, Brechbühl D, Python S, Zumkehr B, Posthaus H, Oevermann A, Summerfield A (2016b) Japanese encephalitis virus tropism in experimentally infected pigs. Vet Res 47:34
Rosen L, Lien JC, Shroyer DA, Baker RH, Lu LC (1989) Experimental vertical transmission of Japanese encephalitis virus by Culex tritaeniorhynchus and other mosquitoes. Am J Trop Med Hyg 40(5):548–556
Samuel MA, Diamond MS (2006) Pathogenesis of West Nile Virus infection: a balance between virulence, innate and adaptive immunity, and viral evasion. J Virol 80(19):9349–9360
Samuel MA, Wang H, Siddharthan V, Morrey JD, Diamond MS (2007) Axonal transport mediates West Nile virus entry into the central nervous system and induces acute flaccid paralysis. Proc Natl Acad Sci U S A 104(43):17140–17145
Sardelis MR, Turell MJ, Dohm DJ, O'Guinn ML (2001) Vector competence of selected North American Culex and Coquillettidia mosquitoes for West Nile virus. Emerg Infect Dis 7(6):1018–1022
Schuh AJ, Ward MJ, Brown AJ, Barrett AD (2013a) Phylogeography of Japanese encephalitis virus: genotype is associated with climate. PLoS Negl Trop Dis 7(8):e2411
Schuh AJ, Guzman H, Tesh RB, Barrett AD (2013b) Genetic diversity of Japanese encephalitis virus isolates obtained from the Indonesian archipelago between 1974 and 1987. Vector Borne Zoonotic Dis 13(7):479–488
Schuh AJ, Ward MJ, Leigh Brown AJ, Barrett AD (2014) Dynamics of the emergence and establishment of a newly dominant genotype of Japanese encephalitis virus throughout Asia. J Virol 88(8):4522–4532
Shi G, Suzuki T (2018) Molecular basis of encapsidation of hepatitis C virus genome. Front Microbiol 9:396
Simmonds P, Becher P, Bukh J, Gould EA, Meyers G, Monath T, Muerhoff S, Pletnev A, Rico-Hesse R, Smith DB, Stapleton JT, ICTV Report Consortium (2017) ICTV virus taxonomy

profile: Flaviviridae. J Gen Virol 98(1):2–3
Smithburn KC, Hughes TP, Burke AW, Paul JH (1940) A neurotropic virus isolated from the blood of a native of Uganda1. Am J Trop Med Hyg 1(4):471–492
Solomon T (2004) Flavivirus encephalitis. N Engl J Med 351(4):370–378
Solomon T, Ni H, Beasley DW, Ekkelenkamp M, Cardosa MJ, Barrett AD (2003) Origin and evolution of Japanese encephalitis virus in Southeast Asia. J Virol 77(5):3091–3098
Su CL, Yang CF, Teng HJ, Lu LC, Lin C, Tsai KH, Chen YY, Chen LY, Chang SF, Shu PY (2014) Molecular epidemiology of Japanese encephalitis virus in mosquitoes in Taiwan during 2005–2012. PLoS Negl Trop Dis 8(10):e3122
Tian HY, Bi P, Cazelles B, Zhou S, Huang SQ, Yang J, Pei Y, Wu XX, Fu SH, Tong SL, Wang HY, Xu B (2015) How environmental conditions impact mosquito ecology and Japanese encephalitis: an eco-epidemiological approach. Environ Int 79:17–24
Turell MJ, Sardelis MR, Dohm DJ, O'Guinn ML (2001) Potential North American vectors of West Nile virus. Ann N Y Acad Sci 951:317–324
Turtle L, Bali T, Buxton G, Chib S, Chan S, Soni M, Hussain M, Isenman H, Fadnis P, Venkataswamy MM, Satishkumar V, Lewthwaite P, Kurioka A, Krishna S, Shankar MV, Ahmed R, Begum A, Ravi V, Desai A, Yoksan S, Fernandez S, Willberg CB, Kloverpris HN, Conlon C, Klenerman P, Satchidanandam V, Solomon T (2016) Human T cell responses to Japanese encephalitis virus in health and disease. J Exp Med 213(7):1331–1352
van den Hurk AF, Ritchie SA, Mackenzie JS (2009) Ecology and geographical expansion of Japanese encephalitis virus. Annu Rev Entomol 54:17–35
Wang H, Liang G (2015) Epidemiology of Japanese encephalitis: past, present, and future prospects. Ther Clin Risk Manag 11:435–448
Wang T, Town T, Alexopoulou L, Anderson JF, Fikrig E, Flavell RA (2004) Toll-like receptor 3 mediates West Nile virus entry into the brain causing lethal encephalitis. Nat Med 10(12):1366–1373
Wang P, Hu K, Luo S, Zhang M, Deng X, Li C, Jin W, Hu B, He S, Li M, Du T, Xiao G, Zhang B, Liu Y, Hu Q (2016) DC-SIGN as an attachment factor mediates Japanese encephalitis virus infection of human dendritic cells via interaction with a single high-mannose residue of viral E glycoprotein. Virology 488:108–119
Ward MP, Schuermann JA, Highfield LD, Murray KO (2006) Characteristics of an outbreak of West Nile virus encephalomyelitis in a previously uninfected population of horses. Vet Microbiol 118(3–4):255–259
WHA (2016) Fact sheet: EXOTIC-Japanese encephalitis. https://wildlifehealthaustralia.com.au/Portals/0/Documents/FactSheets/Exotic/EXOTIC%20-%20Japanese%20Encephalitis%20Mar%202016%20(2.0).pdf
WHO (2019) Japanese encephalitis. Fact sheet. http://www.who.int/mediacentre/factsheets/fs386/en/. Accessed 6-3-2018
Zhang B, Chan YK, Lu B, Diamond MS, Klein RS (2008) CXCR3 mediates region-specific antiviral T cell trafficking within the central nervous system during West Nile virus encephalitis. J Immunol 180(4):2641–2649
Zhang H, Rehman MU, Li K, Luo H, Lan Y, Nabi F, Zhang L, Iqbal MK, Zhu S, Javed MT, Chamba Y, Li JK (2017) Epidemiologic survey of Japanese encephalitis virus infection, Tibet, China, 2015. Emerg Infect Dis 23(6):1023–1024
Zou G, Chen YL, Dong H, Lim CC, Yap LJ, Yau YH, Shochat SG, Lescar J, Shi PY (2011) Functional analysis of two cavities in flavivirus NS5 polymerase. J Biol Chem 286(16):14362–14372

第 8 章 环状病毒

Sushila Maan, Manjunatha N. Belaganahalli, Narender S. Maan, Houssam Attoui, Peter P. C. Mertens

8.1 引言

环状病毒(orbivirus)是 dsRNA 病毒的一个大的种群,属于呼肠孤病毒科,环状病毒属。ICTV 已登记了 27 种环状病毒,包含至少 161 种血清型。本章主要基于对典型环状病毒——“蓝舌病病毒”的研究展开概述。环状病毒的直径为 60~80nm,外观呈球状,含有 10 个基因组片段(Mertens 等,2005b)。环状病毒是“真正的”虫媒病毒(它们在节肢动物媒介和脊椎动物宿主中均可复制),并通过吸血节肢动物作为传播媒介进行传播,包括库蠓、蜱、白蛉和蚊(尽管一些环状病毒的传播媒介尚未确定)。一些环状病毒也可以通过脊椎动物宿主进行垂直传播或通过口腔途径进行水平传播(Batten 等,2014;Menzies 等,2008;Prasad 等,2007)。

总的来说,环状病毒可供选择的宿主有很多种,包括人类、家养和野生的反刍动物、猫科动物、犬科动物、马属动物、有袋动物、树懒、蝙蝠和鸟类。尽管一些环状病毒可以感染人,但一些种类的环状病毒主要还是对饲养动物和野生动物造成重大的经济影响,包括 BTV、AHSV、流行性出血热病毒(epizootic haemorrhagic disease virus,EHDV)和马炎脑病毒(equine encephalosis virus,EEV)。一些感染野生动物的环状病毒也可能影响野生动物种群数量的可持续发展。环状病毒在不同媒介和宿主中复制的能力,以及病毒的数量(血清型),可能有助于解释它们在不同生态系统中普遍存在的原因。因此,它们也可能比其他种类的病毒具有更大的“发生”风险。

将环状病毒引入非疫区和无免疫力易感宿主动物群,可引起高发病率和死亡率。2006 年北欧暴发的由 BTV-8(Elbers 等,2008a;Wilson 等,2007)、欧本南吉病毒(Eubenangee virus)引起的尤金袋鼠猝死综合征,以及沃洛尔和沃里戈病毒(Wallal and Warrego virus)引起的袋鼠流行性失明都证实了这一观点(Hooper 等,1999;Reddacliff 等,1999)。此外,通过血清学已经证实新型环状病毒,如中点环状病毒(middle point orbivirus,MPOV)(Cowled 等,2007)和拉伸环礁湖环状病毒(Stretch Lagoon orbivirus,SLOV)在澳大利亚的家畜和马属动物中存在(Cowled 等,2009),表明这些病毒很可能对家畜和人类健康构成巨大威胁。然而,这些病毒也可能与已经被确认的环状病毒或更进一步地与新的尚未分类的环状病毒共同引发严重的问题。

8.2 环状病毒

《出埃及记》中将埃及的第五次瘟疫描述为一场非常严重的“家畜瘟疫”。这可能是由环状病毒引起的真正意义上的动物流行病的首次记载(Marr 和 Malloy,1996)。

18 世纪晚期,法国生物学家 Francois de Vaillanthad 首次将蓝舌病描述为一种牛和羊的疫病,并将其命名为“tong-sikte”(Gutsche,1979)。然而,Hutcheon(1902)首次对绵羊 tong-sikte 的临床症状进行了详细的科学描述。最初,Spreull(1902)将该病命名为“疟疾卡他热”,但后来,基于感染严重的绵羊的口舌出现青紫现象,建议将该病的名称变更为“bluetongue”蓝舌病(Spreull,1905)。1943 年,发生了第一起被证实的非洲以外地区塞浦路斯蓝舌病,结果造成幼龄绵羊种群的死亡率非常高,虽然有报告称,在 1924 年,塞浦路斯就曾经发生过蓝舌病,但未经确认(Gambles,1949;Polydorou,1985)。

另一种重要的环状病毒疫病是非洲马瘟(AHS),1327 年在也门发生大流行(Moule,1896)。有报道认为该病毒来自非洲(Theal,1899)。虽然 AHS 病在 1719 年被首次确认,并导致 1700 多只动物死亡,但是 AHSV 可能很久以前就存在于南非(Henning,1956)。AHS 的致死率极高(>95%),并被认为在 17 世纪和 18 世纪妨碍了早期在南非定居的荷兰殖民者使用马匹。直到最近，已确认AHSV主要存在于撒哈拉沙漠以南的非洲，斑马是其主要的野生动物宿主。然而,在北非、伊比利亚和阿拉伯半岛也发现了病毒的周期性入侵。1959—1961年,在中东地区和亚洲国家检测到了该病毒,并造成了马的大量死亡(Mellor和Hamblin,2004;Rafyi,1961)。

自1890年以来，流行性出血热(epizootic haemorrhagic disease,EHD)就一直存在于美国东南部的野生反刍动物中，并被伐木工人和猎人称为"黑舌病",但直到1955年,在白尾鹿中发生致死性的流行病时才首次被正式确认（Shope等,1955)。

8.3 分类

1950年以后，多种类似于BTV的病毒被分离到,这些病毒要么来自脊椎动物宿主,要么来自昆虫媒介，通常以首次分离地被命名(Karabatsos,1985)。1971年,Murphy等人和Borden等人得出结论,这些相关的BTV的分类参数(形态学、物理化学和血清学)与其他节肢动物传播的病毒,以及任何其他已确认的病毒种群都不相同，表明它们代表了一个新的特殊种群。通过电子显微镜负染观察，可见环状病毒核心粒子的表面结构是由一连串相互连结的"环状"衣壳组成(Borden等,1971b;Murphy等,1971)。因此,"orbivirus"(在拉丁语中"orbis"意为"环"或"圈")这个名称被用于命名一种类似于BTV的新型节肢动物传播的病毒,并与BTV一起作为环状病毒的"典型代表"病毒(Borden等,1971b)。后来,在1976年,ICTV正式确认了环状病毒属(Fenner,1976)及BTV是对经济产生重大影响病毒的典型。

环状病毒属是呼肠孤病毒科(Reoviridae)15个属中的1种。呼肠孤病毒科的病毒呈二十面体，基因组由9~12个dsRNA线性片段组成，总基因组大小为19~32kb（表8.1)(Attoui等,2005b;Mertens等,2005a)。根据病毒的亚核心"T2"层表面是否存在"螺旋(turrets)"或"刺突(spike)",呼肠孤病毒科的属被分为两个亚科，即棘突呼肠孤病毒亚科(spinareovirinae)(拉丁语spina意为"刺"或"刺突")和光滑呼肠孤病毒亚科(sedoreovirinae)(拉丁语sedo意为"光滑")。属于光滑呼肠孤病毒亚科的环状病毒属是呼肠孤病毒科中最大的属,目前包含22种已被确认的病毒,以及15种尚未被分类的"环状病毒"(Cowled等,2007,2009;Karabatsos,1985;Mertens等,2005b)。通过补体结合试验(complement fixation,CF)、琼脂凝胶免疫扩散试验(agar gel immunodiffusion,AGID)、荧光抗体试验（fluorescent antibody,FA）或ELISA检测，发现每种环状病毒都具有共同的抗原(Gorman等,1983;Mertens等,2005b)。由于它们具有共同的抗原,根据血清学交叉反应,将环状病毒分为不同的"血清群"(Borden等,1971a;Della-Porta,1985;Moore和Lee,1972)，这等同于ICTV目前承认的不同种类病毒(Mertens等,2005b)。然而,在某些情况下,例如,在BTV和EHDV血清群中，在一些血清学试验和血清学交叉反应研究中观察到低水平或"单向"交叉反应(Della-Porta等,1985;Huismans等,1979;Moore,1974),使得个别病毒很难通过血清学鉴定进行诊断和分类(Borden等,1971b;Della-Porta等,1985;Moore,1974;Moore和Lee,1972)。

由于病毒的"种",属于所有生物学分类中最基本的分类范畴，所以为了对呼肠孤病毒科中的不同属和种进行鉴定,ICTV设定了许多不同的"多元参数"(Attoui等,2011;Mertens等,2005b;van Regenmortel和Mahy,2004)。呼肠孤病毒科中决定病毒种类的首要因素是不同毒株在共感染(这一过程被称为重组)过程中,相互间进行基因片段交换的能力,从而产生新的子代病毒。然而,在缺乏呼肠孤毒特定分离株重组数据的情况下,也可单独或联合使用其他参数来鉴定相同种类病毒。因此,不同种类病毒也可通过其他参数进行鉴

表8.1　呼肠孤病毒科的属

属	结构	基因组片段的编号(基因组大小)	种的编号(+未分类的种)	宿主(媒介)
棘突呼肠孤病毒亚科(spinareovirinae)				
水生呼肠孤病毒属(aquareovirus)	螺旋形	11(约23.7kb)	6(+5)	软体动物、有鳍鱼类、甲壳类动物
科罗拉多蜱传热病毒属(coltivirus)	螺旋形	12(约29kb)	2(+1)	哺乳动物(包括人类)(蜱)
昆虫质型多角体病毒属(cypovirus)	螺旋形	10(24.8~33.3kb)[a]	21(+2)	昆虫(鳞翅目、双翅目和膜翅目),已报告出来自淡水水蚤的单一分离株
迪诺维纳呼肠孤病毒属(dinovernavirus)	螺旋形	9(约23.35kb)	1	蚊子
斐济病毒属(fijivirus)	螺旋形	10(约28.7kb)	5	植物(乔木科、百合科)(飞虱)
昆虫非包涵体呼肠孤病毒属(idnoreovirus)	螺旋形	10+1[b](约25.1kb)	5(+1)	昆虫(膜翅目)
真菌呼肠孤病毒属(mycoreovirus)	螺旋形	11或12(约24.4kb)	3	真菌
水稻病毒属(oryzavirus)	螺旋形	10(约26.1kb)	2	植物(禾本科)(飞虱)
正呼肠孤病毒属(orthoreovirus)	螺旋形	10(约23.5kb)	5	鸟类、爬行动物、哺乳动物
光滑呼肠孤病毒亚科(sedoreovirinae)				
甲壳动物呼肠孤病毒属(cardoreovirus)	非螺旋形	12(约20.15kb)	1(+2)	蟹
环状病毒属(orbivirus)	非螺旋形	10(约19.2kb)	22(+14)	哺乳动物(包括人类)、鸟类(库蠓,蚊子、白蛉、蜱)
原生生物微胞藻呼肠孤病毒属(mimoreovirus)	非螺旋形	11(约25.56kb)	1	浮游植物
植物呼肠孤病毒属(phytoreovirus)	非螺旋形	12(约25.1kb)	3(+1)	植物(叶蝉类)
轮状病毒属(rotavirus)	非螺旋形	11(约18.5kb)	5(+2)	鸟类、哺乳动物
东南亚十二RNA病毒属(seadornavirus)	非螺旋形	12(约21kb)	3	哺乳动物(包括人类)(蚊子)
尚未分类呼肠孤病毒		10,11或12	5	

[a] 在某些情况下,仅通过电泳分析估计不同种昆虫质型多角体病毒的基因组大小。

[b] 个别昆虫非包涵体呼肠孤病毒粒子可能包含第11个基因组片段,这取决于其来源宿主黄蜂的性别和染色体倍性。

Reproduced with some modification from Attoui et al.(2009a)

定,包括病毒保守蛋白间的血清学交叉反应;通过序列分析检测保守的RNA和蛋白;或RNA交叉杂交。在每种环状病毒内,更确切地说是在每个属内,保守末端(六核苷酸)存在于所有基因组片段上。在电泳过程中(琼脂糖凝胶中),每种环状病毒基因序列的迁移模式和大小是保守的。相似的宿主范围、临床症状和昆虫媒介种类也有助于鉴别不同种类病毒。每种环状病毒也包含着不同血清型的病毒,这可以通过其在血清中和试验或病毒中和试验进行特异性鉴定,从而确定其病毒种类(Gould 和 Eaton,1990;Mertens 等,2005b)。基于保守基因组片段的交叉杂交或基于 RT-PCR 测定(例如,Seg-1,编码病毒聚合酶或 Seg-3,编码亚核心 T2 蛋白)的序列比较,已用于鉴定相关的环状病毒。最近,根据这些标准已鉴定出 22 种环状病毒,以及 15 种尚未分类的病毒,可能意味着它们是新增的病毒种类(Attoui 等,2011;Cowled 等,2009;Martins 等,2007;Mertens 等,2005b;Vieira Cde 等,2009)。

然而,最近随着更快速和更可靠测序技术的发展,以及相关数据库的建立,"保守"基因的核苷酸序列与之前测定的病毒核苷酸序列的系统发育比较已成为鉴定新病毒的主要方法,包括新环状病毒的分离(Anthony 等,2007;Aradaib,2009;Aradaib 等,2009;Attoui 等,2005a;Belaganahalli,2012;Cowled 等,2007;Maan 等,2011c,2012;Shaw 等,2007)(表 8.2)。这些研究为环状病毒的分类提供了相关信息,并确定了早期用血清学试验无法鉴别的地区型(区域变异体)病毒。编码外衣壳蛋白的易于突变的基因组片段的系统发育树比较也有助于鉴别和提供关于每种环状病毒的不同血清型信息(表 8.2)(Anthony 等,2009c;Maan 等,2007)。

8.4 病毒特性

8.4.1 理化性质

环状病毒颗粒的分子量(Mr)约为 10.8×10^7,"核心"的 Mr 约为 6.7×10^7。病毒粒子和核心在CsCl中的浮力密度分别为 $1.36g/cm^3$ 和 $1.40g/cm^3$。标准沉降常数 S20W 分别为 550S(病毒粒子)和 470S(核心)(Mertens 等,2005b)。环状病毒对洗涤剂和有机溶剂有较强抵抗力,但根据病毒种类的不同,其对洗涤剂的敏感性也不同。十二烷基硫酸钠通常能损坏病毒粒子并破坏其感染性(Gorman,1978,Mertens 等,2005b)。在 pH 值为 8~9 时,环状病毒传染性稳定,但在 pH 值<6.5 或 pH 值>10.2,尤其是在较低 pH 值环境时,病毒粒子传染性显著降低,这是由于外层衣壳蛋白受到破坏(Mertens 等,2005b)。在低 pH 值条件下(pH 值<5.0),病毒粒子逐渐被破坏,当 pH 值降到 3.0 时,病毒粒子完全丧失传染性(Mertens 等,2005b)。根据环状病毒对酸的不稳定性、对有机溶剂的轻度敏感性及其血清学特性,可将其与正呼肠孤病毒和轮状病毒区别开来(Borden 等,1971b;Gorman,1978;Roy,2005)。环状病毒保存在温度 4 ℃,pH 值为 8.0 的 0.1 M Tris-HCl 中,1 年内都可保持高度感染性,但在 60℃时可被迅速灭活。冷冻保存病毒粒子会受到损坏,传染性降低约 90%(Mertens 等,2005b)。使用适当的糖组合(包括海藻糖)进行冻干保存,可保持病毒稳定,防止其在高温下丧失传染性(甚至高达 100℃)。这些病毒特有的理化特性包括:①分段 dsRNA 基因组;②对脱氧胆酸盐和脂质溶剂存在相对耐受;③在酸性 pH 值条件下不稳定。

8.4.2 生物学特性

主要是基于对 BTV 的研究,环状病毒在大多数哺乳动物细胞系内的复制通常发生细胞裂解,并形成 CPE (Karabatsos,1985;Mertens 等,1987)。病毒粒子从被感染的哺乳动物细胞释放出来或通过细胞损伤释放出来,再或者通过出芽方式,形成细胞膜包囊膜病毒颗粒释放出来(Celma 和 Roy,2009;Hyatt 等,1989,1993;Mertens 等,2005b)。环状病毒也可引起细胞死亡和裂解,从哺乳动物细胞内释放出来。然而,BTV 在一些昆虫细胞中培养,BTV 复制通常不会引起细胞裂解或 CPE (Mertens 等,1996,2005b)。它们从昆虫细胞释放出来的机制还不能确定,似乎与病毒小囊膜膜蛋

表8.2 环状病毒种类列表

序号	种类	缩写	血清型编号	年份	脊椎动物宿主	媒介
1	非洲马瘟病毒(African horse sickness virus)	AHSV	9	1989	马和犬	库蠓
2	蓝舌病病毒(bluetongue virus)	BTV	26	1905	家养或野生反刍动物	库蠓
3	张格罗拉病毒(Changuinola virus)	CGLV	12	1960	树懒、犰狳、穿山甲、啮齿动物和人(捕蚊者)	白蛉
4	秦纽达病毒(Chenuda virus)	CNUV	7	1954		
5	乔巴峡病毒(Chobar Gorge virus)	CGV	2	1970	蝙蝠(分离)和家畜(血清学)	蜱
6	科里帕塔病毒(Corriparta virus)	CORV	6	1960	水鸟、人类和家禽(血清学)	库蚊
7	流行性出血热病毒(epizooc haemorrhagic disease virus)	EHDV	7	1955	白尾鹿和牛	库蠓
8	马脑炎病毒(equine encephalosis virus)	EEV	7	1967	马	库蠓
9	欧本南吉病毒(Eubenangee virus)	EUBV	4	1963	袋鼠、沙袋鼠和牛	蚊子和库蠓
10	大岛病毒(Great Island virus)	GIV	34[b]	1971	海鸟、家畜、马和人	蜱
11	伊理病毒(Ieri virus)	IERIV	3	1955	鸟类	蚊子
12	莱邦博病毒(Lebombo virus)	LEBV	1	1956	人、啮齿动物	蚊子
13	奥伦盖病毒(Orungo virus)	ORUV	4	1959	人(分离)、羊、牛、山羊和骆驼(血清学)	蚊子
14	巴尼亚姆病毒(Palyam virus)	PALV	13	1956	家养或野生反刍动物	库蠓、蚊子和蜱
15	秘鲁马病病毒(Peruvian horse sickness virus)	PHSV	1	1997	马	蚊子
16	圣克罗伊河病毒(St Croix River virus)	SCRV	1	1994	蜱?	蜱
17	尤马蒂拉病毒(Umalla virus)	UMAV	4	1969	鸟类	蚊子
18	沃里戈病毒(Warrego virus)	WARV	3	1969	有袋动物(分离)和牛(血清学)	库蠓、蚊子
19	沃洛尔病毒(Wallal virus)	WALV	3	1970	有袋动物	库蠓
20	瓦德麦达尼病毒(Wad Medani virus)	WMV	2	1952	家畜和啮齿动物	蜱
21	王戈尔病毒(Wongorr virus)	WGRV	8	1970	牛和有袋动物	库蠓、蚊子
22	云南环状病毒(Yunnan orbivirus)	YUOV	2	1997	牛、羊和驴	蚊子
23	克麦罗沃病毒(Kemorovo virus)	KEMV	2	1962	人类、鸟类、马、牛、海鸟和小型哺乳动物	蜱
24	帕塔病毒(Pata virus)	PATAV	1	1968	宿主未知	蚊子或库蠓?
25	安达西贝病毒(Andasibe virus)	ANDV	1	1979	宿主未知	蚊子
26	杰潘纳特病毒(Japanaut virus)	JAPV	1	1965	蝙蝠和一些未知宿主	蚊子
27	马图凯亚病毒(Matucare virus)	MATV	1	1963	宿主未知	蜱

(待续)

表 8.2(续)

序号	种类	缩写	血清型编号	年份	脊椎动物宿主	媒介
	暂定种类					
1	科达雅斯病毒(Codajas virus)	COV		1984	宿主未知	蚊子
2	伊费病毒(Ife virus)	IREV		1971	蝙蝠、啮齿动物、鸟类、反刍动物	蚊子
3	伊图皮兰加病毒(Itupiranga virus)	ITUV		1976	宿主未知	蚊子
4	凯马范帕塔病毒(Kammavanpettai virus)	KMPV		1963	鸟类	未知
5	克拉伦登湖病毒(Lake Clarendon virus)	LCV		1981	鸟类、牛(血清学)	蜱
6	布鲁乌布朗库病毒(Breu Branco virus)			1988	未知	蚊子
7	米那库病毒(Minacu virus)			1996	宿主未知	蚊子
8	果洛病毒(Golok virus)			1981	宿主未知	蚊子

[a] 种内原血清型首次分离的年份。根据 Belhouchet 等人(2010)和 Dilcher 等人(2012),本研究鉴别的新种属为序号 24~27 示。克美罗沃病毒被归类为特殊种类。

[b] KEMV 和 TRBV 已从大岛病毒种中删除。

[c] 最近发现的物种(Belaganahalli,2012)。

白 NS3 有关 (由 Seg-10 编码)(Hyatt 等,1989,1993)。

用糜蛋白酶或胰蛋白酶对 BTV 进行处理可产生感染性亚病毒粒子 (其中 VP2 被裂解)(Mertens 等,1987,1996)。这些感染性亚病毒粒子对昆虫细胞系,如 KC 细胞(来自 sonorensis 库蠓)和 C6/36 细胞(来源于白纹伊蚊)的特异性感染显著高于哺乳动物细胞 (BHK-21)(Mertens 等,1987,1996)。与完整纯化的病毒不同,这些 BTV 的感染性亚病毒粒子不具有血凝活性, 也不能凝聚,但对成熟昆虫媒介和昆虫媒介细胞系(KC 细胞)具有很高的感染性(Mertens 等,1996)。

8.5 基因组与编码蛋白

环状病毒基因组由 10 个线性 dsRNA 片段组成,大小为 0.8~3.9kb(线性 dsRNA 片段 1~10),并以精确的等摩尔比进行包装(Huismans 等,1979;Mertens 等,2005b;Verwoerd 等,1970)。相反,利用聚丙烯酰胺凝胶电泳对 dsRNA 基因组片段进行分离时,不仅取决于分子量,更主要的是受到核苷酸序列和 RNA 结构的影响, 但在利用 AGE 进行分离时,只与分子量有关(Maan,2004)。能够产生 AGE“电泳型”,是环状病毒属的一个特性,反映了环状病毒基因组片段的大小/分子量的高度保守性 (Belaganahalli 等,2015;Gonzalez 和 Knudson,1988)。根据 BTV 基因组片段在 1%琼脂糖凝胶中的一致性迁移和变性 dsRNA 在体外翻译,将 BTV 基因组片段分配至其编码的蛋白质(Mertens 等,1984)。

同一种环状病毒分离株的基因组末端通常是有 6 个完全保守的核苷酸序列,但在不同种类环状病毒间的保守性较低(Mertens 和 Diprose,2004;Mertens 和 Sangar,1985;Mertens 等,2005b;Rao等,1983)。每个环状病毒基因组片段的第一个和最后两个核苷酸是反向互补的, 可能参与控制病毒 mRNA 的翻译效率(Roy,1989)。

最近研究已经证实，短的特异性片段保守区域在基因组组装和病毒复制过程中参与不同基因组片段之间的交叉杂交(Boyce 等,2016)。

与其他呼肠孤病毒科成员一样，大多数环状病毒基因组片段被认为是单顺反子的。然而,BTV 由 Seg-10 编码的两个相关蛋白 (NS3 和 NS3a),虽然是同一个 ORF，但起始密码子不同(French 等,1989;Mertens 等,1984,2005b;Wade-Evans,1990;Wade-Evans 等,1992;Wu 等,1992)。小的非结构蛋白由 Seg-9 和 Seg-10 的 ORF 交替翻译而来,分别生成 NS4 和 NS5 蛋白(Belhouchet 等,2010;Ratinier 等,2016;Stewart 等,2015)。

在被环状病毒感染的细胞中发现了 12 种不同的病毒特异性蛋白(Belhouch 等,2011;Gorman 等,1981,1983;Ratinier 等,2011)。BTV 有 7 种结构蛋白(VP1~VP7)(组织结构为 3 层二十面体衣壳）和 5 种非结构蛋白 (NS1~NS5)(Belhouchet 等,2011;Mertens 等,2005b;Stewart 等,2015;Stewart 和 Roy,2010;Verwoerd 等,1972)。VP1~VP7 结构蛋白分别由 BTV 基因组片段-1、-2、-3、-4、-6、-7 和-9 编码,而非结构蛋白(NS1~NS5)分别由 Seg-5、-8、-10、-9 和-10 编码(Belhouchet 等,2011;Huismans,1979;Huismans 等,1979;Mecham 和 Dean,1988;Mertens 等,1984,2005b;Owens 等,2004;Ratinier 等,2011;Verwoerd 等,1972)。

环状病毒结构蛋白在病毒复制过程中发挥着重要的作用,包括与受体结合,并进入细胞;病毒 mRNA 的转录、加帽和甲基化;衣壳形成;基因组衣壳化，以及正链和负链 RNA 合成 (Owens 等,2004)。7 种结构蛋白中的 2 种(VP2 和 VP5)组成了外衣壳。这些外衣壳蛋白(尤其是 VP2)与中和抗体的特异性相互作用决定了病毒的血清型(Huismans 等,1987;Inumaru 和 Roy,1987)。因此,VP2 和 VP5 氨基酸序列均具有高度变异性,尤其是与病毒血清型相关的 VP2(Maan 等,2007)。

VP3 和 VP7 构成了 BTV 核心粒子的内层(次核心)和外层。120 个拷贝的 VP3 形成最内层的亚核衣壳，并且决定了病毒粒子外层的整体大小和形态(Grimes 等,1998;Mertens 和 Diprose,2004)。VP3 也是一种 RNA 结合蛋白，与病毒 RNA 基因组和其他小结构蛋白(VP1、VP4 和 VP6)相互作用。780 个拷贝的 VP7 构成核心外层,在 VP3 亚核心层之外(Grimes 等,1995)。VP7 也是一种免疫优势抗原和血清群特异性抗原,是血清群/种类或型特异性血清学试验的主要靶蛋白(Gumm 和 Newman,1982;Huismans 和 Erasmus,1981)。VP7 还可以结合 dsRNA,并介导进入细胞(在不存在 VP2 和 VP5 的情况下)，导致 BTV 核心粒子对昆虫细胞具有高度特异感染性(Diprose 等,2002;Huismans 和 Erasmus,1981;Mertens 等,1996)。

环状病毒转录酶复合体是由 VP1、VP4 和 VP6(在病毒核心中央空间内包装成 12 个拷贝),以及基因组 dsRNA 共同组成的(Stewart 和 Roy,2010)。VP1 是一种病毒 RNA 依赖性 RdRp,VP4 是一种加帽酶和甲基化酶 (CaP),VP6 是一种解旋酶(Hel)。这些小蛋白均具有高度的保守性(de Waal 和 Huismans,2005;Mertens 和 Diprose,2004;Ramadevi 等,1998;Ramadevi 和 Roy,1998;Stauber 等,1997)。

所有的 BTV 非结构蛋白(存在于被感染细胞中),被认为在病毒复制、细胞内分类和运输、基因组包装、衣壳组装、病毒释放和控制细胞对感染的反应中发挥重要作用(Owens 等,2004;Roy,1996;Kusari 和 Roy,1986;Mertens 等,1984;Sangar 和 Mertens,1983)。Eaton 等人证实 NS1 在被感染细胞中高水平表达,形成小管,附着在细胞骨架的中间丝上,这是环状病毒感染的特征(1988)。NS1 也被鉴定为“小管”蛋白。小管可能在病毒运输、病毒基因表达调节、细胞发病机制和形态发生中直接发挥作用 (Boyce 等,2012;Huismans,1979;Huismans 和 Els,1979;Owens 等,2004;Urakawa 和 Roy,1988)。

NS2 蛋白是一种单股 RNA 结合蛋白,并且是唯一的病毒特异性磷蛋白，构成病毒包涵体(viral inclusion body，VIB)的主要成分，即病毒组装位点(Theron 等，1996a，b；Thomas 等，1990；Uitenweerde 等，1995)。NS2 是一种三磷酸腺苷酶，被认为在 RNA 包装和翻译中发挥重要作用(Butan 和 Tucker，2010；Taraporewala 等，2001)。研究发现 NS2 在细胞质中被蛋白激酶 2 磷酸化，聚合形成 VIB，而未磷酸化的 NS2 蛋白保持分散状态。这表明为了翻译病毒 mRNA，其在 VIB(它们合成的地方)和宿主细胞质之间进行穿梭，NS2 可能在病毒 mRNA 穿梭过程中发挥作用，但具体功能取决于 NS2 的磷酸化状态。

NS3 蛋白和 NS3a 蛋白由 Seg-10 的 ORF 不同的起始位点翻译而来，决定了病毒的毒力、适合的传播媒介及病毒释放，尤其是在感染的昆虫细胞中被大量合成后的释放(Celma 和 Roy，2009，2011；Hyatt 等，1993；Martin 等，1998；O'Hara 等，1998)。NS3 仅在哺乳动物细胞中的翻译较差，并且最初 Seg-10 的翻译只在体外被证实(Mertens 等，1984)。NS3 蛋白和 NS3a 蛋白均为膜蛋白，它们具有两个疏水性跨膜结构域，在 AHSV 中，它是仅次于 VP2 的第二大变异蛋白(Bansal 等，1998；Jensen 和 Wilson，1995；van Niekerk 等，2001)。

最近，发现了两种新的非结构蛋白(NS4 和 NS5)。NS4 由 Seg-9 的一个交替的、重叠的 ORF 编码而来，是核定位信号，尽管其重要性尚未完全明确，但其在抗先天免疫反应中发挥重要作用(Belhouchet 等，2011；Ratinier 等，2011)。研究还表明，在 BTV-8 感染期间，BTV NS4 在宿主抗病毒反应中发挥重要作用(Belhouchet 等，2011；Ratinier 等，2011)。

8.6 结构

研究者已在分子、遗传和结构水平上对 BTV 进行了广泛的研究，并举例说明了环状病毒的最典型特征(Maclachlan 等，2009；Mellor 等，2009；Osburn，1994；Roy，2008；Stuart 和 Grimes，2006)。与经负染在电子显微镜观察到的 BTV 外衣壳特征和模糊外观相反，通过冷冻电镜下观察及对 BTV 的图像分析，其显示出与其他已知呼肠孤病毒明显不同的有序排列的外壳特征(Els 和 Verwoerd，1969；Hewat 等，1992b；Nason等，2004)。

BTV 外衣壳呈二十面体对称性，并展现出由两个不同基序组成的独特结构：VP2 三聚体形成 60 个“三脚蛋白复合体”，VP5 蛋白三聚体形成 120 个球状结构(Hassan 等，2001；Hassan 和 Roy，1999；Nason 等，2004)。VP5 三聚体整齐地位于 BTV 核心外层每个 VP7 蛋白三聚体的六元环中心，而 VP2 蛋白三螺旋基序的“帆状”刺突位于 VP7 蛋白三聚体的 180 之上并覆盖其全部。除了五重轴上的 12 个孔外，VP2 蛋白和 VP5 蛋白在内部外壳周围共同形成一个连续层(Hewat 等，1992a，b)。VP5 蛋白与 VP7 蛋白之间的相互作用具有广泛性，并且可能相对较强，而 VP2 蛋白与 VP7 蛋白之间的相互作用不具有广泛性，并可能较弱，这表明 VP5 蛋白的组装优先于 VP2 蛋白。VP2 蛋白也更容易从粒子表面脱离。然而，经典的“准等价”理论严格意义上并不允许二十面体病毒的结构和组装的三角数为 2 (T=2)(Caspar 和 Klug，1962)。因此，也有人提出(作为替代方法)呼肠孤病毒的内层以 T=1 对称性排列，作为亚核心蛋白的 60 个二聚体 [正呼肠孤病毒的 λ1(Hel)](Reinisch 等，2000)。然而，BTV 的亚核心外壳蛋白 VP3 似乎也被排列成 12 个“碟形”十聚体，在二十面体粒子对称的 12 个五重轴中，它位于每一个轴的中心，它们相互作用并通过它们的 Z 字形外缘组装形成整个亚核心外壳(Grimes 等，1998)。

因此，一种称为“几何准等价”的新原理被用于解释这种组装方式、形成的病毒粒子(Grimes 等，1998)。似乎许多其他 dsRNA 病毒都具有 BTV 亚核心的这一特征，尤其是其他呼肠孤病毒，并且似乎指出了针对特定问题的共同解决方案，涉及

dsRNA 基因组的组装和复制及可能有着共同的祖先。

相反，VP7 蛋白核心表层的 T=13 对称性代表了准等价的一个“典型”示例（Grimes 等，1998；Mertens，2001；Mertens 和 Attoui，2009）。BTV 的核心表层大部分是由核心晶格蛋白 VP7 的 6 个三聚体组成的环状衣壳体。由 5 个 VP7 蛋白三聚体组成的环使得核心层围绕着亚核心，以“完成”和“闭合”二十面体核心层表面外壳。在 AHSV 感染的相关细胞中，VP7 蛋白可组装形成大型、扁平的六角形阵列，这些阵列完全由 VP7 蛋白的六元环组成（Burroughs 等，1994）。

亚核心壳表面/T2 蛋白与病毒基因组片段，以及位于二十面体结构的五重轴上的转录酶复合物（transcriptase complex, TC）相互作用。TC 由 3 个小结构蛋白组成，VP1 蛋白（Pol）（Urakawa 等，1989），VP4 蛋白（CaP）（Ramadevi 等，1998；Ramadevi 和 Roy，1998）及 VP6 蛋白（Hel）（Stauber 等，1997）。病毒 RNA 依赖性 RdRp 和病毒加帽酶（VP4 蛋白）与贯穿五聚体中心的通道内部的基质有关，即核心的五重轴（Patton 和 Spencer，2000；Ramadevi 等，1998）。

TC 似乎能在被包装的 RNA 层中找到合适的位置（Gouet 等，1999）。多种呼肠孤病毒的晶体学和冷冻电镜分析表明，dsRNA 基因组在核心的液晶结构中被部分排列成有序的同心层（Gouet 等，1999；Prasad 等，1996；Reinisch 等，2000）。这将使 RNA 分子保持高度有序且紧密排列的形式，但仍允许粒子内部高度的灵活性和活动性。也有人提出，每个片段可能在核心的五重轴围绕 12 个 RdRp 加帽复合物中的一个，以紧密缠绕的螺旋形式存在（Gouet 等，1999）。

核心的二十面体性质限制了 RdRp 加帽复合物的最大数量为 12，也可能限制了片段的最大数量为 12。事实上，由于大多数呼肠孤病毒含有的片段数量低于 12 个，核心内 12 个潜在位点中的一些可能未被 TC 和 dsRNA 片段占据（Patton 和 Spencer，2000）。

具有转录活性的呼肠孤病毒核心粒子的结构组成表明，每个 RNA 片段/TC 复合物只能合成一种 mRNA，然后通过位于核心结构中相邻五重轴的“通道”输出。核心的每个聚合酶单元的独立功能通过不同 mRNA 的产量被反映出来，其数量大约（但不完全）与重量相等。相反，基因组片段在 RNA 复制过程中完全以等摩尔的水平进行包装（Patton 和 Spencer，2000）。不同 mRNA 的非等摩尔合成反映了每个片段的 RNA 链延伸速率接近恒定，但较小转录本的完成和重新启动更快，导致其在相对摩尔量较大的情况下合成（Mertens 和 Diprose，2004；Patton 和 Spencer，2000）。

尽管它们在高倍镜下具有相似性，但不同种环状病毒之间的某些超微结构特征并不相同。例如，AHSV 的病毒小管（由 NS1 组成）宽 18nm，而 EHDV 的小管宽 54nm，BTV 的小管宽 68nm（Huismans 和 Els，1979）。Gould 和 Hyatt（1994）报告称，一种环状病毒内不同毒株之间也可观察到差异，不同环状病毒之间小管的排列也可不同。研究证明 BTV-1（澳大利亚）的小管成束排列，而 BTV-10（美国）的小管则是平行排列。然而，病毒粒子的直径大小与样品制备方法有关，并且电子显微镜测定出的直径大小被认为是不可靠的诊断特征（Gould 和 Hyatt，1994）。由于小管蛋白 NS1 在每个分布区域内高度保守，并且不随血清型变化而变化，因此，外观或组织上的这些变异可能反映了 BTV 或 EHDV 不同地区型之间存在的较大序列变异，而不是存在于同一地理区域或地区型内的不同血清型之间（Anthony 等，2009b；Maan 等，2010b）。

8.7 感染与复制

针对 BTV 的研究表明，环状病毒能在一些昆虫和哺乳动物细胞中复制（Darpel，2007）。在昆虫中，BTV（和其他环状病毒）作为被吸食血液的一部分被摄入，可以感染中肠上皮细胞（DeMaula

等,2002;Mellor,2000)。在第一轮复制后,病毒通过基板被释放到昆虫的血腔中，在血腔中进行扩散,并可感染包括唾液腺在内的二级器官。此外,在唾液腺中复制的病毒被认为释放到唾液腺导管中,在下一次吸食血液时,病毒可随唾液蛋白一起被注入宿主体内(Anthony,2007;Darpel 等,2011;Mellor,2000;Mellor 和 Boorman,1995;Mellor 等,2000)。BTV 在哺乳动物细胞培养物中的复制通常会导致蛋白质合成中断、细胞周期停滞及病毒通过出芽和裂解方式释放，通常在感染后 48~72 小时造成广泛的 CPE 和细胞死亡（Beaton 等,2002;Owens 等,2004;Shaw 等,2013)。然而,γδT 细胞和树突状细胞可持续被 BTV 感染,宿主细胞不表现出蛋白合成的中止，并可能在脊椎动物宿主中对 BTV 的转移、扩散和持续存在发挥重要作用(Hemati 等,2009;Mertens 和 Attoui,2009;Takamatsu 等,2003)。

昆虫细胞持续不断地被环状病毒感染，与哺乳动物细胞相比，细胞病理损伤明显较轻并且病毒释放较慢。而且不会发生宿主细胞蛋白质合成中断,延长时间后,病毒蛋白质合成和病毒粒子的组装可能会变慢甚至停止,这可能是由免疫反应,如“沉默”导致的。昆虫细胞中的病毒释放机制涉及细胞膜渗透或含有多个病毒粒子的囊泡与细胞膜融合,这可能是由 NS3 介导的(Fu 等,1999;Hyatt 等,1993;Mellor,2000)。

对 BTV 复制的总结见图 8.1。BTV-10 的外衣壳蛋白 VP2 介导病毒粒子吸附在细胞表面,随后通过网格蛋白依赖性的细胞内吞途径内化(Forzan 等,2007;Hassan 和 Roy,1999;Mertens,2001)。然而，近期对 BTV-1 的研究表明,BTV-1 侵入并非仅通过网格蛋白途径，也可由巨胞饮作用介导(Gold 等,2010)。内体低 pH 值环境会使 VP2 解离，并且诱导 VP5 融合蛋白的构象改变和延伸，引起内体膜的渗透，将具有转录活性的核心送入细胞质(Forzan 等,2004)。BTV 核心粒子也可通过核心表面的 VP7 三聚体与细胞（特别是昆虫细胞）结合（Mertens 等,1996;Schwartz-Cornil 等,2008)在核心内部,VP1(RdRp)重复出现并同时将 10 个基因组片段全部转录成多个拷贝的 mRNA

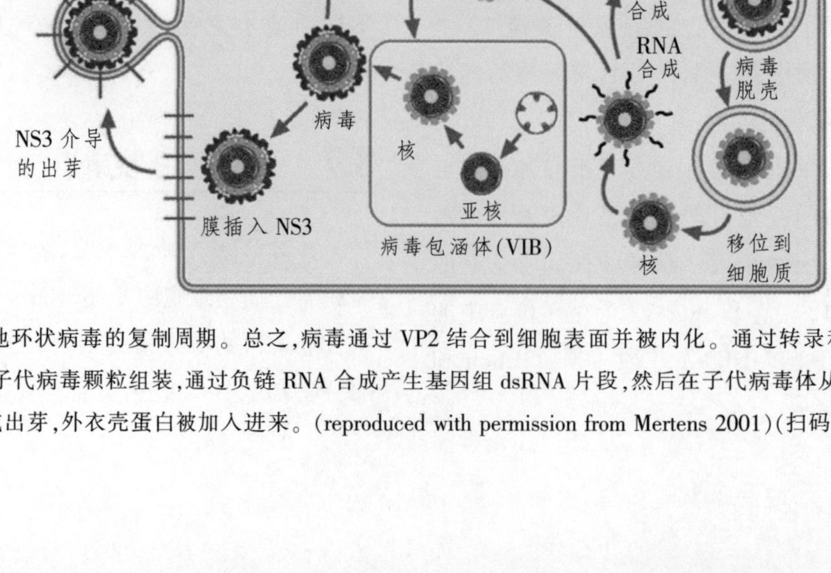

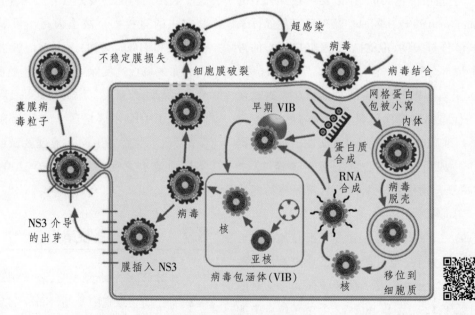

图 8.1 BTV 和其他环状病毒的复制周期。总之,病毒通过 VP2 结合到细胞表面并被内化。通过转录和翻译,产生病毒 mRNA 和蛋白质。子代病毒颗粒组装,通过负链 RNA 合成产生基因组 dsRNA 片段,然后在子代病毒体从细胞中被释放之前,通过膜的破裂或出芽,外衣壳蛋白被加入进来。(reproduced with permission from Mertens 2001)(扫码看彩图)

(Boyce 等,2004;Diprose 等,2001)。新合成的 mRNA 通过活化的 VP4 的鸟苷酸转移酶和甲基化酶催化加帽(Sutton 等,2007),从五重轴上的 VP3 蛋白(T2)层的孔中被挤出(Mertens 和 Diprose,2004)。

环状病毒感染细胞的细胞质中含有被称为 VIB 的大颗粒基质,其含有大量病毒 RNA、非结构蛋白 NS2(ViP)和病毒核心蛋白,新生的子代病毒颗粒(Brookes 等,1993)。在感染晚期,NS2 蛋白主要位于 VIB 的外围,表明 VIB 与宿主细胞质之间发生了交换,可能也表明了其具有传输病毒 mRNA 中的作用。由亲代病毒核心合成的 10 个 mRNA 或正链 RNA 可作为病毒蛋白合成的模板,其也可与 VIB 内的新生病毒颗粒相互作用,作为 RNA 包装期间负链 RNA 合成的模板(Diprose 等,2001)。

在持续感染的昆虫媒介细胞中,VIB 的尺寸逐渐增大,但最终可能不会产生病毒粒子,这表明细胞防御可能会阻止病毒进一步复制(Mertens,2001;Mertens 和 Attoui,2009)。病毒 mRNA 由 VIB 内的亲代或子代核心粒子合成,并作为病毒蛋白质合成的模板。它们也可通过一种尚不清楚了解的组装机制与新生病毒粒子合成一个有 10 个 mRNA(每个片段一个)的群,该组装机制似乎涉及不同片段之间互补的特定短序列(Boyce 等,2016)。包装 mRNA,涉及与解旋酶 VP6、ssRNA 结合 NS2 蛋白、VP1 蛋白和 VP4 蛋白之间的相互作用(Kar 等,2007)。新生粒子被认为通过增加核心表面蛋白 VP7(T13),以及为生成 dsRNA 片段而合成负链 RNA(通过 VP1)的过程而成熟(Boyce 等,2004)。

研究表明,每个 dsRNA 片段分别与位于 VP3 蛋白内部沿五重轴的不同转录复合体(VP1、VP4 和 VP6 蛋白)结合,并在冷冻电镜下形成“花型”(Nason 等,2004;Schwartz-Cornil 等,2008)。

在两种不同的 BTV(血清型或毒株)双重感染的情况下,可能会发生基因组片段的交换/重组,这可能是由不同的亲代病毒粒子形成的不同 VIB 的融合所致(Schwartz-Cornil 等,2008)。一些基因组片段似乎比其他片段交换更加频繁(Gould 和 Hyatt,1994),这可能反映了需要某些片段“配对”的限制条件。尽管 VP3 蛋白亚核心外壳的组装提供的支架决定了病毒粒子其余部分的大小和结构,但相对脆弱且不稳定。然而,VP7 蛋白三聚体的增加会产生更加稳定的核心粒子,会有效地隐蔽来自宿主细胞细胞质的病毒 dsRNA,以防止激活 dsRNA 依赖性的防御机制。然后,新形成的核心转录额外的 mRNA(Schwartz-Cornil 等,2008)。

随着子代粒子离开 VIB 进入宿主细胞质,外层衣壳成分(VP2 蛋白和 VP5 蛋白)的加入,形成了完全成熟的子代病毒粒子。成熟粒子在细胞质内被传输,可能在微导管上,涉及 VP2 蛋白/波形蛋白相互作用(Bhattacharya 等,2007)。由于 NS3 病毒孔蛋白活性介导细胞膜的不稳定作用,病毒粒子可从感染细胞中被释放(Han 和 Harty,2004)。这种机制可能在昆虫细胞中更加重要,其中 NS3/3a 蛋白在昆虫细胞中的合成水平远高于在哺乳动物细胞内的合成水平(Guirakhoo 等,1995)。在一些哺乳动物细胞中(如 BHK-21 细胞),BTV 粒子可通过出芽或使细胞死亡和裂解的方式排出(Mertens,2001;Schwartz-Cornil 等,2008)。

8.8　流行病学

环状病毒虽然对持续感染的媒介生物致病性较弱,但在哺乳动物和节肢动物宿主体内复制并引起各种疾病(Fu 等,1999;Anthony,2007;Attoui 等,2009a)。目前 BTV、EHDV 和 AHSV 是分别引起家养和野生反刍动物及马属动物疾病的最主要的环状病毒,并已在 WOAH 中被列出(Maclachlan 和 Guthrie,2010;WOAH,2010b)。然而,无论在区域性或者全球范围内,其他环状病毒都具有其潜在的重要性(包括 EEV、PALV 和 PHSV)。目前,BTV 在全球范围内广泛分布,血清型也不断增多

(Hofmann 等,2008;Maan 等,2011b)。预计在流行疫区(如南美洲)可能存在许多未被发现的虫媒病毒和其他环状病毒。

8.8.1 蓝舌病病毒

BTV 对多种反刍动物均具有感染性,但发病明显且严重的动物主要见于幼龄改良品种绵羊(例如,无角陶塞特羊),以及某些种的鹿(如短角鹿和白尾鹿)(Mellor,1994)。

报告显示,BTV 可通过注射或经口腔感染某些肉食动物(Alexander 等,1994;Brown 等,1996;Oura 和 El Harrak,2011)。尽管 BTV 主要通过库蠓叮咬传播,但也从公牛精液的单个样本中分离到多种血清型 BTV,表明睾丸是持续感染的无免疫力的易感部位。母牛可通过被 BTV 感染的精子而感染(Bowen 等,1983;Luedke 等,1977)。在感染 BTV-8 的欧洲牛中,已有报告指出 BTV 经口腔和胎盘(垂直传播)等途径传播(Backx 等,2009;Santman-Berends 等,2010b)。

牛、山羊和许多野生反刍动物被 BTV 感染通常不表现出症状或仅表现亚临床症状(Barratt-Boyes 和 MacLachlan,1995;Parsonson,1990;Verwoerd 和 Erasmus,2004),牛通常被视为沉默宿主或贮存宿主,可能通过垂直传播和保护病毒越冬,维持病毒传播(Darpel 等,2009;Luedke 等,1977;Nevill 1971;Santman-Berends 等,2010a,b)。然而,在 2007 年,被 BTV-8 感染的牛中也观察到严重(致死性)临床疾病(其发生频率较低)(Elbers 等,2008a,2009),表现为高热(达到 42℃)、流涕、精神沉郁、大量流涎、跛行、冠状带出血(蹄冠炎)、呼吸加快、口腔病变、口鼻肿胀和舌坏死("蓝舌病")(Gard 等,1987;MacLachlan,1994;MacLachlan 等,2009;Moulton,1961;Verwoerd 和 Erasmus,2004)。牛妊娠早期感染可导致妊娠失败或胚胎死亡,进而导致流产或死产(Osburn,1994;Tabachnick 等,1996)。然而,妊娠后期子宫内感染可导致畸胎效应(包括假犊牛)和出生时已患有病毒血症(Darpel 等,2009)。

将环状病毒传入易感无病的幼龄动物种群中会引起比较高的发病率和死亡率(Elbers 等,2008b;Wilson 等,2007),2006 年在北欧暴发的 BTV-8 证实了这一点(Maan 等,2008)。蓝舌病对经济重要影响是强制限制反刍动物从蓝舌病流行国家或地区运输到无蓝舌病区域(包括它们的种质资源及其他动物产品)(Darpel 等,2007;WOAH,2007,2010a;Osburn,1994)。由于蓝舌病和 AHS 的暴发,这些贸易限制造成了沉重的经济负担。环状病毒的全球性和周期性扩散与其生物媒介的分布一致,后者受气候条件影响(Purse 等,2005;Wilson 等,2009a)。最近在欧洲暴发的 BTV 与该地区的气候变化有关(Purse 等,2005,2008),这表明通过相同的库蠓媒介(例如,AHSV 和 EHDV)传播的其他虫媒病毒(尤其是其他环状病毒)的风险增加(Attoui 等,2009a;Gale 等,2010)。印度曾报告同时感染蓝舌病和小反刍动物兽疫的病例(Maan 等,2018)。环状病毒在新发地区的出现取决于病毒感染的能力及通过常驻媒介种群传播的能力。最近,一些环状病毒已经扩散至全球分布,并且越来越难预见哪些传播媒介能够在全球范围内携带特定的环状病毒。

BTV(库蠓种群)的媒介生物在 18~29℃的活性最强,在这个有效的范围内温度的相对小幅度升高可提高它们的活性和数量,并且能有效提高其作为媒介的效率。蓝舌病的分布基本上限制在美国的北纬 50°至南纬 30°,以及世界其他地区的北纬 40°至南纬 35°。2006 年之后,在欧洲南部和中部地区暴发的 BTV,是其传播受到气候改变影响的最强有力的例证。该地区平均温度的升高导致了拟蚊库蠓(BTV 的主要媒介之一)在该地区的分布发生改变。此外,自 2006 年起,中欧和北欧出现的新型传播媒介种群(超出了拟蚊库蠓的分布范围),其已经作为 BTV 媒介种群在这些地方存在,并使得蓝舌病疫情首次在这些地区传播。

8.8.2 流行性出血热病毒

EHDV 与 BTV 密切相关,它们都是通过相似的库蠓传播。EHDV 在全球范围内流行(Maclachlan 和 Guthrie,2010)。牛是 EHDV 的无症状储存宿主 (Gibbs 和 Lawman 1977;Uren,1986)。然而,1959 年茨城病毒在日本暴发 (由 EHDV-2 毒株引起), 造成了 39 000 多头牛死亡(Anthony,2007)。此后,分别从北美(美国)、南美(圭亚那和苏里南)、以色列、加勒比国家(牙买加、安提瓜和巴布达、格林纳达、特立尼达和多巴哥)、澳大利亚、日本(茨城病毒)和非洲(巴林、苏丹、阿曼、阿尔及利亚和摩洛哥) 分离到 EHDV(Attoui 等,2009a)。

最近在以色列、北非、美国和留尼汪岛等地的牛中暴发 EHDV(2004—2009 年),这一情况增加了 EHDV 对经济上的影响(Anthony 等,2009c)。在临床上也观察到了野生偶蹄动物(白尾鹿、骡鹿、大角羊和叉角羚)发生 EHD(Hoff 等,1973;Hoff 和 Trainer,1974;Karstad 等,1961;Noon 等,2002a,b;Shope 等,1955)。据报道,自 1959 年日本茨城病毒暴发以来, 牛中又暴发了高发病率和高死亡率的 EHDV(Abdy 等,1999;McLaughlin 等,2003;Ohashi 等,2002;Omori 等,1969a,b)。

EHD 的临床症状包括高热、厌食、呼吸困难、面部水肿延伸至颈部和舌部、结膜炎、流涎和流涕,以及皮肤、心脏、胃肠道和黏膜的出血。有时舌可能会发绀,这种疾病在美国当地被称为"黑舌"。死亡动物的典型特征包括出血性腹泻和(或)血尿及脱水(Maclachlan 和 Guthrie,2010;Savini 等,2011)。

8.8.3 非洲马瘟病毒

AHS 是一种古老的疾病, 早在 1657 年就被发现,1719 年首次在非洲南部大暴发, 随后在 1854—1855 年又暴发了数起疫情,造成至少 7 万匹马的死亡(Mellor 和 Hamblin,2004)。AHSV-9 流行最为普遍, 其在非洲以外的所有地区都有流行(Attoui 等,2009a;Mellor 和 Boorman,1995)。唯一的例外是 1987—1990 年在西班牙-葡萄牙流行的 AHSV-4, 是由纳米比亚输入的被感染斑马引起(Mellor 和 Boorman,1995;Sanchez-Vizcaino,2004)。由于 BTV 和 AHSV 的传播媒介相似,最近在欧洲南部、中部和北部暴发的蓝舌病表明,虽然可能因不同地理区域的 BTV 或 AHSV 毒株不同对传播媒介种群的致病力不同, 但该地区仍被视为将来可能暴发任何非洲马病的"风险中心"。

AHS 通常是由最急性转为急性, 在幼龄动物中造成的致死率可超过 95%。马比驴更易感,而非洲驴和斑马可作为病毒的贮存宿主(Maclachlan 和 Guthrie,2010)。AHS 的临床特征表现为高热,食欲缺乏,肺部、皮下和肌间组织水肿并渗出到体腔,以及浆膜表面出血(Maclachlan 和 Guthrie,2010)。

犬似乎在 AHS 的流行病学中不扮演主要角色,但可被高致死性的 AHSV 感染,这主要由于采食了被感染马的尸体(Bevan,1911;McIntosh 1955;Van Rensberg 等,1981)。另有血清学证据表明其他非洲食肉动物普遍被感染(Alexander 等,1995)。除斑马外,其他野生动物或家畜在 AHS 流行病学中均没有实质性作用(Barnard,1997;Davies 和 Otieno,1977)。与某些 BTV 毒株不同, 尚不清楚 AHSV 感染是否会导致妊娠母马的繁殖问题(Maclachlan 和 Guthrie,2010)。

8.8.4 马脑炎病毒

EEV 在马中通常呈轻度感染 (Guthrie 等,2009),EEV 于 1967 年在非洲南部的马中首次被发现,并在非洲呈地方性流行。1967 年以后,EE 在非洲南部呈季节性暴发流行(Theodoridis 等,1979)。由于马妊娠前几个月被 EEV 感染而导致马早期流产,可能会被误诊为不孕(Attoui 等,2009a)。该病在 20 世纪早期被 Arnold Theiler 称为 "马短暂性发热"

(Theiler,1910)。最近以色列首次发生 EEV,引起马发热、不安、厌食、呼吸急促、黏膜充血和面部、颈部、四肢水肿,但死亡率不高(Maclachlan 和 Guthrie,2010;Mildenberg 等,2009)。最近,印度被首次报告了 EEV(Yadav 等,2018)。

8.8.5 秘鲁马瘟病毒

PHSV 是一种蚊媒传播病毒,1997 年在南美洲一匹因脑炎死亡的马中被分离到(Attoui 等,2009b)。该病毒也从牛、羊和犬中被分离出。随后,在 1999 年,从澳大利亚北部地区的患病马中分离到一种与 PHSV 密切相关的病毒(埃尔西病毒)。对埃尔西病毒的流行病学了解甚少,目前尚不清楚在世界上相隔如此遥远的两地为何存在几乎相同的病毒。

8.8.6 巴尼亚姆病毒

PALV 与家畜动物(牛和羊),以及多种分布于世界上热带和亚热带地区的嗜血性节肢动物传播媒介的感染有关。PALV 血清群的致病性仍未得到充分研究,虽然美国农业部限制了尼亚比拉病毒的研究工作(PALV 血清群的一种血清型),但尼亚比拉病毒被认为是家畜的一种潜在病原体(Swanepoel 和 Blackburn,1976;Whistler 和 Swanepoel,1988)。然而,PALV 血清群的若干成员已经从流产的牛胎中被重复分离到。在日本也观察到由初赞病毒引起的犊牛先天性畸形的家畜流行病(1985—1986;Goto 等,1988a,b;Miura 等,1988a,b)。后来也从韩国本地山羊中分离出初赞病毒,表明其具有更广的宿主范围和地理分布,并且将来可能对家畜构成威胁(Yang 等,2008)。尽管从 1987 年至 1996 年多次分离到了初赞病毒,但并没有出现进一步的大规模暴发,而在 20 世纪 90 年代末期,九州地区曾出现过小规模的散发病例(Yamakawa 等,2000)。

8.8.7 欧本南吉病毒

EUBV 属于节肢动物传播的环状病毒,与澳大利亚的尤金袋鼠和袋鼠的猝死及高死亡率有关(Kirkland,2005;Rose 等,2000,2012)。蒂利吉里病毒(Tilligerry virus, TILV)与罗斯河病毒一起被分离到,是澳大利亚纳尔逊湾流行性多发性关节炎的病原体(Marshall 等,1980)。TILV 致病潜力尚未得到充分研究,但在研究病因不明的疾病过程中应考虑此类病毒(Marshall 等,1980)。

8.8.8 沃洛尔病毒和沃里戈病毒

WALV 和 WARV 在澳大利亚东南部的袋鼠中引发流行性失明(Hooper 等,1999;Reddacliff 等,1999)。这两种病毒均于 1994 年在澳大利亚西部灰袋鼠、红袋鼠、东部灰袋鼠,以及大袋鼠等种群中被分离到。这些动物因非化脓性脉络膜炎、视网膜变性、炎症和视神经继发性变性而失明(Hooper 等,1999;Reddacliff 等,1999)。

8.8.9 云南环状病毒

1997 年,中国首次从蚊虫中分离到YUOV(Attoui 等,2005a;Okamoto 等,2010)。YUOV 与里奥哈病毒密切相关,1997 年在秘鲁与 PHSV 同时被分离出(Attoui 等,2009b)。这两种病毒被归为同一种,即 YUOV(Attoui 等,2009b)。1998 年,在澳大利亚,中点环状病毒在健康的牛中被分离出来,并在 2007 年进行了特性鉴定(Cowled 等,2007)。中点环状病毒和 YUOV 的 VP2(T2)精氨酸同源性较高(96%),表明二者属于同一病毒种。YUOV 和 PHSV 都具有一个分布广泛且相互独立的传播形式,但其起源尚未知。

8.8.10 昌金努拉病毒

CGLV 是昌金努拉病毒种中确定的 12 种血清型之一,在轻度发热的患者血液中分离到(Gorman 等,1983;Karabatsos,1985)。在巴拿马的白蛉体内也分离到该病毒，并在啮齿类动物体内检测到了抗体。CGLV 在昆虫细胞系中复制无 CPE,对脑内接种的新生小鼠或仓鼠具有致病性(Karabatsos,1985)。尽管在啮齿类动物和其他哺乳动物中也检测到了该病毒的特异性抗体，但人类感染的患病率还尚未得到研究。

8.8.11 科里帕塔病毒

CORV 于 1960 年从澳大利亚库蚊中被分离到(Doherty 等,1963)。后来,在澳大利亚、非洲和南美洲又分离到血清学相关病毒 (Gonzalez 和 Knudson,1987)。在家畜、野生鸟类、牛、有袋动物、马和人类中检测到了 CORV 的中和抗体(Boughton 等,1990;Doherty 等,1963,1970;Whitehead 等,1968)。

8.8.12 伊理病毒

IERIV 于 1955 年首次从特立尼达常绿季节森林收集的雌性鳞蚊群中被分离到。IERIV 在小鼠脑细胞和 BHK 细胞中被成功分离培养(Spence 等,1967)。这种病毒包含 3 个血清型:IERIV、阿尔戈那姆病毒和戈莫卡病毒。阿尔戈那姆病毒于 1957 年首次从印度采集的蚊群中被分离出来,并且也有报告称，人体内存在抗病毒的中和抗体的证据(Dandawate 和 Shope,1975)。之后,从斯里兰卡采集的蚊子中也分离到了该病毒 (Peiris 等,1994)。戈莫卡病毒于 1970 年从中非共和国赤道森林中采集的蚊子体内被分离到，并且鸟类体内也分离出了该病毒 (Karabatsos,1985)。这表明 IERIV 病毒有可能感染鸟类及人类。

8.8.13 大岛病毒

GIV 的传播媒介是蜱,并且也从鸟类中分离到 GIV(Mertens 等,2000)。1962 年在克麦罗沃地区,从患有脑膜炎患者的血液和脑脊液(及全沟硬蜱体内)中至少分离到 20 株克麦罗沃病毒(Monath 和 Guirakhoo,1996)。在克麦罗沃和中欧(利波夫尼克和特里贝克)等地区,多种 GIV,包括克麦罗沃病毒、利波夫尼克病毒和特里贝克病毒(Tribec virus,TRBV)是造成人类发热和神经感染的病因(Belhouchet 等,2010;Dilcher 等,2012;Libikova 等,1970, 1978)。利波夫尼克病毒和 TRBV 与“中欧脑炎”有关,50%以上的“中欧脑炎”患者的利波夫尼克病毒抗体呈阳性(Attoui 等,2009a;Libikova 等,1978;Monath 和Guirakhoo,1996)。利波夫尼克病毒也被认为与一些慢性神经系统疾病和多发性硬化症有关。1962 年,在克麦罗沃地区，从全沟硬蜱和脑膜炎患者中分离到 20 多株克麦罗沃病毒。该病毒也从鸟类中分离到,其可感染 Vero 细胞或 BHK-21 细胞(Attoui 等,2009a)。

8.8.14 秦纽达病毒

在俄克拉荷马州和得克萨斯州的 “俄克拉荷马州蜱热” 患者中检测到了克麦罗沃相关病毒的抗体。西克斯冈市病毒是秦纽达病毒种(CNUV-1~7)中 7 种蜱传病毒的血清型之一,已从鸟类中分离到(Mertens 等,2000)。在患有“俄克拉荷马州蜱热” 的一些患者体内同时检测到西克斯冈市病毒和 KEMV 的抗体,但没有分离到病毒。然而,因为没有分离到病毒,所以发热的确切原因尚不明确(Fields 等,1985;Monath 和 Guirakhoo,1996)。

8.8.15 莱邦博病毒

LEBV-1 是莱邦博病毒种类中唯一已知的血清型,1968 年在伊巴丹和尼日利亚的一例发热患儿体内分离到(Attoui 等,2009a;Fields 等,1985;Monath 和 Guirakhoo,1996)。该病毒对乳鼠具有致病性，并且已从非洲的啮齿类动物和蚊子中分离到该病毒(曼蚊属和伊蚊属)(Fields 等,1985)。LEBV-1 能在 C6/36 细胞(无细胞病变)中及 Vero 细胞和 LLC-MK2 细胞中进行复制。

8.8.16 奥伦盖病毒

ORUV 有 4 种不同的血清型(ORUV 1~4)，从虐蚊、伊蚊和库蚊(Fields 等,1985)分离到。ORUV 于 1959 年首次从乌干达一例临床表现为发热和腹泻的患者血液中被分离到 (Mohd Jaafar 等,2014)。ORUV 广泛分布于非洲的人类、反刍动物、猴子和蚊子中(Attoui 等,2009a;Fields 等,1985;Monath 和 Guirakhoo,1996)。该病毒在乳鼠和仓鼠中引发致死性脑炎，并且在成年埃及伊蚊的胸腔内接种,可进行复制(Karabatsos,1985)。根据报告,ORUV 与黄热病病毒的混合感染率很高,显示出它们具有相似的地理分布，并且都通过伊蚊传播(Attoui 等,2009a)。

8.8.17 其他环状病毒

SLOV 最初在 2002 年从澳大利亚收集到的蚊群中被分离到。已在马、驴和山羊中检测到了 SLOV 抗体(Cowled 等,2009)。在澳大利亚东部城市附近地区采集到的蚊子中也分离出了该病毒(Jansen 等,2009)。1996 年在巴西首次分离到了米那库病毒(来自南美洲)。小鼠的发病机制研究显示出了 SLOV 广泛的嗜神经性和嗜内脏性(Martins 等,2007)。1988 年，布鲁乌布朗库病毒(也来自南美洲)从蚊群中被分离到,其遗传特征表明该病毒可能代表一种不同的环状病毒(Vieira Cde 等,2009)。根据已发表的数据显示,南美洲存在大量特征不明确的环状病毒，尤其是在亚马孙热带雨林,该地区被公认为是生物多样性中心。初步研究还表明,这些病毒中有几种可能代表了更多新的种类病毒(表 8.1 和表 8.2)。一些起源于蝙蝠的环状病毒(伊费、杰潘纳特、弗莫德和乔巴峡病毒）可能会对人类和家畜构成主要的威胁(Calisher 等,2006)。

在人兽共患病中发现诸多新病毒(Wang 和 Crameri,2014)。它们可以从野生的或驯养的贮存宿主动物传播给人类，也可以通过昆虫媒介或接触感染动物的粪便或组织传播给人类。已从人类中分离出了多种环状病毒(Brown 等,1991;Karabatsos,1985;Libikova 等,1970,1978)。

8.9 生物学多样性

8.9.1 血清学特点

环状病毒是一个多样化的组群,已分化为 27 个不同的种类,也代表了不同的血清学群,其无常见的“通用”抗原。一些环状病毒蛋白是高度保守的,在每个群内都具有共同的抗原/表位,其可通过 CF 试验、FA 试验和 ELISA 检测到(Mertens 等,2005b;Verwoerd 等,1979)。核心表面的“T13”蛋白(BTV 的 VP7)被认为具有“免疫优势”，并且是一些环状病毒血清群的主要特异性抗原(Gumm 和 Newman,1982),因此是一些血清群特异性血清学诊断的主要靶标(Gumm 和 Newman,1982;Huismans 和 Erasmus,1981)。然而,一些关系密切但种类完全不同的环状病毒(例如,在 BTV、EHDV 和 EUBV 之间)通过 ELISA、CF 和琼脂凝胶沉淀试验检测，观察到低水平或单向交叉反应 (Borden 等,1971b;Gorman,1979;Gorman 和 Taylor,1978;Moore,1974;Moore 和 Lee,1972)。这种交叉反应性已被用于衡量不同种类病毒之间的进化距离(Gorman,1986)。然而,由于已被鉴别的部分环状病毒的标准血清难以获得，使得一些病毒分离株还不能通过血清学方法将其归类至已知的血清群/种类中,因此,其仍为该属内的“未分类”病毒(Gorman 等,1983;Karabatsos,1985)。

在许多环状病毒种类/血清群中,根据病毒外表面蛋白(BTV 中的 VP2 蛋白和 VP5 蛋白)的血清学中和试验反应的特异性，划分了多个不同的“血清型”。“病毒中和试验”(根据先前感染或免疫动物的抗毒血清的能力不同，去中和已知血清型的参考毒株)和“血学清学中和试验”是重要的诊断方法。通过这些方法，鉴别了 BTV、AHSV、

CGLV、KEMV、PALV 和 CORV 等血清型（Dandawate，1974；Dandawate 等，1969；Davies 和 Blackburn，1971；Gonzalez 和 Knudson，1987；Gorman，1979；Hamblin 等，1991；House 等，1990；Karabatsos，1985；Libikova 和 Buckley，1971；Myers 等，1971；Travassos da Rosa 等，1984）。

细胞对病毒外衣壳蛋白吸附的能力与 BTV 的 VP2 蛋白主要的中和决定簇有关。然而，BTV 较小的病毒外衣壳蛋白 VP5 也含有一些中和抗原表位，并且可能通过与 VP2 蛋白的构象相互作用影响血清型，在细胞进入过程中参与膜渗透（Cowley 和 Gorman 1989，1990；DeMaula 等，2000；Mertens 等，1989）。BTV 上的 VP2 蛋白是一些蜱传环状病毒的细胞吸附蛋白大小的 2 倍左右，这表明 VP2 蛋白是 BTV 和其他密切相关（库蚊传播）的环状病毒在进化过程中共同经历的产物（Anthony 等，2011；Dilcher 等，2012）。在这种情况下，在 VP2 蛋白氨基末端的一半已经确定了中和表位，其中几个表位似乎在 AHSV、BTV 和 EHDV 之间共有（Bentley 等，2000；Martinez-Torrecuadrada 和 Casal，1995；Martinez-Torrecuadrada 等，2001）。

8.9.2 遗传特性

血清学检测不能用于定量测定不同环状病毒单个基因组片段中的遗传亲缘关系。已用于血清群检测和鉴定的 CF 试验只限于对有限数量的病毒蛋白进行抗原位点比较。血清中和试验确定的相关表位的相关性，仅对应于编码细胞吸附蛋白的部分基因。然而，通过 RNA-RNA 杂交、寡核苷酸定位、特别是测序和系统发育分析对环状病毒基因组片段进行比较，可以准确地和定量地评估每个基因组片段的遗传相关性。这可以用于揭示特定环状病毒毒株和种类之间遗传多样性的范围和程度。

与其他不同环状病毒种类的成员相比，同一种类病毒之间的序列具有高度的保守性，也反映了在 AGE 期间单个环状病毒种类成员的基因组片段迁移的一致模式（Gonzalez 和 Knudson，1988；Mecham 和 Johnson，2005）。但是 EHDV 的 RNA 迁移模式显著不同（Anthony 等，2009a）。与西部地区分离株相比，在东部地区的分离株中有一个较小的 Seg-9（产生了 Seg-7、-8 和-9 的迁移模式，与东部地区分离株中的 BTV 更相似）。

在不同但相关的种类/血清组的成员之间，可能存在相似的电泳型（Gorman 等，1981；Mertens 等，2005b）。然而，目前尚无研究去比较所有环状病毒种类中代表性分离株的聚丙烯胺凝胶电流和（或）AGE 电泳型。

库蠓媒环状病毒（Culicoides-borne orbivirus，CBO）、蚊媒环状病毒（mosquito-borne orbivirus，MBO）和蜱媒环状病毒（tick-borne orbivirus，TBO）的基因组片段大小的变异性在早期已有记载（Attoui 等，2005a，2009b；Belaganahalli 等，2011；Belhouchet 等，2010；Moss 等，1992）。CBO 和 TBO 经过 1%的 AGE 后形成不同但明显保守的电泳型（分别形成 3-3-4 或 2-4-4 模式），与 CBO 和 TBO 对照，MBO 在小、中和大片段中表现出了更多可变的迁移模式，表明基因组片段和蛋白质大小的变异性更大。唯一的例外是 ORUV 与 CBO 有相似的 RNA 电泳迁移模式，这表明它也可以通过库蠓传播。尽管每个血清群/种类具有不同的电泳迁移模式，但 TBO（如 CGV、CNUV 和 WMV）作为一个群，表现出了 2-4-4(2-4-3-1)电泳迁移模式。虽然一些 MBO 的基因组片段也表现出了 2-4-4 电泳迁移模式，但它们的分类和基因组大小与 TBO 不同。因此，根据其独特的 1%AGE 电泳型，可将蜱传播环状病毒与昆虫传播的环状病毒（蚊子和库蠓）区分开来。

环状病毒在流行过程中，通过基因的插入、删除、重整、重组和聚合的组合获得了不同环状病毒种类间的序列差异性和大小异质性（Anthony 等，2009a，c；2011；Cao 等，2008；He 等，2010；Hundley 等，1987；Murao 等，1996；Troupin 等，2011）。TBO 和 CBO 之间的差异说明了这种异质性的程度，其中,TBO 的 OC1 约为 CBO 的 OC1 的一半。因此,CBO 的 OC1 可能是 TBO 的 OC1 基因复制事件的结果,表明 TBO 是 CBO 的祖先 (Belhouchet 等,2010;Dilcher 等,2012)。这也表明可观察到的 OC1 大小的变化可能在某种程度上与传播媒介的特异性变化相关。序列变异不同水平表明选择不同环状病毒基因的作用强度不同。正如 EHDV 表现出的一样,环状病毒 OC1 在种内和种间表现出高度变异性,可能是最高水平的选择压力作用的结果(Anthony 等,2009c)。

8.9.3 病毒种间谱系变化

8.9.3.1 基于亚核心 T2 蛋白的谱系变化

在所有蜱传播的环状病毒(GIV、SCRV、PHSV、YUOV 和 CORV)中,亚核心外壳“T2”蛋白被鉴定为 VP2 而不是 VP3。VP3 蛋白(T2)和 VP2 蛋白(T2)群分别形成两个离散的系统发育谱系,即昆虫媒介群和蜱媒介群(Attoui 等,2009b;Moss 等,1992;Parkes 和 Gould,1996)。对于基因组片段 1、7、8、9 和 10,BTV 和 GIV 的基因组编码分配是等效的。GIV 基因组片段 2、3、4、5 和 6 分别与 BTV 的片段 3、4、6、2 和 5 是同源的 (Attoui 等,2009a;Belhouchet 等,2010;Moss 等,1987,1992)。

环状病毒 T2 蛋白(BTV 的 Seg-3/VP3)的氨基酸序列决定了病毒粒子的构象和结构,并在每种环状病毒中序列具有很高的保守性(>83%)(Attoui 等,2001,2009a;Belaganahalli 等,2011;Belhouchet 等,2010)。最保守的区域是编码 T2 蛋白的基因组片段,可用于各种环状病毒种类(血清群)的鉴别(Attoui 等,2001,2009a)。因此,在测序工作中,优先考虑 T2 基因组片段(Attoui 等,2001,2005a,2009b;Cowled 等,2007,2009;Maan 等,2008;Pritchard 等,1995)。所以,在 GenBank 中 T2 基因序列是几种环状病毒的唯一有效序列(WALV、WGRV、EUBV 和 CORV 的部分序列是有效的)(Belaganahalli 等,2011;Belhouchet 等,2010)。因此,T2 是新型环状病毒测序以确定其病毒种类的理论依据(Attoui 等,2001;Belhouchet 等,2010;Pritchard 等,1995)。

8.9.3.2 基于蛋白 VP1(Pol 或 RdRp)的谱系变化

编码病毒聚合酶(BTV 的 Seg-1/VP1)的基因组片段也非常保守。由于呼肠孤病毒科中的所有病毒均有一个同源基因,因此在科/属/种层面上,它也被选为靶序列用来测序和系统发育比较。用病毒 RdRp 氨基酸序列进行系统发育分析,分别形成螺旋形和非螺旋形病毒两个单独的簇。环状病毒被分为单独的簇,其聚合酶的氨基酸同源性大于 30%,与其他非螺旋状病毒属的氨基酸同源性为 13%~15%,与螺旋状病毒属的氨基酸同源性小于 8%(Attoui 等,2009a)。

已经确定了几种环状病毒属成员的 RdRp 氨基酸序列,包括 BTV、EHDV、AHSV、PALV(Chuzan 毒株)、YUOV、PHSV、EEV 和 SCRV。用 RdRp 序列构建的一个系统发育树表明,SCRV 最接近其他环状病毒起源的祖先(Attoui 等,2009a)。

8.9.3.3 基于外层衣壳蛋白的谱系变化

编码环状病毒外衣壳蛋白的基因具有高度变异性,可能表现出部分或高水平的血清型特异性[如 BTV 的 VP2 蛋白和 VP5 蛋白,以及其他一些病毒的 VP3 蛋白(OC1)]。对来自不同血清型的 BTV Seg-2 进行全核苷酸序列分析,已经证实病毒的变异与血清型相关(Maan 等,2007),这为开发型特异性分子检测方法提供了理论依据(Mertens 等,2007)。对个别 BTV 血清型内的病毒株进行 Seg-2 序列的系统发育分析,也发现了与这些病毒起源地有关的差异。根据这些差异将 BTV 毒株分为“东部”和“西部”两个群或“Seg-2 地区型”,并有证据将其进一步分为地区亚群(Maan 等,2007;Mertens 等,2007)。这些研究和新的检测系统有助于对最近在欧洲入侵或检测到的

11种不同血清型的BTV(1,2,4,6,8,9,11,14,16,25和27),以及它们的输入途径进行检测。

最新的分子系统发育学为环状病毒在病毒种内和种间的分类提供了更好的定量评估(Anthony 2007;Anthony 等,2009c;Attoui 等,2001,2005a;Belaganahalli 等,2011;Belhouchet 等,2010;Maan 等,2007)。

系统发育树(图8.2和图8.3)显示,典型的分支排列反映媒介分组。以前,用环状病毒VP1蛋白(Pol)序列和节肢动物序列(使用线粒体基因)进行的基因序列分析表明,蜱传病毒为昆虫传播病毒提供了家族性根源(Attoui 等,2009a;Belhouchet 等,2010),并且支持了这样一种假设,即环状病毒通过被称为“物种协同进化”的过程与其传播媒介一起进化。对密码子偏嗜性和基因组组成的分析将加强这一假说。

采用遗传距离和成对删除参数对所有环状病毒全部蛋白质的氨基酸序列进行鉴定(表8.3)。这些值可能对这些病毒媒介组群的归类有影响(Belaganahalli,2012)。

8.9.4 遗传进化

有人提出,不同的环状病毒之间是有关联的,它们最初都起源于一个共同的祖先。然而,它们在很长的一段时间内发生了多点突变、插入、删除和复制(串联)。这将环状病毒分成了不同的种类、地区型、血清型、谱系,可通过测序和系统发育分析来识别的其他组群(Attoui 等,2009a)。RNA病毒的进化特点可能是由于其基因组的高度变异性和重组,使其能够在单个病毒基因组内的不同基因组片段中整合不同的正突变。RNA病毒还可以通过生成缺陷性干扰RNA、同源和非同源重组、插入、复制、删除及串联事件产生遗传变异性(Anthony 等,2011;Bonneau 等,2001;Eaton 和 Gould,1987;Mohd Jaafar 等,2008;Qiu 和 Scholthof,2001),并帮助它们适应不同的生态环境(Domingo 和 Holland,1997;Elena 和 Sanjuan,2005;Fields 和 Joklik,1969)。

有关dsRNA病毒(包括环状病毒)的分子进化速率的信息很少。BTV的贝叶斯进化估算表明,Seg-2、-3、-6和-10的平均进化速率为6.9~0.52×10^{-4}置换/(位点·年),低于正链RNA病毒[(24.2~3.3)×10^{-4}置换/(位点·年)],以及呼肠弧病毒科的非媒介传播dsRNA病毒[(19.3~8.4)×10^{-4}置换/(位点·年)]。尽管它们与其他媒介传播的负链ssRNA病毒相似[12.3–0.7×10^{-4}置换/(位点·年)](Carpi 等,2010;Hanada 等,2004)。

对呼肠弧病毒科科罗拉多蜱传热病毒属蜱传病毒的研究表明它们以10^{-8}~10^{-9}置换/(位点·年)的速率进化,这与dsDNA基因组中观察到的速率相似(Attoui 等,2002a,b;Holland 等,1982)。这表明蜱传病毒比其他节肢动物传播的病毒进化得更慢,可能是由于蜱能够长时间休眠的特性,潜在降低了病毒复制频率,进一步有助于降低变异率(Attoui 等,2002b)。

有人认为dsRNA的稳定作用(超过ssRNA),需要宿主的交替,dsRNA基因组在具有转录活性的核心内衣壳化(避免其遗传物质暴露于感染细胞的细胞质中),并且由于在其流行周期中宿主/媒介的持续感染,病毒复制频率降低,共同对病毒基因组片段施加强大的净化选择压力,这可能会影响整体进化速度(Carpi 等,2010)。

病毒mRNA需要进行分子内碱基配对,以维持翻译、包装、复制和其他功能所需的特定结构,这也可能有助于降低中性突变积累的速率。病毒基因组的片段性质本身也可能提供一些保护,防止高突变率的潜在破坏性影响及降低其病毒总体“适应性”。通过将基因组断裂成不同的片段,可以降低靶基因中致死性突变的数量。重组和高选择率(存在瓶颈,例如,在传播给媒介节肢动物和传播来自媒介节肢动物过程中)将迅速移除无活力的子代病毒,并可从病毒种群中特异性地移除含有有害突变的片段。

尽管存在这些限制,但在一些环状病毒种类

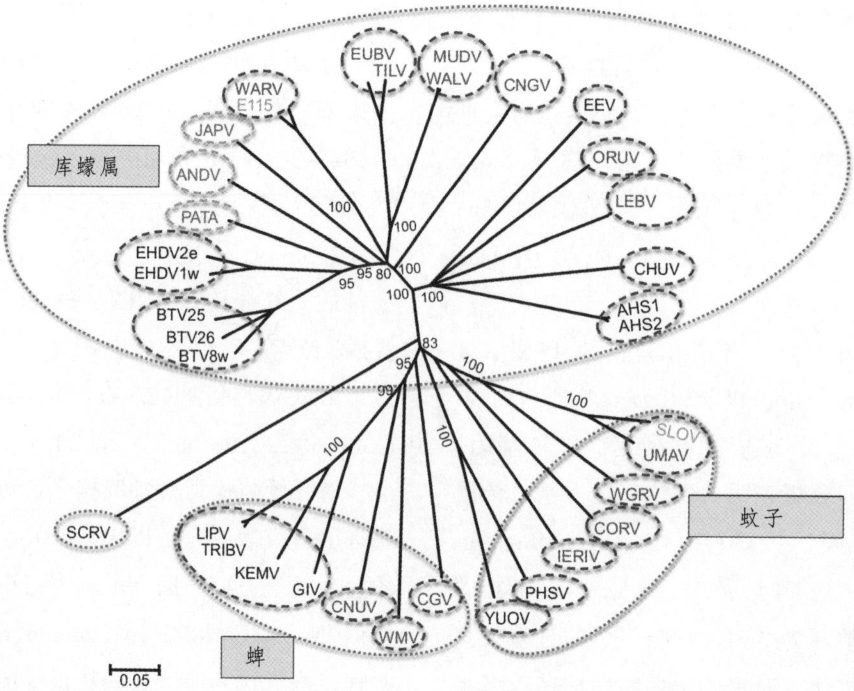

图 8.2 以环状病毒 VP1 氨基酸序列为靶序列，通过邻位相连法建立的系统发育树。使用距离矩阵和 MEGA 5 中的遗传距离算法（1000 次引导复制）生成树（Tamura 等，2011）。节点处的数字表示 1000 次重复后的引导置信度值。病毒名称分别用红色（代表新种类）、绿色（归类为现有种类）和蓝色字体表示，并且在最近进行了测序（Belaganahalli，2012）。“(e)”和“(w)”分别表示东部毒株和西部毒株。（扫码看彩图）

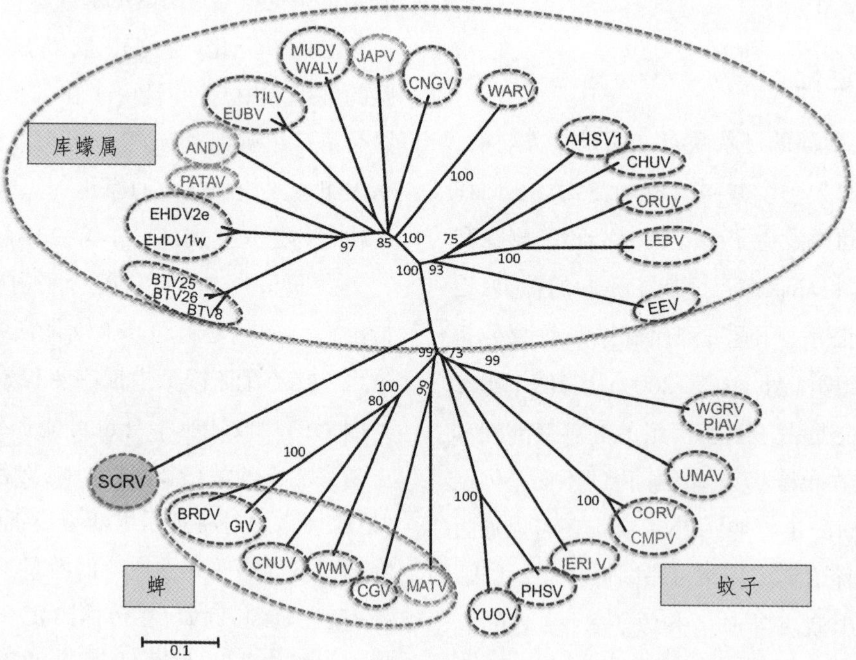

图 8.3 以环状病毒 T2 氨基酸序列为靶序列，通过邻位相连法建立的系统发育树。使用距离矩阵和 MEGA 5 中的遗传距离算法（1000 次引导复制）生成树（Tamura 等，2011）。节点处的数字表示 1000 次重复后的引导置信度值。病毒名称分别用红色（代表新种类）、绿色（归为现有种类）和蓝色字体表示，并在本研究中进行测序（Belaganahalli，2012）。“(e)”和“(w)”分别表示东部毒株和西部毒株。（扫码看彩图）

表8.3 环状病毒分类的建议氨基酸鉴定临界值

蛋白质	种内最小氨基酸同源性(%)	种间最大氨基酸同源性(%)
VP1	79[a](KEMV 和 TRBV 之间)	73.5(BTV 和 EHDV 之间)
OCP1	22.1(Maan 等,2010b)	29.28(PATAV 和 BTV 之间)
T2	87.8(Maan 等,2010b)	82.8[a](GIV 和 KEMV 之间)
Cap(VP4)	80.1(Maan 等,2010b)	66.46(BTV 和 EHDV 之间)
NS1(TuP)	76.6(Maan 等,2010b)	58.8[a](TRBV 和 GIV 之间)
OCP2(VP5)	54.2(Maan 等,2010b)	63.43(BTV 和 EHDV 之间)
T13	79.9[b]	77.6[a](TRBV 和 GIV 之间)
NS2(ViP)	67.7(Maan 等,2010b)	77.6[a](TRBV 和 GIV 之间)
Hel(VP6)	53.1(Maan 等,2010b)	48.7[a](TRBV 和 GIV 之间)
NS3	63.7(van Niekerk 等,2001)	59.29(PATAV 和 BTV 之间)

[a] 数值来自 Belhouchet 等(2010)和 Dilcher 等(2012)。

[b] 数值基于现有 BTV 序列计算而来。

中仍存在高水平的遗传变异(Bonneau 等,2000,2001;de Mattos 等,1996;Gould 和 Pritchard,1991)。一些变化反映了对选择压力的适应，尤其是在蛋白质中，如 BTV 的 VP2 蛋白、VP5 蛋白(靶向宿主免疫系统)和 VP7 蛋白及可能在昆虫和昆虫细胞感染中起作用的 NS3 蛋白 (Bonneau 等,2001;Carpi 等,2010)对选择压力的适应。

虫媒病毒在节肢动物生命周期中的进化动力受到吸血量、体积和 2 次吸食血液的间隔时间影响,同时还受宿主范围、媒介范围,以及潜在的垂直传播影响 (Cook 和 Holmes,2006;Zanotto 等,1996)。蚊传播虫媒病毒的进化速度约为蜱传播虫媒病毒的 2.5 倍(Gould 等,2003;Zanotto 等,1995)。这可能是因为蜱的生命周期较长(2~5 年)、长期不活动及有限的吸血次数(每个生命周期吸 3 次血),导致在蜱中病毒更新率明显变慢,相比之下,蚊子的寿命较短,需要多次进食血液(Gould 等,2003)。此外,病毒在丛林复制循环周期中,蚊媒病毒在媒介和宿主体内的滴度都快速增加。蚊媒病毒的整个复制周期和传播速度远比蜱媒病毒更有活力(即复制周期数更多),这有助于加快进化速率。

库蠓和白蛉生存的生态环境与蚊子相似（它们都是昆虫),因此,认为它们传播病毒的进化速度也可能很相似。通过贝叶斯分析法对环状病毒的分析表明，昆虫传播环状病毒的进化速度比蜱传播的环状病毒的进化速度快 2 倍。

8.10 环状病毒的诊断

8.10.1 病毒学和免疫学(血清学)方法

环状病毒的诊断通常包括病毒分离和血清学鉴定。然而,这些诊断方法速度较慢,有时结果可能还不可靠(Akita 等,1992;Tabachnick 等,1996;Wade-Evans 等,1990;Prasad 等,2000)。

在病毒分离前首先在新生小鼠脑内或鸡胚内进行接种,接着在细胞进行传代培养(Karabatsos,1985)。然而,最新研究表明,用 KC 细胞(来源于北美传播媒介索氏库蠓)直接分离库蠓传播环状病毒的成功率很高。这些细胞对血液或血细胞具有高度耐受性,可能反映了其来源于噬血昆虫,这有利于从低滴度病毒的血液样品中直接分离培养到病毒。通过细胞培养直接分离病毒减少了实验动物的使用,支持了“三 R”原则(替代、优化和减少研究中动物的使用)。

环状病毒的病毒种类特异性(血清群特异性)

诊断最初包括病毒分离、用于病毒抗原检测的CF或FA试验,以及用于抗体检测的CF或AGID试验(Boulanger等,1967;Della-Porta等,1983;Jochim和Chow,1969;Pearson和Jochim,1979)。后来开发了用于病毒特异性抗原(Crafford等,2003;Hawkes等,2000;Hosseini等,1998;Thevasagayam等,1996)和抗体检测的ELISA(Afshar,1994;Afshar等,1987,1997;Crafford等,2011;Hamblin等,1990;Thevasagayam等,1995)。中和试验(血清中和试验或病毒中和试验)可用于病毒血清型的鉴定(Blackburn和Swanepoel,1988;House等,1990;Howell等,1970;Parker等,1975)。

血清学方法,如ELISA,用于地方性流行地区或暴发地区的环状病毒特异性抗体的鉴定,或者用于监测"哨兵"猪群的感染和血清转化(Anderson,1984;Batten等,2009;Crafford等,2011;Laviada等,1992,1995;WOAH,2008)。然而,广泛的疫苗接种(如在北欧BTV-1和BTV-8暴发期间进行疫苗接种)导致无法通过血清学试验监测病毒特异性抗体,因为活疫苗和当前使用的灭活疫苗均可产生针对病毒所有结构和非结构蛋白的特异性抗体。尽管已尝试根据非结构蛋白设计"DIVA"试验(以区分自然感染动物与疫苗接种动物)(Barros等,2009;Laviada等,1995),但获得的结果并不可靠,这可能反映了特定疫苗和野毒株之间蛋白的血清学特性差异,而不是疫苗接种和野毒感染之间的一般差异。

早期基于血清学方法检测和鉴别环状病毒,主要存在的问题是需要资源配备充足(所有病毒的参考毒株和抗病毒血清)、耗费时间(通常需要数周完成),以及最终确认前的大量工作。在疫情暴发时,这些问题及由此产生的确诊延迟难以立即采取预防措施,例如,筛选和使用适当的疫苗。然而,不同种类环状病毒的代表毒株(包括个别种类内部密切相关的分离株)的遗传特性为了解其亲缘关系,以及设计不同诊断试验奠定了基础。

由于BTV、EHDV和AHSV具有重要的经济意义,因此,它们的许多分离株的完整基因组被优先鉴定。这些病毒的基因序列数据库有助于开发快速、可靠的分子诊断方法,用于血清群和血清型的鉴定(Aguero等,2008;Anthony等,2007;Bremer和Viljoen,1998;Fernandez-Pinero等,2009;Maan等,2010a,2011a,2012;Mertens等,2007;Orru等,2006;Shaw等,2007)。OC1序列与血清型的相互关系(Anthony等,2009c;Maan等,2007);RdRp序列与属、种和地区型之间的相互关系(Attoui等,2005a;Belaganahalli等,2011;Belhouchet等,2010);T2序列与种和地区型之间的相互关系(Attoui等,2001,2005a;Nomikou等,2009;Pritchard等,1995),已经用于不同环状病毒的鉴定。因此,在属、种、血清型和地区型层面上,序列数据现已成为环状病毒的系统分类,以及重新分类的重要工具(Attoui等,2005a;Maan等,2007,2010b;Mertens等,2005b)。

8.10.2 分子生物学诊断

分子生物学和测序技术的最新发展有助于开发非常精确、灵敏、快速和可重复的诊断工具,并用于环状病毒的鉴定。核酸杂交技术最初用于环状病毒的检测(Brown等,1988a,b;de Mattos等,1989;Gonzalez和Knudson,1987,1988;Gould,1988;Huismans和Cloete,1987;Koekemoer和Dijk,2004;Mohammed,1996;Roy等,1985;Squire等,1985a,b;Venter等,1991;Zientara等,1998)。在杂交试验中可使用放射性或非放射性标记核酸探针(Akita等,1993)。

在1990年建立了基于RT-PCR的环状病毒RNA检测方法(Dangler等,1990)。后来又对这些方法进行了各种改进(Katz等,1993;MacLachlan等,1994)。在21世纪初,基于RT-PCR,对环状病毒种类和特异血清型开发了更快和更精确的诊断方法(Anthony等,2007;Aradaib等,2003,2005,2009;Billinis等,2001;Maan等,2010a;Zientara等,

2002;Palacios 等,2011;Prasad 等,1999)。

许多“传统”的 RT-PCR 试验需要琼脂糖凝胶电泳来检测和鉴定 cDNA 扩增产物。这使得试验的劳动量增大,且要遵守严格的操作程序,否则会增加 PCR 样本污染的可能性。相反,实时 RT-PCR 试验使用一个密封管装置的检测方法,显著增加了检测敏感性并降低了污染风险。因此,目前已经根据不同环状病毒的不同片段建立许多血清群和血清型特异性实时 RT-PCR 检测方法 (Batten 等,2010;Hoffmann 等,2009;Jimenez-Clavero 等,2006;Monaco 等,2011;Orru 等,2006;Quan 等,2010;Rodriguez-Sanchez 等,2008;Shaw 等,2007;Toussaint 等,2007;Wilson 等,2009b,c)。系统发育研究证实 Seg-2 的序列差异与 BTV 血清型相关,Seg-2 被用于开发 BTV 的血清型特异性 RT-PCR 检测方法(Mertens 等,2007)。这些分析和最近开发的测序技术有助于确认入侵新地理区域的新增的病毒血清型 (Hofmann 等,2008;Maan 等,2011b)。此外,这些数据库还可用于追踪病毒的传播、起源和亲本毒株(Anthony 等,2009a,2010;Batten 等,2008;Maan 等,2008,2010b;Nomikou 等,2009;Oura 2011;Ozkul 等,2009)。

8.11 感染免疫

有关典型的环状病 BTV 的免疫应答研究相对较少,所以关于开发有效疫苗所需的病毒和哺乳动物宿主之间相互作用的理论非常有限。

BTV 感染的反刍动物产生大量抗病毒免疫应答,包括产生针对几种较保守病毒蛋白的特异性血清群抗体,例如,免疫优势蛋白 VP7。

BTV 单一血清型的中和抗体(血清型特异性抗体)通常可对 BTV 同源毒株的攻击提供免疫保护。然而,多个血清型 BTV 的相继感染可产生更广谱的免疫交叉反应中和抗体。但同时接种 3 个不同血清型 BTV 不会产生异型抗体,结果是接种 3 个血清型 BTV,有 2 个血清型进行了复制。因此,接种多价减毒活疫苗可能不会对某些血清型的 BTV 产生免疫应答。

针对 BTV 的细胞免疫应答研究非常有限。BTV 可在体外和体内有效诱导 IFN(MacLachlan 和 Thompson,1985)。然而,尚不清楚 IFN 应答对病毒清除和适应性免疫应答的影响。感染 BTV 或 EHDV 的动物会出现淋巴细胞减少症 (Ellis 等,1990;Quist 等,1997)。

CTL 对于防御细胞内的病原体感染至关重要。

已在被 BTV 感染的小鼠(Jeggo 和 Wardley 1982,1985)和羊中证实了存在抗 BTV CTL(具有血清型交叉反应性)。BTV 感染羊 T 细胞的交叉反应性与中和抗体的交叉反应不相关,这可能是由于 T 细胞通过 MHC 通路识别短的线性肽,而中和抗体主要识别构象表位(Takamatsu 和 Jeggo,1989)。

羊识别 CTL 的主要抗原是 BTV 的 NS1 和 VP2,程度较小。VP5、VP7 和 NS3 蛋白是次要抗原(Andrew 等,1995;Janardhana 等,1999;Jones等,1997)。

8.12 新发和再发

虫媒病毒包括新发和再发环状病毒和其他病原体。但在当今世界,主要差异是虫媒病毒的发生和传播更快且在地理上分布范围更广泛,这主要是由于大规模的全球贸易和运输、节肢动物媒介对城市化的适应、未能控制蚊虫密度的增加和对土地的干扰(Gould 等,2017)。

8.13 监测、预防与控制

监测策略应适用于其目的,并应包括适当的诊断方法。环状病毒的疾病控制和预防,特别是在国际水平上,应根据疾病的严重程度并基于风险/

成本-效益进行评估(Papadopoulos 等,2009)。

8.14 展望

环状病毒属包括通过库蠓、白蛉、蚊子和蜱传播的病毒，代表了其他虫媒病毒相关传播方式的演变范例(Cook 和 Holmes,2006),提供了对新发和再发病毒来源和传播的理解。类似地,环状病毒提供了一个实用的模型来研究媒介传播的分段基因组 dsRNA 病毒的进化及其传播方式。

通过对环状病毒的序列分析，可以得出以下结论:①环状病毒属为单谱系(见图 8.2 和图 8.3),并且 SCRV 是所有其他环状病毒的祖先;②库蠓、蚊子和蜱传播的环状病毒代表单独的系统发育谱系，白蛉传播的病毒群与库蠓传播的病毒在同一群；③选择不同的基因为靶基因构建环状病毒相关的系统发育(树)几乎完全一致,这意味着每个种类内的环状病毒，所有基因组片段和蛋白质均保持一定程度的同源性，这不允许种类间的重组(可能存在功能限制的证据);④在种和属水平上，VP1 基因包含足够的用来鉴别环状病毒的地区型和分类系统遗传信号；⑤在种和属水平的保守性分析表明，环状病毒的 VP1 适用于环状病毒分子进化研究；⑥环状病毒及其传播媒介一起进化,这种现象被称为物种协同进化。

参考文献

Abdy MJ, Howerth EE, Stallknecht DE (1999) Experimental infection of calves with epizootic hemorrhagic disease virus. Am J Vet Res 60(5):621–626

Afshar A (1994) Bluetongue: laboratory diagnosis. Comp Immunol Microbiol Infect Dis 17(3–4):221–242

Afshar A, Thomas FC, Wright PF, Shapiro JL, Shettigara PT, Anderson J (1987) Comparison of competitive and indirect enzyme-linked immunosorbent assays for detection of bluetongue virus antibodies in serum and whole blood. J Clin Microbiol 25(9):1705–1710

Afshar A, Anderson J, Nielsen KH, Pearson JE, Trotter HC, Woolhouse TR, Thevasagayam JA, Gall DE, Lapointe JM, Gustafson GA (1997) Evaluation of a competitive ELISA for detection of antibodies to epizootic hemorrhagic disease virus of deer. J Vet Diagn Invest 9(3):309–311

Aguero M, Gomez-Tejedor C, Angeles Cubillo M, Rubio C, Romero E, Jimenez-Clavero A (2008) Real-time fluorogenic reverse transcription polymerase chain reaction assay for detection of African horse sickness virus. J Vet Diagn Invest 20(3):325–328

Akita GY, Chinsangaram J, Osburn BI, Ianconescu M, Kaufman R (1992) Detection of bluetongue virus serogroup by polymerase chain reaction. J Vet Diagn Invest 4(4):400–405

Akita GY, Glenn J, Castro AE, Osburn BI (1993) Detection of bluetongue virus in clinical samples by polymerase chain reaction. J Vet Diagn Invest 5(2):154–158

Alexander KA, MacLachlan NJ, Kat PW, House C, O'Brien SJ, Lerche NW, Sawyer M, Frank LG, Holekamp K, Smale L et al (1994) Evidence of natural bluetongue virus infection among African carnivores. Am J Trop Med Hyg 51(5):568–576

Alexander KA, Kat PW, House J, House C, O'Brien SJ, Laurenson MK, McNutt JW, Osburn BI (1995) African horse sickness and African carnivores. Vet Microbiol 47(1–2):133–140

Anderson J (1984) Use of monoclonal antibody in a blocking ELISA to detect group specific antibodies to bluetongue virus. J Immunol Methods 74(1):139–149

Andrew M, Whiteley P, Janardhana V, Lobato Z, Gould A, Coupar B (1995) Antigen specificity of the ovine cytotoxic T lymphocyte response to bluetongue virus. Vet Immunol Immunopathol 47(3–4):311–322

Anthony SJ (2007) Genetic studies of epizootic haemorrhagic disease virus (EHDV). DPhil thesis submitted in partial fulfilment for the degree of Doctor of Philosophy, University of Oxford

Anthony S, Jones H, Darpel KE, Elliott H, Maan S, Samuel A, Mellor PS, Mertens PP (2007) A duplex RT-PCR assay for detection of genome segment 7 (VP7 gene) from 24 BTV serotypes. J Virol Methods 141(2):188–197

Anthony SJ, Maan N, Maan S, Sutton G, Attoui H, Mertens PP (2009a) Genetic and phylogenetic analysis of the core proteins VP1, VP3, VP4, VP6 and VP7 of epizootic haemorrhagic disease virus (EHDV). Virus Res 145(2):187–199

Anthony SJ, Maan N, Maan S, Sutton G, Attoui H, Mertens PP (2009b) Genetic and phylogenetic analysis of the non-structural proteins NS1, NS2 and NS3 of epizootic haemorrhagic disease virus (EHDV). Virus Res 145(2):211–219

Anthony SJ, Maan S, Maan N, Kgosana L, Bachanek-Bankowska K, Batten C, Darpel KE, Sutton

G, Attoui H, Mertens PP (2009c) Genetic and phylogenetic analysis of the outer-coat proteins VP2 and VP5 of epizootic haemorrhagic disease virus (EHDV): comparison of genetic and serological data to characterise the EHDV serogroup. Virus Res 145(2):200–210

Anthony SJ, Darpel KE, Maan S, Sutton G, Attoui H, Mertens PP (2010) The evolution of two homologues of the core protein VP6 of epizootic haemorrhagic disease virus (EHDV), which correspond to the geographical origin of the virus. Virus Genes 40(1):67–75

Anthony SJ, Darpel KE, Belaganahalli MN, Maan N, Nomikou K, Sutton G, Attoui H, Maan S, Mertens PP (2011) RNA segment 9 exists as a duplex concatemer in an Australian strain of epizootic haemorrhagic disease virus (EHDV): genetic analysis and evidence for the presence of concatemers as a normal feature of orbivirus replication. Virology 420(2):164–171

Aradaib IE (2009) PCR detection of African horse sickness virus serogroup based on genome segment three sequence analysis. J Virol Methods 159(1):1–5

Aradaib IE, Smith WL, Osburn BI, Cullor JS (2003) A multiplex PCR for simultaneous detection and differentiation of North American serotypes of bluetongue and epizootic hemorrhagic disease viruses. Comp Immunol Microbiol Infect Dis 26(2):77–87

Aradaib IE, Mohamed ME, Abdalla TM, Sarr J, Abdalla MA, Yousof MA, Hassan YA, Karrar AR (2005) Serogrouping of United States and some African serotypes of bluetongue virus using RT-PCR. Vet Microbiol 111(3–4):145–150

Aradaib IE, Mohamed ME, Abdalla MA (2009) A single-tube RT-PCR for rapid detection and differentiation of some African isolates of palyam serogroup orbiviruses. J Virol Methods 161(1):70–74

Attoui H, Stirling JM, Munderloh UG, Billoir F, Brookes SM, Burroughs JN, de Micco P, Mertens PP, de Lamballerie X (2001) Complete sequence characterization of the genome of the St Croix River virus, a new orbivirus isolated from cells of Ixodes scapularis. J Gen Virol 82(Pt 4):795–804

Attoui H, Fang Q, Mohd Jaafar F, Cantaloube JF, Biagini P, de Micco P, de Lamballerie X (2002a) Common evolutionary origin of aquareoviruses and orthoreoviruses revealed by genome characterization of Golden shiner reovirus, Grass carp reovirus, Striped bass reovirus and golden ide reovirus (genus Aquareovirus, family Reoviridae). J Gen Virol 83(Pt 8):1941–1951

Attoui H, Mohd Jaafar F, Biagini P, Cantaloube JF, de Micco P, Murphy FA, de Lamballerie X (2002b) Genus Coltivirus (family Reoviridae): genomic and morphologic characterization of Old World and New World viruses. Arch Virol 147(3):533–561

Attoui H, Mohd Jaafar F, Belhouchet M, Aldrovandi N, Tao S, Chen B, Liang G, Tesh RB, de Micco P, de Lamballerie X (2005a) Yunnan orbivirus, a new orbivirus species isolated from Culex tritaeniorhynchus mosquitoes in China. J Gen Virol 86(Pt 12):3409–3417

Attoui H, Mohd Jaafar F, Belhouchet M, Biagini P, Cantaloube JF, de Micco P, de Lamballerie X (2005b) Expansion of family Reoviridae to include nine-segmented dsRNA viruses: isolation and characterization of a new virus designated Aedes pseudoscutellaris reovirus assigned to a proposed genus (Dinovernavirus). Virology 343(2):212–223

Attoui H, Maan SS, Anthony SJ, Mertens PPC (2009a) Bluetongue virus, other orbiviruses and other reoviruses: their relationships and taxonomy. In: Mellor PS, Baylis M, Mertens PPC (eds) Bluetongue monograph, 1st edn. Elsevier/Academic Press, London, pp 23–552

Attoui H, Mendez-Lopez MR, Rao S, Hurtado-Alendes A, Lizaraso-Caparo F, Jaafar FM, Samuel AR, Belhouchet M, Pritchard LI, Melville L, Weir RP, Hyatt AD, Davis SS, Lunt R, Calisher CH, Tesh RB, Fujita R, Mertens PP (2009b) Peruvian horse sickness virus and Yunnan orbivirus, isolated from vertebrates and mosquitoes in Peru and Australia. Virology 394(2):298–310

Attoui H, Mertens PPC, Becnel J, Belaganahalli M, Bergoin M, Brussaard CP, Chappell JD, Ciarlet M, del Vas M, Dermody TS, Dormitzer PR, Duncan R, Fcang Q, Graham R, Guglielmi KM, Harding RM, Hillman B, Maan S, Makkay A, Marzachì C, Matthijnssens J et al (2011) Reoviridae. Elsevier Academic Press, London

Backx A, Heutink R, van Rooij E, van Rijn P (2009) Transplacental and oral transmission of wild-type bluetongue virus serotype 8 in cattle after experimental infection. Vet Microbiol 138(3–4):235–243

Bansal OB, Stokes A, Bansal A, Bishop D, Roy P (1998) Membrane organization of bluetongue virus nonstructural glycoprotein NS3. J Virol 72(4):3362–3369

Barnard BJ (1997) Antibodies against some viruses of domestic animals in southern African wild animals. Onderstepoort J Vet Res 64(2):95–110

Barratt-Boyes SM, MacLachlan NJ (1995) Pathogenesis of bluetongue virus infection of cattle. J Am Vet Med Assoc 206(9):1322–1329

Barros SC, Cruz B, Luis TM, Ramos F, Fagulha T, Duarte M, Henriques M, Fevereiro M (2009) A DIVA system based on the detection of antibodies to non-structural protein 3 (NS3) of bluetongue virus. Vet Microbiol 137(3–4):252–259

Batten CA, Maan S, Shaw AE, Maan NS, Mertens PP (2008) A European field strain of bluetongue virus derived from two parental vaccine strains by genome segment reassortment. Virus Res 137(1):56–63

Batten CA, Sanders AJ, Bachanek-Bankowska K, Bin-Tarif A, Oura CA (2009) Bluetongue virus:

European Community proficiency test (2007) to evaluate ELISA and RT-PCR detection methods with special reference to pooling of samples. Vet Microbiol 135(3–4):380–383

Batten CA, van Rijn PA, Oura CA (2010) Detection of the European 'field' strain of bluetongue virus serotype 6 by real-time RT-PCR. Vet Microbiol 141(1–2):186–188

Batten C, Darpel K, Henstock M, Fay P, Veronesi E, Gubbins S, Graves S, Frost L, Oura C (2014) Evidence for transmission of bluetongue virus serotype 26 through direct contact. PLoS One 9(5):e96049

Beaton AR, Rodriguez J, Reddy YK, Roy P (2002) The membrane trafficking protein calpactin forms a complex with bluetongue virus protein NS3 and mediates virus release. Proc Natl Acad Sci U S A 99(20):13154–13159

Belaganahalli MN (2012) Full genome sequencing and phylogenetic analyses of different Orbivirus species. PhD thesis, University of London.

Belaganahalli MN, Maan S, Maan NS, Tesh R, Attoui H, Mertens PP (2011) Umatilla virus genome sequencing and phylogenetic analysis: identification of stretch lagoon orbivirus as a new member of the Umatilla virus species. PLoS One 6(8):e23605

Belaganahalli MN, Maan S, Maan NS, Brownlie J, Tesh R, Attoui H, Mertens PP (2015) Genetic characterization of the tick-borne orbiviruses. Viruses 7(5):2185–2209

Belhouchet M, Mohd Jaafar F, Tesh R, Grimes J, Maan S, Mertens PP, Attoui H (2010) Complete sequence of Great Island virus and comparison with the T2 and outer-capsid proteins of Kemerovo, Lipovnik and Tribec viruses (genus Orbivirus, family Reoviridae). J Gen Virol 91(Pt 12):2985–2993

Belhouchet M, Mohd Jaafar F, Firth AE, Grimes JM, Mertens PP, Attoui H (2011) Detection of a fourth orbivirus non-structural protein. PLoS One 6(10):e25697

Bentley L, Fehrsen J, Jordaan F, Huismans H, du Plessis DH (2000) Identification of antigenic regions on VP2 of African horse sickness virus serotype 3 by using phage-displayed epitope libraries. J Gen Virol 81(Pt 4):993–1000

Bevan LEW (1911) The transmission of African horse sickness to the dog by feeding. Vet J 67:402–408

Bhattacharya B, Noad RJ, Roy P (2007) Interaction between Bluetongue virus outer capsid protein VP2 and vimentin is necessary for virus egress. Virol J 4:7

Billinis C, Koumbati M, Spyrou V, Nomikou K, Mangana O, Panagiotidis CA, Papadopoulos O (2001) Bluetongue virus diagnosis of clinical cases by a duplex reverse transcription-PCR: a comparison with conventional methods. J Virol Methods 98(1):77–89

Blackburn NK, Swanepoel R (1988) Observations on antibody levels associated with active and passive immunity to African horse sickness. Trop Anim Health Prod 20(4):203–210

Bonneau KR, Zhang NZ, Wilson WC, Zhu JB, Zhang FQ, Li ZH, Zhang KL, Xiao L, Xiang WB, MacLchlan NJ (2000) Phylogenetic analysis of the S7 gene does not segregate Chinese strains of bluetongue virus into a single topotype. Arch Virol 145(6):1163–1171

Bonneau KR, Mullens BA, MacLachlan NJ (2001) Occurrence of genetic drift and founder effect during quasispecies evolution of the VP2 and NS3/NS3A genes of bluetongue virus upon passage between sheep, cattle, and Culicoides sonorensis. J Virol 75(17):8298–8305

Borden EC, Murphy FA, Nathanson N, Monath TP (1971a) Effect of antilymphocyte serum on Tacaribe virus infection in infant mice. Infect Immun 3(3):466–471

Borden EC, Shope RE, Murphy FA (1971b) Physicochemical and morphological relationships of some arthropod-borne viruses to bluetongue virus—a new taxonomic group. Physicochemical and serological studies. J Gen Virol 13(2):261–271

Boughton CR, Hawkes RA, Naim HM (1990) Arbovirus infection in humans in NSW: seroprevalence and pathogenicity of certain Australian bunyaviruses. Aust N Z J Med 20(1):51–55

Boulanger P, Ruckerbauer GM, Bannister GL, Gray DP, Girard A (1967) Studies on bluetongue. 3. Comparison of two complement-fixation methods. Can J Comp Med Vet Sci 31(7):166–170

Bowen RA, Howard TH, Entwistle KW, Pickett BW (1983) Seminal shedding of bluetongue virus in experimentally infected mature bulls. Am J Vet Res 44(12):2268–2270

Boyce M, Wehrfritz J, Noad R, Roy P (2004) Purified recombinant bluetongue virus VP1 exhibits RNA replicase activity. J Virol 78(8):3994–4002

Boyce M, Celma CC, Roy P (2012) Bluetongue virus non-structural protein 1 is a positive regulator of viral protein synthesis. Virol J 9:178

Boyce M, McCrae MA, Boyce P, Kim JT (2016) Inter-segment complementarity in orbiviruses: a driver for co-ordinated genome packaging in the Reoviridae? J Gen Virol 97(5):1145–1157

Bremer CW, Viljoen GJ (1998) Detection of African horse sickness virus and discrimination between two equine orbivirus serogroups by reverse transcription polymerase chain reaction. Onderstepoort J Vet Res 65(1):1–8

Brookes SM, Hyatt AD, Eaton BT (1993) Characterization of virus inclusion bodies in bluetongue virus-infected cells. J Gen Virol 74(Pt 3):525–530

Brown SE, Gonzalez HA, Bodkin DK, Tesh RB, Knudson DL (1988a) Intra- and inter-serogroup genetic relatedness of orbiviruses. II. Blot hybridization and reassortment in vitro of epizootic haemorrhagic disease serogroup, bluetongue type 10 and Pata viruses. J Gen Virol 69(Pt

1):135–147
Brown SE, Morrison HG, Buckley SM, Shope RE, Knudson DL (1988b) Genetic relatedness of the Kemerovo serogroup viruses: I. RNA-RNA blot hybridization and gene reassortment in vitro of the Kemerovo serocomplex. Acta Virol 32(5):369–378
Brown SE, Morrison HG, Karabatsos N, Knudson DL (1991) Genetic relatedness of two new Orbivirus serogroups: Orungo and Lebombo. J Gen Virol 72(Pt 5):1065–1072
Brown CC, Rhyan JC, Grubman MJ, Wilbur LA (1996) Distribution of bluetongue virus in tissues of experimentally infected pregnant dogs as determined by in situ hybridization. Vet Pathol 33(3):337–340
Burroughs JN, O'Hara RS, Smale CJ, Hamblin C, Walton A, Armstrong R, Mertens PP (1994) Purification and properties of virus particles, infectious subviral particles, cores and VP7 crystals of African horse sickness virus serotype 9. J Gen Virol 75(Pt 8):1849–1857
Butan C, Tucker P (2010) Insights into the role of the non-structural protein 2 (NS2) in Bluetongue virus morphogenesis. Virus Res 151(2):109–117
Calisher CH, Childs JE, Field HE, Holmes KV, Schountz T (2006) Bats: important reservoir hosts of emerging viruses. Clin Microbiol Rev 19(3):531–545
Cao D, Barro M, Hoshino Y (2008) Porcine rotavirus bearing an aberrant gene stemming from an intergenic recombination of the NSP2 and NSP5 genes is defective and interfering. J Virol 82(12):6073–6077
Carpi G, Holmes EC, Kitchen A (2010) The evolutionary dynamics of bluetongue virus. J Mol Evol 70(6):583–592
Caspar DL, Klug A (1962) Physical principles in the construction of regular viruses. Cold Spring Harb Symp Quant Biol 27:1–24
Celma CC, Roy P (2009) A viral nonstructural protein regulates bluetongue virus trafficking and release. J Virol 83(13):6806–6816
Celma CC, Roy P (2011) Interaction of calpactin light chain (S100A10/p11) and a viral NS protein is essential for intracellular trafficking of nonenveloped bluetongue virus. J Virol 85(10):4783–4791
Cook S, Holmes EC (2006) A multigene analysis of the phylogenetic relationships among the flaviviruses (Family: Flaviviridae) and the evolution of vector transmission. Arch Virol 151(2):309–325
Cowled C, Melville L, Weir R, Walsh S, Hyatt A, Van Driel R, Davis S, Gubala A, Boyle D (2007) Genetic and epidemiological characterization of Middle Point orbivirus, a novel virus isolated from sentinel cattle in northern Australia. J Gen Virol 88(Pt 12):3413–3422
Cowled C, Palacios G, Melville L, Weir R, Walsh S, Davis S, Gubala A, Lipkin WI, Briese T, Boyle D (2009) Genetic and epidemiological characterization of Stretch Lagoon orbivirus, a novel orbivirus isolated from Culex and Aedes mosquitoes in northern Australia. J Gen Virol 90(Pt 6):1433–1439
Cowley JA, Gorman BM (1989) Cross-neutralization of genetic reassortants of bluetongue virus serotypes 20 and 21. Vet Microbiol 19(1):37–51
Cowley JA, Gorman BM (1990) Effects of proteolytic enzymes on the infectivity, haemagglutinating activity and protein composition of bluetongue virus type 20. Vet Microbiol 22(2–3):137–152
Crafford JE, Guthrie AJ, van Vuuren M, Mertens PP, Burroughs JN, Howell PG, Hamblin C (2003) A group-specific, indirect sandwich ELISA for the detection of equine encephalosis virus antigen. J Virol Methods 112(1–2):129–135
Crafford JE, Guthrie AJ, Van Vuuren M, Mertens PP, Burroughs JN, Howell PG, Batten CA, Hamblin C (2011) A competitive ELISA for the detection of group-specific antibody to equine encephalosis virus. J Virol Methods 174(1–2):60–64
Dandawate CN (1974) Antigenic relationship among Palyam, Kasba and Vellore viruses—a new serogroup of arboviruses. Indian J Med Res 62(3):326–331
Dandawate CN, Shope RE (1975) Studies on physicochemical and biological properties of two ungrouped arboviruses: Minnal and Arkonam. Indian J Med Res 63(8):1180–1187
Dandawate CN, Rajagopalan PK, Pavri KM, Work TH (1969) Virus isolations from mosquitoes collected in North Arcot district, Madras state, and Chittoor district, Andhra Pradesh between November 1955 and October 1957. Indian J Med Res 57(8):1420–1426
Dangler CA, de Mattos CA, de Mattos CC, Osburn BI (1990) Identifying bluetongue virus ribonucleic acid sequences by the polymerase chain reaction. J Virol Methods 28(3):281–292
Darpel KE (2007) The bluetongue virus 'ruminant host–insect vector' transmission cycle; the role of Culicoides saliva proteins in infection. PhD thesis submitted in partial fulfilment for the degree of Doctor of Philosophy, University of London
Darpel KE, Batten CA, Veronesi E, Shaw AE, Anthony S, Bachanek-Bankowska K, Kgosana L, bin-Tarif A, Carpenter S, Muller-Doblies UU, Takamatsu HH, Mellor PS, Mertens PP, Oura CA (2007) Clinical signs and pathology shown by British sheep and cattle infected with bluetongue virus serotype 8 derived from the 2006 outbreak in northern Europe. Vet Rec 161(8):253–261
Darpel KE, Batten CA, Veronesi E, Williamson S, Anderson P, Dennison M, Clifford S, Smith C,

Philips L, Bidewell C, Bachanek-Bankowska K, Sanders A, Bin-Tarif A, Wilson AJ, Gubbins S, Mertens PP, Oura CA, Mellor PS (2009) Transplacental transmission of bluetongue virus 8 in cattle, UK. Emerg Infect Dis 15(12):2025–2028

Darpel KE, Langner KF, Nimtz M, Anthony SJ, Brownlie J, Takamatsu HH, Mellor PS, Mertens PP (2011) Saliva proteins of vector Culicoides modify structure and infectivity of bluetongue virus particles. PLoS One 6(3):e17545

Davies FG, Blackburn NK (1971) The typing of bluetongue virus. Res Vet Sci 12(2):181–183

Davies FG, Otieno S (1977) Elephants and zebras as possible reservoir hosts for African horse sickness virus. Vet Rec 100(14):291–292

de Mattos CC, de Mattos CA, Osburn BI, Dangler CA, Chuang RY, Doi RH (1989) Recombinant DNA probe for serotype-specific identification of bluetongue virus 17. Am J Vet Res 50(4):536–541

de Mattos CC, de Mattos CA, MacLachlan NJ, Giavedoni LD, Yilma T, Osburn BI (1996) Phylogenetic comparison of the S3 gene of United States prototype strains of bluetongue virus with that of field isolates from California. J Virol 70(8):5735–5739

de Waal PJ, Huismans H (2005) Characterization of the nucleic acid binding activity of inner core protein VP6 of African horse sickness virus. Arch Virol 150(10):2037–2050

Della-Porta AJ (1985) Classification of orbiviruses: a need for supergroups of genera. Prog Clin Biol Res 178:267–274

Della-Porta AJ, Sellers RF, Herniman KA, Littlejohns IR, Cybinski DH, St George TD, McPhee DA, Snowdon WA, Campbell J, Cargill C, Corbould A, Chung YS, Smith VW (1983) Serological studies of Australian and Papua New Guinean cattle and Australian sheep for the presence of antibodies against bluetongue group viruses. Vet Microbiol 8(2):147–162

Della-Porta AJ, Parsonson IM, McPhee DA (1985) Problems in the interpretation of diagnostic tests due to cross-reactions between orbiviruses and broad serological responses in animals. Prog Clin Biol Res 178:445–453

DeMaula CD, Bonneau KR, MacLachlan NJ (2000) Changes in the outer capsid proteins of bluetongue virus serotype ten that abrogate neutralization by monoclonal antibodies. Virus Res 67(1):59–66

DeMaula CD, Leutenegger CM, Jutila MA, MacLachlan NJ (2002) Bluetongue virus-induced activation of primary bovine lung microvascular endothelial cells. Vet Immunol Immunopathol 86(3–4):147–157

Dilcher M, Hasib L, Lechner M, Wieseke N, Middendorf M, Marz M, Koch A, Spiegel M, Dobler G, Hufert FT, Weidmann M (2012) Genetic characterization of Tribec virus and Kemerovo virus, two tick-transmitted human-pathogenic Orbiviruses. Virology 423(1):68–76

Diprose JM, Burroughs JN, Sutton GC, Goldsmith A, Gouet P, Malby R, Overton I, Zientara S, Mertens PP, Stuart DI, Grimes JM (2001) Translocation portals for the substrates and products of a viral transcription complex: the bluetongue virus core. EMBO J 20(24):7229–7239

Diprose JM, Grimes JM, Sutton GC, Burroughs JN, Meyer A, Maan S, Mertens PP, Stuart DI (2002) The core of bluetongue virus binds double-stranded RNA. J Virol 76(18):9533–9536

Doherty RL, Carley JG, Mackerras MJ, Marks EN (1963) Studies of arthropod-borne virus infections in Queensland. III. Isolation and characterization of virus strains from wild-caught mosquitoes in North Queensland. Aust J Exp Biol Med Sci 41:17–39

Doherty RL, Whitehead RH, Wetters EJ, Gorman BM, Carley JG (1970) A survey of antibody to 10 arboviruses (Koongol group, Mapputta group and ungrouped) isolated in Queensland. Trans R Soc Trop Med Hyg 64(5):748–753

Domingo E, Holland JJ (1997) RNA virus mutations and fitness for survival. Annu Rev Microbiol 51:151–178

Eaton BT, Gould AR (1987) Isolation and characterization of orbivirus genotypic variants. Virus Res 6(4):363–382

Eaton BT, Hyatt AD, White JR (1988) Localization of the nonstructural protein NS1 in bluetongue virus-infected cells and its presence in virus particles. Virology 163(2):527–537

Elbers A, Backx A, van der Spek A, Ekker M, Leijs P, Steijn K, van Langen H, van Rijn P (2008a) [Epidemiology of bluetongue virus serotype 8 outbreaks in the Netherlands in 2006]. Tijdschr Diergeneeskd 133(6):222–229

Elbers AR, Popma J, Oosterwolde S, van Rijn PA, Vellema P, van Rooij EM (2008b) A cross-sectional study to determine the seroprevalence of bluetongue virus serotype 8 in sheep and goats in 2006 and 2007 in the Netherlands. BMC Vet Res 4:33

Elbers AR, van der Spek AN, van Rijn PA (2009) Epidemiologic characteristics of bluetongue virus serotype 8 laboratory-confirmed outbreaks in The Netherlands in 2007 and a comparison with the situation in 2006. Prev Vet Med 92(1–2):1–8

Elena SF, Sanjuan R (2005) Adaptive value of high mutation rates of RNA viruses: separating causes from consequences. J Virol 79(18):11555–11558

Ellis JA, Luedke AJ, Davis WC, Wechsler SJ, Mecham JO, Pratt DL, Elliott JD (1990) T lymphocyte subset alterations following bluetongue virus infection in sheep and cattle. Vet Immunol Immunopathol 24(1):49–67

Els HJ, Verwoerd DW (1969) Morphology of bluetongue virus. Virology 38(2):213–219
Fenner F (1976) The classification and nomenclature of viruses. Summary of results of meetings of the International Committee on Taxonomy of Viruses in Madrid, September 1975. J Gen Virol 31(3):463–470
Fernandez-Pinero J, Fernandez-Pacheco P, Rodriguez B, Sotelo E, Robles A, Arias M, Sanchez-Vizcaino JM (2009) Rapid and sensitive detection of African horse sickness virus by real-time PCR. Res Vet Sci 86(2):353–358
Fields BN, Joklik WK (1969) Isolation and preliminary genetic and biochemical characterization of temperature-sensitive mutants of reovirus. Virology 37(3):335–342
Fields BN, Knipe DM, Howley PM (1985) Fields virology, 3rd edn. Lippincott-Raven, Philadelphia, PA
Forzan M, Wirblich C, Roy P (2004) A capsid protein of nonenveloped Bluetongue virus exhibits membrane fusion activity. Proc Natl Acad Sci U S A 101(7):2100–2105
Forzan M, Marsh M, Roy P (2007) Bluetongue virus entry into cells. J Virol 81(9):4819–4827
French TJ, Inumaru S, Roy P (1989) Expression of two related nonstructural proteins of bluetongue virus (BTV) type 10 in insect cells by a recombinant baculovirus: production of polyclonal ascitic fluid and characterization of the gene product in BTV-infected BHK cells. J Virol 63(8):3270–3278
Fu H, Leake CJ, Mertens PP, Mellor PS (1999) The barriers to bluetongue virus infection, dissemination and transmission in the vector, Culicoides variipennis (Diptera: Ceratopogonidae). Arch Virol 144(4):747–761
Gale P, Brouwer A, Ramnial V, Kelly L, Kosmider R, Fooks AR, Snary EL (2010) Assessing the impact of climate change on vector-borne viruses in the EU through the elicitation of expert opinion. Epidemiol Infect 138(2):214–225
Gambles RM (1949) Bluetongue of sheep on Cyprus. J Comp Pathol 59:176–190
Gard GP, Shorthose JE, Weir RP, Erasmus BJ (1987) The isolation of a bluetongue serotype new to Australia. Aust Vet J 64(3):87–88
Gibbs EP, Lawman MJ (1977) Infection of British deer and farm animals with epizootic haemorrhagic disease of deer virus. J Comp Pathol 87(3):335–343
Gold S, Monaghan P, Mertens P, Jackson T (2010) A clathrin independent macropinocytosis-like entry mechanism used by bluetongue virus-1 during infection of BHK cells. PLoS One 5(6):e11360
Gonzalez HA, Knudson DL (1987) Genetic relatedness of corriparta serogroup viruses. J Gen Virol 68(Pt 3):661–672
Gonzalez HA, Knudson DL (1988) Intra- and inter-serogroup genetic relatedness of orbiviruses. I. Blot hybridization of viruses of Australian serogroups. J Gen Virol 69(Pt 1):125–134
Gorman BM (1978) Susceptibility of orbiviruses to low pH and to organic solvents. Aust J Exp Biol Med Sci 56(3):359–367
Gorman BM (1979) Variation in orbiviruses. J Gen Virol 44(1):1–15
Gorman BM (1986) Evolutionary relationships among orbiviruses. Rev Sci Tech Off Int Epiz 5(2):323–332
Gorman BM, Taylor J (1978) The RNA genome of Tilligerry virus. Aust J Exp Biol Med Sci 56(3):369–371
Gorman BM, Taylor J, Walker PJ, Davidson WL, Brown F (1981) Comparison of bluetongue type 20 with certain viruses of the bluetongue and Eubenangee serological groups of orbiviruses. J Gen Virol 57(Pt 2):251–261
Gorman BM, Taylor J, Walker PJ (1983) Orbiviruses. In: Joklik WK (ed) The Reoviridae. Plenum Press, New York, pp 287–357
Goto Y, Miura Y, Kono Y (1988a) Epidemiological survey of an epidemic of congenital abnormalities with hydranencephaly-cerebellar hypoplasia syndrome of calves occurring in 1985/86 and seroepidemiological investigations on Chuzan virus, a putative causal agent of the disease, in Japan. Nippon Juigaku Zasshi 50(2):405–413
Goto Y, Miura Y, Kono Y (1988b) Serologic evidence for the etiologic role of Chuzan virus in an epizootic of congenital abnormalities with hydranencephaly-cerebellar hypoplasia syndrome of calves in Japan. Am J Vet Res 49(12):2026–2029
Gouet P, Diprose JM, Grimes JM, Malby R, Burroughs JN, Zientara S, Stuart DI, Mertens PP (1999) The highly ordered double-stranded RNA genome of bluetongue virus revealed by crystallography. Cell 97(4):481–490
Gould AR (1988) The use of recombinant DNA probes to group and type orbiviruses. A comparison of Australian and South African isolates. Arch Virol 99(3–4):205–220
Gould AR, Eaton BT (1990) The amino acid sequence of the outer coat protein VP2 of neutralizing monoclonal antibody-resistant, virulent and attenuated bluetongue viruses. Virus Res 17(3):161–172
Gould AR, Hyatt AD (1994) The orbivirus genus. Diversity, structure, replication and phylogenetic relationships. Comp Immunol Microbiol Infect Dis 17(3–4):163–188
Gould AR, Pritchard LI (1991) Phylogenetic analyses of the complete nucleotide sequence of the

capsid protein (VP3) of Australian epizootic haemorrhagic disease of deer virus (serotype 2) and cognate genes from other orbiviruses. Virus Res 21(1):1–18

Gould EA, de Lamballerie X, Zanotto PM, Holmes EC (2003) Origins, evolution, and vector/host coadaptations within the genus Flavivirus. Adv Virus Res 59:277–314

Gould E, Pettersson J, Higgs S, Charrel R, de Lamballerie X (2017) Emerging arboviruses: why today? One Health 4:1–13

Grimes J, Basak AK, Roy P, Stuart D (1995) The crystal structure of bluetongue virus VP7. Nature 373(6510):167–170

Grimes JM, Burroughs JN, Gouet P, Diprose JM, Malby R, Zientara S, Mertens PP, Stuart DI (1998) The atomic structure of the bluetongue virus core. Nature 395(6701):470–478

Guirakhoo F, Catalan JA, Monath TP (1995) Adaptation of bluetongue virus in mosquito cells results in overexpression of NS3 proteins and release of virus particles. Arch Virol 140(5):967–974

Gumm ID, Newman JF (1982) The preparation of purified bluetongue virus group antigen for use as a diagnostic reagent. Arch Virol 72(1–2):83–93

Guthrie AJ, Pardini AD, Howell PG (2009) Equine encephalosis. Equine Vet Edu Manual 8:145–150

Gutsche T (1979) There was a man. Timmins, Cape Town, p 4

Hamblin C, Graham SD, Anderson EC, Crowther JR (1990) A competitive ELISA for the detection of group-specific antibodies to African horse sickness virus. Epidemiol Infect 104(2):303–312

Hamblin C, Mellor PS, Boned J (1991) The use of ELISA for the detection of African horse sickness viruses in Culicoides midges. J Virol Methods 34(2):221–225

Han Z, Harty RN (2004) The NS3 protein of bluetongue virus exhibits viroporin-like properties. J Biol Chem 279(41):43092–43097

Hanada K, Suzuki Y, Gojobori T (2004) A large variation in the rates of synonymous substitution for RNA viruses and its relationship to a diversity of viral infection and transmission modes. Mol Biol Evol 21(6):1074–1080

Hassan SS, Roy P (1999) Expression and functional characterization of bluetongue virus VP2 protein: role in cell entry. J Virol 73(12):9832–9842

Hassan SH, Wirblich C, Forzan M, Roy P (2001) Expression and functional characterization of bluetongue virus VP5 protein: role in cellular permeabilization. J Virol 75(18):8356–8367

Hawkes RA, Kirkland PD, Sanders DA, Zhang F, Li Z, Davis RJ, Zhang N (2000) Laboratory and field studies of an antigen capture ELISA for bluetongue virus. J Virol Methods 85(1–2):137–149

He CQ, Ding NZ, He M, Li SN, Wang XM, He HB, Liu XF, Guo HS (2010) Intragenic recombination as a mechanism of genetic diversity in bluetongue virus. J Virol 84(21):11487–11495

Hemati B, Contreras V, Urien C, Bonneau M, Takamatsu HH, Mertens PP, Breard E, Sailleau C, Zientara S, Schwartz-Cornil I (2009) Bluetongue virus targets conventional dendritic cells in skin lymph. J Virol 83(17):8789–8799

Henning MM (1956) African horse sickness, perdesiekte, Pestis equorum. In: Animal diseases in South Africa, 3rd edn. Central News Agency, Johannesburg, pp 785–808

Hewat EA, Booth TF, Loudon PT, Roy P (1992a) Three-dimensional reconstruction of baculovirus expressed bluetongue virus core-like particles by cryo-electron microscopy. Virology 189(1):10–20

Hewat EA, Booth TF, Roy P (1992b) Structure of bluetongue virus particles by cryoelectron microscopy. J Struct Biol 109(1):61–69

Hoff GL, Trainer DO (1974) Observations on bluetongue and epizootic hemorrhagic disease viruses in white-tailed deer: (1) distribution of virus in the blood (2) cross-challenge. J Wildl Dis 10(1):25–31

Hoff GL, Griner LA, Trainer DO (1973) Bluetongue virus in exotic ruminants. J Am Vet Med Assoc 163(6):565–567

Hoffmann B, Eschbaumer M, Beer M (2009) Real-time quantitative reverse transcription-PCR assays specifically detecting bluetongue virus serotypes 1, 6, and 8. J Clin Microbiol 47(9):2992–2994

Hofmann MA, Renzullo S, Mader M, Chaignat V, Worwa G, Thuer B (2008) Genetic characterization of Toggenburg orbivirus, a new bluetongue virus, from goats, Switzerland. Emerg Infect Dis 14(12):1855–1861

Holland J, Spindler K, Horodyski F, Grabau E, Nichol S, VandePol S (1982) Rapid evolution of RNA genomes. Science 215(4540):1577–1585

Hooper PT, Lunt RA, Gould AR, Hyatt AD, Russell GM, Kattenbelt JA, Blacksell SD, Reddacliff LA, Kirkland PD, Davis RJ, Durham PJ, Bishop AL, Waddington J (1999) Epidemic of blindness in kangaroos—evidence of a viral aetiology. Aust Vet J 77(8):529–536

Hosseini M, Hawkes RA, Kirkland PD, Dixon RJ (1998) Rapid screening of embryonated chicken eggs for bluetongue virus infection with an antigen capture enzyme linked immunosorbent assay. J Virol Methods 75(1):39–46

House C, Mikiciuk PE, Berninger ML (1990) Laboratory diagnosis of African horse sickness: comparison of serological techniques and evaluation of storage methods of samples for virus

isolation. J Vet Diagn Invest 2(1):44–50
Howell PG, Kumm NA, Botha MJ (1970) The application of improved techniques to the identification of strains of bluetongue virus. Onderstepoort J Vet Res 37(1):59–66
Huismans H (1979) Protein synthesis in bluetongue virus-infected cells. Virology 92(2):385–396
Huismans H, Cloete M (1987) A comparison of different cloned bluetongue virus genome segments as probes for the detection of virus-specified RNA. Virology 158(2):373–380
Huismans H, Els HJ (1979) Characterization of the tubules associated with the replication of three different orbiviruses. Virology 92(2):397–406
Huismans H, Erasmus BJ (1981) Identification of the serotype-specific and group-specific antigens of bluetongue virus. Onderstepoort J Vet Res 48(2):51–58
Huismans H, Bremer CW, Barber TL (1979) The nucleic acid and proteins of epizootic haemorrhagic disease virus. Onderstepoort J Vet Res 46(2):95–104
Huismans H, van der Walt NT, Cloete M, Erasmus BJ (1987) Isolation of a capsid protein of bluetongue virus that induces a protective immune response in sheep. Virology 157(1):172–179
Hundley F, McIntyre M, Clark B, Beards G, Wood D, Chrystie I, Desselberger U (1987) Heterogeneity of genome rearrangements in rotaviruses isolated from a chronically infected immunodeficient child. J Virol 61(11):3365–3372
Hutcheon D (1902) Malarial catarrhal fever of sheep. Vet Rec 14:629–633
Hyatt AD, Eaton BT, Brookes SM (1989) The release of bluetongue virus from infected cells and their superinfection by progeny virus. Virology 173(1):21–34
Hyatt AD, Zhao Y, Roy P (1993) Release of bluetongue virus-like particles from insect cells is mediated by BTV nonstructural protein NS3/NS3A. Virology 193(2):592–603
Inumaru S, Roy P (1987) Production and characterization of the neutralization antigen VP2 of bluetongue virus serotype 10 using a baculovirus expression vector. Virology 157(2):472–479
Janardhana V, Andrew ME, Lobato ZI, Coupar BE (1999) The ovine cytotoxic T lymphocyte responses to bluetongue virus. Res Vet Sci 67(3):213–221
Jansen CC, Prow NA, Webb CE, Hall RA, Pyke AT, Harrower BJ, Pritchard IL, Zborowski P, Ritchie SA, Russell RC, Van Den Hurk AF (2009) Arboviruses isolated from mosquitoes collected from urban and peri-urban areas of eastern Australia. J Am Mosq Control Assoc 25(3):272–278
Jeggo MH, Wardley RC (1982) Generation of cross-reactive cytotoxic T lymphocytes following immunization of mice with various bluetongue virus types. Immunology 45(4):629–635
Jeggo MH, Wardley RC (1985) Bluetongue vaccine: cells and/or antibodies. Vaccine 3(1):57–58
Jensen MJ, Wilson WC (1995) A model for the membrane topology of the NS3 protein as predicted from the sequence of segment 10 of epizootic haemorrhagic disease virus serotype 1. Arch Virol 140(4):799–805
Jimenez-Clavero MA, Aguero M, San Miguel E, Mayoral T, Lopez MC, Ruano MJ, Romero E, Monaco F, Polci A, Savini G, Gomez-Tejedor C (2006) High throughput detection of bluetongue virus by a new real-time fluorogenic reverse transcription-polymerase chain reaction: application on clinical samples from current Mediterranean outbreaks. J Vet Diagn Invest 18(1):7–17
Jochim MM, Chow TL (1969) Immunodiffusion of bluetongue virus. Am J Vet Res 30(1):33–41
Jones LD, Williams T, Bishop D, Roy P (1997) Baculovirus-expressed nonstructural protein NS2 of bluetongue virus induces a cytotoxic T-cell response in mice which affords partial protection. Clin Diagn Lab Immunol 4(3):297–301
Kar AK, Bhattacharya B, Roy P (2007) Bluetongue virus RNA binding protein NS2 is a modulator of viral replication and assembly. BMC Mol Biol 8:4
Karabatsos N (1985) International catalogue of arboviruses including certain other viruses of vertebrates, 3rd edn. American Society of Tropical Medicine and Hygiene, San Antonio, TX
Karstad L, Winter A, Trainer DO (1961) Pathology of epizootic hemorrhagic disease of deer. Am J Vet Res 22:227–235
Katz JB, Alstad AD, Gustafson GA, Moser KM (1993) Sensitive identification of bluetongue virus serogroup by a colorimetric dual oligonucleotide sorbent assay of amplified viral nucleic acid. J Clin Microbiol 31(11):3028–3030
Kirkland P (2005) Epidemic viral diseases of wildlife—sudden death in tammar wallabies, blind kangaroos, herpesviruses in pilchards-what next? Microbiol Aust 26:82–84
Koekemoer JJ, Dijk AA (2004) African horse sickness virus serotyping and identification of multiple co-infecting serotypes with a single genome segment 2 RT-PCR amplification and reverse line blot hybridization. J Virol Methods 122(1):49–56
Kusari J, Roy P (1986) Molecular and genetic comparisons of two serotypes of epizootic hemorrhagic disease of deer virus. Am J Vet Res 47(8):1713–1717
Laviada MD, Babin M, Dominguez J, Sanchez-Vizcaino JM (1992) Detection of African horse sickness virus in infected spleens by a sandwich ELISA using two monoclonal antibodies specific for VP7. J Virol Methods 38(2):229–242
Laviada MD, Roy P, Sanchez-Vizcaino JM, Casal JI (1995) The use of African horse sickness virus NS3 protein, expressed in bacteria, as a marker to differentiate infected from vaccinated horses.

Virus Res 38(2–3):205–218

Libikova H, Buckley SM (1971) Serological characterization of Eurasian Kemerovo group viruses. II. Cross plaque neutralization tests. Acta Virol 15(1):79–86

Libikova H, Tesarova J, Rajcani J (1970) Experimental infection of monkeys with Kemerovo virus. Acta Virol 14(1):64–69

Libikova H, Heinz F, Ujhazyova D, Stunzner D (1978) Orbiviruses of the Kemerovo complex and neurological diseases. Med Microbiol Immunol 166(1–4):255–263

Luedke AJ, Jones RH, Walton TE (1977) Overwintering mechanism for bluetongue virus: biological recovery of latent virus from a bovine by bites of Culicoides variipennis. Am J Trop Med Hyg 26(2):313–325

Maan S (2004) Complete nucleotide sequence analysis of genome segment 2 from the twenty-four serotypes of bluetongue virus: development of nucleic acid based typing methods and molecular epidemiology. PhD thesis submitted in partial fulfilment for the degree of Doctor of Philosophy, University of London

Maan S, Maan NS, Samuel AR, Rao S, Attoui H, Mertens PP (2007) Analysis and phylogenetic comparisons of full-length VP2 genes of the 24 bluetongue virus serotypes. J Gen Virol 88(Pt 2):621–630

Maan S, Maan NS, Ross-smith N, Batten CA, Shaw AE, Anthony SJ, Samuel AR, Darpel KE, Veronesi E, Oura CA, Singh KP, Nomikou K, Potgieter AC, Attoui H, van Rooij E, van Rijn P, De Clercq K, Vandenbussche F, Zientara S, Breard E, Sailleau C, Beer M, Hoffman B, Mellor PS, Mertens PP (2008) Sequence analysis of bluetongue virus serotype 8 from the Netherlands 2006 and comparison to other European strains. Virology 377(2):308–318

Maan NS, Maan S, Nomikou K, Johnson DJ, El Harrak M, Madani H, Yadin H, Incoglu S, Yesilbag K, Allison AB, Stallknecht DE, Batten C, Anthony SJ, Mertens PP (2010a) RT-PCR assays for seven serotypes of epizootic haemorrhagic disease virus & their use to type strains from the Mediterranean region and North America. PLoS One 5(9):e12782

Maan S, Maan NS, van Rijn PA, van Gennip RG, Sanders A, Wright IM, Batten C, Hoffmann B, Eschbaumer M, Oura CA, Potgieter AC, Nomikou K, Mertens PP (2010b) Full genome characterisation of bluetongue virus serotype 6 from the Netherlands 2008 and comparison to other field and vaccine strains. PLoS One 5(4):e10323

Maan NS, Maan S, Nomikou K, Belaganahalli MN, Bachanek-Bankowska K, Mertens PP (2011a) Serotype specific primers and gel-based RT-PCR assays for 'typing' African horse sickness virus: identification of strains from Africa. PLoS One 6(10):e25686

Maan S, Maan NS, Nomikou K, Batten C, Antony F, Belaganahalli MN, Samy AM, Reda AA, Al-Rashid SA, El Batel M, Oura CA, Mertens PP (2011b) Novel bluetongue virus serotype from Kuwait. Emerg Infect Dis 17(5):886–889

Maan S, Maan NS, Nomikou K, Veronesi E, Bachanek-Bankowska K, Belaganahalli MN, Attoui H, Mertens PP (2011c) Complete genome characterisation of a novel 26th bluetongue virus serotype from Kuwait. PLoS One 6(10):e26147

Maan NS, Maan S, Belaganahalli MN, Ostlund EN, Johnson DJ, Nomikou K, Mertens PPC (2012) Identification and differentiation of the twenty six bluetongue virus serotypes by RT–PCR amplification of the serotype-specific genome segment 2. PLoS One 7(2):e32601

Maan S, Kumar A, Gupta AK, Dalal A, Chaudhary D, Gupta TK, Bansal N, Kumar V, Batra K, Sindhu N, Kumar A, Mahajan NK, Maan NS, Mertens PPC (2018) Concurrent infection of Bluetongue and Peste-des-petits-ruminants virus in small ruminants in Haryana State of India. Transbound Emerg Dis 65(1):235–239

MacLachlan NJ (1994) The pathogenesis and immunology of bluetongue virus infection of ruminants. Comp Immunol Microbiol Infect Dis 17(3–4):197–206

Maclachlan NJ, Guthrie AJ (2010) Re-emergence of bluetongue, African horse sickness, and other orbivirus diseases. Vet Res 41(6):35

MacLachlan NJ, Thompson J (1985) Bluetongue virus-induced interferon in cattle. Am J Vet Res 46(6):1238–1241

MacLachlan NJ, Nunamaker RA, Katz JB, Sawyer MM, Akita GY, Osburn BI, Tabachnick WJ (1994) Detection of bluetongue virus in the blood of inoculated calves: comparison of virus isolation, PCR assay, and in vitro feeding of Culicoides variipennis. Arch Virol 136:1–2), 1–8

Maclachlan NJ, Drew CP, Darpel KE, Worwa G (2009) The pathology and pathogenesis of bluetongue. J Comp Pathol 141(1):1–16

Marr JS, Malloy CD (1996) An epidemiologic analysis of the ten plagues of Egypt. Caduceus 12(1):7–24

Marshall ID, Woodroofe GM, Gard GP (1980) Arboviruses of coastal south-eastern Australia. Aust J Exp Biol Med Sci 58(1):91–102

Martin LA, Meyer AJ, O'Hara RS, Fu H, Mellor PS, Knowles NJ, Mertens PP (1998) Phylogenetic analysis of African horse sickness virus segment 10: sequence variation, virulence characteristics and cell exit. Arch Virol Suppl 14:281–293

Martinez-Torrecuadrada JL, Casal JI (1995) Identification of a linear neutralization domain in the protein VP2 of African horse sickness virus. Virology 210(2):391–399

Martinez-Torrecuadrada JL, Langeveld JP, Meloen RH, Casal JI (2001) Definition of neutralizing sites on African horse sickness virus serotype 4 VP2 at the level of peptides. J Gen Virol 82(Pt 10):2415–2424
Martins LC, Diniz JA, Silva EV, Barros VL, Monteiro HA, Azevedo RS, Quaresma JA, Vasconcelos PF (2007) Characterization of Minacu virus (Reoviridae: Orbivirus) and pathological changes in experimentally infected newborn mice. Int J Exp Pathol 88(1):63–73
McIntosh BM (1955) Horse sickness antibodies in the sera of dogs in enzootic areas. J S Afr Vet Med Assoc 26:269–272
McLaughlin BE, DeMaula CD, Wilson WC, Boyce WM, MacLachlan NJ (2003) Replication of bluetongue virus and epizootic hemorrhagic disease virus in pulmonary artery endothelial cells obtained from cattle, sheep, and deer. Am J Vet Res 64(7):860–865
Mecham JO, Dean VC (1988) Protein coding assignment for the genome of epizootic haemorrhagic disease virus. J Gen Virol 69(Pt 6):1255–1262
Mecham JO, Johnson DJ (2005) Persistence of bluetongue virus serotype 2 (BTV-2) in the southeast United States. Virus Res 113(2):116–122
Mellor PS (1994) Bluetongue. State Vet J 4:7–10
Mellor PS (2000) Replication of arboviruses in insect vectors. J Comp Pathol 123(4):231–247
Mellor PS, Boorman J (1995) The transmission and geographical spread of African horse sickness and bluetongue viruses. Ann Trop Med Parasitol 89(1):1–15
Mellor PS, Hamblin C (2004) African horse sickness. Vet Res 35(4):445–466
Mellor PS, Boorman J, Baylis M (2000) Culicoides biting midges: their role as arbovirus vectors. Annu Rev Entomol 45:307–340
Mellor P, Baylis M, Mertens P (2009) Bluetongue. Elsevier/Academic Press, London
Menzies FD, McCullough SJ, McKeown IM, Forster JL, Jess S, Batten C, Murchie AK, Gloster J, Fallows JG, Pelgrim W, Mellor PS, Oura CA (2008) Evidence for transplacental and contact transmission of bluetongue virus in cattle. Vet Rec 163(7):203–209
Mertens P (2001) Orbiviruses and Bluetongue virus. In: Encyclopedia of life sciences (ELS). John Wiley & Sons, Ltd., Chichester
Mertens P, Attoui H (2009) Orbiviruses and Bluetongue virus. In: Encyclopedia of life sciences (ELS). John Wiley & Sons, Ltd., Chichester
Mertens PP, Diprose J (2004) The bluetongue virus core: a nano-scale transcription machine. Virus Res 101(1):29–43
Mertens PP, Sangar DV (1985) Analysis of the terminal sequences of the genome segments of four orbiviruses. Virology 140(1):55–67
Mertens PP, Brown F, Sangar DV (1984) Assignment of the genome segments of bluetongue virus type 1 to the proteins which they encode. Virology 135(1):207–217
Mertens PP, Burroughs JN, Anderson J (1987) Purification and properties of virus particles, infectious subviral particles, and cores of bluetongue virus serotypes 1 and 4. Virology 157(2):375–386
Mertens PP, Pedley S, Cowley J, Burroughs JN, Corteyn AH, Jeggo MH, Jennings DM, Gorman BM (1989) Analysis of the roles of bluetongue virus outer capsid proteins VP2 and VP5 in determination of virus serotype. Virology 170(2):561–565
Mertens PP, Burroughs JN, Walton A, Wellby MP, Fu H, O'Hara RS, Brookes SM, Mellor PS (1996) Enhanced infectivity of modified bluetongue virus particles for two insect cell lines and for two Culicoides vector species. Virology 217(2):582–593
Mertens PPC, Arella M, Attoui H et al (2000) Family Reoviridae. In: van Regenmortel MHV, Fauquet CM, Bishop DHL et al (eds) Virus taxonomy. Seventh report of the international committee on taxonomy of viruses. Academic Press, New York, pp 395–480
Mertens PPC, Attoui H, Duncan R, Dermody TS (2005a) Reoviridae. In: Fauquet CM, Mayo MA, Maniloff J, Desselberger U, Ball LA (eds) Virus taxonomy. Eighth report of the international committee on taxonomy of viruses. Elsevier/Academic Press, London, pp 447–560
Mertens PPC, Maan S, Samuel A, Attoui H (2005b) Orbiviruses, Reoviridae. In: Fauquet CM, Mayo MA, Maniloff J, Desselberger U, Ball LA (eds) Virus taxonomy. Eighth report of the international committee on taxonomy of viruses. Elsevier/Academic Press, London, pp 466–483
Mertens PP, Maan NS, Prasad G, Samuel AR, Shaw AE, Potgieter AC, Anthony SJ, Maan S (2007) Design of primers and use of RT-PCR assays for typing European bluetongue virus isolates: differentiation of field and vaccine strains. J Gen Virol 88(Pt 10):2811–2823
Mildenberg Z, Westcott D, Bellaiche M, Dastjerdi A, Steinbach F, Drew T (2009) Equine encephalosis virus in Israel. Transbound Emerg Dis 56(8):291
Miura Y, Goto Y, Kubo M, Kono Y (1988a) Isolation of Chuzan virus, a new member of the Palyam subgroup of the genus Orbivirus, from cattle and Culicoides oxystoma in Japan. Am J Vet Res 49(12):2022–2025
Miura Y, Goto Y, Kubo M, Kono Y (1988b) Pathogenicity of Chuzan virus, a new member of the Palyam subgroup of genus Orbivirus for cattle. Nippon Juigaku Zasshi 50(3):632–637
Mohammed ME, Aradaib IE, Mukhtar MM, Ghalib HW, Riemann HP, Oyejide A, Osburn BI (1996) Application of molecular biological techniques for detection of epizootic hemorrhagic

disease virus (EHDV-318) recovered from a sentinel calf in central Sudan. Vet Microbiol 52(3–4):201–208

Mohd Jaafar F, Goodwin AE, Belhouchet M, Merry G, Fang Q, Cantaloube JF, Biagini P, de Micco P, Mertens PP, Attoui H (2008) Complete characterisation of the American grass carp reovirus genome (genus Aquareovirus: family Reoviridae) reveals an evolutionary link between aquareoviruses and coltiviruses. Virology 373(2):310–321

Mohd Jaafar F, Belhouchet M, Belaganahalli M, Tesh RB, Mertens PP, Attoui H (2014) Full-genome characterisation of orungo, lebombo and changuinola viruses provides evidence for co-evolution of orbiviruses with their arthropod vectors. PLoS One 9(1):e86392

Monaco F, Polci A, Lelli R, Pinoni C, Di Mattia T, Mbulu RS, Scacchia M, Savini G (2011) A new duplex real-time RT-PCR assay for sensitive and specific detection of African horse sickness virus. Mol Cell Probes 25(2–3):87–93

Monath TP, Guirakhoo F (1996) Orbiviruses and coltiviruses. In: Fields BN, Knipe DM, Howley PM (eds) Fields virology. Lippincott-Raven, Philadelphia, PA, pp 1735–1766

Moore DL (1974) Bluetongue and related viruses in Ibadan, Nigeria: serologic comparison of bluetongue, epizootic hemorrhagic disease of deer, and Abadina (Palyam) viral isolates. Am J Vet Res 35(8):1109–1113

Moore DL, Lee VH (1972) Antigenic relationship between the virus of epizootic haemorrhagic disease of deer and bluetongue virus. Brief report. Arch Gesamte Virusforsch 37(2):282–284

Moss SR, Ayres CM, Nuttall PA (1987) Assignment of the genome segment coding for the neutralizing epitope(s) of orbiviruses in the Great Island subgroup (Kemerovo serogroup). Virology 157(1):137–144

Moss SR, Jones LD, Nuttall PA (1992) Comparison of the major structural core proteins of tick-borne and Culicoides-borne orbiviruses. J Gen Virol 73(Pt 10):2585–2590

Moule L (1896) Histoire de la Médecine Vétérinaire. Maulde, Paris, p 38

Moulton JE (1961) Pathology of bluetongue of sheep in California. J Am Vet Med Assoc 138:493–498

Murao K, Uyeda I, Ando Y, Kimura I, Cabauatan PQ, Koganezawa H (1996) Genomic rearrangement in genome segment 12 of rice dwarf phytoreovirus. Virology 216(1):238–240

Murphy FA, Borden EC, Shope RE, Harrison A (1971) Physicochemical and morphological relationships of some arthropod-borne viruses to bluetongue virus—a new taxonomic group. Electron microscopic studies. J Gen Virol 13(2):273–288

Myers RM, Carey DE, Reuben R, Jesudass ES, Shope RE (1971) Vellore virus: a recently recognized agent of the Palyam group of arboviruses. Indian J Med Res 59(8):1209–1213

Nason EL, Rothagel R, Mukherjee SK, Kar AK, Forzan M, Prasad BV, Roy P (2004) Interactions between the inner and outer capsids of bluetongue virus. J Virol 78(15):8059–8067

Nevill EM (1971) Cattle and Culicoides biting midges as possible overwintering hosts of bluetongue virus. Onderstepoort J Vet Res 38(2):65–71

Nomikou K, Dovas CI, Maan S, Anthony SJ, Samuel AR, Papanastassopoulou M, Maan NS, Mangana O, Mertens PP (2009) Evolution and phylogenetic analysis of full-length VP3 genes of Eastern Mediterranean bluetongue virus isolates. PLoS One 4(7):e6437

Noon TH, Wesche SL, Cagle D, Mead DG, Bicknell EJ, Bradley GA, Riplog-Peterson S, Edsall D, Reggiardo C (2002a) Hemorrhagic disease in bighorn sheep in Arizona. J Wildl Dis 38(1):172–176

Noon TH, Wesche SL, Heffelfinger J, Fuller A, Bradley GA, Reggiardo C (2002b) Hemorrhagic disease in deer in Arizona. J Wildl Dis 38(1):177–181

O'Hara RS, Meyer AJ, Burroughs JN, Pullen L, Martin LA, Mertens PP (1998) Development of a mouse model system, coding assignments and identification of the genome segments controlling virulence of African horse sickness virus serotypes 3 and 8. Arch Virol Suppl 14:259–279

Ohashi S, Yoshida K, Yanase T, Tsuda T (2002) Analysis of intratypic variation evident in an Ibaraki virus strain and its epizootic hemorrhagic disease virus serogroup. J Clin Microbiol 40(10):3684–3688

OIE (2007) Chapter 2.2.13. Bluetongue. Terrestrial animal health code, pp 131–136. Accessed 23 Dec 2018

OIE (2008) Report reference: OIE, Ref OIE: 6964, report date: 14/04/2008, Australia. http://www.oie.int/wahis/reports/en_imm_0000006964_20080415_200102.pdf

OIE (2010a) Chapter 8.3. Bluetongue. Terrestrial animal health code

OIE (2010b) Chapter 1.2. Criteria for listing diseases. Terrestrial animal health code. Accessed 23 Dec 2018

Okamoto K, Endo Y, Inoue S, Nabeshima T, Nga PT, Guillermo PH, Yu F, Loan do P, Trang BM, Natividad FF, Hasebe F, Morita K (2010) Development of a rapid and comprehensive proteomics-based arboviruses detection system. J Virol Methods 167(1):31–36

Omori T, Inaba Y, Morimoto T, Tanaka Y, Ishitani R (1969a) Ibaraki virus, an agent of epizootic disease of cattle resembling bluetongue. I. Epidemiologic, clinical and pathologic observations and experimental transmission to calves. Jpn J Microbiol 13(2):139–157

Omori T, Inaba Y, Morimoto T, Tanaka Y, Kono M (1969b) Ibaraki virus, an agent of epizootic disease of cattle resembling bluetongue. II. Isolation of the virus in bovine cell culture. Jpn J Microbiol 13(2):159–168

Orru G, Ferrando ML, Meloni M, Liciardi M, Savini G, De Santis P (2006) Rapid detection and quantitation of Bluetongue virus (BTV) using a Molecular Beacon fluorescent probe assay. J Virol Methods 137(1):34–42

Osburn BI (1994) The impact of bluetongue virus on reproduction. Comp Immunol Microbiol Infect Dis 17(3–4):189–196

Oura C (2011) Bluetongue: new insights and lessons learnt. Vet Rec 168(14):375–376

Oura CA, El Harrak M (2011) Midge-transmitted bluetongue in domestic dogs. Epidemiol Infect 139(9):1396–1400

Owens RJ, Limn C, Roy P (2004) Role of an arbovirus nonstructural protein in cellular pathogenesis and virus release. J Virol 78(12):6649–6656

Ozkul A, Erturk A, Caliskan E, Sarac F, Ceylan C, Mertens P, Kabakli O, Dincer E, Cizmeci SG (2009) Segment 10 based molecular epidemiology of bluetongue virus (BTV) isolates from Turkey: 1999–2001. Virus Res 142(1–2):134–139

Palacios G, Cowled C, Bussetti AV, Savji N, Weir R, Wick I, Travassos da Rosa A, Calisher CH, Tesh RB, Boyle D, Lipkin WI (2011) Rapid molecular strategy for orbivirus detection and characterization. J Clin Microbiol 49(6):2314–2317

Papadopoulos O, Mellor PS, Mertens PPC (2009) Bluetongue control strategies. In: Mellor PS, Baylis M, Mertens PPC (eds) Bluetongue monograph, 1st edn. Elsevier/Academic Press, London, pp 429–452

Parker J, Herniman KA, Gibbs EP, Sellers RF (1975) An experimental inactivated vaccine against bluetongue. Vet Rec 96(13):284–287

Parkes H, Gould AR (1996) Characterisation of Wongorr virus, an Australian orbivirus. Virus Res 44(2):111–122

Parsonson IM (1990) Pathology and pathogenesis of bluetongue infections. Curr Top Microbiol Immunol 162:119–141

Patton JT, Spencer E (2000) Genome replication and packaging of segmented double-stranded RNA viruses. Virology 277(2):217–225

Pearson JE, Jochim MM (1979) Protocol for the immunodiffusion test for bluetongue. Proc Am Assoc Vet Lab Diagn 22:463

Peiris JS, Amerasinghe PH, Amerasinghe FP, Calisher CH, Perera LP, Arunagiri CK, Munasingha NB, Karunaratne SH (1994) Viruses isolated from mosquitoes collected in Sri Lanka. Am J Trop Med Hyg 51(2):154–161

Polydorou K (ed) (1985) Bluetongue in Cyprus. Alan R Liss Inc., New York

Prasad BV, Rothnagel R, Zeng CQ, Jakana J, Lawton JA, Chiu W, Estes MK (1996) Visualization of ordered genomic RNA and localization of transcriptional complexes in rotavirus. Nature 382(6590):471–473

Prasad G, Minakshi, Malik YS, Maan S (1999) RT-PCR and its detection limit for cell culture grown bluetongue virus 1 using NS1 gene group specific primers. Indian J Exp Biol 37:1255–1258

Prasad G, Minakshi, Malik YS, Maan S (2000) Trends in bluetongue virus diagnostics: a review. Indian J Anim Sci 70:103–109

Prasad G, Minakshi, Malik YS (2007) Bluetongue. Publisher Indian Council of Agricultural Research, New Delhi, p 163

Pritchard LI, Gould AR, Wilson WC, Thompson L, Mertens PP, Wade-Evans AM (1995) Complete nucleotide sequence of RNA segment 3 of bluetongue virus serotype 2 (Ona-A). Phylogenetic analyses reveal the probable origin and relationship with other orbiviruses. Virus Res 35(3):247–261

Purse BV, Mellor PS, Rogers DJ, Samuel AR, Mertens PP, Baylis M (2005) Climate change and the recent emergence of bluetongue in Europe. Nat Rev Microbiol 3(2):171–181

Purse BV, Brown HE, Harrup L, Mertens PP, Rogers DJ (2008) Invasion of bluetongue and other orbivirus infections into Europe: the role of biological and climatic processes. Rev Sci Tech 27(2):427–442

Qiu W, Scholthof KB (2001) Defective interfering RNAs of a satellite virus. J Virol 75(11):5429–5432

Quan M, Lourens CW, MacLachlan NJ, Gardner IA, Guthrie AJ (2010) Development and optimisation of a duplex real-time reverse transcription quantitative PCR assay targeting the VP7 and NS2 genes of African horse sickness virus. J Virol Methods 167(1):45–52

Quist CF, Howerth EW, Bounous DI, Stallknecht DE (1997) Cell-mediated immune response and IL-2 production in white-tailed deer experimentally infected with hemorrhagic disease viruses. Vet Immunol Immunopathol 56(3–4):283–297

Rafyi A (1961) Horse sickness. Bull Off Int Epizoot 56:216–250

Ramadevi N, Roy P (1998) Bluetongue virus core protein VP4 has nucleoside triphosphate phosphohydrolase activity. J Gen Virol 79(Pt 10):2475–2480

Ramadevi N, Burroughs NJ, Mertens PP, Jones IM, Roy P (1998) Capping and methylation of

mRNA by purified recombinant VP4 protein of bluetongue virus. Proc Natl Acad Sci U S A 95(23):13537–13542
Rao CD, Kiuchi A, Roy P (1983) Homologous terminal sequences of the genome double-stranded RNAs of bluetongue virus. J Virol 46(2):378–383
Ratinier M, Caporale M, Golder M, Franzoni G, Allan K, Nunes SF, Armezzani A, Bayoumy A, Rixon F, Shaw A, Palmarini M (2011) Identification and characterization of a novel non-structural protein of bluetongue virus. PLoS Pathog 7(12):e1002477
Ratinier M, Shaw AE, Barry G, Gu Q, Di Gialleonardo L, Janowicz A, Varela M, Randall RE, Caporale M, Palmarini M (2016) Bluetongue virus NS4 protein is an interferon antagonist and a determinant of virus virulence. J Virol 90(11):5427–5439
Reddacliff L, Kirkland P, Philbey A, Davis R, Vogelnest L, Hulst F, Blyde D, Deykin A, Smith J, Hooper P, Gould A, Hyatt A (1999) Experimental reproduction of viral chorioretinitis in kangaroos. Aust Vet J 77(8):522–528
Reinisch KM, Nibert ML, Harrison SC (2000) Structure of the reovirus core at 3.6 A resolution. Nature 404(6781):960–967
Rodriguez-Sanchez B, Fernandez-Pinero J, Sailleau C, Zientara S, Belak S, Arias M, Sanchez-Vizcaino JM (2008) Novel gel-based and real-time PCR assays for the improved detection of African horse sickness virus. J Virol Methods 151(1):87–94
Rose KA, Kirkland PD, Davies RJ, Cooper DW, Blumstein DT (2000) An epizootic of sudden death in Tammar Wallabies (*Macropus eugenii*). In: Proceedings of the AAZV and IAAAM joint conference, pp 166–170
Rose K, Kirkland P, Davis R, Cooper D, Blumstein D, Pritchard L, Newberry K, Lunt R (2012) Epizootics of sudden death in tammar wallabies (Macropus eugenii) associated with an orbivirus infection. Aust Vet J 90(12):505–509
Roy P (1989) Bluetongue virus genetics and genome structure. Virus Res 13(3):179–206
Roy P (1996) Orbivirus structure and assembly. Virology 216(1):1–11
Roy P (2005) Bluetongue virus proteins and particles and their role in virus entry, assembly, and release. Adv Virus Res 64:69–123
Roy P (2008) Bluetongue virus: dissection of the polymerase complex. J Gen Virol 89(Pt 8):1789–1804
Roy P, Ritter GD Jr, Akashi H, Collisson E, Inaba Y (1985) A genetic probe for identifying bluetongue virus infections in vivo and in vitro. J Gen Virol 66(Pt 7):1613–1619
Sanchez-Vizcaino JM (2004) Control and eradication of African horse sickness with vaccine. Dev Biol (Basel) 119:255–258
Sangar DV, Mertens PPC (1983) Comparison of type-1 bluetongue virus protein synthesis in vivo and in vitro. In: Compans RW, Bishop DHL (eds) Double-stranded RNA viruses. Elsevier, New York, pp 183–191
Santman-Berends IM, Hage JJ, van Rijn PA, Stegeman JA, van Schaik G (2010a) Bluetongue virus serotype 8 (BTV-8) infection reduces fertility of Dutch dairy cattle and is vertically transmitted to offspring. Theriogenology 74(8):1377–1384
Santman-Berends IM, van Wuijckhuise L, Vellema P, van Rijn PA (2010b) Vertical transmission of bluetongue virus serotype 8 virus in Dutch dairy herds in 2007. Vet Microbiol 141(1–2):31–35
Savini G, Afonso A, Mellor P, Aradaib I, Yadin H, Sanaa M, Wilson W, Monaco F, Domingo M (2011) Epizootic haemorrhagic disease. Res Vet Sci 91(1):1–17
Schwartz-Cornil I, Mertens PP, Contreras V, Hemati B, Pascale F, Breard E, Mellor PS, MacLachlan NJ, Zientara S (2008) Bluetongue virus: virology, pathogenesis and immunity. Vet Res 39(5):46
Shaw AE, Monaghan P, Alpar HO, Anthony S, Darpel KE, Batten CA, Guercio A, Alimena G, Vitale M, Bankowska K, Carpenter S, Jones H, Oura CA, King DP, Elliott H, Mellor PS, Mertens PP (2007) Development and initial evaluation of a real-time RT-PCR assay to detect bluetongue virus genome segment 1. J Virol Methods 145(2):115–126
Shaw AE, Bruning-Richardson A, Morrison EE, Bond J, Simpson J, Ross-Smith N, Alpar O, Mertens PP, Monaghan P (2013) Bluetongue virus infection induces aberrant mitosis in mammalian cells. Virol J 10:319
Shope RE, MacNamara LG, Mangold R (1955) Epizootic hemorrhagic disease of deer. N J Outdoor 6:17
Spence L, Anderson CR, Aitken TH, Downs WG (1967) Bushbush, Ieri and Lukuni viruses, three unrelated new agents isolated from Trinidadian forest mosquitoes. Proc Soc Exp Biol Med 125(1):45–50
Spreull J (1902) Report from veterinary surgeon Spreull on the result of his experiments with the malarial catarrhal fever of sheep. Agric J Cape Good Hope 20:469–477
Spreull J (1905) Malarial catarrhal fever (bluetongue) of sheep in South Africa. J Comp Pathol 18:321–337
Squire KR, Chuang RY, Chuang LC, Doi RH, Osburn BI (1985a) Molecular cloning and hybridization studies on bluetongue virus serotype 17. Prog Clin Biol Res 178:355–361
Squire KR, Chuang RY, Chuang LF, Doi RH, Osburn BI (1985b) Detecting bluetongue virus RNA in cell culture by dot hybridization with a cloned genetic probe. J Virol Methods 10(1):59–68

Stauber N, Martinez-Costas J, Sutton G, Monastyrskaya K, Roy P (1997) Bluetongue virus VP6 protein binds ATP and exhibits an RNA-dependent ATPase function and a helicase activity that catalyze the unwinding of double-stranded RNA substrates. J Virol 71(10):7220–7226

Stewart ME, Roy P (2010) Role of cellular caspases, nuclear factor-kappa B and interferon regulatory factors in Bluetongue virus infection and cell fate. Virol J 7:362

Stewart M, Hardy A, Barry G, Pinto RM, Caporale M, Melzi E, Hughes J, Taggart A, Janowicz A, Varela M, Ratinier M, Palmarini M (2015) Characterization of a second open reading frame in genome segment 10 of bluetongue virus. J Gen Virol 96(11):3280–3293

Stuart DI, Grimes JM (2006) Structural studies on orbivirus proteins and particles. Curr Top Microbiol Immunol 309:221–244

Sutton G, Grimes JM, Stuart DI, Roy P (2007) Bluetongue virus VP4 is an RNA-capping assembly line. Nat Struct Mol Biol 14(5):449–451

Swanepoel R, Blackburn NK (1976) A new member of the Palyam serogroup of orbiviruses. Vet Rec 99(18):360

Tabachnick WJ, MacLachlan NJ, Thompson LH, Hunt GJ, Patton JF (1996) Susceptibility of Culicoides variipennis sonorensis to infection by polymerase chain reaction-detectable bluetongue virus in cattle blood. Am J Trop Med Hyg 54(5):481–485

Takamatsu H, Jeggo MH (1989) Cultivation of bluetongue virus-specific ovine T cells and their cross-reactivity with different serotype viruses. Immunology 66(2):258–263

Takamatsu H, Mellor PS, Mertens PP, Kirkham PA, Burroughs JN, Parkhouse RM (2003) A possible overwintering mechanism for bluetongue virus in the absence of the insect vector. J Gen Virol 84(Pt 1):227–235

Tamura K, Peterson D, Peterson N, Stecher G, Nei M, Kumar S (2011) MEGA5: molecular evolutionary genetics analysis using maximum likelihood, evolutionary distance, and maximum parsimony methods. Mol Biol Evol 28(10):2731–2739

Taraporewala ZF, Chen D, Patton JT (2001) Multimers of the bluetongue virus nonstructural protein, NS2, possess nucleotidyl phosphatase activity: similarities between NS2 and rotavirus NSP2. Virology 280(2):221–231

Theal GM (1899) Records of South-Eastern Africa collected in various libraries and archive departments in Europe, vol 3. Government of Cape Colony, Cape Town, p 224

Theiler A (1910) Notes on a fever in horses simulating horse-sickness. Transvaal Agric J 8:581–586

Theodoridis A, Nevill EM, Els HJ, Boshoff ST (1979) Viruses isolated from Culicoides midges in South Africa during unsuccessful attempts to isolate bovine ephemeral fever virus. Onderstepoort J Vet Res 46(4):191–198

Theron J, Huismans H, Nel LH (1996a) Identification of a short domain within the non-structural protein NS2 of epizootic haemorrhagic disease virus that is important for single strand RNA-binding activity. J Gen Virol 77(Pt 1):129–137

Theron J, Huismans H, Nel LH (1996b) Site-specific mutations in the NS2 protein of epizootic haemorrhagic disease virus markedly affect the formation of cytoplasmic inclusion bodies. Arch Virol 141(6):1143–1151

Thevasagayam JA, Mertens PP, Burroughs JN, Anderson J (1995) Competitive ELISA for the detection of antibodies against epizootic haemorrhagic disease of deer virus. J Virol Methods 55(3):417–425

Thevasagayam JA, Wellby MP, Mertens PP, Burroughs JN, Anderson J (1996) Detection and differentiation of epizootic haemorrhagic disease of deer and bluetongue viruses by serogroup-specific sandwich ELISA. J Virol Methods 56(1):49–57

Thomas CP, Booth TF, Roy P (1990) Synthesis of bluetongue virus-encoded phosphoprotein and formation of inclusion bodies by recombinant baculovirus in insect cells: it binds the single-stranded RNA species. J Gen Virol 71(Pt 9):2073–2083

Toussaint JF, Sailleau C, Breard E, Zientara S, De Clercq K (2007) Bluetongue virus detection by two real-time RT-qPCRs targeting two different genomic segments. J Virol Methods 140(1–2):115–123

Travassos da Rosa AP, Tesh RB, Pinheiro FP, Travassos da Rosa JF, Peralta PH, Knudson DL (1984) Characterization of the Changuinola serogroup viruses (Reoviridae: Orbivirus). Intervirology 21(1):38–49

Troupin C, Schnuriger A, Duponchel S, Deback C, Schnepf N, Dehee A, Garbarg-Chenon A (2011) Rotavirus rearranged genomic RNA segments are preferentially packaged into viruses despite not conferring selective growth advantage to viruses. PLoS One 6(5):e20080

Uitenweerde JM, Theron J, Stoltz MA, Huismans H (1995) The multimeric nonstructural NS2 proteins of bluetongue virus, African horse sickness virus, and epizootic hemorrhagic disease virus differ in their single-stranded RNA-binding ability. Virology 209(2):624–632

Urakawa T, Roy P (1988) Bluetongue virus tubules made in insect cells by recombinant baculoviruses: expression of the NS1 gene of bluetongue virus serotype 10. J Virol 62(11):3919–3927

Urakawa T, Ritter DG, Roy P (1989) Expression of largest RNA segment and synthesis of VP1 protein of bluetongue virus in insect cells by recombinant baculovirus: association of VP1 protein with RNA polymerase activity. Nucleic Acids Res 17(18):7395–7401

Uren MF (1986) Clinical and pathological responses of sheep and cattle to experimental infection with five different viruses of the epizootic hemorrhagic disease of deer serogroup. Aust Vet J 63(6):199–201

van Niekerk M, van Staden V, van Dijk AA, Huismans H (2001) Variation of African horse sickness virus nonstructural protein NS3 in southern Africa. J Gen Virol 82(Pt 1):149–158

van Regenmortel MH, Mahy BW (2004) Emerging issues in virus taxonomy. Emerg Infect Dis 10(1):8–13

Van Rensberg IB, De Clerk J, Groenewald HB, Botha WS (1981) An outbreak of African horse sickness in dogs. J S Afr Vet Assoc 52(4):323–325

Venter EH, Viljoen GJ, Nel LH, Huismans H, van Dijk AA (1991) A comparison of different genomic probes in the detection of virus-specified RNA in Orbivirus-infected cells. J Virol Methods 32(2–3):171–180

Verwoerd DW, Erasmus BJ (2004) Bluetongue. In: Coetzer JAW, Tustin RC (eds) Infectious diseases of livestock. Oxford University Press, Cape Town, pp 1201–1220

Verwoerd DW, Louw H, Oellermann RA (1970) Characterization of bluetongue virus ribonuclcic acid. J Virol 5(1):1–7

Verwoerd DW, Els HJ, De Villiers EM, Huismans H (1972) Structure of the bluetongue virus capsid. J Virol 10(4):783–794

Verwoerd DW, Huismans H, Erasmus BJ (1979) Orbiviruses. In: Fraenkel-Conrat H, Wagner RR (eds) Comprehensive virology. Plenum Press, New York, pp 285–345

Vieira Cde M, Nunes MR, da Silva EV, Carvalho VL, Nunes Neto JP, Cruz AC, Casseb SM, Vasconcelos HB, Quaresma JA, Vasconcelos PF (2009) Full-length sequencing and genetic characterization of Breu Branco virus (Reoviridae, Orbivirus) and two related strains isolated from Anopheles mosquitoes. J Gen Virol 90(Pt 9):2183–2190

Wade-Evans AM (1990) The complete nucleotide sequence of genome segment 7 of bluetongue virus, serotype 1 from South Africa. Nucleic Acids Res 18(16):4919

Wade-Evans AM, Mertens PP, Bostock CJ (1990) Development of the polymerase chain reaction for the detection of bluetongue virus in tissue samples. J Virol Methods 30(1):15–24

Wade-Evans AM, Mertens PP, Belsham GJ (1992) Sequence of genome segment 9 of bluetongue virus (serotype 1, South Africa) and expression analysis demonstrating that different forms of VP6 are derived from initiation of protein synthesis at two distinct sites. J Gen Virol 73(Pt 11):3023–3026

Wang LF, Crameri G (2014) Emerging zoonotic viral diseases. Rev Sci Tech 33(2):569–581

Whistler T, Swanepoel R (1988) Characterization of potentially foetotropic Palyam serogroup orbiviruses isolated in Zimbabwe. J Gen Virol 69(Pt 9):2221–2227

Whitehead RH, Doderty RL, Domrow R, Standfast HA, Wetters EJ (1968) Studies of the epidemiology of arthropod-borne virus infections at Mitchell River Mission, Cape York Peninsula, North Queensland. 3. Virus studies of wild birns, 1964–1967. Trans R Soc Trop Med Hyg 62(3):439–445

Wilson A, Carpenter S, Gloster J, Mellor P (2007) Re-emergence of bluetongue in northern Europe in 2007. Vet Rec 161(14):487–489

Wilson A, Mellor PS, Szmaragd C, Mertens PP (2009a) Adaptive strategies of African horse sickness virus to facilitate vector transmission. Vet Res 40(2):16

Wilson WC, Hindson BJ, O'Hearn ES, Hall S, Tellgren-Roth C, Torres C, Naraghi-Arani P, Mecham JO, Lenhoff RJ (2009b) A multiplex real-time reverse transcription polymerase chain reaction assay for detection and differentiation of Bluetongue virus and Epizootic hemorrhagic disease virus serogroups. J Vet Diagn Invest 21(6):760–770

Wilson WC, O'Hearn ES, Tellgren-Roth C, Stallknecht DE, Mead DG, Mecham JO (2009c) Detection of all eight serotypes of Epizootic hemorrhagic disease virus by real-time reverse transcription polymerase chain reaction. J Vet Diagn Invest 21(2):220–225

Wu X, Chen SY, Iwata H, Compans RW, Roy P (1992) Multiple glycoproteins synthesized by the smallest RNA segment (S10) of bluetongue virus. J Virol 66(12):7104–7112

Yadav PD, Albarino CG, Nyayanit DA, Guerrero L, Jenks MH, Sarkale P, Nichol ST, Mourya DT (2018) Equine encephalosis virus in India, 2008. Emerg Infect Dis 24(5):898–901

Yamakawa M, Ohashi S, Kanno T, Yamazoe R, Yoshida K, Tsuda T, Sakamoto K (2000) Genetic diversity of RNA segments 5, 7 and 9 of the Palyam serogroup orbiviruses from Japan, Australia and Zimbabwe. Virus Res 68(2):145–153

Yang DK, Hwang IJ, Kim BH, Kweon CH, Lee KW, Kang MI, Lee CS, Cho KO (2008) Serosurveillance of viral diseases in Korean native goats (Capra hircus). J Vet Med Sci 70(9):977–979

Zanotto PM, Gao GF, Gritsun T, Marin MS, Jiang WR, Venugopal K, Reid HW, Gould EA (1995) An arbovirus cline across the northern hemisphere. Virology 210(1):152–159

Zanotto PM, Gould EA, Gao GF, Harvey PH, Holmes EC (1996) Population dynamics of flaviviruses revealed by molecular phylogenies. Proc Natl Acad Sci U S A 93(2):548–553

Zientara S, Sailleau C, Moulay S, Cruciere C, el-Harrak M, Laegreid WW, Hamblin C (1998) Use of reverse transcriptase-polymerase chain reaction (RT-PCR) and dot-blot hybridisation for the detection and identification of African horse sickness virus nucleic acids. Arch Virol Suppl 14:317–327

Zientara S, Sailleau C, Dauphin G, Roquier C, Remond EM, Lebreton F, Hammoumi S, Dubois E, Agier C, Merle G, Breard E (2002) Identification of bluetongue virus serotype 2 (Corsican strain) by reverse-transcriptase PCR reaction analysis of segment 2 of the genome. Vet Rec 150(19):598–601

第 9 章 马流感病毒

Nitin Virmani, S. Pavulraj, B. C. Bera, Taruna Anand, R. K. Singh,B. N. Tripathi

9.1 引言

马流感(equine influenza,EI)是由马流感病毒(equine influenza virus,EIV)引起的一种具有高度传染性的马急性呼吸道疾病。马病毒性呼吸道感染2/3都是由EIV引起的，造成了重大的经济损失(Mumford等,1998;Singh等,2018)。该病也被称为"马流感"和"纽马克特咳嗽",临床特征表现为发热、厌食、抑郁、呼吸困难、干咳、流涕,继发细菌性肺炎(Gerber,1970)。幼龄马发病率可达60%~90%,确诊感染后死亡率为1%~20%。其中,小马驹营养不良,免疫功能低下,老龄和劳动强度大的马、驴死亡率较高(Paillot等,2006;Waghmare等,2010)。1872年在美国首次记载了EI暴发,被称为"1872年大流行病",但在那个时候病原体还未被分离到(McClure,1998)。1956年在布拉格首次从患有呼吸道症状的马身上分离到EIV，发现了一种只引起马感染的新型流感病毒(Sovinova等,1958),其病毒亚型为H7N7。回顾性血清学分析表明,1956年以前，相同亚型病毒在欧洲马群中也曾经流行过(Beveridge,1965)。1963年晚些时候，美国马群中由不同亚型流感病毒引起呼吸道病,该亚型流感病毒的抗原与H7N7不同,被命名为H3N8。H3N8原型于1963年在迈阿密被分离到(Waddell等,1963)。直到20世纪70年代末,马群中的EI都是由H7N7和H3N8亚型流感病毒引起的。然而,在过去的35年中,H7N7亚型流感病毒已经不再进行传播(Webster,1993),H3N8亚型流感病毒则继续传播，并在全世界范围内引起发病。尽管自20世纪60年代以来,进行了常规疫苗接种,但世界各国仍不断有EI暴发(Bryans等,1966)。EI暴发主要发生在全球幼龄、且未被接种疫苗的动物中,过去几年,在阿根廷、巴西、加拿大、智利、中国、多米尼加共和国、法国、德国、爱尔兰、印度、蒙古、瑞典、阿联酋、英国、美国和乌拉圭等国家经常有EI暴发的报道(WOAH,2019)。一般认为,马流感病毒只发生在马中,但最近观察到EI可进行跨种传播,能够通过马传播给其他新的宿主，如犬科动物(Crawford等,2005;Yamanaka等,2009;Kirkland等,2010)、猪(Tu等,2009)、骆驼(Yondon等,2014)。近期研究表明,猫也易感EIV(Su等,2014)。疫苗接种是预防EI等病毒性疾病最有效的方法之一。控制EI的策略主要是采取疫苗接种,从而产生针对表面糖蛋白,特别是HA的强大抗体反应(Daly等,2011)。HA是A型流感病毒的主要表面蛋白，在自然感染和疫苗接种中是宿主对流感病毒产生免疫反应的关键抗原。尽管进行了疫苗接种,EIV近年来仍在不断传播,成为家畜流行病。这是由于表面糖蛋白不断发生抗原变异。基因内编码HA和NA表位的抗原发生遗传漂移(点突变积累)有助于病毒逃避宿主的体液免疫保护作用,导致免疫失败。因此,对EIV的积极研究已经成为全球科学家关注的焦点，以期破解该病的发病机制、病毒进化,以及为了控制EI积极开发能够产生体液免疫和细胞免疫的疫苗。

9.2 结构与分类

EIV属正黏病毒科,正黏病毒科分为A型、B型、C型和D型流感病毒，托高土病毒和鲑传食病毒(Cox和Subbarao,2000;Lamb和Krug,2001)。根据核蛋白和基质(M)蛋白的抗原差异,流感病毒分为A型、B型和C型。A型流感病毒可感染马、人、家禽、猪,但很少感染犬、海豹、貂和鲸鱼;B型流感病毒只感染人,几乎不感染海豹,而C型流感病毒对人和猪均可感染。以上病毒均可通过气溶胶和飞沫传播。托高土病毒可由蜱传播,极少引起家畜和人发病。传染性鲑鱼贫血病毒是鲑传食病毒属的唯一成员，是一种在大西洋鲑鱼中引起疾病的病毒(MacLachlan,2011)。A型流感病毒属于单股负链RNA病毒，基因组分为8段,D型流感病毒分7段长约13.6 kb,被脂膜包裹。A型流感病毒基因组编码12种不同的蛋白质,即HA、

NA、核蛋白、病毒聚合酶蛋白(PA、PA-X、PB1、PB1-F2 和 PB2)、M1 蛋白、M2 蛋白、NS1 蛋白和核输出蛋白(Chen 等,2001;Jagger 等,2012)。根据表面糖蛋白间的抗原关系,将 A 型流感病毒进一步分为不同的亚型,即棒状三聚体血凝素(H1-H18 亚型)和蘑菇状四聚体神经氨酸酶(N1-N11 亚型)。近期,在秘鲁的一种果蝠体内发现了 H18 和 N11 亚型(Tong 等,2013)。A 型流感病毒形状从球形(直径为 80~120nm)到丝状不等。病毒的脂质双分子层结构含有 2 个糖蛋白,即 HA(负责与宿主细胞上的受体结合)和 NA(指导病毒从被感染细胞释放和传播过程)。两种膜蛋白(MI 和 M2)对 A 型流感病毒的囊膜结构起到支撑作用。A 型流感病毒各片段均被聚合酶蛋白包裹，螺旋状排列的多拷贝核蛋白形成杆状核糖核蛋白(ribonucleo protein,RNP)复合体,这是病毒的核心结构。三种聚合酶蛋白，即 PB1、PB2 和 PA 蛋白间相互作用,形成 RNA 依赖的 RNA 聚合酶复合物,具有控制病毒复制的作用。这种聚合酶蛋白复合物也与核衣壳有关。RNP 是一种具有转录活性的复合物,具有信使 RNA(mRNA)转录和互补 RNA(cRNA)合成的作用,cRNA 是新生病毒 RNA(vRNA)的模板。新生成的负链病毒 RNA 被整合到子代病毒中。NS1 蛋白在 RNA 运输、拼接、转录和破坏宿主免疫应答中起着重要作用，而核输出蛋白则参与了病毒粒子 RNA 的核输出(Paillo 等,2006)。病毒 RNA 依赖的 RNA 聚合酶缺乏校对活性,可能会导致病毒复制出错，进而造成 HA 基因发生抗原漂移和遗传漂移(点突变的积累),从而产生新的毒株。

9.3 流行病学与病毒变异

EI 在欧洲和北美呈地方性流行,具有重要的经济意义(Cullinane 和 Newton,2013)。世界各地经常有关于 EI 的流行报道,如 1986 年和 2003 年在南非(Guthrie,2006)、1987 年和 2008 年在印度(Virmani 等,2008;Singh 等,2018)、1989 年在中国(Guo 等,1992)、1992 年在中国香港(Powell 等,1995)、2007—2008 年在中国、2007 年和 2011 年在蒙古、2012 年在乌拉圭(Acosta 等,2012)、2015 年和 2016 年在巴基斯坦 (Khan 等,2017), 以及 2018—2019 年在欧洲多次报道暴发 EI。在 2007 年前澳大利亚没有发生过 EIV,但 2007 年在澳大利亚暴发了一场大规模的疾病,超过 50 000 匹马受到影响。不过,EI 在澳大利亚流行 4 个月就被彻底根除了(Cowled 等,2009)。在澳大利亚控制和根除 EI 没有进行动物扑杀,主要是通过采取限制运动和严格的生物安全措施有效地控制了 EI。

包括新生小马驹在内的各年龄段马都对 EIV 易感。免疫功能低下的马患病风险更高。2~6 个月龄的马发病率较高(Nyaga 等,1980)。该病通常在 1 周岁赛马的拍卖、运输及其他活动期间暴发(Cullinane 和 Newton,2013;Daly 等,2011)。EIV 传播途径包括直接接触传播和间接接触传播。首次被感染马在接触病毒 24 小时后,就可从鼻分泌物中排出病毒，此过程可持续 10 天 (Myers 和 Wilson,2006)。被感染的马可以通过其强烈的咳嗽将病毒传播到 32 米远的地方(Miller,1965)。EIV 通过气溶胶可随风传播至数公里外，如南非(Huntington,1990) 报道的传播距离超过 8 公里,牙买加报道(Dalglish,1992)传播距离为 3.2 公里,澳大利亚报道传播距离为 5 公里 (Firestone 等,2011)。EIV 通过器械、车辆和人员等污染物进行间接传播(Guthrie 等,1999),用作竞赛和育种的马匹在全球流动促进了流感在全世界的传播(Powell 等,1995;Mumford,1999;Wernery,1999)

流感病毒通过点突变积累、重组和跨物种传播等过程不断变异(Webster,2002)。由于 RNA 聚合酶缺乏校对功能，流感病毒基因组内的突变不断累积。这些突变可能为沉默突变,也可能改变病毒的毒力或致病性。表面糖蛋白点突变的长期积累会造成结构和抗原差异，从而对疫苗的有效性产生不利影响(Nakajima 等,2005)。抗原漂移主要与免疫选择有关，其他因素还包括病毒在新宿主中的适应性。人类A型流感病毒的HA基因2个不同抗原位点至少需要4种抗原变化才能产生显著的抗原漂移(Wilson 和 Cox,1990)。负责编码病毒HA1糖蛋白基因的5个主要抗原位点基因中的任意一个或多个位点突变都会导致循环抗体无法识别接种的疫苗或感染的病毒 (Wiley 和 Skehel,1987;Daly 等,2003)。由于不断暴露于宿主免疫系统之下，流感病毒的抗原HA和NA基因变异很常见(Webster 和 Laver,1975),造成在世界不同地区每5~7年暴发的流感中，就会分离到新的毒株。

流感病毒的分节段性使病毒可在重复感染期进行基因组重排，子代病毒将获得2个亲本的基因组片段并成为新的亚型(抗原转换)。H3N8亚型病毒比人类流感病毒进化速度慢。一些病毒已经传播了超过25年,没有太多的抗原和基因变异(Guo 等,1992);这种现象被称为“进化冻结”(Endo 等,1992)。

H7N7亚型和H3N8亚型EIV均与马的急性呼吸道疾病有关。H7N7于1956年在东欧首次被发现(Sovinova 等,1958),但直到20世纪70年代后才被报道，且认为该病毒在马中已灭绝(Webster 和 Thomas,1993年)。H3N8亚型引发的感染于1963年在佛罗里达被首次报道 (Waddell 等,1963); 自那时起,H3N8亚型流感病毒在世界各地蔓延,目前的EI疫情均是由H3N8亚型流感病毒引起的。EIV比人类流感病毒更稳定,但由于HA和NA基因点突变,仍然周期性地发生抗原漂移,致使病毒抗原性发生重大变化(Cullinane 和 Newton,2013)。A/equine/Miami/63 和 Fontainebleau/76分别是20世纪60年代和20世纪70至80年代早期的EIV原型。HA基因遗传进化树表明,在20世纪80年代，其分化成两种不同的遗传变异分支，根据分离地分别被命名为类美洲系和类欧洲系(Daly 等,1996),但后来这两种病毒不受地域限制共同传播。在整个A-D长度的HA抗原区域中，类美洲系和类欧洲系病毒原型有9个氨基酸残基不相同。两个系病毒感染/接种之间不存在免疫交叉保护,因此EIV疫苗含有两个系病毒的抗原 (Daly 等,1996,2004;Yates 和 Mumford,2000;Mumford 等,2003)。类美洲系毒株进一步演化成三个不同分支,即佛罗里达亚系、肯塔基亚系和阿根廷亚系(Lai 等,2004),其中佛罗里达亚系占主导地位。此外,佛罗里达亚系分化为两个截然不同的抗原分支 (即分支1和分支2)(Bryant 等,2009)。佛罗里达分支1病毒在美国占主导地位,但却在亚洲(日本)、澳大利亚、非洲和欧洲暴发(Cullinane 和 Newton,2013;Yamanaka 等,2008)。佛罗里达分支2病毒在欧洲占主导地位，但在亚洲国家蒙古(2007)和印度(2008—2009)大规模暴发(Virmani 等,2010a;Qi 等,2010;Yondon 等,2013)。2013—2015年,英国、德国和意大利暴发的EIV疫情主要由佛罗里达分支2病毒引起;然而,在英国暴发的EIV,HA的144位(A144V)氨基酸序列发生改变,而来自德国和意大利的EIV,HA的氨基酸则在179位发生变化 (Rash 等,2017),2011—2012年,在智利暴发了佛罗里达分支1病毒，随后蔓延至南美洲的一些国家(Perglione 等,2016)。2015年,马来西亚也暴发了佛罗里达分支1病毒(Gahan 等,2019)。2018—2019年,包括英国、法国、德国、爱尔兰、荷兰、丹麦、瑞典和比利时在内的许多欧洲国家发生了严重的EI疫情,其中一株病毒起源于佛罗里达分支1病毒。

病毒复制完全依赖于宿主细胞系统,因此,流感病毒的宿主适应性受基因组组成和密码子使用模式的影响。很少有研究强调使用同义密码的EIV在宿主的整体适应中的作用。整体EIV基因组密码子的偏嗜性较低,主要受核苷酸组成、免疫压力和自然选择的影响(Kumar 等,2016)。此外,密码子偏嗜性也突显了H7N7亚型灭绝的潜在因素，与H3N8亚型相比,H7N7密码子有更强的偏嗜性、受到突变压力更小、对马宿主tRNA的适应性更低(Kumar 等,2016)。另一项研究则分析了各种变异因素对 EIV 聚合酶基因的影响(Bera 等,2017)。聚合酶基因直接参与病毒复制、转录和宿主传播,在宿主的适应性和发病机制中起重要作用。类似于 EIV 基因组密码子使用模式,聚合酶基因密码子的偏嗜性也较低。自然选择是这些基因密码子使用和变异的主要因素;然而,核苷酸组成、免疫压力、疏水性和芳香族化合物的结构等因素也会影响密码子的使用(Bera 等,2017)。

9.4 宿主

A 型流感病毒的宿主范围实质是由多种基因决定,需要 HA、NA 和宿主体内蛋白的功能平衡。它还取决于受体在宿主体内的分布,特别是与宿主细胞表面半乳糖的连接方式有关。EIV 在进入细胞之前,优先与宿主细胞唾液酸寡糖相连的 N-乙酰神经氨酸 α2,3-半乳糖结合 (Lee 和 Saif,2009),特别是鼻黏膜的上皮细胞、杯状细胞,鼻腺、下呼吸道的气管和细支气管上皮细胞(Scocco 和 Pedini,2008)。在禽类也发现了类似的受体,并在 H3N8 株中分离到了类似的受体。马、驴和骡是 EIV 常规感染的靶向物种,被认为是 El 病毒的分离宿主或终末宿主。然而,在 2004 年,EIV 表现出跨种传播的特性,可传播给犬科动物(Crawford 等,2005)。A 型流感病毒种间传播取决于多种因素,包括不同宿主之间的相互作用,以及新宿主呼吸道上皮细胞上是否存在适于 HA 吸附的受体(α2,3-唾液酸链接)。

正如我们所知,源于犬肾细胞的细胞系,由于细胞表面有 α2,3-唾液酸残基的存在,可支持 EIV 在犬肾细胞增殖 (Kovbasnjuk 和 Spring,2000),在 EIV 的传播和诊断中具有重要的作用,虽然在犬的上呼吸道存在类似的 α2,3-唾液酸残基,然而,在犬中并没有观察到 EIV 感染。2004 年初,在佛罗里达州,EIV 跨越种属屏障,由马传染给犬,其中有 8 只灰犬患有致死性的出血性气管炎、支气管肺炎和胸膜炎,22 只被感染犬的血清呈阳性,并通过分离到病毒获得证实(Crawford 等,2005)。2004—2007 年年底,美国灰犬和宠物犬的发病报道较少(Payungporn 等,2008)。通过对犬流感病毒分离株的 HA 基因进行遗传变异分析发现,各分离株间的关系密切,其共同祖先属于 20 世纪 90 年代在马中流行的 H3N8 亚型佛罗里达亚系毒株。通过回顾性研究发现,H3N8 病毒是导致 2002 年英国猎狐犬发生感染、产生呼吸症状的原因(Daly 等,2008),犬中最早发生的传播可能在 1998—2003 年(Payungporn等,2008)。最近在 2007 年澳大利亚暴发的 EI 疫情中,有 EIV 传播给犬的报道(Kirkland 等,2010)。犬的气管上皮细胞上有优势的 α2,3-唾液酸受体(Daly 等,2008)可能是引起犬类感染的原因,但为什么直到 2004 年才有关于犬类感染的报道尚不清楚,可能是将该病误认为犬窝咳征。不过,目前流行的 EIV 可能会传播给犬(Daly 等,2011)。从犬中共分离到 6 株流感病毒,分别为 A/犬/Jacksonville/2005、A/犬/Miami/2005、A/犬/Florida/242/2003、A/犬/Florida/43/2004、A/犬/Texas/2004 和 A/犬/Iowa/2005 (Crawford 等,2005;Yoon 等,2005)。分离到的 6 株犬流感病毒的 H3 血凝素有 5 个保守氨基酸发生置换,这使其与 EIV 区分开来。发生置换的保守氨基酸分别是 N54K、N835、W222L、1328T 和 N483T,这被认为是犬流感病毒的 H3 血凝素循环的标志。

除了马和犬外,2005 年和 2006 年在中国从患有呼吸症状的猪上分离到 2 株病毒,均属于 H3N8 亚型 EIV。通过对 HA 进行测序研究发现,目前流行的病毒与 20 世纪 90 年代流行的欧洲系病毒非常相似 (Tu 等,2009)。2012 年晚些时候,通过对蒙古的一只骆驼进行鼻拭子检测,结果 H3N8 亚型 EIV 呈阳性。然而,自此之后,没有任何关于骆驼临床感染或病毒在骆驼群中传播的报道(Yondon 等,2014)。

9.5　马流感流行

历史上曾多次报道过类似EI疾病。众所周知，EI首次于1872年在北美暴发，通常被称为“1872年大流行病”，马群中EI发病率高达100%，死亡率为1%~10%(McClure，1998)。1955—1956年，EIV首次在瑞典和东欧被分离到，属于H7N7亚型，命名为A/EQ/Prague/56(Sovinova等，1958；Tumova和Sovinova-Fiserova，1959)。血清学回顾性分析显示，在首次病毒检测之前，该病毒已经在美国和欧洲流行(Doll，1961)，A/eq/Miami/63是在迈阿密被首次报道的H3N8亚型流感病毒(Waddell等，1963)。直到20世纪70年代末，H7N7和H3N8毒株都在马群中流行，并引起发病。在过去的35年里，没有关于H7N7毒株引起EI的报道，认为其已经灭绝（Webster，1993)。近年来，EIV/H3N8佛罗里达亚系分支1和分支2在全世界马群中流行。在欧洲和北美地区呈现地方性流行(Cullinane和Newton，2013)。多个国家报道了EI的流行，比如1986年和2003年在南非(Guthrie，2006)；1987年和2008年在印度（Virmani等，2008)；1989年在中国(Guo等，1992年)；1992年在中国香港（Powell等，1995)；2007—2008年在中国；2007年和2011年在蒙古；2012年在乌拉圭(Acosta等，2012)；2015年和2016年在巴基斯坦(Khan等，2017)，造成了巨大的经济损失。2007年，在澳大利亚暴发了一次大规模EI，5万多匹马受到影响，但目前EI已在该国被根除(Cowled等，2009)。

9.6　印度马流感

印度加尔各答从澳大利亚进口马群后曾出现过类似于EI的报道，并在1915年4月到9月期间传播至全国。由于该病具有传染性，潜伏期短，具有发热症状，呼吸道症状，休息后便可迅速恢复，因而被怀疑为EI。据报道，共有16 921起病例，其中死亡病例893起，但没有分离到病原体(Williams，1924)。Manjrekar等(1965)报道了另一起类似EI的马呼吸道疾病，大约有400匹来自孟买皇家西印度草皮俱乐部的马受到影响。但没有从被感染动物的体内分离到病原体。“纽马克特咳嗽”于1964年10月的最后一周暴发，一直持续到1964年12月。

1987年，印度暴发了EI疫情，当时印度北部地区超过8.3万匹马被感染(Uppal和Yadav，1987)。从法国进口的马是印度此次暴发EI的传染源。在此次暴发的EI中，A/eq/Bhiwani/87和A/eq/udhiana/87(Uppal等，1989)两株病毒被分离、鉴定。用A/equi2/Ludhiana/87全病毒自主研制了佐剂灭活疫苗，并被应用（Gupta等，1993)。从1987—2008年，这二十年间没有再发生过EI疫情(Virmani等，2010a)，不过，在印度一些地区检测到了A/Equine-2抗体，这可能是由接种疫苗引起。

EI于2008年6月份在印度再次暴发，此次EI最早在印度最北部的查谟和克什米尔发生，随后蔓延至全国14个州(Virmani等，2008，2010a，b)。分离到的病毒属于H3N8亚型，根据病毒分离地，将其分别命名为A/eq/Jammu-Katra/08、A/eq/Mysore/08和A/g/Ahamadabad/09。HA基因进化分析证实该病毒与美洲系佛罗里达亚系分支2(H3N8)有关(Virmani等，2010a)。通过对全部分离株的M基因进行系统发育分析发现，与源于亚洲其他分支2病毒的M1和M2氨基酸序列同源性分别为98.41%和99.54%。此外，所有亚洲分离株，包括中国分离株和蒙古分离株的M1和M2蛋白分别有3个和4个特定的氨基酸发生变化(Virmani等，2011)。印度哈里亚纳邦海萨城的国家马研究中心持续开展了对EI血清监测，并采取预防措施，使此病在该国得到控制，在2009年的7月份，报道了最后一例EI。

9.7 发病机制

EI 潜伏期为 18 小时至 5 天，与病毒攻击剂量呈负相关。EIV 与 α2,3-唾液酸受体结合后，在 HA 作用下，对整个呼吸道，尤其是下呼吸道产生病理损伤(Muranaka 等，2010)。在 NA 蛋白作用下，EIV 裂解黏液中的唾液酸，使得 EIV 与细胞受体结合，增加病毒的流动性，并有利于细菌继发感染(Patterson-Kane 等，2008)。病毒被吸附后，通过受体介导内吞作用发生内化作用 (Eierhoff 等，2010)。核内体酸化后，HA 中发生不可逆的构型改变(Sieczkarski 和 Whittaker，2002)。病毒和细胞膜的融合使得病毒 RNP 释放到细胞质中，随后被运送到细胞核中合成 mRNA。在 NS1 和核蛋白等蛋白质作用下，在细胞核合成互补 RNA 和病毒 RNA(Lee 和 Saif，2009)。新合成的病毒蛋白 M1、HA 和 NA，以膜结合形式产生，并在内质网和高尔基体运输过程中发生多种构型改变(Bosch 等，1981)。病毒粒子出芽是通过 HA 和 NA 在细胞膜表面(脂筏结构域)集聚实现(Wang 等，2007)。M1 与 HA 和 NA 细胞质尾部结合，从而与病毒核糖核蛋白对接。M1 聚合作用使得出芽病毒粒子伸长，并将病毒核糖核蛋白定位在一定部位。在 M2 蛋白的作用下，膜与病毒粒子完全分离。最后，成熟的感染性病毒粒子被释放到黏膜管腔表面(Rossman 等，2010；Rossan 和 Lamb，2011)。

病毒感染能够破坏纤毛上皮，降低气管清除率 (Willoughby 等，1992)。黏膜上唾液酸残基与 NA 的相互作用有利于病毒吸附和细菌的继发感染。这是一个很好的病毒-细菌协同作用的例子(Lopez，2007)。

人们对于流感病毒如何杀死细胞的分子机制知之甚少。哺乳动物细胞感染流感病毒会引发细胞凋亡并引起细胞病变(Takizawa 等，1993)。通过利用犬肾细胞对 EIV 体外感染的研究探讨了一些基本的发病机制。被感染的犬肾细胞引起细胞氧化应激，并诱导类似 c-JUN/Ap-1 的应激相关的转录因子生成，这些应激相关的转录因子与细胞凋亡和细胞死亡有关。EIV 还可诱导细胞因子 TGF-β1 表达，并在细胞因子高表达时诱导细胞凋亡。通过激活 JNK/SAPK 途径，介导对 TGF-β 的生长抑制作用。TGF-β 激活 JNK，刺激 c-JUN 的 JNK 磷酸化，诱导细胞凋亡。通过中和抗体抑制 TGF-β1，进而减弱的 EIV 通过下调 c-JUN/API 诱导细胞凋亡。c-JUN 和 TGF-β1 在诱导细胞凋亡中的作用是通过抗氧化剂(N-乙酰半胱胺酸)进行评估，抗氧化剂能够导致抗凋亡基因 BCL2 的过表达。使用抗氧化药物，如卡维地洛，对 EIV 感染引起的犬肾细胞细胞死亡具有很强的抑制作用。流感病毒表面蛋白 HA 激活转录因子 NF-κB，NA 蛋白可直接激活 TGF-β。C-Jun/AP-1 和 NF-κB 激活都是通过活性氧浓度进行介导。这些因子均靶向细胞因子基因，如 TNF-α 和 TGF-β。较高的细胞因子水平对细胞具有细胞毒性作用。也有报道称，流感病毒能够激活与细胞凋亡相关的 Fas 基因。惊人的是，流感病毒的 NS 基因编码蛋白与 Fas 抗原基因编码的 Fas 蛋白有超过 68 个氨基酸的胞质结构域相似，其相似性约 50%。一般认为 NS 蛋白可能在流感病毒诱导的细胞凋亡中起重要作用。EIV 感染也会刺激产生多种与 JNK/SAPK 的激活和凋亡有关的半胱氨酸蛋白酶。通过观察 EIV 感染的犬肾细胞，推测 EIV 在感染的马中也发挥着类似的激活多条信号通路的机制，以及与 EIV 感染有关的细胞凋亡机制(Chengbin 等，2001)。

9.8 病毒感染

有研究表明，雄性动物比雌性动物更易感 EIV，这可能是由于雄性动物的睾酮比雌性动物的雌激素更具有免疫抑制作用 (Nuria 等，2007)。因此，在受到感染之后，雌性动物产生的体液和细胞免疫反应要比雄性动物快。所有品种的马不分年龄大小,均易受到感染。然而,更常见于 2 岁以下的幼马。虽然死亡率不高,但在老马和免疫缺陷的马中常出现流感病毒感染后继发细菌性感染而导致的死亡。

据报道,在印度暴发 EI 时,曾造成 50 匹马死亡,当时患病马处在气候条件恶劣,爬了 14 公里山坡的条件下。50 匹马死亡原因为:感染 EIV 后,继发细菌感染引起死亡(Virmani 等,2010a,b)。EI 可全年暴发; 然而,EI 暴发通常都与一周岁赛马拍卖、动物集市、赛马/繁殖活动、过度拥挤和不受限制的活动有关(Wilson 1993;Virmani 等,2010a,b)。受病毒感染的动物会对同源毒株产生具有保护作用的抗体,且免疫保护作用能够持续 1 年,对异源病毒株可提供部分免疫保护作用。具有部分免疫保护作用的马可能不会受到重度感染, 但可能会向外界排毒, 给未发育成熟的种马带来感染风险,并引起抗原漂移(Paillot,2014)。

9.9 临床症状

马的临床症状及发病严重程度取决于病毒毒株和宿主免疫状态(Mumford 等,1990)。未免疫接种的幼龄马会表现出严重的临床症状 (Daly 等,2011)。临床症状最早于感染后 24 小时出现,其特征为双相热、呼吸困难、持续性干咳、剧烈咳嗽及流涕(Muranaka 等,2012)。初期是浆液性鼻分泌物,3~4 天后变成脓性分泌物。原发性疾病后一周左右观察到第二个发热高峰。其他可变临床症状包括厌食、抑郁、肌痛和呼吸系统相关淋巴结肿大(导致呼吸急促)及肢体水肿。极个别异常病例具有脑炎、心肌炎症状。不过,EIV 在其中的作用需要被阐明(Daly 等,2006)。血液和生化指标的改变具有非特异性,包括贫血、白细胞减少、淋巴细胞减少和轻度低蛋白血症。单纯 EIV 感染的临床症状可能会持续 7~14 天, 但咳嗽可能会持续 21 天(Newton 等,2006)。EI 继发细菌性肺炎通常具有肺实变的特征。在适宜病毒生存的条件下,EI 的发病率通常可达 100%,通常马驹、老马及生活在恶劣环境中的动物多发, 主要表现为继发性细菌性支气管肺炎和间质性肺炎,但很少引起死亡。初期感染的动物通过鼻分泌物排出病毒, 至少可持续 4~6 天。在 EIV 感染的马中未观察到带毒状态。根据免疫状况,接种过疫苗的动物或在上个季节接触过 EI 的动物可能会发生轻度感染。部分受保护的动物在没有明显临床症状情况下仍然可通过鼻分泌物排毒, 因此对于病毒在环境中持续存在起着主要作用。

9.10 病理变化

由于 EIV 感染造成马死亡的病例很少,因此,有关马自然感染 EIV 病理变化的信息极为有限。此外,对 EIV 的致病机制研究,很少用马进行感染实验。自然/实验感染 EIV 的马肉眼可见的病变局限于呼吸器官及其相关淋巴结。通常表现为鼻黏膜充血、鼻分泌物多、急性期伴有咽后淋巴结、肺淋巴结轻微肿胀。感染后 1 周,可见肺部点状出血,气管和支气管有黏性、脓性分泌物伴有间质性肺水肿。感染后 2 周,出现继发性细菌性肺炎引起的肺纤维化/肺实变。

早期感染, 只在上呼吸道发生组织病理学损伤。鼻炎和气管炎特征性组织病理学表现为上皮细胞变性、坏死,纤毛上皮消失,杯状细胞数量减少,固有层淋巴细胞中度浸润。更严重病变可见于气管末端。感染后 7 天,病变扩展至细支气管和肺泡中。上皮细胞明显增生、鳞状上皮化生;2 型肺细胞增殖、化脓性支气管肺炎合并肺水肿;中性粒细胞增生、巨噬细胞和上皮细胞坏死 (Begg 等,2011)。感染后 14 天可观察到严重气管炎、伴有上皮增生的支气管炎、伴有短绒毛的鳞状上皮化生和小杯状细胞。在支气管和细支气管上皮固有层可见中性粒细胞浸润。随后,继发兽疫链球菌、马链球菌、马驹放线杆菌和支气管炎博德特菌等感染引起支气管肺炎, 造成纤毛上皮和杯状细胞损伤。肺泡内中性粒细胞和肺泡巨噬细胞大量浸润，可观察到纤维蛋白渗出、肺水肿及2型肺细胞明显增殖。在呼吸性淋巴结感染的所有阶段都观察到淋巴细胞增生(Muranaka 等,2012)。

为实现对 EIV 致病性的详细研究及对灭活疫苗快速评价，印度国家马属动物研究中心建立了 BALB/c 小鼠模型(Pavulraj 等,2015,2017)。在致病性研究中,BALB/c 小鼠经鼻内接种佛罗里达亚系分支2的 EIV(H3N8),并对建立的感染和相关参数进行检测。感染 EIV 的小鼠表现为体重下降、嗜睡、呼吸困难等临床症状,感染后3~5天,可进一步观察到呼吸道病变,包括鼻炎、气管炎、支气管炎、细支气管炎、肺泡炎及弥漫性间质性肺炎。利用透射电子显微镜、免疫组织化学法及通过滴定和 qRT-PCR 对病毒进行定量，可观察到病毒在上、下呼吸道感染情况。血清学调查显示血清乳酸脱氢酶和血清阳性转化率较高。自然感染和试验感染 EIV 的马整个病程发展模式、病理变化，以及被感染小鼠的鼻和肺洗脱液中病毒培养增殖情况一致。随后,利用建立的 EIV 小鼠模型，通过血清学、临床体征、肉眼可见病理变化及组织病理学病变分级、免疫组织化学、病毒定量对灭活 EIV 疫苗效能进行评价。接种2次疫苗的小鼠产生具有免疫保护作用的血凝抑制实验(hemagglutination inhibition,HAI) 和单向辐射溶血(single radial hemolysis,SRH)抗体,并产生 Th1/Th2 平衡反应。接种过疫苗的小鼠对 EIV 攻毒具有明显的免疫保护作用，具体表现为血清抗体效价显著增高,临床症状轻微,恢复快,肉眼可见病理变化及组织病理学病变等级变低，病毒抗原分布强度较低,病毒只在呼吸道复制,在鼻洗脱液中检测到的病毒少且存在时间短。接种疫苗的小鼠仅出现肺实质充血,未见肺实变。本研究发现,BALB/c 小鼠可作为研究 EIV(H3N8)感染的小型动物模型,在病毒致病机制研究、疫苗疗效评价、候选疫苗的初步筛选和抗 EIV 治疗等方面具有巨大的潜力。

9.11 细胞免疫

EIV 感染或接种 EIV 疫苗诱导的细胞免疫应答反应尚未进行详细研究。当 EIV 感染具备完善免疫防御能力的马时,在血清中检测不到 EIV 特异性 SRH 抗体,这与免疫系统的另一个支持性免疫作用,即 CMI 有关。前期研究表明,加入氢氧化铝或磷酸铝佐剂的全 EIV 疫苗不能激活细胞毒性 CTL (Hannant 等,1994)。疫苗的佐剂对激活 CMI 或体液免疫应答具有重要作用。众所周知,氢氧化铝对 Th2 反应具有激活作用，诱导体液免疫应答，而对 CMI 不具有诱导作用(Lindblad, 2004)。基于免疫刺激复合物基质的 EIV 佐剂疫苗可诱导以 T 细胞增殖为特征的较强的 CMI 反应，并且会增加 EIV 诱导的特异性 CTL 百分比(Paillot 和 Prowse,2012)。近期有关接种 Duvaxyn IE-T Plus 疫苗后激活 CMI 的反应被报道(Paillot 等,2013),其中合成 γ-干扰素的阳性细胞率被认为是 CMI 标记。在第1次免疫接种前、后3周,第1次和第2次加强免疫后2周，分别对产生 EIV 特异性 γ-干扰素的细胞百分率进行测定。与对照组和免疫前小马相比，免疫接种后小马体内产生特异性 γ-干扰素的细胞百分率有所增加。此外，据报道,给小鼠接种羧乙烯聚合物 974p 疫苗会激活以强烈的 γ-干扰素反应和体液免疫为特征的 CMI 反应。虽然在 EIV 感染期间,CTL 反应的保护作用还不清楚，但这仍是开发更好疫苗的重要组成部分。

9.12 人兽共患

EIV 通过与细胞表面 α2,3-唾液酸受体结合进入细胞,然后开始复制。与之相比,人的呼吸道上皮细胞含有 α2,6-唾液酸受体。鉴于人类中暴发的高致病性禽流感 H5N1，细胞表面虽然缺乏特定受体，但并不排除 EIV 跨种传播的可能性。

然而,H5 亚型 HPAI 的 HA 裂解位点上具有多个特殊的碱性氨基酸，而 H3 亚型流感病毒的 HA 蛋白不具备这个特点。为证明人可能感染马 H3N8 亚型流感病毒，很少在人类志愿者中进行 EIV 攻毒实验研究。虽然人感染病毒后出现阳性血清,但未观察到临床症状(Kasel 等,1965,1969;Alford 等,1967)。最近,在艾奥瓦州和马接触的人中,检测到了 EIV 的血清学阳性反应。然而,并没有分离到病毒(Larson 等,2015)。值得注意的是,虽然观测到有支持人感染 EIV 的证据,但其结果仍需谨慎解释。由于已经存在另一种人类 H3 亚型流感病毒抗体或人类疫苗接种的抗体，所以 H3N8 亚型 EIV 血清学应答不应被混淆。

9.13 治疗

目前还没有特异性治疗 EI 的方法,现有治疗方法主要是根据临床症状给予支持性治疗，如充分休息。治疗方法可选择广谱抗生素(避免继发性细菌并发症)、抗组胺药和非甾体抗炎药(减少呼吸道炎症反应)。良好的营养水平和充足的饮水状态有助于受感染动物迅速恢复。充足的休息(发病后休息 3~4 周）对于动物快速康复很有必要。初步研究表明，通过口服抗病毒制剂金刚烷胺和金刚乙胺(针对 M2 离子通道蛋白的跨膜结构域)对患有 EI 的马具有良好的治疗效果;此外,扎那米韦和奥司他韦(NA 抑制剂)在体外表现出抗 EIV 活性。虽然对于马的抗病毒治疗的研究很少,但研究结果仍然是充满希望的。由于成本、劳动力、特殊场地等原因，使得在马身上进行临床试验或攻毒试验难度大,最终使得相关研究进展缓慢。

9.14 疫情控制

受感染动物体外病毒持续时间取决于病毒毒株及动物免疫水平。动物向体外排毒可持续 3~8 天（Ragni-Alunni 和 Zande,2009;Daly 等,2007 Paillot 等,2008)，很少超过 12 天（Paillot 等，2013)。在流感暴发(全面接种)早期诊断之后,无论其以往接种计划如何，都要对所有动物进行免疫接种，这对于缓解疫情十分重要（Nuria 等，2007)。2007 年,澳大利亚暴发 EI,控制疫情基本策略包括限制行动、公众意识和沟通,根据风险划分区域,实验室检测、疾病追踪和监测、加强预防传播的生物安全措施和紧急接种疫苗。通过采取以上措施,澳大利亚在 5 个月内就实现了对 EI 的控制和消除,其中限制行动在 EI 的控制和消除中发挥了巨大的作用。疫苗接种在控制疫情中的作用需要进一步深入研究(Cowled 等,2009)。EIV 极易被阳光、紫外线、热、冷和常见的消毒剂灭活(Yadav 等,1993)。因此,在 EI 暴发期间,可用消毒剂对种马进行常规消毒，防止病毒通过污染物进行传播。

9.15 预防与疫苗

1979 年疫情暴发后,英国等国家自 1981 年开始对赛马进行强制接种疫苗(Elton 和 Bryant，2011)。1966 年初,Bryans 及其同事（Bryans 等，1966)根据人类流感疫苗的经验,研发了针对 EI 的多价灭活全病毒疫苗，其他工作者又对其进行了升级研发(Petermann 等,1970;Burki 和 Sibalin，1973;Frerichs 等,1973)。目前可用的 EI 疫苗包括油佐剂、氢氧化铝和高分子胶的全病毒灭活疫苗或含有免疫刺激复合物，还有植物皂苷胶束的亚单位疫苗(表 9.1)。含有磷酸铝或氢氧化物等佐剂的疫苗可诱导长达 16~20 周的抗体反应。免疫计划为初次免疫接种 2 个剂量疫苗，间隔 4~6 周再接种 2 个剂量疫苗加强免疫 1 次，然后每年重复进行免疫接种或根据测定的血清 SRH/血凝抑制试验抗体效价后再进行重复接种。用同源病毒对接种疫苗的动物进行攻击通常会产生高水平体液免疫应答及最高水平的保护性免疫，而异源病毒则不能(Mumford 和 Wood,1992)。SRH 抗体水平在 120~154mm^2 时能够对小马在实验雾化气溶胶

表 9.1 目前可用的 EI 疫苗

类型	疫苗	公司	疫苗毒株	疫苗性质	佐剂
全病毒	Duvaxyn™ IE T Plus	Elanco Animal Health	A/eq/Suffolk/89(H3N8) A/eq/Newmarket/93(H3N8) A/eq/Prague/56(H7N7)	灭活	羧乙烯聚合物 934P 和氢氧化铝
	Calvenza®-03 EIV	Boehringer Ingelheim Animal Health	A/eq/Newmarket/2/93 (H3N8),A/eq/Kentucky/2/95 (H3N8) A/eqoHIO/03(H3N8)	灭活	Carbimmune 佐剂
	Equilis Prequenza (2013 年更新)	MSD Animal Health	A/eq/Newmarket/2/93 (H3N8),A/eq/South Africa/4/03(H3N8)		免疫刺激复合物基质
	Equip F	Zoetis	A/eq/Borlänge/91(H3N8) A/eq/Kentucky98(H3N8) A/eq/Newmarket/77(H7N7)	灭活	植物皂苷磷酸铝
	Fluvac Innovator	Zoetis	A/eq/Kentucky97(H3N8)	灭活	MetaStim®
	Equip™ F	Pfizer Ltd.	A/eq/Newmarket/77 (H7N7), A/eq/Borlänge/91(H3N8), A/eq/Kentucky/98(H3N8)	灭活	植物皂苷
亚单位	Equilis Prequenza Te	MSD Animal Health	A/eq/South Africa/4/03(H3N8),A/eq/Newmarket/2/93(H3N8) 破伤风类毒素	亚单位 HA	精制皂素
病毒重组载体疫苗	PROTEQ FLU™	Merial Animal Health Ltd.	A/eq/Ohio/03 (H3N8),A/eq/Richmond/1/07(H3N8)	灭活重组金丝雀痘-HA 亚单位	高分子胶
EIV 修饰的活疫苗	Flu Avert® I.N.	Intervet/Schering-Plough Animal Health (美国)	A/eq/Kentucky/91(H3N8)	减毒冷适应全 EIV	NA

攻击条件下提供完全的免疫保护 (Mumford 和 Wood,1992),这一点通过 EI 自然暴发期间的实地研究得到了证实。感染前 SRH 抗体水平>140mm² 的动物在感染过程中不会产生任何临床症状或用基于核蛋白抗原的 ELISA 检测血清呈阴性 (Townsend 等,1999;Newton 等,2000)。有足够的证据证明 SRH 抗体水平>80mm² 可预防临床感染 (Mumford 等,1988,1990)。防止 EIV 感染与抗体应答/效价绝对相关 (Park 等,2003;Paillot 等,2013)。根据人工接种 EIV 小马驹的研究结果显示,SRH 抗体效价越高,通过 ELISA 检测到的经鼻孔排出的病毒越少,同样通过 qRT-PCR 检测到的经鼻孔排出病毒 RNA 拷贝数越少。另一方面,SRH 抗体效价较低的小马,其排出的病毒量

及其 RNA 拷贝数增加。在接种/免疫的动物中，针对不同病毒的保护可能需要很高的 SRH 抗体效价，才能对抗原相似的病毒提供类似的免疫保护水平(Nuria 等，2007；Paillot 等，2013)。

即使 SRH 抗体已经在宿主中消失，自然感染后的保护性免疫仍可持续较长时间(Hannant 等，1988 a,b)。自然感染病毒的宿主可产生持久性免疫力，其特点包括黏膜局部生成 IgA 抗体、IgGa 和 IgGb(Nelson 等，1998)。另一方面，全病毒灭活疫苗免疫后引起以 IgGa、IgGb 和 IgG(T)抗体为特征的免疫反应，在黏膜部位几乎没有 IgA 应答(Wilson 等，2001)。综上所述，高效的病毒感染是产生特异性黏膜 IgA 抗体的必要条件 (Lunn 等，1999)。此外，除了上述所指近期开发的疫苗外，由于抗原的处理和递呈不同，以 CTL 反应为特征的 CMI 通常只在自然感染病毒后被观察到(Hannant 等，1989)。

免疫保护不全或免疫应答低是普遍现象，大多数免疫的马不能产生最佳免疫反应，仍然容易受到感染。虽然马体内免疫反应不佳的确切原因尚不清楚，但建议在免疫之后检测马体内 SRH 抗体水平，获得最佳的群体免疫保护水平(Gildea 等，2011)。此外，它可能为确定不良免疫反应和优化疫苗免疫接种时间奠定基础。在母源抗体存在的情况下，早期接种疫苗会增加马后期患有 EI 的风险。免疫接种应在母源抗体下降后 6 个月进行(Nuria 等，2007)。尽管进行了疫苗接种，由于疫苗与暴发毒株不匹配，2004 年在克罗地亚(Barbic 等，2009)、2005 年在意大利 (Vito Martella 等，2007)发生过几次疫苗失效事件。疫苗的效力、佐剂、疫苗接种计划及抗原漂移监测是目前研究的重点。

参考文献

Acosta R, Rimondi A, Mino S, Tordoya MS, Fernandez A, Betancor G, Larrauri H, Larrauri H, Nuñez A, Barrandeguy M (2012) Outbreak of equine influenza among Thoroughbred horses in Maronas, Montevideo (Uruguay) during March and April 2012. J Equine Vet 32:7

Alford RH, Kasel JA, Lehrich JR, Knight V (1967) Human responses to experimental infection with influenza a/equi 2 virus. Am J Epidemiol 86:185–192

Animal Health Trust, UK (2019) Equine influenza outbreaks reported in 2019. https://www.aht.org.uk/wp-content/uploads/2019/04/Equiflunet-outbreaks-to-15-April2019.pdf

Barbic L, Madic J, Turk N, Daly J (2009) Vaccine failure caused an outbreak of equine influenza in Croatia. Vet Microbiol 133:164–171

Begg AP, Reece RL, Hum S, Townsend W, Gordon A, Carrick J (2011) Pathological changes in horses dying with equine influenza in Australia. Aust Vet J 89:19–22

Bera BC, Virmani N, Kumar N, Anand T, Pavulraj S, Rash A, Elton D, Rash N, Bhatia S, Sood R, Singh RK, Tripathi BN (2017) Genetic and codon usage bias analyses of polymerase genes of equine influenza virus and its relation to evolution. BMC Genomics 18(1):652. https://doi.org/10.1186/s12864-017-4063-1

Beveridge WI (1965) Some topical comments on influenza in horses. Vet Rec 77:42

Bosch FX, Garten W, Klenk HD, Rott R (1981) Proteolytic cleavage of influenza virus hemagglutinins: primary structure of the connecting peptide between HA1 and HA2 determines proteolytic cleavability and pathogenicity of avian influenza viruses. Virology 113(2):725–735

Bryans JT, Doll ER, Wilson JC, McCollum WH (1966) Immunization for equine influenza. J Am Vet Med Assoc 148:413–417

Bryant NA, Rash AS, Russell CA, Ross J, Cooke A, Bowman S, MacRae S, Lewis NS, Paillot R, Zanoni R, Meier H, Griffiths LA, Daly JM, Tiwari A, Chambers TM, Newton JR, Elton DM (2009) Antigenic and genetic variations in European and North American equine influenza virus strains (H3N8) isolated from 2006 to 2007. Vet Microbiol 138:41–52

Burki F. Sibalin M (1973) Conclusions and questions arising from a study of serology and immunology of equine influenza. In: Proceedings of the 3rd International Conference on Equine Infectious Diseases, Eds: J.T. Bryans and H. Gerber, Karger Basel, New York, pp 510–526

Chen W, Calvo PA, Malide D, Gibbs J, Schubert U, Bacik I, Basta S, O'Neill R, Schickli J, Palese P, Henklein P, Bennink JR, Yewdell JW (2001) A novel influenza A virus mitochondrial protein that induces cell death. Nat Med 7:1306–1312

Chengbin L, Zimmer SG, Lu Z, Holland RE Jr, Dong Q, Chambers TM (2001) The involvement of a stress-activated pathway in equine influenza virus-mediated apoptosis. Virology 287:202–213

Cowled B, Ward MP, Hamilton S, Garner G (2009) The equine influenza epidemic in Australia: Spatial and temporal descriptive analyses of a large propagating epidemic. Prev Vet Med 92:60–70

Cox NJ, Subbarao K (2000) Global epidemiology of influenza: past and present. Annu Rev Med 51(1):407–421

Crawford PC, Dubovi EJ, Castleman WL, Stephenson I, Gibbs EPJ, Chen L, Smith C, Hill RC, Ferro P, Pompey J, Bright RA (2005) Transmission of equine influenza virus to dogs. Science 310(5747):482–485

Cullinane A, Newton JR (2013) Equine influenza—a global perspective. Vet Microbiol 167:205–214

Dalglish RA (1992) The international movement of horses—the current infectious disease situation. In: Short CR (ed) Proceedings of the 9th International Conference of Racing Analysts and Veterinarians. Louisiana State University, New Orleans, LA, pp 37–53

Daly JM, Lai AC, Binns MM, Chambers TM, Barrandeguy M, Mumford JA (1996) Antigenic and genetic evolution of equine H3N8 influenza A viruses. J Gen Virol 77:661–761

Daly JM, Yates PJ, Browse G, Swann Z, Newton JR, Jessett D, Davis-Poynter N, Mumford JA (2003) Comparison of hamster and pony challenge models for evaluation of effect of antigenic drift on cross-protection afforded by equine influenza vaccines. Equine Vet J 35:458–462

Daly JM, Yates PJ, Newton JR, Park A, Henley W, Wood JL, Davis-Poynter N, Mumford JA (2004) Evidence supporting the inclusion of strains from each of the two co-circulating lineages of H3N8 equine influenza virus in vaccines. Vaccine 22:4101–4109

Daly JM, Whitwell KE, Miller J, Dowd G, Cardwell JM, Smith KC (2006) Investigation of equine influenza cases exhibiting neurological disease: coincidence or association? J Comp Pathol 134(2):231–235

Daly JM, Sindle T, Tearle J, Barquero N, Newton JR, Corning S (2007) Equine influenza vaccine containing older H3N8 strains offers protection against A/eq/South Africa/4/03 (H3N8) strain in a short-term vaccine efficacy study. Equine Vet J 39:446–450

Daly JM, Blunden AS, Macrae S, Miller J, Bowman SJ, Kolodziejek J, Nowotny N, Smith KC (2008) Transmission of equine influenza virus to English foxhounds. Emerg Infect Dis 14:461–464

Daly JM, MacRae S, Newton JR, Wattrang E, Elton DM (2011) Equine influenza: a review of an unpredictable virus. Vet J 189(1):7–14

Doll ER (1961) Influenza of horses. Am Rev Respir Dis 83:48

Eierhoff T, Hrincius ER, Rescher U, Ludwig S, Ehrhardt C (2010) The epidermal growth factor receptor (EGFR) promotes uptake of influenza A viruses (IAV) into host cells. PLoS Pathog 6:e1001099

Elton D, Bryant N (2011) Facing the threat of equine influenza. Equine Vet J 43(3):250–258

Endo A, Pecoraro R, Sugita S, Nerome K (1992) Evolutionary pattern of the H3 haemagglutinin of equine influenza viruses: multiple evolutionary lineages and frozen replication. Arch Virol 123:73–87

Firestone SM, Schemann KA, Toribio Jenny-Ann LML, Ward MP, Dhand NK (2011) A case-control study of risk factors for equine influenza spread onto horse premises during the 2007 epidemic in Australia. Prev Vet Med 100:53–63

Frerichs GN, Burrows R, Frerichs CC (1973) Serological response of horses and laboratory animals to equine influenza vaccines. In: Bryans JT, Gerber H (eds) Proceedings of the 3rd International Conference on Equine Infectious Diseases. Karger Basel, New York, pp 503–509

Gahan J, Garvey M, Asmah Abd Samad R, Cullinane A (2019) Whole genome sequencing of the first H3N8 equine influenza virus identified in Malaysia. Pathogens 8(2):62. https://doi.org/10.3390/pathogens8020062

Gerber H (1970) Clinical features, sequelae and epidemiology of equine influenza. In: Proceedings of the Second International Conference on Equine Infectious Diseases, Paris, 1969. S. Karger, Basel, pp 63–80

Gildea S, Arkins S, Cullinane A (2011) Management and environmental factors involved in equine influenza outbreaks in Ireland 2007–2010. Equine Vet J 43:608–617

Guo Y, Wang M, Zheng S, Wang P, Ji W, Chen Q (1991) Aetiologic study on an influenza-like epidemic in horses in China. Acta Virol 35:190–195

Guo Y, Wang M, Kawaoka Y, Gorman O, Ito T, Saito T, Webster RG (1992) Characterization of a new avian-like influenza A virus from horses in China. Virology 188:245–255

Gupta AK et al (1993) Single radial immunodiffusion potency test for standardization of indigenous influenza vaccine. Indian J Exp Biol 31:944–947

Guthrie AJ (2006) Equine influenza in South Africa, 2003 outbreak. In: Proceedings of the 9th International Congress of World Equine Veterinary Association. International Veterinary Information Service, Ithaca, NY

Guthrie AJ, Stevens KB, Bosman PP (1999) The circumstances surrounding the outbreak and spread of equine influenza in South Africa. Rev Sci Tech 18:179–185

Hannant D, Jessett DM, O'Neill T, Sundquist B, Mumford JA, Powell DG (1988a) Nasopharyngeal, tracheobronchial, and systemic immune responses to vaccination and aerosol infection with equine-2 influenza A virus (H3N8). In: Equine infectious diseases. V: Proceedings of the Fifth International Conference, vol 5. University Press of Kentucky, Lexington, KY, pp 66–73

Hannant D, Mumford JA, Jessett DM (1988b) Duration of circulating antibody and immunity following infection with equine influenza virus. Vet Rec 122(6):125–128

Hannant D, Mumford JA (1989) Cell mediated immune responses in ponies following infection with equine influenza virus (H3N8): the influence of induction culture conditions on the properties of cytotoxic effector cells. Vet Immunol Immunopathol 21:327–337

Hannant D, Jessett DM, O'Neill T, Livesay GJ, Mumford JA (1994) Cellular immune responses stimulated by inactivated virus vaccines and infection with equine influenza virus (H3N8). In: Nakajima H, Plowright W (eds) Equine infectious diseases. VII. Proceedings of Seventh International Conference on Equine Infectious Diseases. R & W Publications Ltd, Newmarket, pp 169–174

Huntington PJ (1990) Equine influenza—the disease and its control. Technical Report Series No. 184. Department of Agriculture and Rural Affairs, Victoria

Jagger BW, Wise HM, Kash JC, Walters KA, Wills NM, Xiao YL, Dunfee RL, Schwartzman LM, Ozinsky A, Bell GL, Dalton RM, Lo A, Efstathiou S, Atkins JF, Firth AE, Taubenberger JK, Digard P (2012) An overlapping protein-coding region in influenza a virus segment 3 modulates the host response. Science 337:199–204

Kasel JA, Alford RH, Knight V, Waddell GH, Sigel MM (1965) Experimental infection of human volunteers with equine influenza virus. Nature 206:41–43

Kasel JA, Couch RB (1969) Experimental infection in man and horses with Influenza A viruses. Bull World Health Organ 41:447–452

Khan A, Mushtaq MH, Ahmad MUD, Nazir J, Farooqi SH, Khan A (2017) Molecular Epidemiology of a novel re-assorted epidemic strain of equine influenza virus in Pakistan in 2015–16. Virus Res 240:56–63

Kirkland PD, Finlaison DS, Crispe E, Hurt AC (2010) Influenza virus transmission from horses to dogs, Australia. Emerg Infect Dis 16:699–702

Kovbasnjuk ON, Spring KR (2000) The apical membrane glycocalyx of MDCK cells. J Membr Biol 176:19–29

Kumar N, Bera BC, Greenbaum BD, Bhatia S, Sood R, Selvaraj P, Anand T, Tripathi BN, Virmani N (2016) Revelation of influencing factors in overall codon usage bias of equine influenza viruses. PLoS One 11(4):e0154376. https://doi.org/10.1371/journal.pone.0154376

Lai AC, Rogers KM, Glaser A, Tudor L, Chambers T (2004) Alternate circulation of recent equine-2 influenza viruses (H3N8) from two distinct lineages in the United States. Virus Res 100:159–164

Lamb RA, Krug RM (2001) Orthomyxoviridae: the viruses and their replication. In: Knipe DM, Howle PM (eds) Fields virology. Lippincott Williams & Wilkins, Philadelphia, pp 1487–1532

Larson KR, Heil GL, Chambers TM, Capuano A, White SK, Gray GC (2015) Serological evidence of equine influenza infections among persons with horse exposure, Iowa. J Clin Virol 67:78–83

Lee CW, Saif YM (2009) Avian influenza virus. Comp Immunol Microbiol Infect Dis 32:301–310

Lindblad EB (2004) Aluminium adjuvants—In retrospect and prospect. Vaccine 22:3658–3668

Lopez A (2007) Respiratory system. In: McGavin D, Zachary JF (eds) Pathologic basis of veterinary disease, 4th edn. Mosby Elsevier, St Louis, MO, p 470

Lunn DP, Soboll G, Schram BR, Quass J, McGregor MW, Drape R, Macklin MD, McCabe DE, Swain WF, Olsen CW (1999) Antibody responses to DNA vaccination of horses using the influenza virus hemagglutinin gene. Vaccine 17:2245–2258

MacLachlan N, Dubovi EJ (eds) (2011) Fenner's veterinary virology, 4th edn. Academic, San Diego

Manjrekar SL, Gorhe DS, Paranjape VL (1965) Observation on the coughing outbreaks 'Newmarket Cough' in the race horses in Bombay. Indian Vet J 48:460–464

Martella V, Elia G, Decaro N, di Trani L, Lorusso E, Campolo M, Desario C, Parisi A, Cavaliere N, Buonavoglia C (2007) An outbreak of equine influenza virus in vaccinated horses in Italy is due to an H3N8 strain closely related to recent North American representatives of the Florida sub-lineage. Vet Microbiol 121:56–63

McClure JP (1998) The epizootic of 1872: horses and disease in a nation in motion. New York History 79:4–22

Miller WC (1965) Equine influenza—further observations on "coughing" outbreak 1965. Vet Rec 77:455

Mumford JA (1999) Control of influenza from an international perspective. In: Proceedings of the Eighth International Conference on Equine Infectious Diseases. R & W Publications, Newmarket, pp 11–24

Mumford JA, Wood J (1992) Establishing an acceptability threshold for equine influenza vaccines. Dev Biol Stand 79:137–146

Mumford JA, Wood JM, Folkers C, Schild GC (1988) Protection against experimental infection with influenza virus A/equine/Miami/63 (H3N8) provided by inactivated whole virus vaccines

containing homologous virus. Epidemiol Infect 100(3):501–510
Mumford J, Hannant D, Jessett DM (1990) Experimental infection of ponies with equine influenza (H3N8) viruses by intranasal inoculation or exposure to aerosols. Equine Vet J 22:93
Mumford EL, Traub-Dargatz JL, Salman MD, Collins JK, Getzy DM, Carman J (1998) Monitoring and detection of acute viral respiratory tract disease in horses. J Am Vet Med Assoc 213:385–390
Mumford J, Cardwell J, Daly J, Newton R (2003) Efforts to preempt an equine influenza epidemic. Vet Rec 152:405–406
Muranaka M, Yamanaka T, Katayama Y, Hidari K, Kanazawa H, Suzuki T, Oku K, Oyamada T (2010) Distribution of influenza virus sialoreceptors on upper and lower respiratory tract in horses and dogs. J Vet Med Sci 73:125–127
Muranaka M, Yamanaka T, Katayama Y, Niwa H, Oku K, Matsumura T, Oyamada T (2012) Time-related pathological changes in horses experimentally inoculated with equine influenza A virus. J Equine Sci 23(2):17–26
Myers C, Wilson WD (2006) Equine influenza virus. Vet Clin North Am Equine Pract 5:187–196
Nakajima K, Nobusawa E, Nagy A, Nakajima S (2005) Accumulation of amino acid substitutions promotes irreversible structural changes in the hemagglutinin of human influenza A H3 virus during evolution. J Virol 79(10):6472–6477
Nelson KM, Schram BR, McGregor MW, Sheoran AS, Olsen CW, Lunn DP (1998) Local and systemic isotype-specific antibody responses to equine influenza virus infection versus conventional vaccination. Vaccine 16:1306–1313
Newton JR, Lakhani KH, Wood JLN, Baker DJ (2000) Risk factors for equine influenza serum antibody titres in young Thoroughbred racehorses given an inactivated vaccine. Prev Vet Med 46:129–141
Newton JR, Daly JM, Spencer L, Mumford JA (2006) Description of the outbreak of equine influenza (H3N8) in the United Kingdom in 2003, during which recently vaccinated horses in Newmarket developed respiratory disease. Vet Rec 158:185–192
Nuria B, Janet M, Daly J, Newton R (2007) Risk factors for influenza infection in vaccinated racehorses: lessons from an outbreak in Newmarket, UK in 2003. Vaccine 25:7520–7529
Nyaga PN et al (1980) Epidemiology of equine influenza, risk by age, breed and sex. Comp Immunol Microbiol Infect Dis 3:67
OIE, Equine influenza (2019). https://www.oie.int/fileadmin/Home/eng/Health_standards/tahm/3.05.07_EQ_INF.pdf
Paillot R (2014) A systematic review of recent advances in equine influenza. Vaccines (Basel) 2014(2):797–831. https://doi.org/10.3390/vaccines2040797
Paillot R, Prowse L (2012) ISCOM-matrix-based equine influenza (EIV) vaccine stimulates cell-mediated immunity in the horse. Vet Immunol Immunopathol 145(1–2):516–521
Paillot R, Hannant D, Kydd JH, Daly JM (2006) Vaccination against equine influenza: Quid novi? Vaccine 24:4047–4061
Paillot R, Grimmett H, Elton D, Daly JM (2008) Protection, systemic IFN gamma γ, and antibody responses induced by an ISCOM-based vaccine against a recent equine influenza virus in its natural host. Vet Res 39(3):1
Paillot R, Prowse L, Montesso F, Huang CM, Barnes H, Escala J (2013) Whole inactivated equine influenza vaccine: efficacy against a representative clade 2 equine influenza virus, IFN gamma synthesis and duration of humoral immunity. Vet Microbiol 162:396–407
Park AW, Wood JL, Newton JR, Daly J, Mumford JA, Grenfell BT (2003) Optimising vaccination strategies in equine influenza. Vaccine 21:2862–2870
Patterson-Kane JC, Carrick JB, Axon JE, Wilkie I, Begg AP (2008) The pathology of bronchointerstitial pneumonia in young foals associated with the first outbreak of equine influenza in Australia. EquineVet J 40:199–203
Pavulraj S, Bera BC, Joshi A, Anand T, Virmani M, Vaid RK, Shanmugasundaram K, Gulati BR, Rajukumar K, Singh R, Misri J (2015) Pathology of equine influenza virus (H3N8) in murine model. PLoS One 10(11):143094
Pavulraj S, Virmani N, Bera BC, Joshi A, Anand T, Virmani M, Singh R, Singh RK, Tripathi BN (2017) Immunogenicity and protective efficacy of inactivated equine influenza (H3N8) virus vaccine in murine model. Vet Microbiol 210:188–196
Payungporn S, Crawford PC, Kouo TS, Chen LM, Pompey J, Castleman WL, Dubovi EJ, Katz JM, Donis RO (2008) Influenza A virus (H3N8) in dogs with respiratory disease, Florida. Emerg Infect Dis 14:902–908
Perglione CO, Gildea S, Rimondi A, Miño S, Vissani A, Carossino M et al (2016) Epidemiological and virological findings during multiple outbreaks of equine influenza in South America in 2012. Influenza Other Respir Viruses 10:37–46
Petermann HG, Fayet MT, Fontaine M. Fontaine MP (1970) Vaccination against equine influenza. In: Bryans JT (ed) Proceedings of the 2nd International Conference On Equine Infectious Diseases. Karger Basel, New York, pp 105–110
Powell DG, Watkins KL, Li PH, Shortridge KF (1995) Outbreak of equine influenza among horses in Hong Kong during 1992. Vet Rec 136:531–536

Qi T, Guo W, Huang WQ, Li HM, Zhao LP, Dai LL, He N, Hao XF, Xiang WH (2010) Genetic evolution of equine influenza viruses isolated in China. Arch Virol 155:1425–1432
Ragni-Alunni R, van de Zande S (2009) Efficacy of Equilis Prequenza te against Recent Equine Influenza Isolates. Equitana; Essen, Germany: 2009. Animal studies with individual challenge method
Rash A, Morton R, Woodward A, Maes O, McCauley J, Bryant N, Elton D (2017) Evolution and divergence of H3N8 equine influenza viruses circulating in the United Kingdom from 2013 to 2015. Pathogens 6:6
Rossman JS, Lamb RA (2011) Influenza virus assembly and budding. Virology 411:229–236
Rossman JS, Jing X, Leser GP, Lamb RA (2010) Influenza virus M2 protein mediates ESCRT-independent membrane scission. Cell 142(6):902–913
Scocco P, Pedini V (2008) Localization of influenza virus sialoreceptors in equine respiratory tract. Histol Histopathol 23:973–978
Sieczkarski SB, Whittaker GR (2002) Influenza virus can enter and infect cells in the absence of clathrin-mediated endocytosis. J Virol 76:10455–10464
Singh RK, Dhama K, Karthik K, Khandia R, Munjal A, Khurana SK, Chakraborty S, Malik YS, Virmani N, Singh R, Tripathi BN, Munir M, van der Kolk JH (2018) A comprehensive review on equine influenza virus: etiology, epidemiology, pathobiology, advances in developing diagnostics, vaccines, and control strategies. Front Microbiol 9:1941. https://doi.org/10.3389/fmicb.2018.01941
Sovinova O, Tumova B, Pouska F, Nemec J (1958) Isolation of a virus causing respiratory disease in horses. Acta Virol 2:52
Su S, Wang L, Fu X, He S, Hong M, Zhou P, Lai A, Gray G, Li S (2014) Equine influenza A(H3N8) virus infection in cats. Emerg Infect Dis 20:2096–2099
Takizawa T, Matsukawa S, Higuchi Y, Nakamura S, Nakanishi Y, Fukuda R (1993) Induction of programmed cell death (apoptosis) by influenza virus infection in tissue culture cells. J Gen Virol 74:2347–2355
Tong S, Zhu X, Li Y, Shi M, Zhang J et al (2013) New world bats harbor diverse influenza A viruses. PLoS Pathog 9(10):e1003657. https://doi.org/10.1371/journal.ppat.1003657
Tu J, Zhou H, Jiang T, Li C, Zhang A, Guo X, Zou W, Chen H, Jin M (2009) Isolation and molecular characterization of equine H3N8 influenza viruses from pigs in China. Arch Virol 154:887–890
Tumova B, Sovinova-Fiserova O (1959) Properties of influenza viruses. A/Asia/57 and A-equi/Praha/56. Bull World Health Organ 20:445
Townsend HGG, Cook A, Watts TC (1999) Efficacy of a cold-adapted, modified-live virus influenza vaccine: a double-blind challenge trial. Proc Am Assoc Equine Pract 45:41–42
Uppal PK, Yadav MP (1987) Outbreak of equine influenza in India. Vet Rec 121:569–570
Uppal PK, Yadav MP, Oberoi MS (1989) Isolation of A/Equi-2 virus during 1987 equine influenza epidemic in India. Equine Vet J 21(5):364–366
Virmani N, Singh BK, Gulati BR, Kumar S (2008) Equine influenza outbreak in India. Vet Rec 163:607–608
Virmani N, Bera BC, Singh BK, Shanmugasundaram K, Gulati BR, Barua S, Vaid RK, Gupta AK, Singh RK (2010a) Equine influenza outbreak in India (2008–09): virus isolation, sero-epidemiology and phylogenetic analysis of HA gene. Vet Microbiol 143:224–237
Virmani N, Bera BC, Gulati BR, Karuppusamy S, Singh BK, Kumar Vaid R, Kumar S, Kumar R, Malik P, Khurana SK, Singh J, Manuja A, Dedar R, Gupta AK, Yadav SC, Chugh PK, Narwal PS, Thankur VL, Kaul R, Kanani A, Rautmare SS, Singh RK (2010b) Descriptive epidemiology of equine influenza in India (2008–2009): temporal and spatial trends. Vet Ital 46(4):449–458
Virmani N, Bera BC, Shanumugasundaram K, Singh BK, Gulati BR, Singh RK (2011) Genetic analysis of the matrix and non-structural genes of equine influenza virus (H3N8) from epizootic of 2008–2009 in India. Vet Microbiol 152:169–175
Waddell GH, Teigland MB, Sigel MM (1963) A new influenza virus associated with equine respiratory disease. J Am Vet Med Assoc 143:587–590
Waghmare SP, Mode SG, Kolte AY, Babhulkar N, Vyavahare SH, Patel A (2010) Equine influenza: an overview. Vet World 3:194–197
Wang X, Hinson ER, Cresswell P (2007) The interferon-inducible protein viperin inhibits influenza virus release by perturbing lipid rafts. Cell Host Microbe 2:96–105
Webster RG, Laver WG (1975) Antigenic variation of influenza viruses. In: Kilbournc ED (ed) The influenza viruses and influenza. Academic, New York, pp 269–314
Webster RG, Thomas TL (1993) Efficacy of equine influenza vaccines for protection against A/Equine/Jilin/89 (H3N8)—a new equine influenza virus. Vaccine 11:987–993
Webster RG (1993) Are equine 1 influenza viruses still present in horses? Equine Vet J 25:537–538
Webster RG (2002) The importance of animal influenza for human disease. Vaccine 20:16–20
Wernery R, Yates PJ, Wernery U, Mumford JA (1999) Equine influenza outbreak in a polo club in Dubai, United Arab Emirates in 1995/96. In: Wernery U, Wade JF, Mumford JA, Kaaden OR (eds) Proc. of the 8th Int. Conference on equine infectious diseases. Dubai, 1998, pp. 342–346.

Wiley DC, Skehel JJ (1987) The structure and function of the hemagglutinin membrane glycoprotein of influenza virus. Annu Rev Biochem 56:365–394

Williams AJ (1924) Analogies between influenza of horses and influenza of man. Section of epidemiology and state medicine. Proc R Soc Med 17:47–58

Willoughby R, Ecker G, McKee S, Riddolls L, Vernaillen C, Dubovi E, Lein D, Mahony JB, Chernesky M, Nagy E et al (1992) The effects of equine rhinovirus, influenza virus and herpesvirus infection on tracheal clearance rate in horses. Can J Vet Res 56:115–121

Wilson WD (1993) Equine influenza. Vet Clin North Am Equine Pract 9(2):257–282

Wilson IA, Cox NJ (1990) Structural basis of immune recognition of influenza virus hemagglutinin. Annu Rev Immunol 8:737–771

Wilson WD, Mihalyi JE, Hussey S, Lunn DP (2001) Passive transfer of maternal immunoglobulin isotype antibodies against tetanus and influenza and their effect on the response of foals to vaccination. Equine Vet J 33:644–650

Yadav MP, Uppal PK, Mumford JA (1993) Physico-chemical and biological characterization of A/Equl-2 virus isolated from 1987 equine influenza epidemic in India. Int J Anim Sci 8:99–99

Yamanaka T, Niwa H, Tsujimura K, Kondo T, Matsumura T (2008) Epidemic of equine influenza among vaccinated racehorses in Japan in 2007. J Vet Med Sci 70:623–625

Yamanaka T, Nemoto M, Tsujimura K, Kondo T, Matsumura T (2009) Interspecies transmission of equine influenza virus (H3N8) to dogs by close contact with experimentally infected horses. Vet Microbiol 139:351–355

Yates P, Mumford JA (2000) Equine influenza vaccine efficacy: the significance of antigenic variation. Vet Microbiol 74:173–177

Yondon M, Heil GL, Burks JP, Zayat B, Waltzek TB, Jamiyan BO, McKenzie PP, Krueger WS, Friary JA, Gray GC (2013) Isolation and characterization of H3N8 equine influenza A virus associated with the 2011 epizootic in Mongolia. Influenza Other Respir Viruses 7:659–665

Yondon M, Zayat B, Nelson MI, Heil GL, Anderson BD, Lin X, Halpin R, McKenzie PP, White SK, Wentworth DE, Gray GC (2014) Equine influenza A (H3N8) virus isolated from Bactrian camel, Mongolia. Emerg Infect Dis 20:2144–2147

Yoon KJ, Cooper VL, Schwartz KJ, Harmon KM, Kim WI, Janke BH, Strohbehn J, Butts D, Troutman J (2005) Influenza virus infection in racing greyhounds. Emerg Infect Dis 11:1974–1976

第 10 章 施马伦贝格病毒

S. B. Sudhakar, P. N. Gandhale, F. Singh, A. A. Raut,
A. Mishra, D. D. Kulkarni, V. P. Singh

10.1 引言

施马伦贝格病毒（Schmallenberg virus，SBV）是一种新出现的病毒，于2011年夏秋两季在荷兰和德国边境地区的奶牛中首次被发现，临床表现为体温升高、腹泻、产奶量降低的特征性综合征。常见病原体检测结果均为阴性，随后运用宏基因组工具确定了该综合征的病原体，并通过细胞培养技术对病毒进行分离。施马伦贝格病毒首次从德国施马伦贝格镇的组织样本中分离得到，因此将其命名为施马伦贝格病毒(Hoffmann 等，2012)。自此，SBV在欧洲各国传播开来。SBV是一种新型虫媒病毒，属布尼亚病毒科(bunyaviridae)，正布尼亚病毒属(orthobunyavirus)辛波血清群，可通过库蠓叮咬传播和胎盘传播，造成致畸效应，但对于公共卫生影响却可以忽略不计。目前，来自亚洲、非洲和澳大利亚及WOAH报道的辛波血清群病毒尚未被列为法定报告病毒。该病毒主要对家养反刍动物造成影响，导致羔羊、山羊及犊牛先天畸形、死产及流产（Hoffmann 等，2012；Bayrou 等，2014；Peperkamp 等，2015）。已经证实SBV在野生反刍动物中也流行(EFSA，2014)。在北欧，大规模SBV暴发造成了重大经济损失，因此，对家畜及其产品的跨境贸易实施了限制(Hoffmann 等，2012)。

自2011年首次发现SBV以来，获得了大量关于病毒起源、发生、流行病学、分子病毒学、临床症状、发病机制、诊断、疫苗开发、血清阳性率、进一步暴发流行及再次发生的可能性，以及该病造成的经济影响等信息，本章将对一些实验的研究分析和迄今为止在科学期刊和基于网络报告工具获得研究数据进行整理。

10.2 病毒分离

2011年，在德国出现了一种未知的奶牛疾病综合征，发病率为20%~70%，几天就能痊愈。同时，荷兰也发出现了严重腹泻的类似病例(Hoffmann 等，2012；Tarlinton 等，2012；Bilk 等，2012；Elbers 等，2012；EFSA，2012；Beer 等，2013；Tarlinton 和 Daly，2013)。

2011年下半年，荷兰、德国和比利时在新生羔羊中出现流产和死产现象，同时在山羊和牛中也出现流产和死产现象，并伴有难产发生(van den Brom 等，2012)。在2012年2月、3月中旬、5月和8月，比利时、丹麦、法国、德国、意大利、卢森堡、荷兰、西班牙、瑞士和英国也报道了各种类似疫情的发生。在绵羊养殖场发生严重的传染，随后在牛和山羊养殖场发生传染。通过对早期出现的先天畸形病例进行详细调查，排除了所有可能的病原体，进而通过宏基因组分析发现了一种新型病毒，被分离、鉴定，并根据其首次分离地被命名为施马伦贝格病毒。通过宏基因组和全序列分析发现，SBV与在日本牛身上发现的3种病毒有相似之处，这3种病毒分别是阿伊诺病毒(Aino virus)、沙蒙达病毒(Shamonda virus)和阿卡巴纳病毒(Akabane virus)，均属于布尼亚病毒科，正布尼亚病毒属(Hoffmann 等，2012；van den Brom 等，2012)。

通过对1961—2010年反刍动物SBV流行的回顾性研究表明，在德国发生这次疫情前，没有任何证据证明SBV蛋白和RNA的存在(Gerhauser 等，2014)。在土耳其进行了另一项研究，通过应用ELISA检测抗体发现，2006年1头水牛、2007年12头牛和1头水牛抗体呈阳性。然而，直到2012年6月，才检测到SBV的核酸(Azkur 等，2013)。此外，2012年，在莫桑比克也发现了血清呈阳性的绵羊、山羊和牛(Blomstrom 等，2014)。

10.3 临床症状

该病在成年牛中临床症状不明显或出现亚临床症状(Hoffmann 等，2012；Schulz 等，2014)。潜伏期为1~4天，病毒感染期非常短(1~6天)，随后食欲减退，在病毒增殖期间体温升高（超过 40°C），身体健康状况恶化，产奶量减少（高达 50%），发生腹泻，几天后即可痊愈（Hoffmann 等，2012；Laloy 等，2015；Lechner 等，2017）。SBV 感染无性别差异（Wernike 等，2013b）。

该病除了增加绵羊和山羊流产及后代先天畸形的风险，可能没有其他临床症状。然而，有些羊可能表现出非常轻的临床症状，如腹泻、嗜睡、发热、精神沉郁、流涕（Wernike 等，2013c；Helmer等，2013）。

SBV 感染会引起生殖障碍，母畜在妊娠期间被感染，可将病毒传播至胎儿，导致胎儿发生先天性关节弯曲–积水性无脑综合征的严重先天性异常。临床表现为早产、木乃伊胎、死胎、跖骨不成比例、四肢弯曲、僵硬、严重斜颈、关节强直、头骨扁平及短颌变形（van den Brom 等，2012；Gelagay 等，2018）。在出生前不止一个胎儿被病毒感染时，其中只有一个可能出现临床症状或关节卷曲病，另一个可能出现与神经系统相关的疾病。在另一种情况下，双胞胎中的一个可能畸形，而另一个可能存活，或者仅表现出生长发育迟缓（van den Brom 等，2012；Wernike 等，2014）。

神经系统相关疾病表现为黑蒙、共济失调和（或）行为异常、侧卧、丧失吸吮能力、偶尔痉挛（“失声综合征”）、四肢无力、麻痹、游泳状和绕圈运动。被感染的幼畜主要表现为脊柱多发畸形（斜颈、脊柱前凸、后凸、脊柱侧凸）。犊牛最易发生斜颈，羔羊则是脊柱侧弯（Bayrou 等，2014；Peperkamp 等，2014）。除畸形外，新生犊牛的体重明显减轻（Bayrou 等，2014）。

10.4 病理变化

病变的大体特征为关节挛缩、短颌变形、脊柱和中枢神经系统畸形，包括小脑发育不全、无脑畸形、脑积水、脑穿通畸形、小脊髓炎和皮下水肿（小腿）（van den Brom 等，2012）。中枢神经系统特征性病变表现为脑白质形成空腔，大脑皮质、小脑皮层、脑干核和脊髓灰质柱的神经元丢失。羔羊受影响程度比犊牛严重（Peperkamp 等，2014；Laloy 等，2017）。此外，当羔羊和犊牛在子宫内被感染时，被感染关节的肌腱变短，相关肌肉会出现颜色变化和质量下降（Bayrou 等，2014）。

10.5 危害

疾病的危害性由多种因素决定，例如，先天性异常羔羊的数量、泌乳量减少期和感染发生时所处的妊娠期（Wuthrich 等，2016）。该病对成年动物带来的直接影响为体温升高、腹泻、产奶量下降、不孕不育、反复配种、流产和具有致死性的难产。在某些妊娠阶段，有可能出现新生胎儿畸形和液化的并发症。垂直传播对妊娠前期的影响最为明显（Hoffmann 等，2012；Elbers 等，2012；Bilk 等，2012；EFSA，2012；Tarlinton 等，2012；Wernike 等，2013a；Tarlinton 和 Darly，2013）。2012 年，通过观察发现，病毒传播到一个未受感染的宿主群体中可造成直接危害。SBV 感染引起的其他危害包括用于治疗产犊和产羔并发症产生的费用，除此之外，还需支付后备种畜费用，以此补偿丧失繁殖能力和未能出售的后备种畜，以及流通受限带来的损失（Alarcon 等，2014）。

虽然这种病毒对反刍动物的直接经济影响有限，但是对国际贸易产生了重大的经济影响，由于受疾病影响国家的家畜产品出口受到限制，如胚胎、精液、活畜，继而对出口业影响巨大。这种贸易限制对欧洲经济造成了重大损失。在 2011 年和 2012 年，纯种动物的出口量下降了 20%；2012 年，牛精液贸易量下降了 11%~26%，相当于 890 万剂量（EFSA，2014）。

在个体农场中，受影响损失的幼畜数量很少，多时超过 50%（Helmer 等，2013）。总的来说，SBV 对羊场的影响最大，羊场的流产率、羔羊死亡率、难产率、畸形率增加及生育率降低。

在养牛场，因产奶量下降及恢复生产所造成的经济损失远大于小牛的先天性畸形所造成的损失(EFSA,2012;Beer 等,2013)。该病毒对家养山羊的影响小于对绵羊的影响。然而,由于幼畜的高死亡率和产奶量下降，使得受影响的山羊农场产生了高达 50%的经济损失(Helmer 等,2013)。2016 年,Wuthrich 等在一个感染 SBV 的标准化农场进行了一项研究，经计算得出，经济损失平均为 1338 欧元(1 欧元≈7.5 元)。在全国的平均经济损失可能较低，但在不同农场之间也存在经济损失波动很大的情况,因此,个别农场也可能遭受更大的经济损失。

10.6 结构与分类

在布尼亚病毒科中的最大的属,即正布尼亚病毒属被分为 18 个血清群,其中辛波血清群包括 SBV 在内的至少 25 个病毒,根据病毒中和交叉试验和血凝抑制交叉实验结果,被分为 7 种病毒[辛波病毒(Simbu virus)、赤羽病毒(Akabane virus)、奥洛普切病毒(Oropouche virus)、沙门达病毒(Shamonda virus)、萨苏伯里病毒(Sathuperi virus)、舒尼病毒(Shuni virus)和曼扎尼拉病毒(Manzanilla virus)](Goller 等,2012;Yanase 等,2012;Hoffmann 等,2012;Plyusnin 等,2012)。

SBV 呈球形(直径 80~120mm,表面有短的凸起),有囊膜,分三节段、单股负链 RNA 病毒。基因组可分为大(L)、中(M)和小(S)3 个基因片段,在片段末端互补的非编码碱基形成狭长结构。这些复合物与少数几个拷贝的 L 聚合酶和多个拷贝核衣壳蛋白相结合,形成 RNP,即感染性病毒颗粒(Tilston 等,2017)。L 片段编码 RNA 依赖性 RdRp,负责病毒的复制和转录(Kraatz 等,2018)。

M 片段编码前体多聚蛋白，被细胞蛋白酶裂解成两个表面糖蛋白(Gn 和 Gc),以及一个非结构蛋白(NSm)。Gn 和 Gc 共同形成异质二聚体混合物,负责病毒进入细胞,并且通过中和抗体证实其具有抗原决定簇的作用,但 NSm 蛋白作用尚不清楚,其有可能参与病毒粒子的组装。一般来说,在 S、M、L 片段中,M 片段的 RNA 具有高度变异性。基因自然重组是新病毒株出现的原因,新出现毒株的宿主范围、毒力和抗原性都有可能发生变化。

S 片段以 ORF 重叠的方式编码核衣壳蛋白和另一个小的非结构蛋白。核衣壳蛋白主要功能是将整个病毒基因组包裹在衣壳内并防止被细胞降解,这是病毒 RNA 转录和复制所必需的。核衣壳蛋白是介导病毒粒子和受感染宿主细胞间相互作用的重要蛋白。因此,核衣壳蛋白被广泛应用于 SBV 的分子鉴定和血清学鉴定(Bilk 等,2012),同时,核衣壳蛋白也是 SBV 与补体结合的主要抗原蛋白(Goller 等,2012;Yanase 等,2012),对宿主的先天性免疫应答起到调控作用(Elliott 等,2013)。

10.7 理化特性

SBV 在 50~60℃的温度下持续加热 30 分钟可失活,2%戊二醛、1%次氯酸钠、甲醛和 70%乙醇等常见消毒剂会直接影响其毒力,使其不能够在传播媒介或者宿主体外长期存活(WOAH,2017)。

10.8 遗传多样性

病毒被发现后不久,Hoffmann 等于 2012 年对 SBV 全基因组进行测序,S (830 个核苷酸)、M (4415 个核苷酸)和 L(6865 个核苷酸)片段与其他正布尼亚病毒进行比较之后发现，其 S 基因片段与沙门达病毒的同源性是 97%;M 基因片段与艾罗病毒的同源性是 71%;L 基因片段与赤羽病毒的同源性是 69%,所有检测样本均采自于日本牛。根据这些观察结果,将 SBV 归类于辛波血清群的类沙门达病毒。

随后在 2012 年,Yanase 等认为 SBV 可能是萨苏伯里病毒和沙门达病毒重组得到的,S 片段和 L 片段来自沙门达病毒,M 片段来自道格拉斯病毒和萨苏伯里病毒。之后,辛波血清群中 5 个病毒属的 9 种病毒[即沙门达病毒属(桑戈病毒、皮顿病毒和沙门达病毒),萨苏伯里病毒属(道格拉斯病毒和萨苏伯里病毒),舒尼病毒属(舒尼病毒和艾罗病毒)、赤羽病毒属(萨博病毒)和辛波病毒属(辛波病毒)]的几乎所有病毒的全基因组序列都被获得。通过遗传进化分析发现,SBV 属于萨苏伯里病毒属,其可能不是重组病毒,可能是沙门达病毒的祖先, 因沙门达病毒本身就是重组病毒,S 片段和 L 片段来自 SBV,M 片段来自未知病毒。血清学检测也支持该结论, 通过血清学检测发现 SBV 血清能够中和道格拉斯病毒和萨苏伯里病毒,但不能中和沙门达病毒(Goller 等,2012)。

10.9　流行病学

10.9.1　发病率

自 SBV 首次在欧洲西北部出现(Hoffmann 等,2012)以来,已经蔓延到欧洲大部分地区。现已在德国、荷兰和比利时(血清复阳率高达 99.8%)、英国、法国(血清复阳率高到 90%)、意大利、卢森堡、西班牙、意大利、丹麦、爱沙尼亚、埃塞俄比亚(血清复阳率高达 56.6%)、北爱尔兰、瑞士、挪威、奥地利、瑞典、芬兰、波兰和土耳其发现了 SBV 的感染 (Elbers 等,2012;Azkur 等,2013;Afonso 等,2014;Gelagay 等,2018)。2011—2013 年首次流行后,德国(Wernike 等,2015a)、比利时(Delooz 等,2016)、英格兰和威尔士(APHA,2017)报道了SBV 可能循环流行。非洲、亚洲、澳大利亚和中东也发现了辛波血清群中的正布尼亚病毒。非洲国家的血清学检测显示 SBV 抗体呈阳性。基于非洲多地发现了辛波血清群病毒, 且存在着血清学交叉反应,Mathew 等人于 2015 年推测 ELISA 检测血清抗体呈阳性的病毒可能是辛波血清群中的其他病毒,而不是 SBV。同样,萨苏伯里病毒(辛波血清群)首次在印度杂鳞库蚊中被分离到,随后在尼日利亚的奶牛和库蠓中又被分离到 (Dandawat 等,1969;Causey 等,1972)。辛波血清群的另一成员凯卡勒病毒在印度克里希纳区的三带喙库蚊中被分离到。由于存在很强的双向免疫交叉反应,凯卡勒病毒和阿伊诺病毒被认为是同一种病毒或同一种病毒的变种。

在牛中,幼年牛的患病率低于成年牛(Gelagay 等,2018)。然而,Elbers 等人于 2012 年在荷兰没有发现与年龄相关的发病率差异。

10.9.2　病毒感染与流行

迄今为止,还没有国家报道人被 SBV 感染的病例,因此,SBV 公共卫生风险可以不予考虑。辛波血清群除奥罗普切病毒能够导致人类产生严重流感样症状外, 其他病毒不存在成为人兽共患病的可能性。

尽管如此,SBV 越过欧洲边界开始蔓延到欧洲以外的地区。病毒媒介、储存宿主和易感宿主都进一步促进了 SBV 的传播和(或)成为地方性流行病,同时也决定了 SBV 的持续存在。每一批引入的新动物都会有新的易感宿主, 新易感宿主与现有种群的混合比例大小决定了该病的持续时间和年际间的流行周期。这也取决于牧群更新率、免疫接种水平和免疫持续时间。根据不同的农场管理和生产体系,更新率可能有所不同。在农业企业中,正常的更新率为 20%或 25%,导致每年在牧场中产生大量易感宿主。

此外,在引进雄性种用动物之前,如果雌性种用动物已经被感染, 预计不会产生不利影响。不过,在妊娠早期受到感染则会导致胚胎死亡,母畜反复受孕。尽管如此,母畜在下一次妊娠时,还是可能正常生产的。当动物身体状况和气象条件对传播媒介和病毒有利时, 特别是在疫区边界出现大量易感宿主,病毒就会再次出现。

10.10　传播

10.10.1　易感动物

自 SBV 首次发现以来,对多种动物中的病毒

RNA 和(或)抗体进行调查研究,结果发现家养反刍动物(牛、绵羊和山羊),多种野生动物(羊驼、安纳托利亚水牛、麋鹿、野牛、马鹿、扁角鹿、狍子、梅花鹿、麂、麂皮、摩弗伦羊和野猪)和动物园中的动物(紫羚、野猪、野牛、刚果水牛、欧洲野牛、白肢野牛、好望角大羚羊、大种弯角羚、细纹斑马、驼鹿、沼水羚、努比亚山羊、波斯野驴、驯鹿、马羚、弯角大羚羊、林羚和牦牛)均对 SBV 易感(EFSA, 2014)。家养反刍动物中,山羊的易感性要低于牛和绵羊。

有报道表明,犬可能会感染 SBV,妊娠的母犬感染 SBV 会出现致畸效应,但这种现象可能较为罕见(Sailleau 等,2013)。也有猪被感染的报道,病毒能够诱导猪的血清转阳,但没有涉及家畜的流行(Poskin 等,2014)。也有骆驼被感染的相关报道(Wernike 等,2012)。

10.10.2 水平传播

SBV 不能通过直接接触传播,也不能通过口服传播。对牛(Hoffimann 等,2012;Wernike 等,2013b)、绵羊(Wernike 等,2013c;Martinelle 等,2017)进行皮下接种病毒,对山羊(Laloy 等,2015)进行皮内接种病毒(Martinelle 等,2017)都能导致病毒血症发生。

10.10.3 传播媒介

SBV 通过库蠓叮咬传播,尤其是通过不显库蠓复组进行传播,但其他库蠓(雪翅库蠓、丘夫库蠓、苏格兰库蠓、灰黑库蠓)也具有传播能力,它们在日出和日落前 1 小时非常活跃(Hoffmann 等,2012;Bilk 等,2012;Elbers 等,2012;Tarlinton 等,2012;Beer 等,2013;Wernike 等,2013;Tarlinton 和 Daly,2013)。

10.10.4 垂直传播

SBV 能够穿过胎盘,对发生垂直传播具有重要意义。胎盘感染发生在第一个胎盘及其附属物出现到胎儿具有免疫力前,牛发生在受孕后 30~150 天(Bayrou 等,2014),绵羊和山羊则发生在受孕后 28~56 天(Helmer 等,2013;Laloy 等,2017)。临床表现取决于胎儿日龄(Bayrou 等,2014),在母畜妊娠早期受感染会导致胎儿死亡、低生育能力和死产,但是在妊娠后期感染,胎儿的免疫系统对病毒有抵抗能力,尽管如此,也有可能发生木乃伊化、死产和流产的情况(Helmer 等,2013)。被感染的后代不发生病毒血症,也没有证据表明病毒能够从受感染的后代传播给传播媒介(EFSA, 2014)。

10.10.5 精液传播

病毒能够进入到受感染动物的精液中,并且已经在精液的血浆和细胞中检测到了病毒核酸(Kesik 和 Larska, 2016)。含有 SBV 的精液不太可能感染胚胎,如果母畜被病毒感染,可能是由于发生了虫媒传播(Schulz 等,2014)。

10.11 感染免疫

关于感染后的免疫反应及免疫持续时间的研究并不多。牛被病毒感染后立即产生先天免疫反应(Wernike 等,2013b)。然而,SBV 可以通过在转录水平上抑制干扰素的产生,特别是通过 NS 蛋白介导作用,改变了宿主的先天免疫反应。NS 蛋白类似于毒力因子,可能通过抑制细胞代谢产生拮抗干扰素的作用。因此,宿主的先天免疫反应被破坏,病毒就能够有效进行复制(Elliott 等,2013)。牛在被感染后 8~14 天发生血清转阳,阳性血清可在自然感染的牛中持续 3 年以上,因此,有希望能够产生长期免疫能力。然而,在实验感染牛中,免疫持续至少 8 周,能够有效预防再次感染发生(Wernike 等,2013b;Elbers 等,2014;Schulz 等,2014;Wernike 等,2015b)。绵羊在被感染后 6~22 天发生血清转阳,且至少持续 15 个月,而山羊在感染后 7~14 天发生血清转阳(Wernike 等,2013c;Poskin 等,2015;Laloy 等,2015)。犊牛获得母源抗体的唯一方法是初乳喂养，由于反刍动物不会发生胎盘移位，抗体水平可以维持 5~6 个月(Elbers 等,2014)。此外,胎儿能够产生中和抗体,已经在死产或流产的犊牛和羔羊体内检测到中和抗体。然而,如果胎儿具有免疫能力,在子宫内发生胎儿感染,出生后胎儿不会表现出临床症状。

10.12 诊断

临床诊断一般通过观察临床症状进行，但不同动物的临床表现也不尽相同。用于检测 SBV 感染的各种实验室操作规程已经被介绍，包括实时 RT-PCR、中和实验和间接免疫荧光实验、ELISA 和通过细胞培养进行病毒的分离和鉴定。

10.12.1 被检样本

SBV 在被感染的胎儿体内持续时间要比被感染的成年动物更长(4~6 天),在畸形新生儿体内能够检测到病毒(Laloy 等,2017 年)。相当长一段时间内,在急性感染动物的肠系膜淋巴结、脾脏和精液中能够检测到病毒核酸。胎盘和羊水也是用于检测病毒的首选材料(Bilk 等,2012)。

从急性感染的成年动物中采集到的 EDTA 抗凝全血和血清的样本,应进行妥善包装,然后冷链运输到指定的实验室。从流产胎儿或新生动物上采集到的样本可以用于组织病理学（固定包括脊髓在内的中枢神经系统)、血清学和病毒学的研究。病死动物的最理想的样本应该是脑样本的大脑、小脑,以及其他中枢神经系统、血液,活体动物最理想的样本应该是初乳前的血液、血清和胎粪,样本采集后应妥善包装，冷链运输到指定的实验室(WOAH,2017)。

10.12.2 核酸检测

实时定量 RT-PCR 是用于临床样本中病毒核酸(病毒基因组的 L 片段或 S 片段)检测的有效方法。为此,FLI 建立了 SBV 的 RT-qPCR 检测方法(Bilk 等,2012;Hoffman 等,2012)。

10.12.3 病毒分离

用于病毒分离的样本可采自成年患病动物体温升高时的血液、死亡的胎儿及其大脑(Laloy 等,2017)。SBV 可用动物和人的细胞系进行分离培养,例如,Vero 细胞、羊脉络丛细胞、牛胎主动脉内皮细胞、293T 细胞，犬肾细胞、BHK-21 细胞、BSR 细胞和 KC 细胞(库蠓属幼虫),在上述大多数细胞系中都能产生细胞病变效应。在这些细胞系中,SBV 对羊脉络丛细胞最敏感 (Hoffmann 等,2012;Wernike 等 2013c;Ichmann 等,2017)。

10.12.4 血清学检测

病毒中和实验、免疫荧光和 ELISA 方法常用于 SBV 感染动物的血清阳性率,以及动物个体的抗体水平的检测(Loeffen 等,2012)。病毒中和实验是 SBV 诊断的金标准,其敏感性和特异性可达 100%(Loeffen 等,2012)。尽管有关报道表明辛波血清群内的病毒存在着免疫交叉反应，但 ELISA 方法则是用于 SBV 抗体检测的最为灵敏、特异、有效的一种检测方法,并且获得批准应用,能够用于 SBV 抗体的检测。可以通过对大量牛奶的抗体检测，来监测群体暴露在病毒中的水平 (Hoffmann 等,2012;Tarlinton 等,2012;EFSA 2012;Elbers 等,2012;Bilk 等,2012;Beer 等,2013;Tarlinton 和 Daly，2013)。ELISA 和 RT-PCR 检测试剂盒在市场均有销售,其中竞争性 ELISA 试剂盒可用于多种动物的血清或血浆病毒抗体的检测(ID Screen® Schmallenberg virus Competition Multi-species,IDvet Laboratories,Montpellier,France)。

10.12.5 鉴别诊断

由于 SBV 感染成年动物中没有特征性的临床症状,因而该病的鉴别诊断很重要。所有引起体温升高、腹泻、产奶量低、死产和流产等潜在的致病因素都应该被考虑,例如,牛病毒性腹泻病毒,边界病毒和其他瘟病毒,牛疱疹病毒 1 型和其他疱疹病毒,牛流行热病毒,流行性出血热病毒,口蹄疫病毒，蓝舌病病毒、裂谷热病毒等病毒，以及山藜芦、羽扇豆等有毒物质。而在调查先天性畸形致病因素时，也必须考虑到正布尼亚病毒属的卡奇谷病毒感染、遗传（如蜘蛛羊综合征）、有毒物质、营养不良（如缺乏妊娠蛋白、锰）等因素。

10.13 预防与控制

目前还没有用于治疗 SBV 感染的特效药物，因此，支持性疗法是对病毒进行干预的唯一可靠选择。为了降低该病带来的直接影响，可以通过使用有效的疫苗，或者在传播媒介活跃期对动物采取避孕的措施来减轻其直接影响。不同种灭活疫苗已经被成功研制并通过了测试（Wernike 等，2013c；Merial，2013），其中两种疫苗 SBV_{vax} 和 Bovilis SBV 已经获得进入英国、法国市场的临时许可。根据生产厂家的使用说明，大型动物间隔 28 天免疫接种 2 次，绵羊仅免疫接种 1 次。在 3 周内即可产生免疫保护。在繁殖季节前接种疫苗是预防感染最有效的措施。接种疫苗后，就会在未受病毒感染的宿主体内产生 SBV 抗体，一旦动物妊娠，这种抗体抑制病毒通过胎盘感染胎儿，避免引起畸形。此外，疫苗也能够保护易感动物不被感染，还可以防止传播媒介进一步传播病毒（Tarlinton 等，2012）。DNA 免疫能够刺激机体产生多种 Th 细胞和抗体的免疫应答反应。但到目前为止，针对该病还没有研制出有效的 DNA 疫苗。为了研制 SBV DNA 疫苗，以核蛋白和假定的 GC 胞外区基因作为靶基因进行了研制（Boshra 等，2017）。

通过采取适当措施控制蚊媒传播是控制 SBV 的切实可行的方法，例如，通过在蚊媒生长的自然栖息地使用杀虫剂/杀幼虫剂和对蚊媒有致病作用的病原体，以及通过环境处理，清除幼虫繁殖地；成年蚊的控制则可通过对畜舍或动物本身使用杀虫剂或驱虫剂（如拟除虫菊酯和寄主利它素）（Carpenter 等，2008）。

此外，可以通过改良育种体系保护幼畜或血清阳性率低的畜群避免 SBV 感染。繁殖期可以安排在晚秋，库蠓少的时候。为了使妊娠期动物获得免疫保护，可以将易感动物转移至疫区，动物很快就能获得免疫力（Helmer 等，2013）。犊牛的母源抗体在 5~6 个月就会消失，但在成年牛的特异性抗体至少维持 2 年（Elbers 等，2014）。由于该疫病并非法定报告传染病，因此没有实施行动限制。

参考文献

Afonso A, Abrahantes JC, Conraths F et al (2014) The Schmallenberg virus epidemic in Europe 2011–2013. Prev Vet Med 116:391–403

Alarcon P, Hasler B, Raboisson D et al (2014) Application of integrated production and economic models to estimate the impact of Schmallenberg virus for various sheep production types in the UK and France. Vet Rec Open 1:e000036

Animal & Plant Health Agency (APHA) (2017) Disease surveillance in England and Wales, February 2017. Vet Rec 180:243–246

Azkur AK, Albayrak H, Risvanli A et al (2013) Antibodies to Schmallenberg virus in domestic livestock in Turkey. Trop Anim Health Prod 45:1825–1828

Bayrou C, Garigliany MM, Sarlet M et al (2014) Natural intrauterine infection with Schmallenberg virus in malformed newborn calves. Emerg Infect Dis 20:1327–1330

Beer M, Conraths FJ, van der Poel WH (2013) Schmallenberg virus—a novel orthobunyavirus emerging in Europe. Epidemiol Infect 141:1–8

Bilk S, Schulze C, Fischer M et al (2012) Organ distribution of Schmallenberg virus RNA in malformed newborns. Vet Microbiol 159:236–238

Blomstrom AL, Stenberg H, al SI (2014) Serological screening suggests presence of Schmallenberg virus in cattle, sheep and goat in the Zambezia Province, Mozambique. Transbound Emerg Dis 61:289–292

Boshra HY, Charro D, Lorenzo G et al (2017) DNA vaccination regimes against Schmallenberg virus infection in IFNAR–/– mice suggest two targets for immunization. Antivir Res

141:107–115
van den Brom R, Luttikholt SJ, Lievaart-Peterson K et al (2012) Epizootic of ovine congenital malformations associated with Schmallenberg virus infection. Tijdschr Diergeneeskd 137:106–111
Carpenter S, Mellor PS, Torr SJ (2008) Control techniques for Culicoides biting midges and their application in the UK and northwestern Palaearctic. Med Vet Entomol 22:175–187
Causey OR, Kemp GE, Causey CE (1972) Isolations of Simbu-group viruses in Ibadan, Nigeria 1964–69, including the new types Sango Shamonda, Sabo and Shuni. Ann Trop Med Parasitol 66:357–362
Dandawate CN, Rajagopalan PK, Pavri KM et al (1969) Virus isolations from mosquitoes collected in North Arcot district, Madras state and Chittoor district, Andhra Pradesh between November 1955 and October 1957. Indian J Med Res 57(8):1420–1426
Delooz L, Saegerman C, Quinet C et al (2016) Resurgence of Schmallenberg virus in Belgium after 3 years of epidemiological silence. Transbound Emerg Dis 64(5):1641–1642
EFSA (2012) Scientific report of European Food Standards Agency: Schmallenberg virus: analysis of the epidemiological data and assessment of impact. EFSA J 10:2768
EFSA (2014) Schmallenberg virus: state of art. EFSA J 12(5):3681
Elbers AR, Loeffen WL, Quak S et al (2012) Seroprevalence of Schmallenberg virus antibodies among dairy cattle, the Netherlands, winter 2011–2012. Emerg Infect Dis 18:1065–1071
Elbers AR, Stockhofe-Zurwieden N, van der Poel et al (2014) Schmallenberg virus antibody persistence in adult cattle after natural infection and decay of maternal antibodies in calves. BMC Vet Res 10:103
Elliott RM, Blakqori G, van Knippenberg IC (2013) Establishment of a reverse genetics system for Schmallenberg virus, a newly emerged orthobunyavirus in Europe. J Gen Virol 94:851–859
Gelagay A, Endrias Z, Gebremedhinc E et al (2018) Seroprevalence of Schmallenberg virus in dairy cattle in Ethiopia. Acta Trop 178:61–67
Gerhauser I, Weigand M, Hahn K et al (2014) Lack of Schmallenberg virus in ruminant brain tissues archived from 1961 to 2010 in Germany. J Comp Pathol 150:151–154
Goller KV, Hoper D, Schirrmeier H et al (2012) Schmallenberg virus as possible ancestor of Shamonda virus. Emerg Infect Dis 18:1644–1646
Hechinger S, Wernike K, Beer M (2014) Single immunization with an inactivated vaccine protects sheep from Schmallenberg virus infection. Vet Res 45:79
Helmer C, Eibach R, Tegtmeyer PC et al (2013) Survey of Schmallenberg virus (SBV) infection in German goat flocks. Epidemiol Infect 141:2335–2345
Hoffmann B, Scheuch M, Hoper D et al (2012) Novel orthobunyavirus in cattle, Europe. Emerg Infect Dis 18:469–472
Ilchmann A, Armstrong AA, Clayton RF et al (2017) Schmallenberg virus, an emerging viral pathogen of cattle and sheep and a potential contaminant of raw materials, is detectable by classical in-vitro adventitious virus assays. Biologicals 49:28–32
Kesik MJ, Larska M (2016) Detection of Schmallenberg virus RNA in bull semen in Poland. Pol J Vet Sci 19(3):655–657
Kraatz F, Wernike K, Reiche S et al (2018) Schmallenberg virus non-structural protein NSm: intracellular distribution and role of non-hydrophobic domains. Virology 516:46–54
Laloy E, Breard E, Trapp S et al (2017) Fetopathic effects of experimental Schmallenberg virus infection in pregnant goats. Vet Microbiol 211:141–149
Laloy E, Riou M, Barc C et al (2015) Schmallenberg virus: experimental infection in goats and bucks. BMC Vet Res 11:221
Lechner I, Wuthrich M, Meylan M et al (2017) Association of clinical signs after acute Schmallenberg virus infection with milk production and fertility in Swiss dairy cows. Prev Vet Med 146:121–129
Loeffen W, Quak S, de Boer-Luijtze E et al (2012) Development of a virus neutralisation test to detect antibodies against Schmallenberg virus and serological results in suspect and infected herds. Acta Vet Scand 54:44
Martinelle L, Poskin A, Dal Pozzo F et al (2017) Three different routes of inoculation for experimental infection with Schmallenberg virus in sheep. Transbound Emerg Dis 64(1):305–308
Mathew C, Klevar S, Elbers AR (2015) Detection of serum neutralizing antibodies to Simbu serogroup viruses in cattle in Tanzania. BMC Vet Res 11:208
Merial (2013) Merial receives approval for new vaccine to prevent Schmallenberg disease in livestock. http://www.merial.com/EN/PressRoom/PressRelease/Pages/MerialApprovalSchmallenbergVaccine.aspx. Accessed 5 Nov 2014
Merck Animal Health (2013) Veterinary medicines directorate grants provisional marketing authorisation to MSD animal health for first vaccine targeting Schmallenberg virus. http://www.merck-animal-health.com/news/2013-12-18.aspx. Accessed 5 Nov 2014
OIE (2017): Schmallenberg virus: Technical Fact Sheet
Peperkamp K, Van Schaik G, Vellema P (2014) Risk factors for malformations and impact on reproductive performance and mortality rates of Schmallenberg virus in sheep flocks in the

Netherlands. PLoS One 9:e100135
Peperkamp NH, Luttikholt SJ, Dijkman R et al (2015) Ovine and bovine congenital abnormalities associated with intrauterine infection with Schmallenberg virus. Vet Pathol 52:1057–1066
Plyusnin A, Beaty BJ, Elliott RM (2012) Virus taxonomy: ninth report of the International Committee on taxonomy of viruses. Elsevier Academic Press, London, pp 725–741
Poskin A, Van Campe W, Mostin L (2014) Experimental Schmallenberg virus infection of pigs. Vet Microbiol 170:398–402
Poskin A, Verite S, Comtet L et al (2015) Persistence of the protective immunity and kinetics of the isotype specific antibody response against the viral nucleocapsid protein after experimental Schmallenberg virus infection of sheep. Vet Res 46:119
Rodrigues FM, Singh PB, Dandawate CN et al (1977) Kaikalur virus a new arthropod borne virus belonging to the Simbu group isolated in India from *Culex tritaeniorhynchus* (Giles). Indian J Med Res 66(5):719–725
Sailleau C, Boogaerts C, Meyrueix A (2013) Schmallenberg virus infection in dogs, France, 2012. Emerg Infect Dis 19(11):1896–1898
Schulz C, Wernike K, Beer M et al (2014) Infectious Schmallenberg virus from bovine semen, Germany. Emerg Infect Dis 20:338–340
Tarlinton R, Daly J (2013) Testing for Schmallenberg virus. Vet Rec 172:190
Tarlinton R, Daly J, Dunham S et al (2012) The challenge of Schmallenberg virus emergence in Europe. Vet J 194:10–18
Tilston LNL, Shi X, Elliott RM, Acrani GO (2017) The Potential for Reassortment between Oropouche and Schmallenberg Orthobunyaviruses. Viruses 9(8):220
Wernike K, Breithaupt A, Keller M (2012) Schmallenberg virus infection of adult type I interferon receptor knockout mice. PLoS One 7:e40380
Wernike K, Eschbaumer M, Schirrmeier H et al (2013b) Oral exposure, reinfection and cellular immunity to Schmallenberg virus in cattle. Vet Microbiol 165:155–159
Wernike K, Hoffman B, Conraths FJ et al (2015a) Schmallenberg virus recurrence, Germany, 2014. Emerg Infect Dis 21:1202–1204
Wernike K, Hoffmann B, Bréard E et al (2013a) Schmallenberg virus experimental infection of sheep. Vet Microbiol 166:461–466
Wernike K, Holsteg M, Saßerath M et al (2015b) Schmallenberg virus antibody development and decline in a naturally infected dairy cattle herd in Germany, 2011–2014. Vet Microbiol 181:294–287
Wernike K, Holsteg M, Schirrmeier H et al (2014) Natural infection of pregnant cows with Schmallenberg virus a follow up study. PLoS One 9:e98223
Wernike K, Nikolin VM, Hechinger S et al (2013c) Inactivated Schmallenberg virus prototype vaccines. Vaccine 31:3558–3563
Wuthrich M, Lechner I, Aebi M et al (2016) A case control study to estimate the effects of acute clinical infection with the Schmallenberg virus on milk yield, fertility and veterinary costs in Swiss dairy herds. Prev Vet Med 126:54–65
Yanase T, Kato T, Aizawa M et al (2012) Genetic reassortment between Sathuperi and Shamonda viruses of the genus orthobunyavirus in nature: implications for their genetic relationship to Schmallenberg virus. Arch Virol 157:1611–1616

第 11 章 克里米亚 - 刚果出血热病毒

Ashwin Ashok Raut, Pradeep N. Gandhale, D. Senthil Kumar,
Naveen Kumarl, Atual Paterita, Anamika Mishra,
Diwakar D. Kulkarni

11.1 引言

克里米亚-刚果出血热病毒(Crimean-Congo haemorrhagic fever virus,CCHFV)是一种蜱传人兽共患病毒，可引发急性出血热，人类死亡率高达40%。野生动物和家畜被感染属无症状感染。该病毒于1944—1945年第二次世界大战结束时,在帮助克里米亚农民的俄罗斯士兵体内首次被发现，这种疾病被命名为克里米亚出血热。在1953—1968年的阿斯特拉罕、1963—1971年的苏联罗斯托夫和1953—1971年的保加利亚,CCHFV持续流行。1967年,在哺乳小鼠大脑内接种分离到了病毒，这种技术当时被称为新生小白鼠技术(Hoogstraal,1979)。刚果热病毒于1959年在一名非洲比属刚果患者的血液中分离到。在1969年，美苏两国通过合作研究证实，这种病原体与刚果热病毒的抗原完全相同(Casals,1969)，于是在1973年,用上述两个名字合在一起重新命名该病毒,为克里米亚-刚果出血热病毒。

CCHFV是已知与人类健康有关的蜱传病毒中分布范围最广泛的病毒,目前已被东欧、亚洲、中东和非洲确定为公共卫生问题。该病毒主要存在于璃眼蜱属硬蜱，通过卵巢传播和处于发育期的蜱传播。人除了被蜱叮咬传播外,还可以通过接触携带病毒动物的血液和肉传播。CCHFV感染人群的高致死率，以及动物的无症状感染使得该人兽共患病难以控制。院内传播及其内部的人与人的传播进一步加剧了CCHFV对人类的危害(Watts等,2005)。蜱的数量不断增加,以及气候变化、不断扩大的农业和贸易活动使得蜱在全球扩散,这些都是造成CCHFV传入到土耳其、希腊和印度等地区的主要原因。CCHFV被认为是生物恐怖主义可能利用的生物战剂(Sidwell和Smee,2003)。伊拉克已经对CCHFV作为生物武器的潜力进行了研究,并对病毒的潜在传播方式进行了气溶胶化(Zilinskas,1997;Bronze等,2002)。

目前还没有被批准使用的CCHFV疫苗,只能通过具有支持性和预防性的策略对CCHFV的传播进行控制(Papa等,2015),加之CCHFV高致死率的特点，其已被列为生物安全4级病原微生物。在CCHFV感染初期能够做出快速、准确的诊断是有效防控CCHFV及其医院感染的关键。CCHFV属于生物安全4级病原微生物,不能在常规临床实验室中进行分离，而分子生物学诊断方法因其快速、可靠，越来越多地用于临床样本检测。CCHFV具有强大遗传变异性和潜在的病毒重组可能性，这成为阻碍疫苗研制和疾病诊断的主要障碍之一。

11.2 基因组

布尼亚病毒科被分为5个属,CCHFV属于布尼亚病毒科内罗毕病毒属(Schmaljohn和Nichol,2007)。内罗毕病毒属有32个成员通过硬蜱传播,其中,CCHFV是传播最为广泛、最重要的一种病毒(WHO,2011)。CCHFV病毒粒子有囊膜，直径为80~120nm,呈现多形性。

CCHFV基因组是分节段、单链RNA，有17 100~22 800个核苷酸。基因组分为小(S)、中(M)和大(L)(图11.1)三个节段。三个片段末端的互补碱基对形成三个独立非共价闭合环状结构保护RNA。互补的非编码区(NCR)作为病毒启动子区域位于三个片段的5’端和3’端;S、M和L在内罗毕病毒属中是保守序列。在内罗毕病毒属中的两端NCR分别有9个保守的核苷酸序列(5’-UCUCAAAGA和3’-AGAGUUUCU)。完整的NCR序列是病毒粒子复制、转录、衣壳化和组装的前提。NCR全序列在病毒基因片段之间存在差异,是结合病毒RNA依赖RNA聚合酶(L蛋白)的必要条件，负责启动病毒基因组的复制和转录(Zivcec等,2016)。

S片段长度约为1.6kb,类似于其他布尼亚病毒，具有一个 ORF，负链编码病毒核蛋白（NP）。核蛋白将病毒 RNA 基因组包装成核糖核蛋白颗粒，形成螺旋结构（图 11.1）。S 片段的第二个 ORF 是核蛋白 ORF 的反义链，编码一个非结构蛋白，这表明该片段具有双义特性（Barnwal 等，2015）。S 片段为双义编码，在其他布尼亚病毒之中同样也存在，不过在 CCHFV 中，它是通过重叠的编码区域进行编码，而在其他布尼亚病毒中，是作为转录终止信号被基因区间所分隔（Albarino 等，2007）。M 片段长度约为 5.4kb，比布尼亚病毒属的其他成员长。它编码包膜糖蛋白 G_N 和 G_C，以及非结构蛋白（NSm）（Bergeron 等，2007）。糖蛋白 G_N 和 G_C 不仅能够介导病毒进入细胞（图 11.1），还能够与中和抗体结合。L 片段编码生成的 RNA 依赖性 RdRp，长度为 4000 个氨基酸，分子量为 448 kDa。L 片段有一个单独的 ORF，长度为 12kb，比其他布尼亚病毒 L 片段长（图 11.1）（Honig 等，2004）。

11.3 遗传变异

与其他分节病毒基因组一样，CCHFV 片段间重组的可能性很大（Hewson 等，2004）。相较于 S 和 L 片段，M 片段基因重组更常见。但通过大量的基因组序列分析，发现 S 片段也存在基因重组。

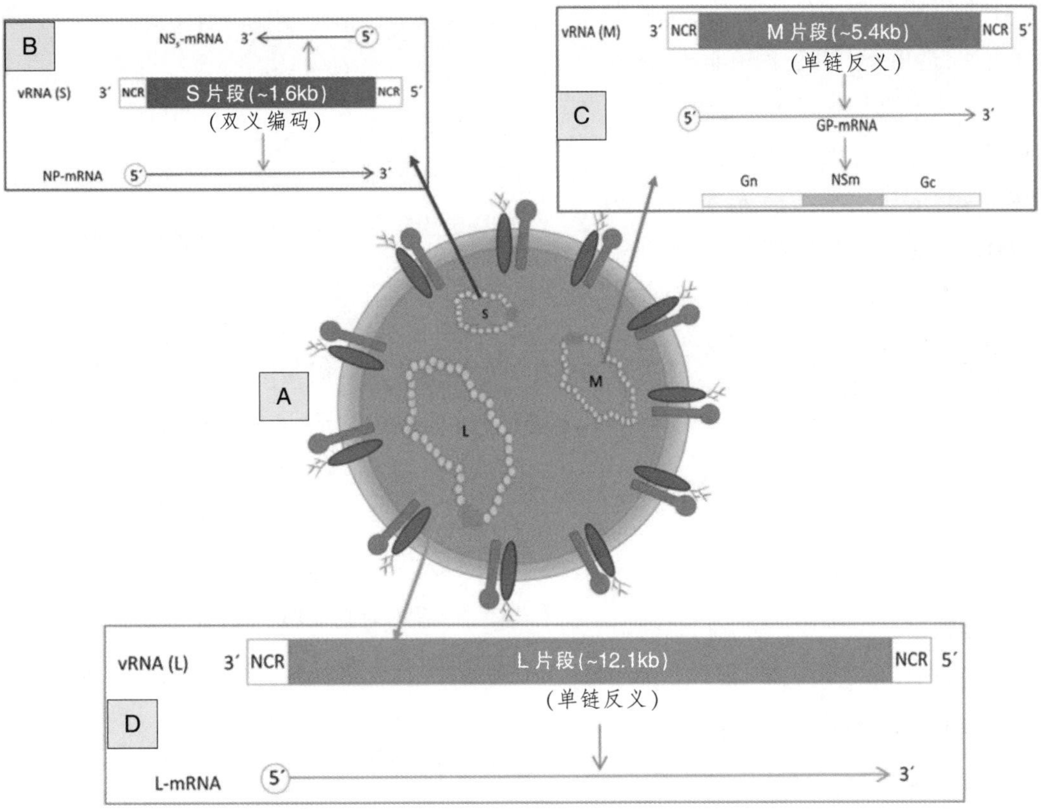

图 11.1 CCHFV 病毒粒子结构和基因组结构：(A)CCHFV 呈球形，被包含两种糖蛋白[Gn（37kDa）和 Gc（75kDa）]的脂质双分子层所围绕。病毒通过糖蛋白 Gn 与未知受体的融合进入细胞。病毒核心由 3 个 vRNA 片段组成，这 3 个片段被病毒核蛋白（NP）包裹。vRNA、NP 和 RdRp 通过非共价结合共同形成 3 个环状基因组 RNP，每个病毒粒子中有 3 个 vRNA 片段。(B)S 片段：一个单链双义 vRNA，长度约 1.6kb，具有保守的 3’和 5’NCR，图中标为阴影长方体。S 片段反义链编码结构核蛋白，正义链编码 NS 蛋白。(C)M 片段：单股反义 vRNA，长度约 5.4kb，具有保守的 3’和 5’NCR，图中标为阴影长方体。M 片段编码单个前体糖蛋白，经蛋白水解酶的作用，可裂解成两个成熟的糖蛋白 Gn 和 Gc。(D)L 片段：单链反义 vRNA，长度约 12.1kb，具有 3’和 5’ NCR，图中标为阴影长方体。L 片段编码 RNA 依赖性 RdRp。（扫码看彩图）

然而,M 片段和 L 片段发生基因重组的证据尚不明确(Lukashev,2005)。因此,从世界不同地区分离得到的 CCHFV 表现出庞大的遗传多样性。全球范围内,S、M、L 片段核苷酸序列的遗传变异系数分别为 20%、31%和 22%。N 蛋白、GPC 蛋白和 L 蛋白的氨基酸变异系数分别为 8%、27%和 10%(Deyde 等,2006)。然而,RNA 片段总长度、ORF 长度和 RNA 聚合酶核心区域等重要基序具有高保守性。据估计,S、M、L 片段的进化率为 1.09×10^{-4}、1.52×10^{-4} 和 0.58×10^{-4} 替换/位点/年 (Carroll 等,2010)。

目前 S 片段已被大量测序,NCBI 数据库收录了 62 个 S 片段全长序列,根据 S 片段的遗传进化分析,可将 CCHFV 分成 7 个进化分支。进化分支Ⅰ包含西非分离株,进化分支Ⅱ包含中非分离株,进化分支Ⅲ包含南非和西非分离株, 进化分支Ⅳ包含中东和亚洲分离株, 进化分支Ⅴ包含欧洲和土耳其分离株, 进化分支Ⅵ包含希腊分离株(Hewson 等,2004)。进化分支Ⅳ被划分为亚洲亚群 1 和亚洲亚群 2, 并与印度分离株单独形成一个群,其中亚洲亚群 2 包括中国、乌兹别克斯坦和塔吉克斯坦等远东地区分离株。印度分离株与塔吉克斯坦分离株核苷酸和氨基酸同源性上分别为 98.5%和 99.5%。邻国巴基斯坦的分离株和亚洲亚群 1 共同形成一个群(Yadav 等,2013)(图 11.2)。

利用 L 片段构建的系统进化树与利用 S 片段构建的系统进化树相似。M 片段对研究病毒的遗传多样性具有重要意义,虽然已知的 M 片段基因序列数量有限,但利用 M 片段构建的系统进化树与利用 S 片段构建的系统进化树存在明显差异。中国分离株 79121 和 7001 聚集到西非进化分支中,南非分离株 SPU415/85 和 SPU97/85 则聚集到进化分支Ⅳ中。另外,在进化分支Ⅶ中,以 M 片段为靶基因只检测到了毛里塔尼亚分离株。进化分支Ⅴ和Ⅲ中,集群模式也发生了变化。这些研究清楚地表明,M 片段基因重组频繁远高于 S 片段和 L 片段。

从CCHFV 遗传进化时间轴来看,松弛分子钟模型更适合于 CCHFV 的遗传变异分析, 而非传统的恒定分子钟模型。大量的序列遗传变异分析表明,虽然 CCHFV 在 20 世纪 40 年代被发现,但它确是一种古老的病毒。据估计,这些病毒最近的共同祖先起源于公元前 1599—1100 年。基于 S、M 和 L 片段分析, 病毒的共同祖先可追溯到 3198 年、3560 年和 7358 年以前。在具有多重序列代表的谱系中,进化分支Ⅳ历史更为久远,进化分支Ⅵ则是 CCHFV 最早的分支之一(Carroll 等,2010)。

总之,以病毒的地理分布为研究对象,开展病毒遗传多样性研究,发现相关病毒在南非、西非、伊拉克等地理位置不相邻的区域存在。另一方面,在同一地理区域也发现了不同遗传谱系, 就有如病毒进化分支Ⅰ和Ⅲ同时在西非地区存在一样,这主要是由于家畜的交易流通、候鸟携带感染的蜱或本身被感染候鸟迁徙引起。这也表明 CCHFV 可能是比现代病毒学中已知病毒更为古老的一种病毒(Chen,2013)。

CCHFV 的这种遗传多样性在任何蜱传病毒中并不常见, 这是因为蜱传病毒无论是以节肢动物作为传播媒介, 还是在脊椎动物宿主增殖中都必须保持适应性,这被称之为双重过滤。然而,双重过滤对 CCHFV 可能是无效的,这是因为 CCHFV 既能够经蜱卵传播,也能通过发育期的蜱进行传播, 而宿主对病毒的维持并不发挥关键作用。1976—2006 年,在伊朗、塔吉克斯坦和阿富汗发现了三个 M 片段重组的 CCHFV, 证明了该病毒在蜱-脊椎-蜱循环中具有高度的适应能力(Chen,2013)。无论在自然界中引起 CCHFV 遗传多样性驱动因素是什么, 都可根据其巨大的遗传多样性(包括所有三个片段在内的全基因组序列)和关键毒株全基因组序列推断出 CCHFV 的分子流行病学特点。如仅用一个片段序列或不完整基因组序列进行推断, 可能会限制血清学诊断和疫苗的研发、应用。

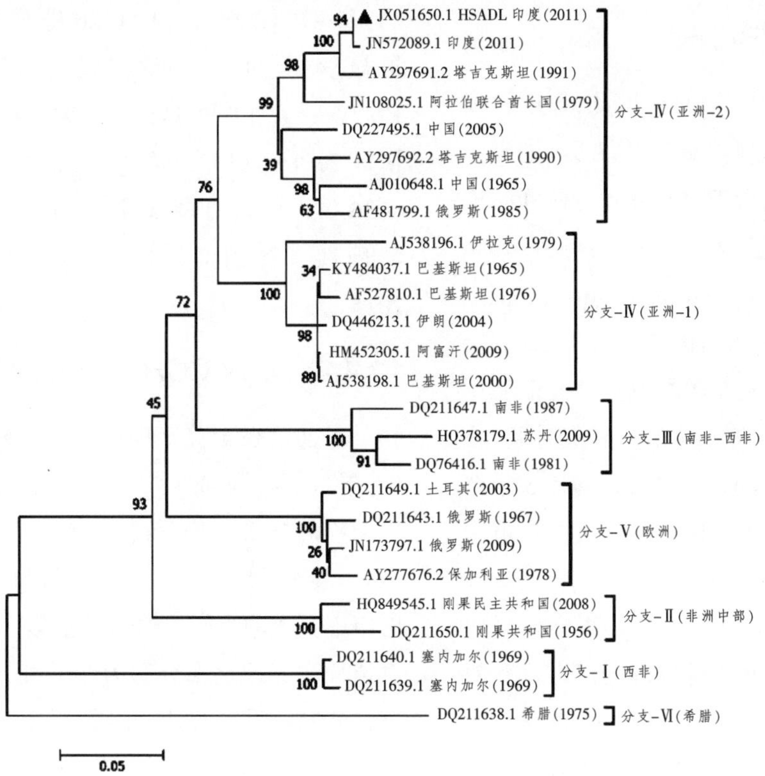

图 11.2　以印度流行的毒株作为参考毒株构建系统进化树，根据 CCHFV 的 S 片段全序列进行绘制。将 CCHFV 基因序列导入到 MEGA5 软件中，并将所有基因序列进行比对分析，利用最大似然树法，设定为 1000 个重复构建系统进化树。系统进化树用 26 个 S 片段全序列构建。括号表示病毒遗传进化分支及其地理分布。

11.4　流行病学

动物感染 CCHFV 无症状，但人感染 CCHFV 会引起严重的出血热。家畜的病毒血症虽然是短暂的，是一种低强度疾病，但对病毒在蜱的整个生命周期中的传播、增殖和维持起着重要的作用，因此从流行病学的角度来看，家畜的病毒血症意义很大。CCHFV 在所有蜱传病毒中引起疾病地理分布范围最广，在南纬 50°与北纬平行的至少 45 个国家都有相关报道（WHO，东南亚区域办事处，2014；CDC，2018）。参与病毒-蜱生命周期的脊椎动物和无脊椎动物的范围，以及大量的环境因素都在蜱的生命周期中发挥着重要作用，由于合法或非法动物贸易和鸟类迁徙，病毒大量传播到以前未受影响的地区，并通过多种途径传播给人类，以及各种教育、文化和行为特征都使得该病的流行病学更具有复杂性和独特性（Papa 等，2015）。根据一些出现地方性流行国家的血清学流行病学研究，各种家畜都表现为无症状感染。在地方性流行中，蜱-脊椎动物-蜱中的 CCHFV 循环大多与动物中蜱的数量及动物血清阳性率一致，因此，大量感染蜱的动物更有可能是 CCHFV 血清阳性（Adam 等，2013；Ibrahim 等，2015）。一般来说，与小型反刍动物相比，牛感染璃眼蜱属蜱的概率更高，因此它们是 CCHFV 低水平循环中最敏感的指标（Camicas 等，1990）。在家畜中检测 CCHFV 抗体对于确定 CCHFV 疫源地的本地化热点地区、增加人类感染的风险，以及传播病毒的初步证据都具有重要意义（Spengler 等，2016）。

11.4.1 印度 CCHFV 流行

印度周边国家 CCHFV 流行，因此，印度也面临很大的风险。由于印度与毗邻国家边界的复杂性，再加上动物/人跨境活动，CCHFV 传播到印度次大陆的风险显而易见。尽管第一例克里米亚–刚果出血热（Crimean–Congo haemorrhagic fever，CCHF）确诊病例是在 2011 年被报道的（Mishra 等，2011），但在 2010 年 2 月收集的人样本中就已经证明了 CCHFV 的存在（Mourya 等，2012）。从那时起，古吉拉特邦、拉贾斯坦邦和北方邦就有蜱通过家畜传播或医院感染而导致 CCHF 零星暴发的报道。

在实验室首次确认 CCHF 暴发之前，就有几项研究在不同动物中检测到了抗 CCHFV 抗体。1976 年，Shanmugam 等利用琼脂凝胶沉淀试验对来自山羊（186 份）、绵羊（149 份）、马（282 份）、牛（25 份）、公牛（12 份）和水牛（2 份）的 656 份血清进行抗体检测分析发现，山羊、马和绵羊的血清阳性率分别为 16.1%、1.1%和 0.7%。但 1986 年，Rodrigues 等对来自查谟和克什米尔的绵羊（38 份）、山羊（75 份）、牛（66 份）、水牛（23 份）、马（16 份）、驴（6 份）、骡子（64 份）和骆驼（3 份）血清进行抗体检测分析，并没有发现 CCHFV 感染的证据。以前，在拉贾斯坦邦（34 份）、马哈拉施特拉邦（132 份）和西北孟加拉邦的屠宰场中的家畜血清中检测到了 CCHFV 抗体，其中来自拉贾斯坦邦的水牛（1 份）、山羊（73 份）和绵羊（6 份）呈阳性（Mourya 等，2012）。此外，在拉贾斯坦邦的加纳巴拉特普尔凯奥拉德奥·盖纳国家公园采集的 10 种候鸟的 61 份血清样本中没有检测到 CCHFV 阳性血清（Ghosh 等，1978）。

第一次报道之后，多项 CCHFV 血清学检测研究证明，在印度检测到动物有抗 CCHFV 抗体。通过对古吉拉特邦的蜱和动物进行 CCHFV 筛查，305 份家畜血清样本中有 66 份抗 CCHFV 抗体阳性（Mourya 等，2012）。2014 年，Mourya 等对来自古吉拉特邦 15 个区牛（711 只）、山羊（279 只）和绵羊（236 只），共 1226 份血清进行检测，结果显示牛、山羊和绵羊的血清阳性率分别为 12.09%、41.21%和 33.62%。通过横断式调查法对横跨 22 个州和 1 个联邦地区的牛、山羊和绵羊的 5636 份血清进行调查发现，牛和山羊/绵羊的血清阳性率分别为 5.43%和 10.99%（Mourya 等，2015）。

11.4.2 全球 CCHFV 流行

目前已在全球范围内开展了对水牛、骆驼、牛、羊、山羊、马、猪、犬、鸡、鸵鸟等家养动物和野生动物的抗 CCHFV 抗体调查。其中研究最多的是牛，其次是绵羊和山羊。到目前为止，南非和津巴布韦报道的牛样本血清数量最多，共计 9430 份血清样本，阳性率分别为 28%和 45%（Swanepoel 等，1987）。另外有几个国家也报道了动物中血清阳性率很高的情况，其中牛（阿富汗）、绵羊（阿富汗）、山羊（土耳其）、马（伊拉克）、驴（塔吉克斯坦）和骆驼的血清阳性率分别为 79.1%、75.0%、66.0%、58.8%、39.5%、26%（Spengler 等，2016）。

非生物因素的变化，如季节、平均气温升高、归一化植被指数和栖息地的碎片化，也可能增加人的 CCHF 发病率和动物 CCHFV 血清阳性率。2012 年，Vescio 等发现，病例报道年度前秋季气温升高，CCHF 零病例报道的可能性增加。

关于伴侣动物的研究非常有限，因此还无法进行解读。不过，据报道，5.9%的犬体内都有 CCHFV 抗体存在（Shepherd 等，1987）。人类 CCHF 病例与来源于家畜、犬及其寄生蜱之间的关系也得到了研究，其中来自绵羊身上的伊文斯扇头蜱为 CCHFV 阳性，而犬身上的血红扇头蜱却为阴性（Nabeth 等，2004）。伴侣动物和相关虫媒载体之间的数据越多，越有助于评估伴侣动物在 CCHFV 生态学中的作用。

11.4.3 野生动物CCHFV流行

在野兔（3%~22%）、水牛（10%~20%）和犀牛

(40%~68%)中均发现较高的血清阳性率。在未成熟蜱活动的高峰期，由于受到伊氏边缘璃眼蜱幼虫和蜱蛹的传染，西欧刺猬和长耳刺猬的血清检测呈阳性。虽然在地方性流行过程中刺猬的作用取决于它的物种特性，但长耳刺猬在实验感染后能够形成病毒血症，说明长耳刺猬既有可能是CCHFV 自然宿主，也有可能是蜱的 CCHFV 的来源。然而，欧洲刺猬(西欧刺猬)在实验感染后并没有形成病毒血症，这可能是因为其对病毒易感性降低或者病毒被清除(Spengler 等，2016)。一些研究也报道了哺乳动物类翼手目的 CCHFV 或相关内罗毕病毒属的抗体。Tkachenko 等在 1969 年和Saidi 等在 1975 记录了在法国两种蝙蝠、伊朗大鼠耳蝠和常见夜蝠的血清中发现了抗 CCHFV 抗体。最近，通过利用 CCHFV 糖蛋白间接免疫荧光试验，对来自刚果、加蓬、加纳、德国和巴拿马的16 种蝙蝠，共 1135 份血清进行检测分析，10%的蝙蝠血清中检测到了抗体，其中穴居蝙蝠的血清阳性率为 3.6%~42.9%，叶栖蝙蝠的血清阳性率为0.6%~7.1%，这说明它们可能参与了 CCHFV 生命周期和地理分布(Muller 等，2016)。

11.4.4 鸟类 CCHFV 流行

许多鸟类是璃眼属蜱的宿主，能够高效的远距离传播蜱，因此蜱被 CCHFV 感染后，蜱再感染鸟类是造成病毒区域传播的主要因素。虽然CCHF 病例的增加与白嘴鸦种群数量的增加也有关联，但 CCHFV 感染与鸟类体内是否存在抗体之间的关系尚不明确。尽管进行了大量的调查，但大多数野生鸟类的血清学调查未显示出任何感染CCHFV 的血清学证据。形成鲜明对比的是，感染CCHFV 的蜱能够在许多物种身上产生病毒血症而存活下来，而 CCHFV 可以在无病毒血症的红嘴犀鸟中进行复制，并感染其他蜱。虽然家养鸡对CCHFV 感染耐受，但鸡、鸭血清中 CCHFV 阳性率为 0.2%。在鸟类中没有检测到抗 CCHFV 抗体，这可能是由检测方法敏感性不高造成的，大多数鸟类的血清学检测都是用琼脂凝胶扩散沉淀试验(agar gel diffusion precipitation，AGDP)方法，而AGDP试验的灵敏性比反向间接血凝抑制试验和免疫荧光抗体试验要低。然而，鸟类缺乏抗体反应并不普遍，携带 CCHFV 的鸵鸟似乎是一个例外，而且也可能将其传播到人类身上(Spengler 等，2016)。

11.5 传播

CCHF 的地理分布与硬蜱(尤其是璃眼蜱属)的地理分布一致。尽管已经从 30 多种蜱中分离到了 CCHFV，但只有少数蜱携带 CCHFV，而且其在全球的分布仅与璃眼蜱属相关(Estrada-Pena 等，2007)。璃眼蜱属喜好干燥的气候环境和干旱类型的植被，同时大量的哺乳动物在其生命周期具有重要的作用。鉴于璃眼蜱属的宿主动物分布广泛，再加上许多国家适宜的气候和生态条件，出现CCHF 的区域可能会随着时间的推移而增加。

蜱将病毒传播给哺乳动物，导致哺乳动物患上短暂性病毒血症，但是蜱终生携带病毒，且具有传染性；也因此蜱成了天然病毒库。人并不是蜱的传染源，只是偶然成为病毒的终末宿主(Bente 等，2013)。在地方性流行的周期性传播中，血清阳性率与带毒生物传播病毒的能力、宿主易感性，以及特定物种携带蜱的数量相关。璃眼蜱属是最主要的病毒携带者，地方性流行也只发生在这些传播媒介存在的地方，家畜中流行则主要发生在繁殖期和活动频繁时期(Spengler 等，2016)。

人被 CCHFV 感染主要通过蜱叮咬或者裸露皮肤直接接触被压碎的感染蜱两种途径，也可通过皮肤和烫伤，以及直接接触被感染动物的血液或组织进行传播。虽然只有少数研究报道表明可通过气溶胶和性传播，但仍需进一步研究 (Al-Abri 等，2017)。在对人类 CCHF 病例进行调查时，常将家畜及其近亲动物当成 CCHFV 的传染源。一些实例已经证明在某些地方性流行的地区，绵羊是 CCHFV 非常重要的宿主，并且在流行病学上，通常与人类病例有关。地方性流行地区的家畜屠宰工人及其协助人员仍然具有很高的感染CCHFV风险。因此，家畜中CCHFV血清阳性率的增加与接触活畜人群的CCHF病例相关，特别是那些处理被感染家畜的血液和器官的人群（Spengler等，2016）。

医护人员由于接触患者血液或分泌物引起的院内感染导致死亡率很高。医护人员由于缺乏充足的用于避免液体飞溅和针头刺伤的个人防护设施而引发院内感染，缺乏快速、灵敏的早期诊断方法进一步增加医护人员院内感染风险（Al-abri等，2017）。

11.6 感染免疫

11.6.1 潜伏期

传播途径和被感染病毒数量多少导致了该病的潜伏期3天到1周不等，但院内感染潜伏期要比蜱叮咬或直接接触家畜引发感染的潜伏期短（Vorou等，2007）。病毒数量和侵入途径可能会影响疾病的严重程度。蜱叮咬感染的致死率比院内传播感染的致死率相对要低；这可能是由于蜱叮咬接种的病毒量较低（Chinikar等，2010；Whitehouse，2004）。

11.6.2 动物临床症状

该病引起动物无症状感染。已有报道，CCHFV能在实验接种的绵羊和牛体内进行复制，引起动物轻度体温升高，但并不表现出临床症状。病毒血症期相对较短，且在病毒血症期出现不久后，就可在在体内检测到病毒。疫区中成年家畜（牛、绵羊和山羊）抗体的阳性率可高达50%以上。

11.6.3 人类临床症状

人类在感染初期症状表现不明显，特征性临床症状通常表现为突发高热、头痛、肌痛、虚弱、恶心和呕吐，可持续3天甚至更长时间。从无症状感染、轻度感染到重症感染和死亡的各类临床表现都有相关报道（Bodur等，2012）。重症感染6天后会出现出血热症状。1/3被感染患者表现为脾大和肝大等。实验室检测CCHF患者，主要是白细胞减少、血小板减少、肝酶升高和出血时间延长（Vorou等，2007；Cevik等，2008；Hatipoglu等，2010）。被感染3周后，患者可康复，但嗜睡、头晕和虚弱等症状可能会持续一年以上（Whitehouse，2004）。CCHF患者因受地理区域和病毒侵入途径影响，病死率为5%~40%（Yilmaz等，2009）。导致死亡的重要原因是重度贫血、脱水和休克引起的多器官功能衰竭。该病的预后和转归取决于血液中病毒的数量、血小板数、肝酶水平、出血时间、纤维蛋白原水平及胃肠道出血情况（Cevik等，2008；Hatipoglu等，2010）。

11.6.4 发病机制

由于CCHFV必须在生物安全4级实验室进行处理，并且缺乏合适的动物模型，所以有关CCHFV的发病机制了解甚少。CCHFV发病机制涉及病毒入侵、扩散、血管内皮损伤和渗漏，这主要是由内皮细胞和免疫细胞受到病理损伤而引起的。

11.6.5 病毒入侵

上皮细胞是阻挡病毒进入机体的首要屏障，但蜱叮咬可突破这道障碍。蜱叮咬后，病毒附着在内皮细胞基底外侧腔室的附着蛋白上，病毒进入血液循环系统进行传播（Connolly-Andersen，2010）。病毒在表面糖蛋白和人细胞表面受体的相互作用介导下进入细胞，细胞表面受体可能是人体内的核蛋白（Xiao等，2011）。

11.6.6 传播

CCHFV通过蜱叮咬或其他途径进入机体后，在巨噬细胞和树突状细胞中定居，并进行复制，随着血液循环到局部淋巴结、脾脏和其他器官（Connolly-Andersen，2010）。肝脏被认为是CCHFV

重要的靶器官，因为肝脏缺失基膜后使得病毒更容易进入肝细胞(Connolly-Andersen,2010)。在感染后期大脑也会受到影响(Bente 等,2010),这是因为细胞因子的释放增加了血管通透性，破坏了脑脊液屏障,导致病毒在脑脊液中进行传播(Kang 和 McGavern,2010)。

11.6.7 内皮细胞损伤和血管渗漏

根据出血及血管通透性增强等病理特征,内皮细胞似乎是 CCHFV 的一个主要靶细胞(Bodur 等,2010b;Burt 等,1997;Ozturk 等,2010)。内皮细胞在启动炎症反应方面具有非常重要的作用,如趋化性、白细胞黏附、血管通透性、炎症部位血细胞渗出，以及通过释放化学介质启动先天免疫反应和适应性免疫反应(Connolly-Andersen 等,2011)。机体的内在凝血级联反应通过病毒直接激活,也可通过病毒损伤的内皮细胞或宿主的可溶性介质间接激活(Connolly-Andersen 等,2011)。CCHFV 感染内皮细胞引起的炎症反应，造成 E 选择素、血管细胞黏附分子-1 和细胞间黏附分子-1 等可溶分子表达上调，进而增加细胞间的黏附作用(Connolly-Andersen 等,2011)。这些可溶性黏附分子可以作为对血管内皮细胞激活、血管损伤和疾病严重程度评价的生物标志物 (Bodur 等,2010b;Ozturk 等,2010)。释放细胞因子是促进 CCHFV 发病机制和疾病进展的主要因素之一。炎症因子、白介素 IL-1、IL-6、IL-8、IL-10 和 TNF-α 在调节致病机制和免疫应答中发挥着重要作用。虽然细胞因子主要是由巨噬细胞和树突状细胞分泌的，但被感染的内皮细胞也能够释放细胞因子(Connolly Andersen 等,2011)。细胞间黏附分子-1 在内皮细胞上的表达依次被细胞因子，以及 CCHFV 感染的树突状细胞释放可溶性介质调控(Schnittler 和 Feldmann,2003; Conndly-Andersen 等,2011)。TNF-α 是激活内皮细胞的关键细胞因子(Connolly-Andersen,2010)。血清促炎性细胞因子 IL-6、IL-8、IL-10 和 TNF-α 水平的升高被认为是 CCHF 患者预后的重要指标(Bente 等,2010;Saksida 等,2010)。在几例重症病例中,免疫失调、细胞因子分泌亢进(细胞因子血症)会导致血管通透性增加、血管舒张、多器官衰竭和休克(Connolly-Andersen,2010)。因此,CCHFV 的发病机制主要包括免疫失调,以及由细胞因子风暴而导致全身血管衰竭。CCHF 患者的病毒载量、血管内皮细胞激活的数量和细胞因子释放之间呈正相关。

11.6.8 免疫损伤

非特异性免疫应答,即先天免疫应答,是机体抵御病毒的第一道防线。CCHFV 可麻痹先天性免疫系统，延迟靶向病毒复制和病毒清除的特异性免疫反应(Bente 等,2010;Saksida 等,2010;Peyrefitte 等,2010)。病毒通过多种机制抑制免疫反应。IFN-α 和 IFN-β 在抑制病毒传播方面发挥着重要作用。病毒复制过程中产生的分子激活了先天免疫系统,诱导产生 IRF-3,随即诱导 Ⅰ 型干扰素产生(Andersson 等,2008)。干扰素的诱导延迟是 CCHFV 调控宿主防御机制的重要策略之一。已有报道表明 IFN 诱导表达的 MxA 蛋白与核衣壳蛋白可发生相互作用,并减少病毒的复制(Andersson 等,2008)。麻痹 IFN 反应机制似乎是人类特有的，因为 CCHFV 无法感染具有免疫活性的小鼠，但在免疫缺陷小鼠体内会迅速发病并导致其死亡(Bente 等,2010)。特异性免疫应答主要是由抗原刺激机体产生特异性抗体。在 CCHFV 感染期,通常检测不到抗体,并且感染患者能够存活的重要指标就是其体内有充足的抗病毒抗体，这与其他病毒性疾病的情况一样，特异性抗体水平与病毒载量呈负相关 (Saksida 等,2010)。据推测，CCHFV 感染后产生抗体的低应答反应可能是由于抗原提呈细胞，即树突状细胞和巨噬细胞的发育成熟受阻 (Bente 等,2010;Peyrefitte 等,2010)。淋巴细胞和自然杀伤细胞是机体抗病毒的先天性免疫应答和适应性免疫应答的重要组成部分,在病毒感染细胞的检测和裂解中发挥重要作用。类似于自然杀伤细胞等免疫细胞的耗竭，T淋巴细胞和B淋巴细胞耗竭是CCHFV感染患者常见的特征（Yilmaz等，2009）。通过CCHFV小鼠模型的研究发现，在感染早期分泌大量细胞因子，导致感染后第一天自然杀伤细胞、T淋巴细胞和B淋巴细胞增加，随后，由于淋巴细胞耗竭及其过度凋亡导致淋巴细胞减少（Bente等，2010；Geisbert等，2000）。因此CCHFV复制使得先天免疫反应和适应性免疫反应调控异常。

11.6.9 噬红细胞作用

单核细胞和巨噬细胞的过度激活造成过度吞噬，导致血红细胞数显著减少，并伴有发热、脾大和骨髓噬血细胞现象，因此该病又称噬血细胞综合征。CCHFV感染患者一半会发生严重的出血（Tasdelen Fisgin等，2008），因此，这也被认为是人类感染CCHFV发病机制之一。

11.6.10 止血和凝血功能障碍

广泛性出血是CCHF的重要临诊特征之一，首先是由于凝血功能受损和内皮细胞遭破坏，其次是免疫介导细胞损伤（Chen和Cosgriff，2000）。因此，弥漫性血管内凝血（Van Gorp等，1999）、血小板减少和凝血酶原时间延长（Cevik等，2008）是评估疾病预后的重要指标。

11.6.11 组织病理学

组织病理学病变主要表现为，在肝脏多发性坏死病灶内，有不同程度的肝细胞坏死、库普弗细胞增生、脂肪变性和门静脉周围单核细胞浸润。坏死区通常有明显的出血特征，且与肝内康氏小体有关（Burt等，1997）。肝脏损伤与丙氨酸转氨酶和天冬氨酸转氨酶升高相关（Cevik等，2008；Hatipoglu等，2010）。在脾脏中发生淋巴衰竭伴有淋巴细胞坏死和细胞核破裂（Bente等，2010）。在重症CCHFV感染患者中，常被报道会出现肝细胞和脾细胞的凋亡（Rodrigues等，2012）。也有发生肺泡受损、肺泡内出血和肠道出血等病理变化的报道（Burt等，1997）。CCHFV抗原多见于肝细胞、库普弗细胞和肝、脾内皮细胞中（Bente等，2010）

综上所述，该疾病的发病机制是血管内皮细胞损伤和免疫损伤共同作用的结果（图11.3）。在整个发病过程中依次出现的一系列病理变化过程，即细胞因子分泌增加、内皮细胞激活和血管扩张造成多器官衰竭和休克。I型干扰素诱导延迟、抗原提呈细胞成熟受阻及抗体应答反应低，最终导致病毒复制失控。

11.7 诊断

CCHF早期的非特异性临床症状与其他病毒性出血热相似，因此，确诊主要依赖于实验室检测。CCHFV属于生物安全4级级病原微生物，病毒分离只能在具备最高级生物防护设施的实验室中进行。当生物样本被灭活后，可以在非生物安全4级实验室中进行分子检测。

在实验室可以通过病毒分离，RT-PCR、RT-qPCR方法检测病毒基因组，抗原捕获ELISA方法检测病毒，ELISA方法检测抗体实现对CCHFV的确诊。大多数情况下，通过采集血清和血浆用于临床样本的病毒分离和分子检测。最好利用EDTA抗凝血作为检测样本进行检测，可大大提高PCR检测效率（Drosten等，2003）。组织、唾液和尿液也可以作为临床样本用于分子生物学检测。被病毒感染的血清样本在室温环境下可存活数天，因此，其被认为是流行病学调查的首选材料（Bodur等，2010a）。

在病毒血症期间，被感染后12天内都可以分离到病毒，在感染后的第一周（病毒最高滴度出现在感染后1~6天）进行病毒分离的成功概率更大（Shepherd等，1988），然而，对血清样本的基因组检测可以持续至18天。据观察，病毒RNA在致死病例中随时间推移而增加，在非致死病例中却随

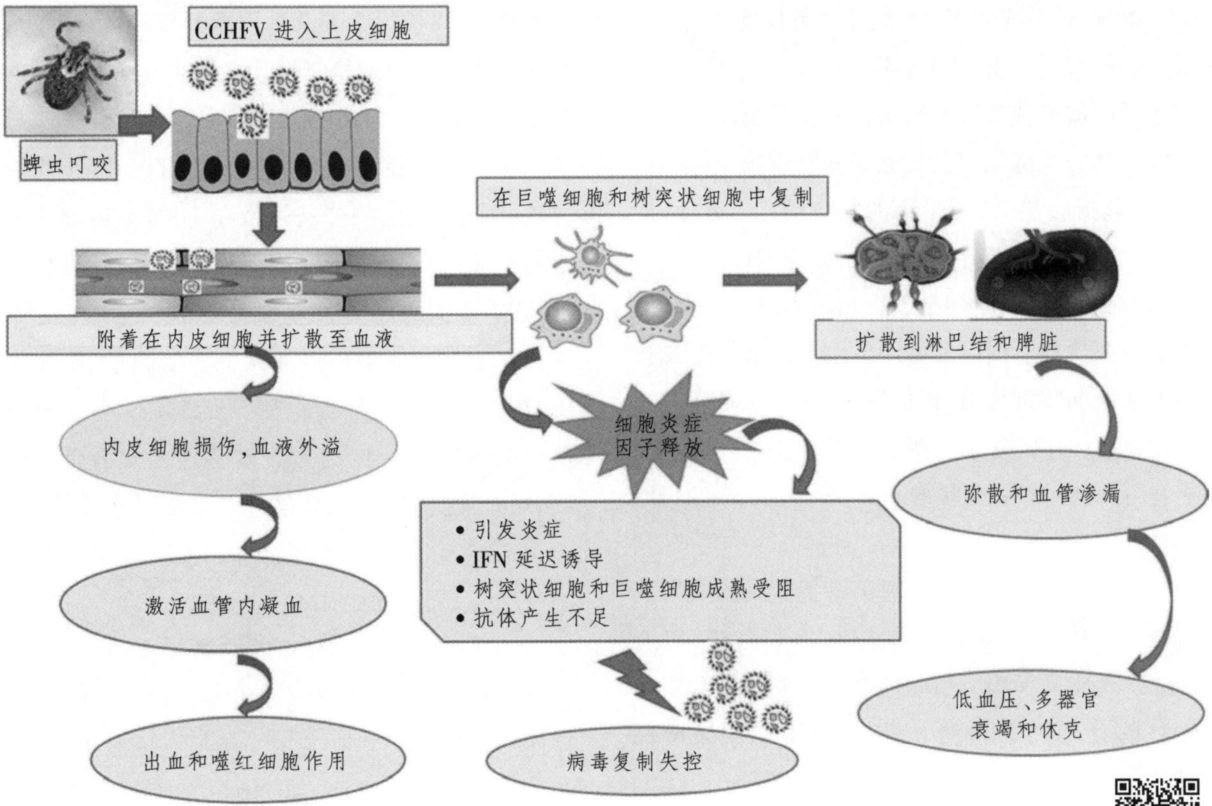

图 11.3 在 CCHEV 发病过程中依次发生的一系列病理变化。(扫码看彩图)

时间推移而减少(Saksida 等,2010)。

11.7.1 病毒分离

病毒分离被认为是诊断 CCHF 的黄金标准，然而,由于 CCHFV 检测需要在密封的生物安全 4 级防护设施中进行,病毒分离方法的敏感性较低，以及其相关风险因素存在，使得病毒分离不适用于常规诊断。大多数情况下,血液、器官分泌液等临床样本用于乳鼠颅内接种或细胞（Vero、LLC-MK2、SW-13、CER 和 BHK-21 细胞）培养分离。CCHFV 接种细胞不产生确切的细胞病变效应。因此，确诊要通过免疫荧光实验或分子生物学试验进行检测。在乳鼠体内进行病毒分离平均需要 7.7 天,而细胞培养只需要 3.3 天,但通过接种乳鼠进行病毒分离比用细胞对病毒分离培养的灵敏度高(Shepherd 等,1986)。

11.7.2 抗原检测

抗原捕获 ELISA 检测方法可以替代病毒分离用于 CCHFV 抗原检测，该检测方法可在数小时内得出结果,且检测样品预先灭活后,不需要在特殊防护设施中进行操作，但该方法的检测灵敏度低于病毒分离和 RT-PCR（Shepherd 等,1988; Burt,2011)。抗原捕获 ELISA 检测方法是利用抗核蛋白基因开发的单克隆抗体作为捕获抗体,可用于急性感染期疾病快速检测,尤其适用于重病毒血症患者死亡病例的检测，但该检测方法只适用于血清抗体阳性样本(非致死性病例)的检测,这是其局限性之一(Saijo 等,2005)。

11.7.3 分子生物学诊断

与传统 CCHFV 检测方法相比,RT-PCR 和

RT-qPCR 等分子生物学检测方法具有快速、灵敏、特异的优点，且无须生物安全 4 防护设施，在数小时内就可获得检测结果，有利于及时采取防控措施和控制感染。然而，分子诊断取决于样本的状态，包括收集时间、采集的材料/器官、运输、冷链维护等。在世界范围内流行的 CCHFV 不同毒株序列的高度变异性是 RT-PCR 检测方法所面临的重要挑战。国际质量评估也支持 RT-PCR 检测方法所面临的挑战，评估发现 44 个参考实验室应用的 CCHFV 分子检测方法均未达最佳标准，突出了改善这一现状的重要性（Escadafal 等，2012）。

1990 年，首次报道了基于琼脂糖凝胶结果或 DNA 印迹检测的传统套式 PCR 检测方法，该方法易产生交叉污染，且耗时较长。但利用两套引物建立的套式 PCR 检测方法，提高了检测灵敏性（Burt 等，1998）。

以 S 片段为靶基因建立的 RT-qPCR 方法，通过特异的引物和探针实现对 CCHFV 所有流行株的检测，扩大了检测范围，且该检测方法具有高度的灵敏性和特异性，实现了交叉污染最小化甚至零交叉污染（表 11.1）。基于化学反应、SYBR Green、TaqMan 和分子信标的许多 RT-qPCR 检测方法已经被建立，这些检测方法均以 CCHFV 的 NP 基因为靶基因。第一个基于 SYBR Green 建立的 RT-qPCR 检测方法的敏感性为在每毫升血浆中可检测到 2779 个拷贝数的病毒基因组（Drosten 等，2003）。首个被建立的 TaqMan RT-qPCR 方法是参考世界内流行的 19 株 CCHFV 分离株进行了引物和探针设计（Yapar 等，2005）。此外，针对巴尔干地区流行毒株，也建立了特异性 qPCR 检测方法（Duh 等，2006）。后来，基于特异性 TaqMan MGB 蛋白质探针的 qPCR 检测方法也被建立（Garrison 等，2007）。基于探针的多重 qPCR 检测方法也被建立，该方法可用于所有 CCHFV 毒株的检测（Wolfel 等，2007）。设计实时 PCR 检测方法应用简单探针技术就可以实现对非洲南部 CCHFV 分离株的检测及基因型分型（Kondiah 等，2010）。由于大多数检测都无法实现对高度变异的 AP92 基因型的 CCHFV 检测，Atkinson 等于 2012 年基于单探针技术建立一种新的 RT-qPCR 检测方法，实现包括 AP92 基因型 CCHFV 在内的所有 CCHFV 已知毒株的检测。此外，以印度流行的 CCHFV 毒株的保守 N 基因为靶基因，基于分子信标建立的实时 RT-PCR 检测方法，可用于印度 CCHFV 毒株的特异性检测（图 11.4）。基于分子信标的 RT-PCR 检测方法敏感性比先前介绍的 TaqMan 检测方法高 10 倍，该方法具有经济、快速、特异、灵敏的特点，可替代 CCHFV 现有的检测方法（Kamboj 等，2014）（表 11.1）。

为了实现对 CCHFV 的监测和基因型分型，WHO 的一个参考实验室利用 CCHFV 的 20 种特异性捕获探针成功开发了一种低密度基因芯片，通过对过去 20 年确诊的 CCHF 临床病例进行检测，对基因芯片进行了验证（Wolfel 等，2009）。一种 DNA 高通量测序的基因芯片已经被研制，可以同时对包括 CCHFV 在内的各种病毒性出血热病毒进行检测和鉴别诊断（Filippone 等，2013）。为了能够在现场使用，一种非 PCR 的基因检测的滚环扩增技术被开发，该方法的应用，使 CCHFV 诊断变得便捷、灵敏，尤其适用于发展中国家（Ke 等，2011）。

11.8 预防和控制

到目前为止，还没有获得许可的 CCHF 疫苗或获批的治疗方法，这主要是因为多年来缺乏合适的动物模型、生物安全 4 级实验室，以及有关 CCHF 的研究较少。

11.8.1 支持性疗法

对于确诊 CCHF 的患者，最有效的处理措施就是给予及时的支持性治疗。通过对多种生理指标进行实验室检测（例如，白细胞计数，肾功能检

表 11.1　CCHFV 实时 PCR 检测实验

序号	引物序列	探针	化学物质	参考文献
1	5'-ATGCAGGAACCATTAARTCTTGGA 3'-CTAATCATAATCTGACAACATTTC 和 CTAATCATGTCTGACAGCATCTC	—		Drosten 等 (2002)
2	CCHF L1-GCTTGGGTCAGCTCTACTGG CCHF D1-TGCATTGACACGGAAACCTA	CCHF S1 FAM-AGAAGGGGC TTGAGTGGTT-DABCYL	SYBR Green	Duh 等 (2006)
3	5'-CAAGGGGTACCAAGAAAATGAAGAAGGC 3'-GCCACAGGGATTGTTCCAAAGCAGAC	FAM-ATCTACATGCACCCTG CTGTGTTGACA-TMARA	TaqMan	Wolfel 等 (2007)
4	5'-GGAGTGGTGCAGGGAATTTG 3'-CAGGGCGGGTTGAAAGC	6FAM-CAAGGCAAGGTACAT CAT-MGBNFQ	TaqMan	Garrison 等 (2007)
6	5'-AGTGTTCTCTTGAGTGCTA 3'-CCACAAGTCCATTTCCTT	6-FAM-CGCGATCATCTCATC TTTGTTGTTCACCTCGATC GCG-BHQ-1	分子信标	Kamboj 等 (2014)

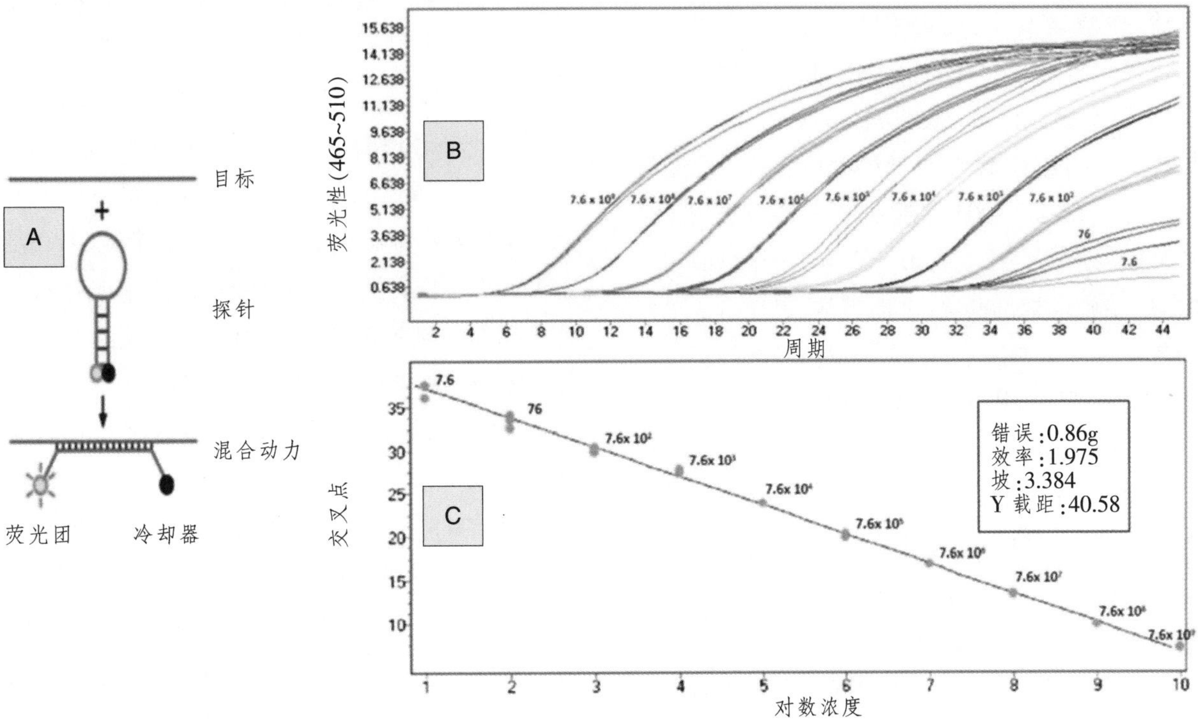

图 11.4　基于分子信标的 RT-qPCR 检测可实现 CCHFV 快速诊断。(A)分子信标化学图示。(B)扩增曲线。(C)使用 10 倍连续稀释 CCHFV S 段 IVT-RNA 进行分子信标探针 RT-qPCR 检测的标准曲线。该检测方法对 IVT-RNA 扩增曲线从 7.6 到 7.6×10^9 个拷贝数。RT-qPCR 检测方法的敏感性为 7.6~76 个拷贝数 RNA。(Adapted from kamboj et al. 2014)

查，血清电解质和转氨酶、血液凝血参数），进而实现对住院患者密切监测。输血，血浆或血小板的支持疗法越早进行有效性就越高。CCHF 患者若伴有难治性的血小板减少症，建议静脉注射免疫球蛋白。

11.8.2 抗病毒治疗

目前还没有被批准用于治疗 CCHF 的特异抗病毒药物，不过，利巴韦林是目前唯一用于治疗 CCHF 的药物。它是一种广谱合成嘌呤核苷类似物，其确切作用机制尚不明确。然而，利巴韦林在体内和体外均以浓度依赖的方式抑制 CCHFV 的复制。有时通过口服利巴韦林可预防接触性传染，但疗效仍不确定。值得注意的是，通过 Meta 分析发现，利巴韦林对患者治疗是无效的，因为与未接受治疗的患者相比，使用利巴韦林治疗并没有提高生存率和缩短住院时间（Ascioglu 等，2011）。联合用药（利巴韦林和类皮质激素）用于 CCHF 患者治疗，特别是 CCHFV 早期感染患者的治疗是有效的（Jabbari 等，2006）。近期，一种被批准用于治疗流感的新药，法匹拉韦（T-705）比目前用于 CCHF 治疗的标准药物（利巴韦林）表现出了更好的疗效（Oestereich 等，2014）。此外，特异性免疫球蛋白也得到了应用，特别是在保加利亚，对恢复期 CCHF 患者肌内注射特异性免疫球蛋白，已达到预防和治疗目的（Christova 等，2009）。在土耳其进行的一项研究中，应用健康供体的 CCHFV 高免球蛋白进行治疗，尤其是对病毒载量高的患者进行治疗是一种有效的治疗手段（Kubar 等，2011）。此外，为了确定和采取对 CCHF 患者有效的抗病毒治疗方法，急需进行大量安慰剂对照实验、前瞻性随机对照实验。

11.8.3 疫苗

对 CCHF 疫苗的探索始于 20 世纪 60 年代左右，1970 年利用感染了 CCHFV 乳鼠的脑组织研制了第一种疫苗（传统的俄罗斯/保加利亚疫苗）（Papa 等，2011）。尽管这种疫苗分别于 1970 年、1974 年得到苏联卫生部和保加利亚批准，但由于其免疫原性差且髓鞘碱性蛋白可能引起自身免疫反应，因此并未获得国际监管机构批准（Hemachudha 等，1987；Mousavi-Jazi 等，2012）。几十年来，由于缺乏适宜的动物模型，疫苗的开发也受到严重阻碍。最近开发 CCHF 候选疫苗的方法包括一种携带编码 CCHFV 糖蛋白基因的 DNA 疫苗。接种这种 DNA 疫苗的小鼠大约有 50%都能诱导产生中和抗体（Spik 等，2006）。另一种候选疫苗则运用转基因烟草叶片表达 CCHFV 糖蛋白，表达蛋白可诱导小鼠体内产生 IgG 和 IgA 两种抗体（Ghiasi 等，2011）。然而，由于没有合适实验动物模型，并没有对这两种疫苗进行致病性保护实验。2010 年开发的 A129（Ⅰ型干扰素受体缺陷）小鼠模型和 STAT-1 基因敲除小鼠模型，使 CCHFV 致病机制研究和进行疫苗攻毒保护实验成为可能。随后，开发了一种表达 CCHFV M 全基因的重组人血清 5 型腺病毒活载体疫苗，该疫苗可诱导机体产生体液免疫应答和细胞免疫应答，但是不能对 STAT-1 基因敲除小鼠提供免疫保护（Sahib，2010）。最近，证明用昆虫细胞表达 Gn 或 Gc 胞外域的亚单位佐剂疫苗，能够诱导机体产生中和抗体，但在对 STAT-2 基因敲除小鼠的攻毒保护实验中没有起到免疫保护作用（Kortekaas 等，2015）。该疫苗失败的原因可能是由于昆虫细胞中对蛋白的糖基化修饰与哺乳动物细胞不同。另一种基于转录能力强大的病毒样颗粒疫苗也已经被研制，虽然能够诱导机体产生效价较高的中和抗体，但对 A129 小鼠只能提供 40%的攻毒保护（Hinkula 等，2017）。迄今为止，最有希望的 CCHF 候选疫苗是用改进的痘苗病毒载体表达编码糖蛋白的 M 基因的重组活载体疫苗。该疫苗在小鼠模型中能够诱导机体产生体液免疫反应和细胞免疫反应，也能够为 CCHFV 感染小鼠提供 100%的保护作用（Buttigieg 等，2014）。此外，使用相同的痘苗病毒载体表达

NP 蛋白的疫苗不能在攻毒实验中提供免疫保护作用(Dowall 等,2016)。然而,到目前为止,还没有一种疫苗进入临床试验。

11.8.4 预防与控制

有效预防 CCHFV 感染的方法可分为两大类:①控制患有 CCHF 的动物和蜱;②降低感染人的风险。蜱-动物-蜱的循环通常不易被发现,家畜发生无症状感染,除此之外,带毒蜱广泛分布,目前仍未开发出用于动物的疫苗,因此,使用杀螨剂控制蜱是控制动物感染的有效方法。

在没有批准用于人类的疫苗情况下,培养人们如何减少接触病毒、提高安全意识尤为重要,特别是动物饲养员、屠宰场工人、农民和卫生工作者。一般来说,为了尽量减少与蜱的接触,建议使用批准的驱虫剂,穿着浅色防护服,避免前往蜱流行地区。医护人员需要频繁接触患有 CCHF 的患者的血液和组织,建议戴手套、穿防护服、戴口罩,避免与感染 CCHFV 患者有密切的身体接触,从而减少接触风险。此外,应采取适当措施,防止非法跨界动物运输,从而限制 CCHF 的跨地域传播。

由于 CCHFV 的传播媒介蜱广泛分布,由此造成 CCHFV 传播的地理区域范围扩大。由于气候变化的影响,使蜱种群大量增殖,从而促进 CCHFV 的传播和发生。密集的城市和郊区畜牧业靠近人群,以及管理措施不到位,因蜱叮咬和接触被感染动物血液、肉类而促进了动物-人之间的传播。通过监测动物,预防家畜被蜱感染,以及在高危人群中采取适当的预防措施可以有效控制 CCHFV 感染。

参考文献

Adam I, Mahmoud MA, Aradaib IE (2013) A seroepidemiological survey of Crimean Congo hemorrhagic fever among cattle in North Kordofan State, Sudan. Virol J 10:178

Al-Abri SS, Abaidani A, Fazlalipour M, Mostafavi E, Leblebicioglu H, Pshenichnayaf N, Memish ZA, Hewson R, Peterseni E, Mala P, Nguyenj TMN, Malik MR, Formenty P, Jeffries R (2017) Current status of Crimean-Congo haemorrhagic fever in the World Health Organization Eastern Mediterranean Region: issues, challenges, and future directions. Intl J Infect Dis 58:82–89

Albariño CG, Bird BH, Nichol ST (2007) A shared transcription termination signal on negative and ambisense RNA genome segments of Rift Valley fever, sandfly fever Sicilian, and Toscana viruses. J Virol 81:5246–5256

Andersson I, Karlberg H, Mousavi-Jazi M, Martı'nez-Sobrido L et al (2008) Crimean-Congo hemorrhagic fever virus delays activation of the innate immune response. J Med Virol 80(8):1397–1404

Ascioglu S, Leblebicioglu H, Vahaboglu H, Chan KA (2011) Ribavirin for patients with Crimean-Congo haemorrhagic fever: a systematic review and meta-analysis. J Antimicrob Chemother 66:1215–1222

Atkinson B, Chamberlain J, Logue CH, Cook N, Bruce C, Dowall SD, Hewson R (2012) Development of a real-time RT-PCR assay for the detection of Crimean-Congo hemorrhagic fever virus. Vector Borne Zoonotic Dis 12:786–793

Barnwal B, Karlberg H, Mirazimi A, Tan YJ (2015) Non-structural protein of Crimean-Congo hemorrhagic fever virus disrupts mitochondrial membrane potential and induces apoptosis. J Biol Chem 291:582–592

Bente DA, Alimonti JB, Shieh WJ, Camus G et al (2010) Pathogenesis and immune response of Crimean-Congo hemorrhagic fever virus in a STAT-1 knockout mouse model. J Virol 84:11089–11100

Bente DA, Forrester NL, Watts DM, McAuley AJ, Whitehouse CA, Bray M (2013) Crimean-Congo hemorrhagic fever: history, epidemiology, pathogenesis, clinical syndrome and genetic diversity. Antivir Res 100(1):59–189

Bergeron E, Vincent MJ, Nichol ST (2007) Crimean–Congo hemorrhagic fever virus glycoprotein processing by the endoprotease SKI-1/S1P is critical for virus infectivity. J Virol 81(23):13271–13276

Bodur H, Akinci E, Ascioglu S, Ongürü P, Uyar Y (2012) Subclinical infections with Crimean-Congo hemorrhagic fever virus, Turkey. Emerg Infect Dis 18:640–642

Bodur H, Akinci E, Ongürü P, Carhan A, Uyar Y, Tanrici A, Cataloluk O, Kubar A (2010a) Detection of Crimean–Congo hemorrhagic fever virus genome in saliva and urine. Int J Infect

Dis 14:247–249
Bodur H, Akinci E, Ongürü P et al (2010b) Evidence of vascular endothelial damage in Crimean-Congo hemorrhagic fever. Int J Infect Dis 14:e704–e707
Bronze MS, Huycke MM, Machado LJ, Voskuhl GW, Greenfield RA (2002) Viral agents as biological weapons and agents of bioterrorism. Am J Med Sci 323(6):316–325
Burt FJ (2011) Laboratory diagnosis of Crimean–Congo hemorrhagic fever virus infections. Future Virol 6:831–841. Review on laboratory diagnosis of Crimean-Congo hemorrhagic fever virus
Burt FJ, Leman PA, Smith JF, Swanepoel R (1998) The use of a reverse transcription-polymerase chain reaction for the detection of viral nucleic acid in the diagnosis of Crimean–Congo haemorrhagic fever. J Virol Methods 70:129–137
Burt FJ, Swanepoel R, Shieh WJ, Smith JF et al (1997) Immunohistochemical and in situ localization of Crimean-Congo hemorrhagic fever (CCHF) virus in human tissues and implications for CCHF pathogenesis. Arch Pathol Lab Med 121:839–846
Buttigieg KR, Dowall SD, Findlay-Wilson S, Miloszewska A, Rayner E, Hewson R et al (2014) A novel vaccine against Crimean-Congo haemorrhagic fever protects 100% of animals against lethal challenge in a mouse model. PLoS One 9:e91516
Camicas JL, Wilson ML, Cornet J, Digoutte J, Calvo M, Adam F, Gonzalez JP (1990) Ecology of ticks as potential vectors of Crimean-Congo hemorrhagic fever virus in Senegal: epidemiological implications. Arch Virol 1:303–322
Carroll SA, Bird BH, Rollin PE, Nichol ST (2010) Ancient common ancestry of Crimean-Congo hemorrhagic fever virus. Mol Phylogenet Evol 55:1103–1110
Casals J (1969) Antigenic similarity between the virus causing Crimean hemorrhagic fever and Congo virus. Proc Soc Exp Biol Med 131(1):233–236
Centers for Disease Control and Prevention 2018, Crimean-Congo Hemorrhagic Fever (CCHF) distribution map. https://www.cdc.gov/vhf/crimean-congo/outbreaks/distribution-map.html. Accessed 16 Mar 2018
Cevik MA, Erbay A, Bodur H, Gülderen E et al (2008) Clinical and laboratory features of Crimean-Congo hemorrhagic fever: predictors of fatality. Int J Infect Dis 12:374–379
Chen JP, Cosgriff TM (2000) Hemorrhagic fever virus-induced changes in hemostasis and vascular biology. Blood Coagul Fibrinolysis 11:461–483
Chen S (2013) Molecular evolution of Crimean-Congo hemorrhagic fever virus based on complete genomes. J Gen Virol 94:843–850
Chinikar S, Ghiasi SM, Moradi M, Goya MM et al (2010) Phylogenetic analysis in a recent controlled outbreak of Crimean-Congo haemorrhagic fever in the south of Iran, December 2008. Euro Surveill 15(47):19720
Christova I, Di Caro A, Papa A, Castilletti C, Andonova L, Kalvatchev N et al (2009) Crimean-Congo hemorrhagic fever, southwestern Bulgaria. Emerg Infect Dis 15:983–985
Connolly-Andersen AM (2010) Pathogenesis of an emerging pathogen Crimean-Congo hemorrhagic fever. Thesis, Stockholm
Connolly-Andersen AM, Moll G, Andersson C, Akerström S et al (2011) Crimean-Congo hemorrhagic fever virus activates endothelial cells. J Virol 85:7766–7774
Deyde VM, Khristova ML, Rollin PE, Ksiazek TG, Nichol ST (2006) Crimean-Congo hemorrhagic fever virus genomics and global diversity. J Virol 80:8834–8842
Dowall SD, Buttigieg KR, Findlay-Wilson SJ, Rayner E, Pearson G, Miloszewska A et al (2016) A Crimean-Congo hemorrhagic fever (CCHF) viral vaccine expressing nucleoprotein is immunogenic but fails to confer protection against lethal disease. Hum Vaccin Immunother 12(2):519–517
Drosten C, Göttig S, Schilling S, Asper M, Panning M, Schmitz H, Günther S (2002) Rapid detection and quantification of RNA of Ebola and Marburg viruses, Lassa virus, Crimean-Congo hemorrhagic fever virus, Rift Valley fever virus, dengue virus, and yellow fever virus by real-time reverse transcription-PCR. J Clin Microbiol 40:2323–2330. The first one-step real-time-RT-PCR for detection of Crimean-Congo hemorrhagic fever virus
Drosten C, Kummerer BM, Schmitz H, Gunther S (2003) Molecular diagnostics of viral haemorrhagic fevers. Antivir Res 57:61–87
Duh D, Saksida A, Petrovec M, Dedushaj I, Avsic-Zupanc T (2006) Novel one-step real-time RT-PCR assay for rapid and specific diagnosis of Crimean–Congo hemorrhagic fever encountered in the Balkans. J Virol Methods 133:175–179
Escadafal C, Olschläger S, Avšič-Županc T, Papa A, Vanhomwegen J, Wolfel R, Mirazimi A, Teichmann A, Donoso-Mantke O, Niedrig M (2012) First international external quality assessment of molecular detection of Crimean–Congo hemorrhagic fever virus. PLoS Negl Trop Dis 6:1706–1714
Estrada-Pena A, Zatansever Z, Gargili A, Aktas M, Uzun R, Ergonul O, Jongejan F (2007) Modeling the spatial distribution of Crimean-Congo hemorrhagic fever outbreaks in Turkey. Vector Borne Zoonotic Dis 7(4):667–678
Filippone C, Marianneau P, Murri S, Mollard N, Avsic-Zupanc T, Chinikar S, Desprès P, Caro V, Gessain A, Berthet N, Tordo N (2013) Molecular diagnostic and genetic characterization of

highly pathogenic viruses: application during Crimean-Congo haemorrhagic fever virus outbreaks in Eastern Europe and the Middle East. Clin Microbiol Infect 19:118–128
Garrison AR, Alakbarova S, Kulesh DA, Shezmukhamedova D, Khodjaev S, Endy TP, Paragas J (2007) Development of a TaqMan minor groove binding protein assay for the detection and quantification of Crimean–Congo hemorrhagic fever virus. Am J Trop Med Hyg 77:514–520
Geisbert TW, Hensley LE, Gibb TR, Steele KE et al (2000) Apoptosis induced in vitro and in vivo during infection by Ebola and Marburg viruses. Lab Investig 80:171–186
Ghiasi SM, Salmanian AH, Chinikar S, Zakeri S (2011) Mice orally immunized with a transgenic plant expressing the glycoprotein of Crimean-Congo hemorrhagic fever virus. Clin Vaccine Immunol 18(12):2031–2037
Ghosh SN, Sokhey J, Dandawate CN, Gupta NP, Obukhova VR, Gaidamovich SY (1978) Arthropod-borne virus activity in migratory birds, Ghana Bird Sanctuary, Rajasthan, state, 1973. Indian J Med Res 68:192–196
van Gorp EC, Suharti C, ten Cate H, Dolmans WM et al (1999) Review: infectious diseases and coagulation disorders. J Infect Dis 180:176–186
Hatipoglu CA, Bulut C, Yetkin MA, Ertem GT et al (2010) Evaluation of clinical and laboratory predictors of fatality in patients with Crimean-Congo haemorrhagic fever in a tertiary care hospital in Turkey. Scand J Infect Dis 42:516–521
Hemachudha T, Griffin DE, Giffels JJ, Johnson RT, Moser AB, Phanuphak P (1987) Myelin basic protein as an encephalitogen in encephalomyelitis and polyneuritis following rabies vaccination. N Engl J Med 316:369–374
Hewson R, Gmyl A, Gmyl L, Smirnova SE, Karganova G, Jamil B, Hasan R, Chamberlain J, Clegg C (2004) Evidence of segment reassortment in Crimean-Congo haemorrhagic fever virus. J Gen Virol 85:3059–3070
Hinkula J, Devignot S, Akerstrom S, Karlberg H, Wattrang E, Bereczky S et al (2017) Immunization with DNA plasmids coding for Crimean-Congo hemorrhagic fever virus capsid and envelope proteins and/or virus-like particles induces protection and survival in challenged mice. J Virol 91(10):e02076–6
Honig JE, Osborne JC, Nichol ST (2004) Crimean-Congo hemorrhagic fever virus genome L RNA segment and encoded protein. Virology 321:29–35
Hoogstraal H (1979) The epidemiology of tick borne Crimean-Congo hemorrhagic fever in Asia, Europe and Africa. J Med Entomol 15:307–417
Ibrahim AM, Adam I, Osman BT, Aradaib IE (2015) Epidemiological survey of Crimean Congo hemorrhagic fever virus in cattle in East Darfur State, Sudan. Ticks Tick Borne Dis 6(4):439–444
Jabbari A, Besharat S, Abbasi A, Moradi A, Kalavi K (2006) Crimean-Congo hemorrhagic fever: case series from a medical center in Golestan province, Northeast of Iran (2004). Indian J Med Sci 60:327–329
Kamboj A, Pateriya AK, Mishra A, Ranaware P, Kulkarni DD, Raut AA (2014) Novel molecular beacon probe-based real-time RT-PCR assay for diagnosis of Crimean-Congo hemorrhagic fever encountered in India. Biomed Res Int 2014:496219
Kang SS, McGavern DB (2010) Microbial induction of vascular pathology in the CNS. J Neuroimmune Pharmacol 5:370–386
Ke R, Zorzet A, Göransson J, Lindegren G, Sharifi-Mood B, Chinikar S, Mardani M, Mirazimi A, Nilsson M (2011) Colorimetric nucleic acid testing assay for RNA virus detection based on circle-to-circle amplification of padlock probes. J Clin Microbiol 49:4279–4285
Kondiah K, Swanepoel R, Paweska JT, Burt FJ (2010) A simple-probe real-time PCR assay for genotyping reassorted and non-reassorted isolates of Crimean–Congo hemorrhagic fever virus in southern Africa. J Virol Methods 169:34–38
Kortekaas J, Vloet RP, McAuley AJ, Shen X, Bosch BJ, de Vries L et al (2015) Crimean-Congo hemorrhagic fever virus subunit vaccines induce high levels of neutralizing antibodies but no protection in STAT1 knockout mice. Vector Borne Zoonotic Dis 15:759–754
Kubar A, Haciomeroglu M, Ozkul A, Bagriacik U, Akinci E, Sener K et al (2011) Prompt administration of Crimean-Congo hemorrhagic fever (CCHF) virus hyperimmunoglobulin in patients diagnosed with CCHF and viral load monitorization by reverse transcriptase-PCR. Jpn J Infect Dis 64:439–433
Lukashev AN (2005) Evidence for recombination in Crimean-Congo hemorrhagic fever virus. J Gen Virol 86:2333–2338
Mishra AC, Mehta M, Mourya DT, Gandhi S (2011) Crimean-Congo haemorrhagic fever in India. Lancet 378(9788):372
Mourya DT, Yadav PD, Shete A, Majumdar TD, Kanani A, Kapadia D, Chandra V, Kachhiapatel AJ, Joshi PT, Upadhyay KJ, Dave P, Raval D (2014) Serosurvey of Crimean-Congo hemorrhagic fever virus in domestic animals, Gujarat, India, 2013. Vector Borne Zoonotic Dis 14(9):690–692
Mourya DT, Yadav PD, Shete AM, Gurav YK, Raut CG, Jadi RS, Pawar SD, Nichol ST, Mishra AC (2012) Detection, isolation and confirmation of Crimean-Congo hemorrhagic fever virus in

human, ticks and animals in Ahmadabad, India, 2010–2011. PLoS Negl Trop Dis 6(5):e1653
Mourya DT, Yadav PD, Shete AM, Sathe PS, Sarkale P, Pattnaik B, Sharma G, Upadhyay KJ, Gosavi S, Patil DY, Chaubal GY, Majumdar TD, Katoch VM (2015) Cross-sectional serosurvey of Crimean-Congo hemorrhagic fever virus IgG in livestock, India, 2013–2014. Emerg Infect Dis 21(10):1837–1839
Mousavi-Jazi M, Karlberg H, Papa A, Christova I, Mirazimi A (2012) Healthy individuals' immune response to the Bulgarian Crimean-Congo hemorrhagic fever virus vaccine. Vaccine 30:6225–6229
Müller MA, Devignot S, Lattwein E, Corman VM, Maganga GD, Gloza-Rausch F, Binger T, Vallo P, Emmerich P, Cottontail VM, Tschapka M, Oppong S, Drexler JF, Weber F, Leroy EM, Drosten C (2016) Evidence for widespread infection of African bats with Crimean-Congo hemorrhagic fever-like viruses. Sci Rep 6:26637
Nabeth P, Cheikh DO, Lo B, Faye O, Vall IOM, Niang M, Wague B, Diop D, Diallo M, Diallo B, Diop OM, Simon F (2004) Crimean-Congo hemorrhagic fever, Mauritania. Emerg Infect Dis 10(12):2143–2149
Oestereich L, Rieger T, Neumann M, Bernreuther C, Lehmann M, Krasemann S et al (2014) Evaluation of antiviral efficacy of ribavirin, arbidol, and T-705 (favipiravir) in a mouse model for Crimean-Congo hemorrhagic fever. PLoS Negl Trop Dis 8(5):e2804
Ozturk B, Kuscu F, Tutuncu E, Sencan I et al (2010) Evaluation of the association of serum levels of hyaluronic acid, sICAM-1, sVCAM-1, and VEGF-A with mortality and prognosis in patients with Crimean-Congo hemorrhagic fever. J Clin Virol 47:115–119
Papa A, Mirazimi A, Köksal I, Estrada-Pena A, Feldmann H (2015 Mar) Recent advances in research on Crimean-Congo hemorrhagic fever. J Clin Virol 64:137–143
Papa A, Papadimitriou E, Christova I (2011) The Bulgarian vaccine Crimean-Congo haemorrhagic fever virus strain. Scand J Infect Dis 43:225–229
Peyrefitte CN, Perret M, Garcia S, Rodrigues R et al (2010) Differential activation profiles of Crimean-Congo hemorrhagic fever virus and Dugbe virus-infected antigen-presenting cells. J Gen Virol 91:189–198
Rodrigues FM, Padbidri VS, Ghalsasi GR, Gupta NP, Mandke VB, Pinto BD, Hoon RS, Bapat MB, Mohan Rao CV (1986) Prevalence of Crimean haemorrhagic-Congo virus in Jammu and Kashmir state. Indian J Med Res 84:134–138
Rodrigues R, Paranhos-Baccala G, Vernet G, Peyrefitte CN (2012) Crimean-Congo hemorrhagic fever virus-infected hepatocytes induce ER-stress and apoptosis crosstalk. PLoS One 7:e29712
Sahib MM (2010) Rapid development of optimized recombinant adenoviral vaccines biosafety level 4 viruses. University of Manitoba, Winnipeg
Saidi S, Casals J, Faghih MA, Faghih AA (1975) Crimean hemorrhagic fever-Congo (CHF-C) virus antibodies in man, and in domestic and small mammals, in Iran. Am J Trop Med Hyg 24:353–357
Saijo M, Tang Q, Shimayi B, Han L, Zhang Y, Asiguma M, Tianshu D, Maeda A, Kurane I, Morikawa S (2005) Antigen capture enzyme-linked immunosorbent assay for the diagnosis of Crimean–Congo hemorrhagic fever using a novel monoclonal antibody. J Med Virol 77:83–88
Saksida A, Duh D, Wraber B, Dedushaj I et al (2010) Interacting roles of immune mechanisms and viral load in the pathogenesis of Crimean-Congo hemorrhagic fever. Clin Vaccine Immunol 17:1086–1093
Schmaljohn CS, Nichol ST (2007) Bunyaviruses. In: Knipe DM, Howley PM (eds) Fields virology, 5th edn. Lippincott Williams & Wilkins, Philadelphia, pp 1741–1790
Schnittler HJ, Feldmann H (2003) Viral hemorrhagic fever—a vascular disease? Thromb Haemost 89:967–972
Shanmugam J, Smirnova SE, Chumakov MP (1976) Presence of antibody to arboviruses of the Crimean Haemorrhagic Fever-Congo (CHF-Congo) group in human beings and domestic animals in India. Indian J Med Res 64:1403–1413
Shepherd AJ, Swanepoel R, Gill DE (1988) Evaluation of enzyme-linked immunosorbent assay and reversed passive hemagglutination for detection of Crimean–Congo hemorrhagic fever virus antigen. J Clin Microbiol 26:347–353
Shepherd AJ, Swanepoel R, Leman PA, Shepherd SP (1986) Comparison of methods for isolation and titration of Crimean–Congo hemorrhagic fever virus. J Clin Microbiol 24:654–656
Shepherd AJ, Swanepoel R, Shepherd SP, McGillivray GM, Searle LA (1987) Antibody to Crimean-Congo hemorrhagic fever virus in wild mammals from southern Africa. Am J Trop Med Hyg 36(1):133–142
Sidwell RW, Smee DF (2003) Viruses of the Bunya- and Togaviridae families: potential as bioterrorism agents and means of control. Antivir Res 57(1–2):101–111
Spengler JR, Bergeron É, Rollin PE (2016) Seroepidemiological studies of Crimean-Congo Hemorrhagic fever virus in domestic and wild animals. PLoS Negl Trop Dis 10(1):e0004210
Spik K, Shurtleff A, McElroy AK, Guttieri MC, Hooper JW, SchmalJohn C (2006) Immunogenicity of combination DNA vaccines for Rift Valley fever virus, tick-borne encephalitis virus, Hantaan virus, and Crimean Congo hemorrhagic fever virus. Vaccine 24(21):4657–4656

Swanepoel R, Shepherd AJ, Leman PA, Shepherd SP, McGillivray GM, Erasmus MJ, Searle LA, Gill DE (1987) Epidemiologic and clinical features of Crimean-Congo hemorrhagic fever in Southern Africa. Am J Trop Med Hyg 36:120–132

Tasdelen Fisgin N, Fisgin T, Tanyel E, Doganci L et al (2008) Crimean-Congo hemorrhagic fever: five patients with hemophagocytic syndrome. Am J Hematol 83:73–76

Tkachenko E, Khanun K, Berezin V (1969) Serological investigation of human and animal sera in agar gel diffusion and precipitation (AGDP) test for the presence of antibodies of Crimean hemorrhagic fever and Grand Arbaud viruses. Mater Nauch Sess Inst Polio Virus Entsef 2:265

Vescio FM, Busani L, Mughini-Gras L, Khoury C, Avellis L, Taseva E, Rezza G, Christova I (2012) Environmental correlates of Crimean-Congo haemorrhagic fever incidence in Bulgaria. BMC Public Health 12:1116

Vorou R, Pierroutsakos IN, Maltezou HC (2007) Crimean-Congo hemorrhagic fever. Curr Opin Infect Dis 20:495–500

Watts DM, Flick R, Peters CJ, Shope R (2005) Bunyaviral fevers: Rift Valley fever and Crimean-Congo hemorrhagic fever. In: Guerrant RL, Walker DH, Weller PF (eds) Tropical infectious diseases: principles, pathogens, and practice. Elsevier Churchill Livingstone, Philadelphia, PA, pp 756–760

Whitehouse CA (2004) Crimean-Congo hemorrhagic fever. Antivir Res 64:145–160

Wolfel R, Paweska JT, Petersen N, Grobbelaar AA, Leman PA, Hewson R, Courbot MCG, Papa A, Günther S, Drosten C (2007) Virus detection and monitoring of viral load in Crimean–Congo hemorrhagic fever virus patients. Emerg Infect Dis 13:1097–1100

Wölfel R, Paweska JT, Petersen N, Grobbelaar AA, Leman PA, Hewson R, Georges-Courbot MC, Papa A, Heiser V, Panning M, Günther S, Drosten C (2009) Low density macroarray for rapid detection and identification of Crimean–Congo hemorrhagic fever virus. J Clin Microbiol 47:1025–1030

World Health Organization (2011) WHO Fact Sheet No. 208. WHO, Geneva. http://www.who.int/mediacentre/factsheets/fs208/en/

World Health Organization, Regional Office for South-East Asia (2014) A brief guide to emerging infectious diseases and zoonoses. WHO, New Delhi

Xiao X, Feng Y, Zhu Z, Dimitrov DS (2011) Identification of a putative Crimean-Congo hemorrhagic fever virus entry factor. Biochem Biophys Res Commun 411(2):253–258

Yadav PD, Cherian SS, Zawar D, Kokate P, Gunjikar R, Jadhav S, Mishra AC, Mourya DT (2013) Genetic characterization and molecular clock analyses of the Crimean-Congo hemorrhagic fever virus from human and ticks in India, 2010–2011. Infect Genet Evol 14:223–231

Yapar M, Aydogan H, Pahsa A, Besirbellioglu BA, Bodur H, Basustaoglu AC, Guney C, Avci IY, Sener K, Setteh MH, Kubar A (2005) Rapid and quantitative detection of Crimean–Congo hemorrhagic fever virus by one-step real-time reverse transcriptase-PCR. Jpn J Infect Dis 58:358–362

Yilmaz GR, Buzgan T, Irmak H, Safran A, Uzun R, Cevik MA, Torunoglu MA (2009) The epidemiology of Crimean-Congo hemorrhagic fever in Turkey, 2002–2007. Int J Infect Dis 13(3):380–386

Zilinskas RA (1997) Iraq's biological weapons. The past as future? JAMA 278(5):418–424

Zivcec M, Scholte FE, Spiropoulou CF, Spengler JR, Bergeron E (2016) Molecular insights into Crimean-Congo hemorrhagic fever virus. Viruses 8(4):106

第12章 猪繁殖与呼吸综合征病毒

Tridib Kumar Rajkhowa

12.1 引言

猪繁殖与呼吸综合征(porcine reproductive and respiratory syndrome,PRRS)是一种能引起各年龄段猪的急性呼吸道疾和种猪繁殖障碍的疫病。该病于 20 世纪 80 年代末在美国养猪场中首次被发现,随后在欧洲出现(Botner 等,1997;Done 等,1996)。由于在发病初期不能确定病原体,最初 PRRS 被称为"猪神秘病"。病猪最初表现为急性型疾病,包括厌食、嗜睡、发热、呼吸困难、四肢发绀,耳朵常呈蓝紫色,因此,也被称为"猪蓝耳病"。种猪繁殖障碍通常发生在急性发病期或急性发病期后不久,主要表现为死产、木乃伊胎、弱仔和断奶前仔猪高死亡率(Meulenberg,2000;Rowland,2007)。这种疾病也被称之为"猪流行性流产与呼吸综合征"或"猪繁殖障碍和呼吸综合征"(Christianson 等,1992;Collins 等,1992;Wensvoort 等,1991)。

PRRSV 于 1991 年在荷兰首次被分离,命名为莱利斯塔病毒 (Terpstra 等,1991;Wensvoort 等,1991)。不久之后,美国通过病毒分离和实验室感染,确定了 PRRS 病原体(Benfield 等,1992;Collins 等,1992)。在两大洲同时出现 PRRSV 的起源尚不清楚,通常认为小鼠乳酸脱氢酶增高病毒和 PRRSV 有着共同的祖先 (Carman 等,1995;Hanada 等,2005;Plagemann,2003)。不过,关于 PRRSV 的分化时间及其出现在猪群前的发展史,还未达成共识(Forsberg,2005;Hanada 等,2005)。

之后,PRRSV 在全球大多数生产猪的国家被检测到。PRRSV 主要引起仔猪严重呼吸道疾病的亚临床感染,种猪繁殖障碍(Baron 等,1992;Dea 等,1992;Hopper 等,1992;Jiang 等,2000;Plana-Duran 等,1992;Shimizu 等,1994)。1996 年前后,美国暴发了几次异常严重的 PRRS,导致种猪发生高流产率和高死亡率。近期,高致病性强毒株出现,导致中国及其周边国家暴发重大疫情(Halbur 和 Bush,1997;Hurd 等,2001;Normile,2007;Tian 等,2007)。自 PRRS 出现以来,已经给养猪业造成了重大的经济损失,在美国,PRRSV 每年造成的经济损失约为 5 亿美元(1 美元≈6.7 元)(Neumann 等,2005;Holtkamp 等,2013)。尽管已经采取了相关防控措施,但它仍将继续影响着全球养猪产业发展。

12.2 病毒结构

PRRSV 属于套式病毒目,动脉炎病毒科(Gorbalenya 等,2006)。动脉炎病毒科包括马动脉炎病毒、猴出血热病毒和乳酸脱氢酶增高病毒。这些病毒的宿主各不相同,PRRSV 感染猪,马动脉炎病毒感染马和驴,猴出血热病毒感染猴子,乳酸脱氢酶增高病毒感染小鼠 (Snijder 和 Meulenberg,1998)。根据 PRRSV 抗原性和基因组的差异,分为两个不同的基因型:欧洲型(Ⅰ型)和北美型(Ⅱ型)(Meng 等,1995)。这两个基因型在易感猪群中引起疾病的症状相似(Kapur 等,1996;Meng 等,1995;Meng,2000;Meulenberg,2000;Nelsen 等,1999;Ropp 等,2004;Stadejek 等,2006)。

PRRS 病毒粒子呈球形,直径为 50~60nm,外表面相对光滑(图 12.1)。病毒基因组被核衣壳蛋白包裹,外绕脂质双层囊膜,表面糖蛋白和细胞表面膜蛋白相互融合。PRRSV 基因组长度约为 15kb;是单股正链 RNA 病毒,3'端有多聚腺苷酸(poly A)尾巴;包含 11 个 ORF(表 12.1)。多顺反子基因组 5' 端包含两个阅读框,ORF la 和 1b,负责编码两个非结构蛋白 PP la 和 PP lab,非结构性的表达依赖于 ORF la/ORF 1b 重叠区域的 a-1 核糖体阅读框架移码信号。两种多聚蛋白 ppla 和 ppl ab 合成后,经 nspla、nsplp、nsp2 和 nsp4 剪切,至少形成 14 个非结构蛋内。由于非结构蛋白 2 (nsp2)中心区域的高度变异性,使得 ORF la 大小也不一致。近期在 ORF 1a 中心区域发现了一个新的 ORF(TF)和-1/-2 程序化的核糖体阅读框架移码信号,表达 nsp2TF 和 nsp2N 两种新的蛋白质(Fang 等,2012;Li 等,2014)。病毒基因组 3'端包

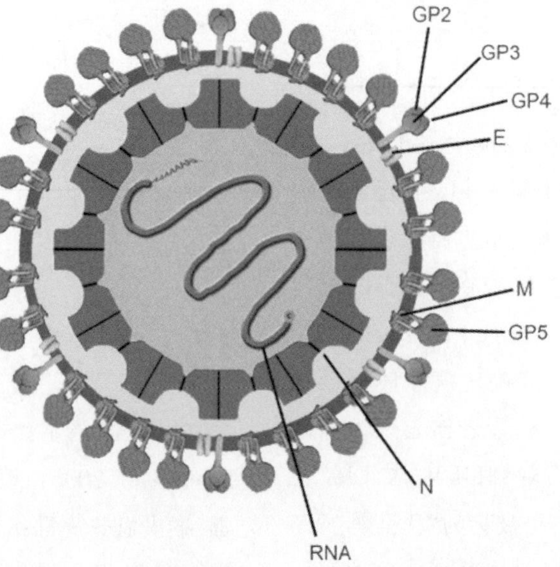

图 12.1 PRRS 病毒粒子图示。(扫码看彩图)

含 8 个相对较小的基因，除 PRRSV 2 型的 ORF 4/ORF 5 外，这些基因的 5'和 3'末端序列与邻近基因重叠，3'端基因编码四种膜糖蛋白(GP2a、GP3、GP4 和 GP5)、三种非糖基化膜蛋白(E、ORF 5a 和 M)和核衣壳(N)蛋白(图 12.2)。

12.3 分子流行病学

20 世纪 80 年代末 90 年代初，PRRSV 几乎是在北美(基因 2 型)和西欧(基因 1 型)同时出现的(Keffaber，1989；Wensvoort 等，1991)。从两大洲分离到 PRRSV 的核苷酸同源性只有 55%~70%(Allende 等，1999；Nelsen 等，1999)。根据两个谱系的遗传进化距离，提出假设，这两个谱系是从一个非常遥远的共同祖先各自进化而来的(Plagemann，2003)。乳酸脱氢酶增高病毒被认为与小动脉病毒科中 PRRSV 的亲缘关性最近，推测，PRRSV 有一个类似于乳酸脱氢酶增高病毒的祖先在啮齿动物中流行(Plagemann，2003)。PRRSV 祖先最初在啮齿动物中传播，后来传播至亚欧野猪身上。野猪作为中间宿主，把病毒从啮齿动物传播到家养猪群中。进一步推测，PRRSV 两种基因型差异性起始于野猪从欧洲引入到美国之时(Plagemann，2003；Mayer 和 Brisbin，1991)，PRRSV 引入后，分别在亚欧大陆和北美大陆野猪种群中开始进化，而后又分别进入家猪种群中进行进化(Plagemann，2003)。

12.3.1 1 型 PRRSV 的变异与进化

20 世纪 90 年代早期在西欧首次记载了 1 型 PRRSV 流行(Wensvoort 等，1991)。莱利斯塔病毒是 1 型 PRRSV 原始毒株，最先在荷兰被分离到，与同期在比利时、法国、德国、英国、荷兰和西班牙分离到的毒株共同形成一个高度同源的种群 (类似莱利斯塔)(Drew 等，1997；Forsberg 等，2002；Le Gall 等，1998；Suarez 等，1996)。不过，类似莱利斯塔种既不能代表基因 1 型 PRRSV 的多样性，也不能视为目前衍生出的大部分基因 1 型 PRRSV 的祖先病毒类型。类莱利斯塔病毒被分离出来后，样本范围不断扩大，在系统进化分析中表现出良好的遗传进化分支的差异性，这表明它们不可能是类莱利斯塔病毒 PRRSV“的后代”。因此证实了 1 型 PRRSV 在西欧“最初”病毒流行之前就已经存在。

根据 ORF 5 序列差异和系统发育分类，1 型 PRRSV 被分为 4 个亚型。近期，ORF 7 大小多态

表 12.1　PRRSV 蛋白及其功能

基因和表达蛋白			蛋白大小		
序号	基因	蛋白	基因Ⅰ型(LV,基因库号96262)	基因Ⅲ型(VR-2332,基因库号AY150564)	已知或预测功能
1	ORF1a	nsp1α	180	180	包含类木瓜半胱氨酸蛋白酶(PLPα);锌指蛋白;亚基因组 mRNA 合成调节因子;潜在 IFN 拮抗剂
		nsp1β	205	203	包含 PLPβ 蛋白酶;潜在干扰素拮抗剂
		nsp2	1078	1196	包含 PLP2 蛋白酶;去泛素化蛋白酶;潜在干扰素拮抗剂;参与细胞膜修饰形成复制复合体
2	ORF 1a'-TF	nsp2TFc	902	1019	包含 PLP2 区
		nsp2TN*	733	850	包含 PLP2 区
3	ORF 1a	nsp3	230	230	跨膜区参与细胞膜修饰形成复制复合体
		nsp4	203	204	主要丝氨酸蛋白酶(SP);细胞凋亡诱导物;潜在干扰素拮抗剂
		nsp5	170	170	可能参与膜修饰的 TM 蛋白
		nsp6	16	16	?
		nsp7α	149	149	重组 nsp7 具有强抗原性
		nsp7β	120	110	
		nsp8	45	45	nsp9 的 N 端区
4	ORF 1b	nsp9	685	685	依赖于 RNA 的 RNA 聚合酶
		nsp10	442	441	RNA 核苷水解酶/解螺旋酶;包含假定的锌结合结构域
		nsp11	224	223	尿苷的特异性核糖核酸内切酶
		nsp12	152	153	?
5	ORF 2a	GP2a	249	256	小糖基化结构蛋白;为病毒感染所必需;以 GP3-4 的多聚体形式整合到病毒粒子中;病毒黏附蛋白
6	ORF 3	E	70	73	小非糖基化和十八烷基化结构蛋白;为病毒感染所必需;以多聚体形式整合到病毒粒子中;具有类离子通道特性,在病毒囊膜上起到病毒孔蛋白作用
7	ORF 4	GP3	265	254	小糖基化结构蛋白;为病毒感染所必需;具有高度抗原性并且可能参与病毒中和反应;以 GP2a 和 GP4 多聚体复合物的形式与病毒粒子整合;GP3 的一个亚群可以分泌为非病毒相关的可溶性蛋白

(待续)

表 12.1(续)

	基因和表达蛋白		蛋白大小		
序号	基因	蛋白	基因Ⅰ型(LV,基因库号 M96262)	基因Ⅲ型(VR-2332,基因库号 AY150564)	已知或预测功能
8	ORF 5	GP4	183	178	小糖基化结构蛋白;为病毒感染所必需;以 GP2a-3-4 多聚体复合物的形式与病毒粒子整合;病毒吸附蛋白,可能参与病毒中和反应
9	ORF 6	GP5	201	200	主要糖基化结构蛋白;具有数量不定的潜在 N-糖基化位点的跨膜蛋白;最可变的结构蛋白;与 M 蛋白联结形成二硫化物异源二聚体
10	ORF 7	ORF5a	43	41	小非糖基化、疏水性结构蛋白;为病毒存活所必需;以多聚体复合物的形式整合到病毒粒子中
11	ORF 8	M	173	174	高度保守的非糖基化结构蛋白;GP5-M 异源二聚作是病毒感染所必需的;在病毒装配和出芽过程中发挥关键作用
12	ORF 9	N	128	123	非糖基化磷酸化结构蛋白;病毒衣壳蛋白组成成分;高度抗原性;潜在 IFN 拮抗剂

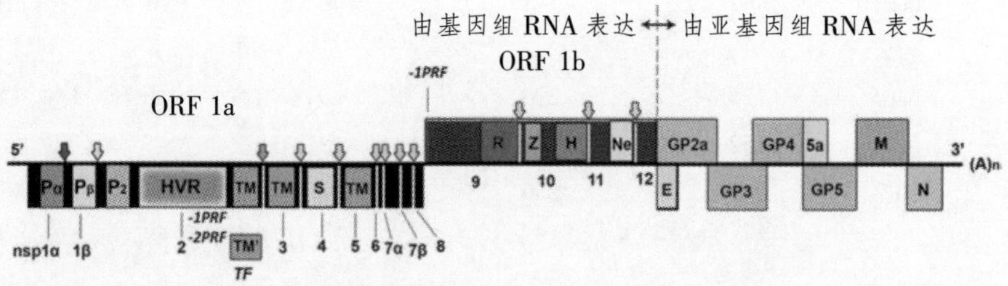

图 12.2 PRRS 病毒基因组结构。

性被推荐为一种稳定和独立的分型指标(Stadejek 等,2008),也证实了三个先前确定的亚型。这包括泛欧洲亚型Ⅰ、亚型Ⅱ(主要来自白俄罗斯、立陶宛和俄罗斯)和亚型Ⅲ(主要来自白俄罗斯,核蛋白大小分别为 128、125 和 124 个氨基酸)(Stadejek 等,2008)。尽管基因 1 型 PRRSV 主要分布在欧洲,但现在也被引入到 5 个非欧洲国家,美国(Fang 等,2007;Ropp 等,2004)、加拿大(Dewey 等,2000)、韩国 (Lee 等,2010)、中国和泰国(Thanawongnuwech 等,2004)。

12.3.2 2 型 PRRSV 的遗传变异与进化

2 型 PRRSV 的最早分离株 VR-2332 于 20 世纪 80 年代晚期在北美暴发流行时被分离到(Collins 等,1992;Hill,1990;Keffaber,1989)。从 1989 年到 20 年代 90 年代早期,在美国分离到的 PRRSV 大部分毒株与 VR-2332 有着很近的亲缘相关 (Kapur 等,1996;Meng 等,1995;Wesley 等,

1998;Shi 等,2010)。1996 年晚夏,在美国艾奥瓦州和其他州多次暴发了种猪以高流产率和高死亡率为特征的 PRRS。无论是单独接种 Ingelvac PRRS 改良式活毒疫苗，还是同时接种 Ingelvac PRRS 和 PrimePac PRRS 疫苗，仍旧发生 PRRSV 暴发流行(Bush 等,1999;Halbur 和 Bush,1997;Key 等,2001)。两种减毒活疫苗之所以没有产生免疫保护作用,是因为当时流行的 PRRSV 毒株与两种活疫苗毒株基因不同（Key 等,2001)。2001 年末,在美国明尼苏达州发现了 PRRSV 变异的强毒株(Han 等,2006),与 PRRSV 的 VR-2332 株相比，这些分离株（MN184 A、B 和 C)ORF 复制酶 NSP2 分别在 VR-2332 对应的位置 324-434 位、486 位和 505-523 位发生了不连续的 11 个、1 个和 19 个氨基酸缺失(Han 等,2006)。为何会突然出现这些毒株仍然是个谜。

2 型 PRRSV 最初在北美流行，并被鉴定,因此被称之为北美型 PRRSV。近期，根据 2 型 PRRSV 序列分析和构建的系统发育树,其形成了三个分支，这三个分支主要由北美分离株和亚洲地区高度变异的分离株集群(一个分支为PrimePac PRRS 疫苗株,一个分支为加拿大、泰国分离株序列，一个分支为 VR-2332 株和 Ingelvac PRRS 改良式活毒疫苗株）和高度多样化的几个亚洲集群(Wang 等,2008)。近期以 ORF 5 所有高质量序列为基础(n=8500)(Shi 等,2010),运用贝叶斯法对 2 型 PRRSV 分离株进行研究,2 型 PRRSV 被分为 9 个谱系,包括 5 大集群(n>1000)和 4 个小集群，相邻谱系遗传距离大于 10%。在已定义的 9 个谱系中,在北美发现了 7 个,剩下的 2 个谱系仅在亚洲国家中被发现(Shi 等,2010),7 个北美谱系在亚洲和欧洲国家也经常出现，造成局部疫情暴发(An 等,2007;Cha 等,2004;Chen 等,2006;Kang 等,2004;Kim 等,2009;Madsen 等,1998;Shi 等,2010;Thanawongnuwech 等,2004)。在所有谱系中,含有 Ingelvac PRRS 改良式活毒疫苗株的谱系 5 在世界分布最为广泛，已经传播到北美大陆以外的 8 个国家。

12.3.3 高致病性 PRRSV

高致病性 PRRS 临床主要表现为高热、皮肤发红和各年龄段猪高死亡率。印度在 2013 年米佐拉姆猪群中首次暴发 PRRS 疫情（Rajkhowa 等，2015a)，后来在 2015 年、2016 年和 2017 年也暴发过几次疫情，后来确定暴发的这些疫情也是由高致病性 PRRS 引起的(Rajkhowa 等,2015a,b,2016,2018;Gogoi 等,2017)。

12.4 传播、感染及发病机制

PRRSV 的传播路径可能与距离相关(局部传播),也可能与距离无关。当传播路径与距离相关时,即使在两个农场之间没有互动交流,也可从一个猪群传播到周边猪群中，现在认为作为传播媒介的气溶胶(Brockmeier 和 Lager,2002)、动物、昆虫(Otake 等,2003),以及污染物(Dee 等,2003;O-take 等,2003)等是造成局部传播的原因。当传播路径与距离无关时，两地之间互动交流是前提条件,这种传播途径与猪的运输、受精(Christopher-Hennings 等,2008),应用疫苗的毒力有关。此外，在给药和接种疫苗时如果不勤换针头，也会加剧传播(Otake 等,2003)。

完全分化的猪肺泡巨噬细胞是PRRSV 主要的靶细胞，但 PRRSV 对单核细胞系不易感(Duan 等,1997)。据报道，树突状细胞也能够支持 PRRSV 的复制(Loving 等,2007)。通过研究发现，只有非洲绿猴肾细胞 MA-104 及其衍生细胞 MARC-145 能够完全支持 PRRSV 体外复制。PRRSV 通过标准网格蛋白介导的内吞作用进入宿主细胞,在核内体酸化和膜融合后,病毒基因组被释放到细胞质中(Nauwynck 等,1999)。CD163 是介导病毒内化和裂解的主要受体（Welch 和 Calvert,2010)。通过小结构蛋白 GP2a 和 GP4 与 CD163 间相互作用介导病毒进入易感宿主细胞

(Das 等,2010;Welch 和 Calvert,2010)。病毒通过受体介导的内吞作用进入细胞,进行裂解,翻译复制酶多聚蛋白 ppa-nsp2TF、ppa-nsp2N、pp1a 和 pp1ab。这些多聚蛋白被病毒自身蛋白酶裂解,至少产生 14 个非结构蛋白,这些蛋白被组装成一个复制转录复合物。复制转录复合物首先参与合成负链 RNA,产生单链全长负链 RNA 和亚基因组长度的负链 RNA。随后,亚基因组 mRNA 作为正链亚基因组 mRNA 的合成模板,正链亚基因组 mRNA 参与基因组 3’端近 1/4 处结构蛋白基因的表达。新生成的 RNA 基因组被包装成核衣壳,核衣壳被光滑的细胞内膜以出芽的方式包装成新的病毒粒子,新包装的病毒粒子通过胞吐从细胞中释放出来。

PRRSV 感染可分为三个阶段,急性感染期、持续感染期和消失。在免疫学、病毒学和临床疾病方面,每个阶段都具有其独特性,急性感染期,肺是主要感染部位,表现为急性呼吸道症状。PRRSV 主要在肺和上呼吸道的巨噬细胞和树突状细胞中复制,病毒感染后 6~12 小时引发病毒血症。尽管有血清抗体的存在,但病毒血症仍要持续数周。第二阶段的主要特征是病毒在淋巴器官定居和复制,包括淋巴器官,扁桃体和淋巴结,但不包括脾脏,继而导致持续感染(Wills 等,1997;Allende 等,2000;Rowland 等,2003)。在这个阶段,在血液和肺部中检测不到该病毒,猪也不表现出明显的临床症状。然而,病毒在该区域淋巴结内仍继续复制,并通过口鼻分泌物和精液传播给初生猪。在感染最后阶段病毒复制逐渐衰减,直至病毒在宿主中消失。病毒感染后,其复制可持续进行 250 天(Wills 等,2003),因此 PRRSV 可导致“终身”感染。

12.5 感染免疫

PRRSV 感染后 7~9 天,猪体内快速产生大量抗蛋白 N 抗体,少量抗蛋白 M 抗体,这些抗体均是非中和抗体(Loemba 等,1996),中和抗体通常在 PRRSV 感染后 28 天产生,主要是抗 GP5 的抗体。一般情况下,中和抗体对同源毒株具有中和活性,而对异源毒株的交叉中和活性较低甚至没有(Vu 等,2011;Zhou 等,2012)。

12.5.1 先天性免疫

先天免疫系统是宿主抵御病毒感染的第一道防线。充分激活宿主抗病毒感染的先天免疫反应,尤其是启动强大后天免疫反应对抗细胞内病原体,是机体抑制病毒在体内复制、防御病毒入侵黏膜组织的关键(Koyama 等,2008)。PRRSV 通过免疫调节引起先天性免疫应答和适应性免疫应答低下,导致病毒不能被完全清除(Albina 等,1998a,b;Renukaradhya 等,2010)。自然杀伤细胞是一种先天淋巴细胞亚群,非特异性清除体内病毒感染细胞,PRRSV 感染对自然杀伤细胞的杀伤活性具有显著抑制作用,最早可在感染后第 2 天显示出抑制作用,可持续 3~4 周(Renukaradhya 等,2010)。PRRSV 感染最显著的特征之一是不能引起任何重要的炎性细胞因子表达,尤其是Ⅰ型干扰素、白细胞介素-1(IL-1)和 TNF-α(genini 等,2008;love 等,2007;Lee 等,2004)。无论是在体内,还是在体外,在 PRRSV 感染肺泡巨噬细胞位置,只能引起少量 TNF-α 的产生。研究证明,Ⅰ型 IFN 的表达下调,尤其是 TNF-α 表达下调,被认为是 PRRSV 发病机制的关键环节,因为 TNF-α 对 PRRSV 复制具有抑制作用(Albina 等,1998a,b;Le Bon 等,2001)。

PRRSV 对发育成熟和未成熟的猪树突状细胞均具有感染性(Loving 等,2007;Flores-Mendoza 等,2008;Chang 等,2008;Charerntantanakul 等,2006b;Wang 等,2007;Park 等,2008),可引起细胞凋亡,CD11b/c、CD14、CD80/86 及 MHC 1 型、2 型分子表达下调;减少 T 细胞的同种异体的刺激;IL-10、IL-12 和 TNF-α 表达上调(Genini 等,2008;Flores-Mendoza 等,2008;Wang 等,2007;Park 等,

2008)。据观察,PRRSV 可能通过改变巨噬细胞和树突状细胞的细胞因子产生、调节抗原提呈分子的表达,抑制先天性和特异性免疫反应。最终自然杀伤细胞和免疫系统中致敏细胞的活化延迟,而导致中和抗体产生、淋巴增殖反应,以及 IFN-γ 对抗 PRRSV 反应均延迟 (Flores-Mendoza 等,2008;Chang 等,2008;Butler 等,2008)。初期机体对 PRRSV 先天性免疫应答反应很弱,导致病毒在感染猪体内长期存在。

12.5.2 获得性免疫

PRRSV 感染后的免疫保护抗体水平存在着争议。自然感染后至少 3 个月免疫水平达到高峰,且不能完全抵御 PRRSV 再次感染,特别是 PRRSV 异源毒株引起的感染(Murtaugh 等,2002;Zuckermann 等,2007)。由于流行的 PRRSV 野毒株具有广泛的遗传性和抗原高度变异性,使得体液免疫和细胞免疫对流行的 PRRSV 野毒株起不到保护作用。

一般在 PRRSV 感染后 5~14 天可以检测到抗体,感染后 4 周左右,抗体效价迅速达到最高水平(Albina 等,1998b;Batista 等,2004;Diaz 等,2005;Labarque 等,2000)。IgM 在感染后 14~21 天达到峰值,然后迅速下降,直至检测不到。特异性抗体 IgG(IgG1 和 IgG2)在感染后第 3 周或第 4 周达到最高值,并在几个月内都保持较高水平(Labarque 等,2000;Loemba 等,1996;Mulupuri 等,2008)。然而,即使在 PRRSV 持续感染期间,抗体水平和浆细胞数也会下降,且再次接触病毒通常也不会引起免疫记忆反应。虽然 PRRSV 特异性抗体在幼龄动物中产生过程表现为正常,但在保护期却出现了不同的情况。被 PRRSV 感染的巨噬细胞对抗体依赖-补体介导的裂解是耐受的,而且极有可能对抗体依赖细胞介导的细胞毒作用也是耐受的,因为病毒蛋白在这些细胞表面并不表达,或者至少用抗体检测不到这些蛋白表达(Costers 等,2006)。

相反,利用 MARC-145 细胞和猪肺泡巨噬细胞实现了 PRRSV 与血清抗体和纯化抗体在体外的中和反应,说明抗体与游离病毒粒子间能发生相互作用。然而,利用体外病毒中和实验检测血清中的中和抗体时,发现在感染 PRRSV 的幼畜体内,病毒中和抗体的出现速度缓慢且数量不多。PRRSV 的中和抗体产生之所以延迟,推测可能是因为 GP5(27-77 氨基酸)上存在一个优势的“诱骗性”抗原表位,在早期诱导发生不具有保护作用的强免疫应答,封闭或削弱了主要的中和抗原表位免疫应答反应(Ostrowski 等,2002)。另一种关于病毒逃逸中和抗体,以及病毒在体内持续存在的解释是,由于膜糖蛋白的 N 端糖基化反应,引起的类似于人免疫缺陷病毒和灵长类免疫缺陷病毒的“聚糖屏蔽” 效应(Wei 等,2003)。此外,病毒进入靶细胞的抗体依赖性增强作用(Cancel-Tirado 等,2004)、先天性免疫应答抑制 (Sang 等,2011)和正常 B 细胞发育受阻(Butler 等,2014)都被证明是导致获得性免疫对 PRRSV 感染抵抗力减弱的原因。

12.6 临床症状

PRRS 的临床症状主要是由急性病毒血症和胎盘传播引起的繁殖障碍,PRRSV 进入非免疫猪群,就会在临床上引起 PRRSV 流行。PRRS 地方流行毒株对同源 PRRSV 毒株的感染具有免疫保护作用,只在易感的亚种群中表现出临床症状(Zimmerman 等,2006)。

PRRS 流行的第一阶段可持续 2 周甚至更久,由于急性病毒血症,各年龄段感染均表现出厌食、嗜睡、体温升高(40.5~41.6℃)、呼吸困难,出现短暂的皮肤充血或肢体发绀。第二阶段表现为母猪繁殖障碍,以及断奶前仔猪死亡率高,这个阶段可持续 1~4 个月。

被感染的成年动物表现为厌食症、高热和嗜睡(Keffaber,1989;Loula,1991),有时也可观察到皮下和后肢水肿、神经症状，以及耳和外阴发绀(Hopper 等,1992;Rossow 等,1998)。PRRSV 对成年猪的感染尤其妊娠后期的母猪感染，是不可治愈的。母猪繁殖障碍包括受孕率、产仔率低,活仔猪数少,弱仔、死胎、木乃伊胎数量多(Zimmeberman,2006)。病毒对哺乳期母猪与妊娠期母猪的健康状况具有同样影响，表现为食欲不振和发热引起无乳症，进而导致仔猪饥饿及其因缺少母源抗体保护而导致其他疾病发生,如大肠杆菌病,严重时,断奶前仔猪死亡率可超过 80%(Christianson 等,1992)。公猪感染,除具有急性呼吸症状外,还存在性欲降低和精液质量差,即精子活力低和发育缺陷(Prieto 和 Castro,2005;Zimmerman 等,2006)。

PRRSV 感染的早产猪和正常足月出生的猪在断奶前的死亡率都较高,临床表现为精神沉郁、腹泻、消瘦、姿势异常、呼吸急促和困难,但肌肉震颤、四肢划动、贫血、多发性关节炎及脑膜炎等情况不常见(Zimmerman 等,2006)。新生仔猪临床症状还表现为双眼肿胀、结膜炎、耳朵发蓝、皮肤瘀血、腹泻、被毛粗乱和注射后大量出血(Rossow,1998)。PRRSV 感染断奶后仔猪表现出不同的临床症状(Rossow 等,1994),表现为厌食、嗜睡、皮肤充血、呼吸急促、呼吸困难、不咳嗽、被毛粗乱、死亡率高(Zimmerman 等,2006)。

PRRSV 一旦传入农场,将在易感猪群中的保育猪、育肥猪、后备母猪、后备公猪中定期或偶然发生典型的急性型 PPRSV 流行(Stevenson 等,1993;Zimmerman 等,2006)。保育猪在冬季感染 PRRSV,导致死亡率高,这表明环境温度低、温度波动幅度大、通风率低和相对湿度高等环境因素可能引起病毒的再循环和动物机体免疫力降低(Stevenson 等,1993)。猪的繁育水平主要取决于被感染猪的数量及其被感染时所处的妊娠阶段,母猪感染,会出现零星流产、不规律的发情、不孕和妊娠后期繁殖障碍(Zimmerman 等,2006)。

12.7 病理变化

PRRSV 的特征性病变是仔猪呼吸道综合征，感染 4~28 天仔猪,甚至更大的仔猪,在肺部和淋巴结中可观察到肉眼可见的病理学变化，在显微镜下观察到病理组织学变化，大多数病毒在肺部和淋巴结中复制。PRRSV 毒株毒力决定了感染猪病变的严重程度和病变分布。病理变化提示可能感染了 PRRSV，但还不能依据病理变化进行确诊,由于其他病毒、细菌也可引起类似病理变化。虽然 PRRSV 特征性病变和病理组织学变化可以作为诊断 PRRSV 强有力的依据,但最终确诊还是要取决于 PRRSV 的检测(Mengeling 和 Lager,2000;Zimmerman 等,2006)。

3~28 日龄的猪被 PRRSV 感染会发展成为间质性肺炎,尤其是在 10~14 日龄的猪病理变化更为严重,表现为肺实质有弹性、不塌陷、稍硬、有斑点或散在红褐色斑、湿润(Zimmerman 等,2006)。由于巨噬细胞、淋巴细胞、浆细胞浸润,水肿,充血,以及Ⅱ型肺细胞增生,可在显微镜下观察到肺泡间隔增厚。肺泡可能含有坏死的细胞碎片、巨噬细胞和渗出液，而淋巴细胞和浆细胞则在呼吸道和血管周围形成管套。淋巴结病理变化为棕褐色至灰色,表现为肿大和水肿,通常是正常淋巴结的数倍(Mengeling 和 Lager,2000)。病理组织学病变主要在生发中心,表现为坏死和消失。淋巴皮质可能含有小的囊性空腔，囊性空腔是由内皮细胞不规律排列而成的,且含有蛋白质的液体、淋巴细胞和多核原核细胞。在心脏可能引发轻度至中度的多灶性淋巴细胞性血管炎和血管周围心肌炎。在脑内可发展为淋巴性的脑白质炎或累及小脑、大脑和脑干的脑炎。淋巴细胞和巨噬细胞在血管周围形成不连续分节的管套,以及出现多灶性神经胶质过多症。在肾小球和肾小管周可见轻度的淋巴组织细胞聚集。受到影响的血管内皮细胞肿胀,皮下蛋白液渗出，淋巴细胞和巨噬细胞在血管壁内及其周围聚集。PRRSV 感染的同一窝猪中有数量不等的正常猪、弱仔，死胎、发生自溶的胎，木乃伊胎。总体上，胎儿病变表现为肾周围水肿、脾韧带水肿、肠系膜水肿、腹腔和胸腔积液，有时会在弱仔和死胎脐带中出现节段性出血肿大(Zimmerman 等，2006)。感染母猪子宫内可见子宫肌层水肿、子宫内膜的血管周围伴有淋巴组织细胞形成的管套。小血管内可能有不连续的淋巴组织细胞性血管炎，子宫内膜上皮与胎盘滋养层之间可能轻度分离。公猪被 PRRSV 感染后 7~25 天，可观察到睾丸的精小管萎缩，这与生殖细胞的凋亡和衰竭有关(Zimmerman 等，2006)。

12.8 诊断

根据种猪的繁殖障碍或任何年龄段猪的呼吸道症状，可对 PRRS 病毒感染初步诊断。PRRS 感染通常不会引起特征性的肉眼可见的病理学变化或微观病理组织学变化，而且与其他病原体感染引起的病变相似或表现不明显。因此，PRRS 的最终确诊需要与猪细小病毒(porcine parvovirus)、伪狂犬病毒(pseudorabies virus)、血凝性脑脊髓炎病毒 (haemagglutinating encephalomyelitis virus)、猪圆环病毒 2 型(porcine circovirus type 2)、猪肠道病毒(porcine enterovirus)、猪流感病毒(swine inffuenza virus)、典型猪瘟（猪霍乱）病毒[classical swine fever(hog cholera)virus]、猪巨细胞病毒(porcine cytomegalovirus)和钩端螺旋体病(leptospirosis) 进行鉴别诊断(Keffaber，1989；Halbur 等，1993，1995；Paton 等，1992a，b)。当临床症状和病理学变化出现 PRRS 特征时，只有对病毒抗原、病毒基因组或从临床样本中分离出的病毒进行检测，才能确诊。另外，发病期间血清抗体水平升高对诊断也能起到支持作用。

12.8.1 抗原检测

利用荧光抗体和免疫组化对冷冻组织切片进行检测，能够检测出组织中 PRRSV 抗原。尽管荧光抗体检测具有特异性，但荧光抗体检测容易受样品质量影响且敏感性较差。相比之下，通过免疫组化检测福尔马林固定组织中的病毒更为有效。对冷冻组织切片中的病毒检测，免疫组化比荧光抗体具有更好的敏感性。因此，可以通过联合运用免疫组化与荧光抗体检测实现对 PRRS 特征性微观病变的检测，实现最终确诊。检测的首选组织样本为肺、淋巴结、脾脏、胸腺和扁桃体(Halbur 等，1996；Rossow 等，1998)。

RT–PCR 检测方法具有高度敏感性和高度特异性，已被广泛用于临床标本中 PRRSV 基因组的检测(Benson 等，2002；Horter 等，2002)。RT–PCR 检测方法不需要利用细胞对病毒进行分离培养，就可实现对病毒的 RNA 检测，检测耗时比病毒分离培养短。目前以 PRRSV 基因 ORF 7、ORF 6、ORF 5 或 ORF 1b 为靶基因，已经开发多种 RT–PCR 的检测方法，可直接用于临床样本检测(Benson 等，2002；Chen 和 Plagemann，1995；Christopher –Hennings 等，1995a；Cook 和 Spatz，1998；Egli 等，2001；Gilbert 等，1997；Legeay 等，1997；Mardassi 等，1994；Oleksiewicz 等，1998；Shin 等，1997；Spagnuolo Weaver 等，1998；Suarez 等，1994；Rajkhowa 等，2016)。为了提高检测敏感性，一些检测方法采用“巢式”PCR 检测(RT–nPCR)。近期，自动荧光 PCR 检测，例如，TaqManTMPCR (Egli 等，2001；Spagnuolo Weaver 等，2000)或“分子信标”PCR(Carlson 等，2002)，已经开发用于临床样本中的 PRRSV 检测。目前 PCR 检测方法已经被广泛地应用于 PRRS 诊断和协助种群监测(Bierk 等，2001；Dee 等，2001；Horter 等，2002；Kleiboeker 等，2002)。

12.8.2 抗体检测

ELISA、间接免疫荧光抗体试验、血清病毒中和试验和免疫过氧化物酶单层细胞试验均已被用于 PRRSV 特异性抗体检测。用于北美株和欧洲株的 PRRSV 特异性抗体检测的 ELISA 方法，具有快速、敏感和特异的优点(Albina 等，1992；

Edwards 等,1994;Nodelijk 等,1996;O'Connor 等,2002;Takikawa 等,1996)。多种 ELISA 方法已经被介绍,包括使用检测样本与阳性标本比值(S/P)的系统间接 ELISA(Yoon 等,1995)、直接使用 OD 值的间接 ELISA(Albina 等,1992;Cho 等,1996;Takikawa 等,1996)和液相阻断 ELISA(Ferrin 等,2002;Houben 等,1995;Zhou 等,2001)。

免疫荧光抗体试验具有较高的特异性(99.5%),但对于个体动物检测的敏感性还不清楚(Yoon 等,1992a,b),与 ELISA 相比,免疫荧光抗体试验的优点在于能够测定抗体效价,可以在感染后 2~3 个月检测出特异性抗体 (Frey 等,1992;Yoon 等,1995;Bautista 等,1993)。免疫过氧化物酶单层细胞试验也被认为是一种具有高度特异性和高度灵敏性的检测方法,在 PRRS 感染后 7~15 天,就能够检测到特异性抗体,甚至感染后 2~3 个月也能检测到(Frey 等,1992;Ohlinger 等,1992;Wensvoort 等,1991;Yoon 等,1995)。检测中使用的毒株与感染猪的毒株之间的亲缘关系可能会对免疫过氧化物酶单层细胞试验检测结果产生影响 (Wensvoort 等,1992;Yoon 等,1997)。

血清病毒中和试验不如免疫荧光抗体试验和 ELISA 敏感性高 (Benson 等,2002),这是由于 PRRS 病毒的中和抗体要在病毒感染后 1~2 个月才出现 (Frey 等,1992;Goyal 和 Collins,1992;Minehart 等,1992;Nelson 等,1994;Yoon 等,1995)。同时,由于血清病毒中和试验耗时费力更适用于研究,不适用于常规的诊断检测。

12.9 病毒分离

PRRS 病毒仅在两种类型的细胞中复制:猪肺泡巨噬细胞和某些非洲猴肾细胞(Bautista 等,1993;Dea 等,1992;Paton 等 1992a,b;Wensvoort 等,1991;Yoon 等,1992b)。猪肺泡巨噬细胞用于分离欧洲株 PRRSV,而北美株 PRRSV 在 MA-104/MAC145 等非洲猴肾细胞中生长良好(Dewey 等,2000;Wensvoort 等,1991)。PRRSV 可从多种临床样本中分离到,包括血清、血浆、外周血单核细胞(即白膜层)、骨髓、扁桃体、肺、淋巴结、胸腺、脾、心、脑、肝、睾丸、附睾、输精管、尿道球腺、阴茎组织、口咽刮片、鼻甲、鼻拭子、胎盘、唾液、尿液、粪便和精液(Baron 等,1992;Christianson 等,1992;Christopher Hennings 等,1995b,1998,2001;Dea 等,1992;Done 等,1992;Goyal 和 Collins,1992;Horter 等,2002;Ohlinger 等,1992;Paton 等,1992a,b;Rossow 等,1994,1998)。急性感染动物可采集精液、肺和支气管肺泡灌洗液进行病毒分离培养。老年动物的病毒血症持续时间短,且在组织中 PRRSV 存留时间可能比在血清中存留时间长(Christopher-Hennings 等,2001)。对于持续感染动物来说,扁桃体、口咽刮片和支气管肺泡灌洗液比血清和肺更适合用于病毒分离(Christopher-Hennings 等,2001;Horter 等,2002)。而对于妊娠晚期流产和早产样本采集,选择新生的弱仔和未吃初乳的仔猪比选择木乃伊胎、流产胎或死胎更适合用于病毒分离(Done 等,1992)。

12.10 疫苗

PRRSV 出现不久,就开始针对 PRRSV 进行疫苗研制、批准和接种。不过,尽管有些疫苗在实验环境和现场条件下都有效,但始终未能实现彻底根除病毒。这很可能是由于疫苗对遗传变异的野毒株不具有免疫保护力。虽然目前批准的疫苗已被广泛使用,但全球仍在努力开发一种全新的、更有效的 PRRSV 疫苗。

目前有两种类型的 PRRSV 疫苗可供现场使用:改良式活毒疫苗或弱毒疫苗,以及灭活病毒疫苗或灭活疫苗。改良式活毒疫苗的研制是将强毒株通过体外细胞连续传代培养,使病毒毒力致弱,而灭活病毒疫苗是运用化学、物理方式将强毒进行灭活,并与免疫佐剂联合应用,但大多数改良式活毒疫苗并不使用佐剂。基于改良式活毒疫苗和灭活病毒疫苗的Ⅰ型和Ⅱ型疫苗已在全球内开发使用。

改良式活毒疫苗能够提供最有效的免疫保护,

能够防止同源毒株攻毒引起的病毒血症及其在肺和淋巴组织中的复制(Sattler 等,2018)。因此,通过接种疫苗能够很好地预防同源毒株引起的呼吸道疾病、生殖障碍和胎盘传染的发生 (Labarque 等,2004;Labarque 等,2003;Martelli 等,2007 年;Nielsen 等,1997;Scortti 等,2006;Zuckermann 等,2007)。然而,改良式活毒疫苗对异源病毒株的攻毒免疫保护作用效果较差。PRRSV 氨基酸或核苷酸同源性并不能作为预测 PRRSV 不同毒株之间是否存在免疫交叉保护作用的一个有效参数 (Diaz 等,2006;Labarque 等,2003,2004;Mengeling 等,2003;Nodeljk 等,2001;Okuda 等,2008;van Woensel 等,2008;Zuckermann 等,2007)。接种改良式活毒疫苗后产生的免疫应答与自然感染产生的免疫应答存在着同样的问题,随着病毒中和抗体的产生,细胞免疫受到抑制。接种改良式活毒疫苗产生的病毒中和抗体通常只能使用疫苗毒作为检测抗原进行病毒中和实验才能检测到,这在一定程度上解释了这些疫苗的种特异性免疫保护作用(Charerntantanakul 等,2006a;Meier 等,2003;Okuda 等,2008;Scortti 等,2006b;Zuckermann 等,2007)。

改良式活毒减毒疫苗不能在巨噬细胞中增殖,也不会导致疾病的产生,然而,接种的改良式活毒疫苗仍然会在体内进行一定程度的复制,并且经常能在接种动物的血液和肺部(大多数为 PRRSV Ⅱ 型)中检测到改良式活毒疫苗的病毒复制(Beilage 等,2009;Diaz 等,2006;Labarque 等,2004;Scortti 等,2006b)。此外,感染病毒的妊娠母猪会将病毒传播给胎儿,而感染病毒的公猪的精液中也存在病毒 (Christopher Henning 等,1997;Nielsen 等,1997;Scortti 等,2006)。由于在疫苗病毒复制过程中可能会出现毒力增强的变异株,并传播给阴性动物,构成了极大安全风险。20 世纪 90 年代末丹麦因为给 PRRSV 阴性猪进行疫苗接种而导致 PRRSV 相关的繁殖障碍大规模暴发(Nielsen 等,2001;Nielsen 等,2002)。PRRSV 毒力的减弱和返祖与非结构蛋白的变化有关,但具体的毒力决定因子尚未确定,这就限制了针对性地对新型安全减毒 PRRSV 疫苗的研发 (Kim 和 Yoon,2008;Kwon 等,2008,2009;Nielsen 等,2001)。总之,改良式活毒疫苗可产生适应性免疫反应,但其与自然感染引起的免疫应答高度相似,因而只对同源病毒具有免疫保护作用。

从安全性方面考虑,灭活病毒疫苗可能是替代改良式活毒疫苗一个很好的选择。然而,通过使用灭活病毒疫苗发现,对灭活病毒疫苗的接种动物进行攻毒时,体内只能产生中度记忆性抗体应答反应,而不能产生特异性中和抗体。曾有报道称,如果用强毒疫苗株进行攻毒,灭活病毒疫苗可对生殖性能提供有效的免疫保护,但是在同源毒株感染条件下,灭活病毒疫苗一般不会影响病毒血症的形成,组织内的病毒复制和病毒逃逸(Nielsen 等,1997;Nilubol 等,2004;Plana-Duran 等,1997;Scorti 等,2007;Zuckermann 等,2007)。灭活病毒疫苗不能特异激活分泌 IFN-γ 的细胞,这表明此种类型疫苗不会刺激机体产生适应性的细胞免疫反应 (Meier 等,2003;Nilubol 等,2004;Piras 等,2005;Zuckermann 等,2007)。

12.11 展望

自 20 世纪 80 年代末 PRRS 出现以来,对全球养猪产业造成了毁灭性打击。尽管近 40 年来对 PRRSV 进行了详尽的研究,但 PRRSV 发病机制和免疫特性仍然没有完全弄清楚。PRRSV 在巨噬细胞内进行复制会导致长期的病毒血症,持续数月的持续感染,在持续感染期间,PRRSV 新的变异毒株就会出现。这种病毒通过几种免疫逃逸机制,逃避天然免疫和获得性免疫,特别是 Ⅰ 型 IFN-α 的表达下调,似乎对获得性免疫的产生具有干扰作用。通过对 PRRSV 毒株的观察,发现其具有遗传多样性,因而利用单一毒株为基础的疫苗预防在抗原和遗传方面具有多样性的野生株,似乎并不能实现对全球猪的免疫保护目的。因此,开发具有高免疫原性、广谱保护性、安全性的 PRRS 疫苗仍将是未来研究者所面临的挑战。

参考文献

Albina E, Carrat C, Charley B (1998a) Interferon-alpha response to swine arterivirus (PoAV), the porcine reproductive and respiratory syndrome virus. J Interf Cytokine Res 1(18):485–490

Albina E, Leforban Y, Baron T, Plana Duran J, Vannier P (1992) An enzyme linked immunosorbent assay (ELISA) for the detection of antibodies to the porcine reproductive and respiratory syndrome (PRRS) virus. Ann Rech Vet 23:167–176

Albina E, Piriou L, Hutet E, Cariolet R, L'Hospitalier R (1998b) Immune responses in pigs infected with porcine reproductive and respiratory syndrome virus (PRRSV). Vet Immunol Immunopathol 61:49–66

Allende R, Laegreid WW, Kutish GF, Galeota JA, Wills RW, Osorio FA (2000) Porcine reproductive and respiratory syndrome virus: description of persistence in individual pigs upon experimental infection. J Virol 74:10834–10837

Allende R, Lewis TL, Lu Z, Rock DL, Kutish GF, Ali A, Doster AR, Osorio FA (1999) North American and European porcine reproductive and respiratory syndrome viruses differ in nonstructural protein coding regions. J Gen Virol 80:307–315

An TQ, Tian ZJ, Xiao Y, Li R, Peng JM, Wei TC, Zhang Y, Zhou YJ, Tong GZ (2009) Origin of highly pathogenic porcine reproductive and respiratory syndrome virus, China. Emerg Infect Dis 16(2):365–367

An TQ, Zhou YJ, Liu GQ, Tian ZJ, Li J, Qiu HJ, Tong GZ (2007) Genetic diversity and phylogenetic analysis of glycoprotein 5 of PRRSV isolates in mainland China from 1996 to 2006: coexistence of two NA-subgenotypes with great diversity. Vet Microbiol 123(1–3):43–52

Baron T, Albina E, Leforban Y, Madec F, Guilmoto H, Plana Duran J, Vannier P (1992) Report on the first outbreaks of the porcine reproductive and respiratory syndrome (PRRS) in France. Diagnosis and viral isolation. Ann Rech Vet 23:161–166

Batista L, Pijoan C, Dee S, Michael O, Molitor T, Joo HS, Xiao Z, Murtaugh M (2004) Virological and immunological responses to porcine reproductive and respiratory syndrome virus in a large population of gilts. Can J Vet Res 68:267–273

Bautista EM, Goyal SM, Yoon IJ, Joo HS, Collins JE (1993) Comparison of porcine alveolar macrophages and CL2621 for the detection of porcine reproductive and respiratory syndrome (PRRS) virus and anti-PRRS antibody. J Vet Diagn Invest 5:163–165

Beilage EG, Nathues H, Meemken D, Harder TC, Doherr MG, Grotha I, Greiser Wilke I (2009) Frequency of PRRS live vaccine virus (European and North American genotype) in vaccinated and non-vaccinated pigs submitted for respiratory tract diagnostics in North-Western Germany. Prev Vet Med 92:31–37

Benfield DA, Nelson E, Collins JE, Harris L, Goya SM, Robison D, Christianson WT, Morrison RB, Gorcyca DE, Chladek DW (1992) Characterization of swine infertility and respiratory syndrome (SIRS) virus (isolate ATCC-VR2332). J Vet Diagn Investig 4:127–133

Benson J, Yaeger M, Christopher-Hennings J, Lager K, Yoon KJ (2002) A comparison of virus isolation, immunohistochemistry, fetal serology and reverse-transcription polymerase chain reaction assay for the identification of porcine reproductive and respiratory syndrome virus transplacental infection in the fetus. J Vet Diagn Investig 14:8–14

Bierk MD, Dee SA, Rossow KD, Otake S, Collins JE, Molitor TW (2001) Transmission of porcine reproductive and respiratory syndrome virus from persistently infected sows to contact controls. Can J Vet Res 65:261–266

Botner A, Strandbygaard B, Sorensen KJ, Have P, Madsen KG, Madsen ES, Alexandersen S (1997) Appearance of acute PRRS-like symptoms in sow herds after vaccination with a modified live PRRS vaccine. Vet Rec 141(19):497–499

Brockmeier SL, Lager KM (2002) Experimental airborne transmission of porcine reproductive and respiratory syndrome virus and Bordetella bronchiseptica. Vet Microbiol 89(4):267–275

Bush E, Corso B, Zimmerman J, Swenson S, Pyburn D, Burkgren T (1999) Update on the acute PRRS investigative study. Swine Health Prod 7:179–180

Butler JE, Lager KM, Golde W, Faaberg KS, Sinkora M, Loving C, Zhang YI (2014) Porcine reproductive and respiratory syndrome (PRRS): an immune dysregulatory pandemic. Immunol Res 59:81–108

Butler JE, Wertz N, Weber P, Lager KM (2008) Porcine reproductive and respiratory syndrome virus subverts repertoire development by proliferation of germline encoded B cells of all isotypes bearing hydrophobic heavy chain CDR3. J Immunol 180:2347–2356

Cancel-Tirado SM, Evans RB, Yoon KJ (2004) Monoclonal antibody analysis of porcine reproductive and respiratory syndrome virus epitopes associated with antibody-dependent enhancement and neutralization of virus infection. Vet Immunol Immunopathol 102:249–262

Carlson DL, Fang Y, Nelson EA, Christopher-Hennings J. (2002) Discriminating between PRRSV isolates and vaccine with quantitative, real-time RT-PCR. Proceedings of the American

Association of Veterinary Laboratory Diagnosticians, pp 57–67

Carman S, Sanford SE, Dea S (1995) Assessment of seropositivity to porcine reproductive and respiratory syndrome (PRRS) virus in swine herds in Ontario—1978 to 1982. Can Vet J 36:776–777

Cha SH, Chang CC, Yoon KJ (2004) Instability of the restriction fragment length polymorphism pattern of open reading frame 5 of porcine reproductive and respiratory syndrome virus during sequential pig-to-pig passages. J Clin Microbiol 42(10):4462–4467

Chang HC, Peng YT, Chang HL, Chaung HC, Chung WB (2008) Phenotypic and functional modulation of bone marrow-derived dendritic cells by porcine reproductive and respiratory syndrome virus. Vet Microbiol 129:281–293

Charerntantanakul W, Platt R, Johnson W, Roof M, Vaughn E, Roth JA (2006a) Immune responses and protection by vaccine and various vaccine adjuvant candidates to virulent porcine reproductive and respiratory syndrome virus. Vet Immunol Immunopathol 109:99–115

Charerntantanakul W, Platt R, Roth JA (2006b) Effects of porcine reproductive and respiratory syndrome virus-infected antigen-presenting cells on T cell activation and antiviral cytokine production. Viral Immunol 19:646–661

Chen J, Liu T, Zhu CG, Jin YF, Zhang YZ (2006) Genetic variation of Chinese PRRSV strains based on ORF5 sequence. Biochem Genet 44(9–10):425–435

Chen Z, Plagemann PG (1995) Detection of related positive-strand RNA virus genomes by reverse transcription/polymerase chain reaction using degenerate primers for common replicase sequences. Virus Res 39:365–375

Cho HJ, Deregt D, Joo HS (1996) An ELISA for porcine reproductive and respiratory syndrome: production of antigen of high quality. Can J Vet Res 60:89–93

Christianson WT, Collins JE, Benfield DA, Harris L, Gorcyca DE, Chladek DW, Morrison RB, Joo HS (1992) Experimental reproduction of swine infertility and respiratory syndrome in pregnant sows. Am J Vet Res 53:485–488

Christopher-Hennings J, Holler LD, Benfield DA, Nelson EA (2001) Detection and duration of porcine reproductive and respiratory syndrome virus in semen, serum, peripheral blood mononuclear cells, and tissues from Yorkshire, Hampshire and Landrace boars. J Vet Diagn Investig 13:133–142

Christopher-Hennings J, Nelson EA, Althouse GC, Lunney J (2008) Comparative antiviral and proviral factors in semen and vaccines for preventing viral dissemination from the male reproductive tract and semen. Anim Health Res Rev 9:59–69

Christopher-Hennings J, Nelson EA, Hines RJ, Nelson JK, Swenson SL, Zimmerman JJ, Chase CCL, Yaeger MJ, Benfield DA (1995a) Persistence of porcine reproductive and respiratory syndrome virus in serum and semen of adult boars. J Vet Diagn Investig 7:456–464

Christopher-Hennings J, Nelson EA, Nelson JA, Hines RJ, Swenson SL, Hill HT, Zimmermann JJ, Katz JB, Yaeger MJ, Chase CCL, Benfield DA (1995b) Detection of porcine reproductive and respiratory syndrome virus in boar semen by PCR. J Clin Microbiol 33:1730–1734

Christopher-Hennings J, Nelson EA, Nelson JK, Benfield DA (1997) Effects of a modified-live virus vaccine against porcine reproductive and respiratory syndrome in boars. Am J Vet Res 58:40–45

Christopher-Hennings J, Nelson EA, Nelson JK et al (1998) Identification of porcine reproductive and respiratory syndrome virus in semen and tissues from vasectomized and nonvasectomized boars. Vet Pathol 35:260–267

Collins JE, Benfield DA, Christianson WT, Harris L, Hennings JC, Shaw DP, Goyal SM, McCullough S, Morrison RB, Joo HS et al (1992) Isolation of swine infertility and respiratory syndrome virus (isolateATCCVR-2332) in North America and experimental reproduction of the disease in gnotobiotic pigs. J Vet Diagn Investig 4(2):117–126

Cook CA, Spatz SJ (1998) Development of a commercial diagnostic assay for the detection of PRRSV in body fluids. Proceedings of the American Association of Swine Practitioners, pp 267–274

Costers S, Delputte PL, Nauwynck HJ (2006) Porcine reproductive and respiratory syndrome virus-infected alveolar macrophages contain no detectable levels of viral proteins in their plasma membrane and are protected against antibody-dependent, complement-mediated cell lysis. J Gen Virol 87:2341–2351

Das PB, Dinh PX, Ansari IH, de Lima M, Osorio FA, Pattnaik AK (2010) The minor envelope glycoproteins GP2a and GP4 of porcine reproductive and respiratory syndrome virus interact with the receptor CD163. J Virol 84:1731–1740

Dea SA, Bilodeau R, Athanassious R (1992) PRRS in Quebec: virology and pathologic aspects. Am Assoc Swine Pract Newsl 4:2–7

Dee S, Deen J, Rossow K, Weise C, Eliason R, Otake S, Joo HS, Pijoan C (2003) Mechanical transmission of porcine reproductive and respiratory syndrome virus throughout a coordinated sequence of events during warm weather. Can J Vet Res 67(1):12–19

Dee SA, Bierk MD, Deen J, Molitor TW (2001) An evaluation of test and removal for the elimination of porcine reproductive and respiratory syndrome virus from 5 swine farms. Can J Vet

Res 65:22–27

Dewey C, Charbonneau G, Carman S, Hamel A, Nayar G, Friendship R, Eernisse K, Swenson S (2000) Lelystad-like strain of porcine reproductive and respiratory syndrome virus (PRRSV) identified in Canadian swine. Can Vet J 41(6):493–494

Diaz I, Darwich L, Pappaterra G, Pujols J, Mateu E (2005) Immune responses of pigs after experimental infection with a European strain of porcine reproductive and respiratory syndrome virus. J Gen Virol 86:1943–1951

Diaz I, Darwich L, Pappaterra G, Pujols J, Mateu E (2006) Different European-type vaccines against porcine reproductive and respiratory syndrome virus have different immunological properties and confer different protection to pigs. Virology 351:249–259

Done SH, Paton DJ, White ME (1996) Porcine reproductive and respiratory syndrome (PRRS): a review, with emphasis on pathological, virological and diagnostic aspects. Br Vet J 152:153–174

Done SH, Paton DS, Edwards S (1992) Porcine reproductive and respiratory syndrome ("blue-eared" pig disease). Pig Vet J 28:9–23

Drew TW, Lowings JP, Yapp F (1997) Variation in open reading frames 3, 4 and 7 among porcine reproductive and respiratory syndrome virus isolates in the UK. Vet Microbiol 55(1–4):209–221

Duan X, Nauwynck HJ, Pensaert MB (1997) Effects of origin and state of differentiation and activation of monocytes/macrophages on their susceptibility to porcine reproductive and respiratory syndrome virus (PRRSV). Arch Virol 142:2483–2497

Edwards S, Robertson IB, Wilesmith JW (1994) PRRS ("blue-eared pig disease") in Great Britain. Am Assoc Swine Pract Newsl 4:32–36

Egli C, Thur B, Liu L (2001) Quantitative TaqMan RT-PCR for the detection and differentiation of European and North American strains of porcine reproductive and respiratory syndrome virus. J Virol Methods 98:63–75

Fang Y, Schneider P, Zhang WP, Faaberg KS, Nelson EA, Rowland RR (2007) Diversity and evolution of a newly emerged North American Type 1 porcine arterivirus: analysis of isolates collected between 1999 and 2004. Arch Virol 152(5):1009–1017

Fang Y, Treffers EE, Li Y, Tas A, Sun Z, van der Meer Y, de Ru AH, van Veelen PA, Atkins JF, Snijder EJ, Firth AE (2012) Efficient−2 frameshifting by mammalian ribosomes to synthesize an additional arterivirus protein. PNAS 109:E2920–E2928

Ferrin N, Fang Y, Carroll J (2002) Validation of a blocking ELISA for antibodies against PRRSV. Proceedings of the International Pig Veterinary Society Congress, p 365

Flores-Mendoza L, Silva-Campa E, Reséndiz M, Osorio FA, Hernández J (2008) Porcine reproductive and respiratory syndrome virus infects mature porcine dendritic cells and up-regulates interleukin-10 production. Clin Vaccine Immunol 15:720–725

Forsberg R (2005) Divergence time of porcine reproductive and respiratory syndrome virus subtypes. Mol Biol Evol 22:2131–2134

Forsberg R, Storgaard T, Nielsen HS, Oleksiewicz MB, Cordioli P, Sala G, Hein J, Botner A (2002) The genetic diversity of European type PRRSV is similar to that of the North American type but is geographically skewed within Europe. Virology 299(1):38–47

Frey ML, Eernisse KA, Landgraf JG (1992) Diagnostic testing for SIRS virus at the National Veterinary Services Laboratories (NVSL). Am Assoc Swine Pract Newsl 4:31

Genini S, Delputte PL, Malinverni R, Cecere M, Stella A, Nauwynck HJ (2008) Genome-wide transcriptional response of primary alveolar macrophages following infection with porcine reproductive and respiratory syndrome virus. J Gen Virol 89:2550–2564

Gilbert SA, Larochelle R, Magar R (1997) Typing of porcine reproductive and respiratory syndrome viruses by a multiplex PCR assay. J Clin Microbiol 35:264–267

Gogoi A, Rajkhowa TK, Singh YD, Ravindran R, Arya RS, Hauhnar L (2017) Epidemiology of porcine reproductive and respiratory syndrome (PRRS) outbreak in India. Indian J Vet Pathol 41(1):31–37

Gorbalenya AE, Enjuanes L, Ziebuhr J, Snijder EJ (2006) Nidovirales: evolving the largest RNA virus genome. Virus Res 117:17–37

Goyal SM (1993) Porcine reproductive and respiratory syndrome. J Vet Diagn Investig 5:656–664

Goyal SM, Collins J (1992) SIRS serology and virus isolation. Minnesota Board An Health Newsl 37:2

Halbur PG, Bush EJ (1997) Update on abortion storms and sow mortality. Swine Health Prod 5:73

Halbur PG, Paul PS, Janke BH. (1993) Viral contributors to the porcine respiratory disease complex. Proceedings of the American Association of Swine Practitioners. pp 343–350

Halbur P, Andrew J, Paul P (1995) Strain variation of PRRS virus: field and research experiences. Proceedings of the American Association of Swine Practitioners, pp 200–204

Halbur PG, Paul PS, Frey ML, Landgraf J, Eernisse K, Meng XJ, Andrews JJ, Lum MA, Rathje JA (1996) Comparison of the antigen distribution of two US porcine reproductive and respiratory syndrome virus isolates with that of the Lelystad virus. Vet Pathol 33:159–170

Han J, Wang Y, Faaberg KS (2006) Complete genome analysis of RFLP 184 isolates of porcine reproductive and respiratory syndrome virus. Virus Res 122(1–2):175–182

Hanada K, Suzuki Y, Nakane T, Hirose O, Gojobori T (2005) The origin and evolution of porcine

reproductive and respiratory syndrome viruses. Mol Biol Evol 22:1024–1031
Hill H 1990 Overview and history of mystery swine disease (swine infertility respiratory syndrome). Proceedings of the Mystery Swine Disease Committee Meeting, October 6, Denver, CO, pp 29–30
Holtkamp DJ, Kliebenstein JB, Neumann EJ, Zimmerman J, Rotto HF, Yoder TK, Wang C, Yeske P, Mowrer CL, Haley CA (2013) Assessment of the economic impact of porcine reproductive and respiratory syndrome virus on United States pork producers. J Swine Health Prod 21:72–84
Hopper SA, White MEC, Twiddy N (1992) An outbreak of blue-eared pig-disease (porcine reproductive and respiratory syndrome) in 4 pig herds in Great Britain. Vet Rec 131:140–144
Horter DC, Pogranichniy RC, Chang CC, Evans RB, Yoon KJ, Zimmerman JJ (2002) Characterization of the carrier state in porcine reproductive and respiratory syndrome virus infection. Vet Microbiol 86:213–218
Houben S, Callebaut P, Pensaert MB (1995) Comparative study of a blocking enzyme-linked immunosorbent assay and the immunoperoxidase monolayer assay for the detection of antibodies to the porcine reproductive and respiratory syndrome virus in pigs. J Virol Methods 51:125–128
Hu H, Li X, Zhang Z, Shuai J, Chen N, Liu G, Fang W (2009) Porcine reproductive and respiratory syndrome viruses predominant in southeastern China from 2004 to 2007 were from a common source and underwent further divergence. Arch Virol 154(3):391–398
Hurd HS, Bush EJ, Losinger W, Corso B, Zimmerman JJ, Wills R, Swenson S, Pyburn D, Yeske P, Burkgren T (2001) Outbreaks of porcine reproductive failure: report on a collaborative field investigation. J Swine Health Prod 9:103–108
Jiang P, Chen PY, Dong YY, Cai JL, Cai BX, Jiang ZH (2000) Isolation and genome characterization of porcine reproductive and respiratory syndrome virus in P. R. China. J Vet Diagn Investig 12:156–158
Kang SY, Yun SI, Park HS, Park CK, Choi HS, Lee YM (2004) Molecular characterization of PL97-1, the first Korean isolate of the porcine reproductive and respiratory syndrome virus. Virus Res 104(2):165–179
Kapur V, Elam MR, Pawlovich TM, Murtaugh MP (1996) Genetic variation in porcine reproductive and respiratory syndrome virus isolates in the midwestern United States. J Gen Virol 77(6):1271–1276
Keffaber K (1989) Reproductive failure of unknown etiology. Am Assoc Swine Pract Newsl 1(2):1–9
Key KF, Haqshenas G, Guenette DK, Swenson SL, Toth TE, Meng XJ (2001) Genetic variation and phylogenetic analyses of the ORF5 gene of acute porcine reproductive and respiratory syndrome virus isolates. Vet Microbiol 83(3):249–263
Kim HK, Yang JS, Moon HJ, Park SJ, Luo Y, Lee CS, Song DS, Kang BK, Ann SK, Jun CH, Park BK (2009) Genetic analysis of ORF5 of recent Korean porcine reproductive and respiratory syndrome viruses (PRRSVs) in viremic sera collected from MLV-vaccinating or non-vaccinating farms. J Vet Sci 10(2):121–130
Kim WI, Yoon KJ (2008) Molecular assessment of the role of envelope associated structural proteins in cross neutralization among different PRRS viruses. Virus Genes 37:380–391
Kleiboeker SB, Lehman JR, Fangman TJ (2002) Concurrent use of reverse transcription-polymerase chain reaction testing of oropharyngeal scrapings and paired serological testing for detection of porcine reproductive and respiratory syndrome virus infection in sows. J Swine Health Prod 10(6):251–258
Koyama S, Ishii KJ, Coban C, Akira S (2008) Innate immune response to viral infection. Cytokine 43:336–341
Kwon B, Ansari IH, Pattnaik AK, Osorio FA (2008) Identification of virulence determinants of porcine reproductive and respiratory syndrome virus through construction of chimeric clones. Virology 380:371–378
Labarque G, Reeth KV, Nauwynck H, Drexler C, Van Gucht S, Pensaert M (2004) Impact of genetic diversity of European-type porcine reproductive and respiratory syndrome virus strains on vaccine efficacy. Vaccine 22:4183–4190
Labarque G, Van Gucht S, Van Reeth K, Nauwynck H, Pensaert M (2003) Respiratory tract protection upon challenge of pigs vaccinated with attenuated porcine reproductive and respiratory syndrome virus vaccines. Vet Microbiol 95:187–197
Labarque GG, Nauwynck HJ, Van Reeth K, Pensaert MB (2000) Effect of cellular changes and onset of humoral immunity on the replication of porcine reproductive and respiratory syndrome virus in the lungs of pigs. J Gen Virol 81:1327–1334
Le Bon A, Schiavoni G, D'Agostino G, Gresser I, Belardelli F, Tough DF (2001) Type I interferons potently enhance humoral immunity and can promote isotype switching by stimulating dendritic cells in vivo. Immunity 14:461–470
Le Gall A, Legeay O, Bourhy H, Arnauld C, Albina E, Jestin A (1998) Molecular variation in the nucleoprotein gene (ORF7) of the porcine reproductive and respiratory syndrome virus (PRRSV). Virus Res 54(1):9–21

Lee C, Kim H, Kang B, Yeom M, Han S, Moon H, Park S, Song D, Park B (2010) Prevalence and phylogenetic analysis of the isolated type I porcine reproductive and respiratory syndrome virus from 2007 to 2008 in Korea. Virus Genes 40(2):225–230

Lee SM, Schommer SK, Kleiboeker SB (2004) Porcine reproductive and respiratory syndrome virus field isolates differ in in vitro interferon phenotypes. Vet Immunol Immunopathol 102:217–231

Legeay O, Bounaix S, Denis M (1997) Development of a RT-PCR test coupled with a microplate colorimetric assay for the detection of a swine Arterivirus (PRRSV) in boar semen. J Virol Methods 68:65–80

Li Y, Treffers EE, Napthine S, Tas A, Zhu L, Sun Z, Bell S, Mark BL, van Veelen PA, van Hemert MJ, Firth AE, Brierley I, Snijder EJ, Fang Y (2014) Transactivation of programmed ribosomal frameshifting by a viral protein. PNAS 111:E2172–E2181

Loemba HD, Mounir S, Mardassi H, Archambault D, Dea S (1996) Kinetics of humoral immune response to the major structural proteins of the porcine reproductive and respiratory syndrome virus. Arch Virol 141:751–761

Loula T (1991) Mystery pig disease. Agri Prac 12:23–34

Loving CL, Brockmeier SL, Sacco RE (2007) Differential type I interferon activation and susceptibility of dendritic cell populations to porcine arterivirus. Immunology 120:217–229

Madsen KG, Hansen CM, Madsen ES, Strandbygaard B, Botner A, Sorensen KJ (1998) Sequence analysis of porcine reproductive and respiratory syndrome virus of the American type collected from Danish swine herds. Arch Virol 143(9):1683–1700

Mardassi H, Wilson L, Mounir S, Dea SA (1994) Detection of porcine reproductive and respiratory syndrome virus and efficient differentiation between Canadian and European strains by reverse transcription and PCR amplification. J Clin Microbiol 32:2197–2203

Martelli P, Cordioli P, Alborali LG, Gozio S, De Angelis E, Ferrari L, Lombardi G, Borghetti P (2007) Protection and immune response in pigs intradermally vaccinated against porcine reproductive and respiratory syndrome (PRRS) and subsequently exposed to a heterologous European (Italian cluster) field strain. Vaccine 25:3400–3408

Mayer J, Brisbin I (1991) Wild pigs of the United States: their biology, history, morphology and current status. University of Georgia Press, Athens, GA

Meier WA, Galeota J, Osorio FA, Husmann RJ, Schnitzlein WM, Zuckermann FA (2003) Gradual development of the interferon-gamma response of swine to porcine reproductive and respiratory syndrome virus infection or vaccination. Virology 309:18–31

Meng XJ (2000) Heterogeneity of porcine reproductive and respiratory syndrome virus: implications for current vaccine efficacy and future vaccine development. Vet Microbiol 74:309–329

Meng XJ, Paul PS, Halbur PG, Morozov I (1995) Sequence comparison of open reading frames 2 to 5 of low and high virulence United States isolates of porcine reproductive and respiratory syndrome virus. J Gen Virol 76(12):3181–3188

Mengeling WL, Lager KM (2000) A brief review of procedures and potential problems associated with the diagnosis of porcine reproductive and respiratory syndrome. Vet Res 3:61–69

Mengeling WL, Lager KM, Vorwald AC, Koehler KJ (2003) Strain specificity of the immune response of pigs following vaccination with various strains of porcine reproductive and respiratory syndrome virus. Vet Microbiol 93:13–24

Meulenberg JJ (2000) PRRSV, the virus. Vet Res 31:11–21

Minehart MJ, Nelson EA, Harley DJ (1992) Comparison of virus neutralization (VN) and indirect fluorescent antibody (IFA) for the detection of antibodies to porcine reproductive and respiratory syndrome (PRRS) virus. Proceedings of the Conference for Research Workers in Animal Disease, p 66

Mulupuri P, Zimmerman JJ, Hermann J, Johnson CR, Cano JP, Yu W, Dee SA, Murtaugh MP (2008) Antigen-specific B-cell responses to porcine reproductive and respiratory syndrome virus infection. J Virol 82:358–370

Murtaugh MP, Xiao Z, Zuckermann F (2002) Immunological responses of swine to porcine reproductive and respiratory syndrome virus infection. Viral Immunol 15:533–547

Nauwynck HJ, Duan X, Favoreel HW, Van Oostveldt P, Pensaert MB (1999) Entry of porcine reproductive and respiratory syndrome virus into porcine alveolar macrophages via receptor-mediated endocytosis. J Gen Virol 80:297–305

Nelsen CJ, Murtaugh MP, Faaberg KS (1999) Porcine reproductive and respiratory syndrome virus comparison: divergent evolution on two continents. J Virol 73:270–280

Nelson EA, Christopher-Hennings J, Benfield DA (1994) Serum immune responses to the proteins of porcine reproductive and respiratory syndrome (PRRS) virus. J Vet Diagn Investig 6:410–415

Neumann EJ, Kliebenstein JB, Johnson CD, Mabry JW, Bush EJ, Seitzinger AH, Green AL, Zimmerman JJ (2005) Assessment of the economic impact of porcine reproductive and respiratory syndrome on swine production in the United States. J Am Vet Med Assoc 227:385–392

Nielsen HS, Oleksiewicz MB, Forsberg R, Stadejek T, Bøtner A, Storgaard T (2001) Reversion of a live porcine reproductive and respiratory syndrome virus vaccine investigated by parallel

mutations. J Gen Virol 82:1263–1272
Nielsen J, Botner A, Bille-Hansen V, Oleksiewicz MB, Storgaard T (2002) Experimental inoculation of late term pregnant sows with a field isolate of porcine reproductive and respiratory syndrome vaccine-derived virus. Vet Microbiol 84:1–13
Nielsen TL, Nielsen J, Have P, Baekbo P, Hoff-Jorgensen R, Botner A (1997) Examination of virus shedding in semen from vaccinated and from previously infected boars after experimental challenge with porcine reproductive and respiratory syndrome virus. Vet Microbiol 54:101–112
Nilubol D, Platt KB, Halbur PG, Torremorell M, Harris DL (2004) The effect of a killed porcine reproductive and respiratory syndrome virus (PRRSV) vaccine treatment on virus shedding in previously PRRSV infected pigs. Vet Microbiol 102:11–18
Nodelijk G, de Jong MC, van Leengoed LA, Wensvoort G, Pol JM, Steverink PJ, Verheijden JH (2001) A quantitative assessment of the effectiveness of PRRSV vaccination in pigs under experimental conditions. Vaccine 19:3636–3644
Nodelijk G, Wensvoort G, Kroese B (1996) Comparison of a commercial ELISA and an immunoperoxidase monolayer assay to detect antibodies against porcine respiratory and reproductive syndrome virus. Vet Microbiol 49:285–295
Normile D (2007) China, Vietnam grapple with 'rapidly evolving' pig virus. Science 317:1017
O'Connor M, Fallon M, O'Reilly PJ (2002) Detection of antibody to porcine reproductive and respiratory syndrome (PRRS) virus: reduction of cut-off value of an ELISA, with confirmation by immunoperoxidase monolayers assay. Irish Vet J 55:73–75
Ohlinger VF, Haas B, Sallmüller A (1992) In vivo and in vitro studies on the immunobiology of PRRS. Am Assoc Swine Pract Newsl 4:24
Okuda Y, Kuroda M, Ono M, Chikata S, Shibata I (2008) Efficacy of vaccination with porcine reproductive and respiratory syndrome virus following challenges with field isolates in Japan. J Vet Med Sci 70:1017–1025
Oleksiewicz MB, Botner A, Madsen KG, Storgaard T (1998) Sensitive detection and typing of porcine reproductive and respiratory syndrome virus by RT-PCR amplification of whole viral genes. Vet Microbiol 64:7–22
Ostrowski M, Galeota JA, Jar AM, Platt KB, Osorio FA, Lopez OJ (2002) Identification of neutralizing and nonneutralizing epitopes in the porcine reproductive and respiratory syndrome virus GP5 ectodomain. J Virol 76:4241–4250
Otake S, Dee SA, Moon RD, Rossow KD, Trincado C, Pijoan C (2003) Evaluation of mosquitoes, Aedes vexans, as biological vectors of porcine reproductive and respiratory syndrome virus. Can J Vet Res 67(4):265–270
Park JY, Kim HS, Seo SH (2008) Characterization of interaction between porcine reproductive and respiratory syndrome virus and porcine dendritic cells. J Microbiol Biotechnol 18:1709–1716
Paton DJ, Brown IH, Scott AC (1992a) Isolation of a Lelystad virus-like agent from British pigs and scanning electron microscopy of infected macrophages. Vet Microbiol 33:195–201
Paton DJ, Drew TW, Brown IH (1992b) Laboratory diagnosis of porcine reproductive and respiratory syndrome. Pig Vet J 29:188–192
Piras F, Bollard S, Laval F, Joisel F, Reynaud G, Charreyre C, Andreoni C, Juillard V (2005) Porcine reproductive and respiratory syndrome (PRRS) virus-specific interferon-gamma(+) T-cell responses after PRRS virus infection or vaccination with an inactivated PRRS vaccine. Viral Immunol 18:381–389
Plagemann PG (2003) Porcine reproductive and respiratory syndrome virus: origin hypothesis. Emerg Infect Dis 9:903–908
Plana-Duran J, Bastons M, Urniza A, Vayreda M, Vila X, Mane H (1997) Efficacy of an inactivated vaccine for prevention of reproductive failure induced by porcine reproductive and respiratory syndrome virus. Vet Microbiol 55:361–370
Plana-Duran J, Vayreda M, Vilarrasa J, Bastons M, Rosell R, Martinez M, San Gabriel A, Pujols J, Badiola JL, Ramos JA, Mariano D (1992) Porcine epidemic abortion and respiratory syndrome (mystery swine disease). Isolation in Spain of the causative agent and experimental reproduction of the disease. Vet Microbiol 33:203–211
Prieto C, Castro JM (2005) Porcine reproductive and respiratory syndrome virus infection in the boar: a review. Theriogenology 63:1–16
Rajkhowa TK, Catherine V, Singh YD, Ravindran R, Arya RS (2018) Genetic variation of Indian highly pathogenic porcine reproductive and respiratory syndrome viruses after introduction in 2013. Indian J Anim Sci 88(10):1118 1126
Rajkhowa TK, Gogoi A, Hauhnar L, Lalrohlua I (2015a) Molecular detection, epidemiology and clinico-pathological studies on first outbreak of porcine reproductive and respiratory syndrome (PRRS) in pig population of Mizoram, India. Indian J Anim Sci 85(4):343–347
Rajkhowa TK, Mohan Rao GJ, Gogoi A, Hauhnar L (2016) Indian porcine reproductive and respiratory syndrome virus bears discontinuous deletion of 30 amino acids in nonstructural protein 2. Virusdisease 27:287–293. https://doi.org/10.1007/s13337-016-0341-9
Rajkhowa TK, Mohan Rao GJ, Gogoi A, Hauhnar L, Lalrohlua I (2015b) Porcine reproductive and respiratory virus (PRRSV) from first outbreak of India shows close relationship with the

highly pathogenic variant of China. Vet Q 35(4):186–193. https://doi.org/10.1080/01652176.2015.1066043

Renukaradhya GJ, Alekseev K, Jung K, Fang Y, Saif LJ (2010) Porcine reproductive and respiratory syndrome virus-induced immunosuppression exacerbates the inflammatory response to porcine respiratory coronavirus in pigs. Viral Immunol 23:457–466

Ropp SL, Wees CE, Fang Y, Nelson EA, Rossow KD, Bien M, Arndt B, Preszler S, Steen P, Christopher-Hennings J, Collins JE, Benfield DA, Faaberg KS (2004) Characterization of emerging European-like porcine reproductive and respiratory syndrome virus isolates in the United States. J Virol 78(7):3684–3703

Rossow KD (1998) Porcine reproductive and respiratory syndrome. Vet Pathol 35:1–20

Rossow KD, Bautista EM, Goyal SM, Molitor TW, Murtaugh MP, Morrison RB, Benfield DA, Collins JE (1994) Experimental porcine reproductive and respiratory syndrome virus infection in one-, four-, and 10-week-old pigs. J Vet Diagn Investig 6:3–12

Rowland RR (2007) The stealthy nature of PRRSV infection: the dangers posed by that ever-changing mystery swine disease. Vet J 174(3):451

Rowland RRR, Lawson S, Rossow K, Benfield DA (2003) Lymphoid tissue tropism of porcine reproductive and respiratory syndrome virus replication during persistent infection of pigs originally exposed to virus in utero. Vet Microbiol 96:219–235

Sang Y, Rowland RR, Blecha F (2011) Interaction between innate immunity and porcine reproductive and respiratory syndrome virus. Anim Health Res Rev 12:149–167

Sattler T, Pikalo J, Wodak E, Revilla-Fernández S, Steinrigl A, Bagó Z, Entenfellner F, Claude JB, Pez F, Francillette M, Schmoll F (2018) Efficacy of live attenuated porcine reproductive and respiratory syndrome virus 2 strains to protect pigs from challenge with a heterologous Vietnamese PRRSV 2 field strain. Vet Res 14:133

Scortti M, Prieto C, Alvarez E, Simarro I, Castro JM (2007) Failure of an inactivated vaccine against porcine reproductive and respiratory syndrome to protect gilts against a heterologous challenge with PRRSV. Vet Rec 1(61):809–813

Scortti M, Prieto C, Martinez-Lobo FJ, Simarro I, Castro JM (2006a) Effects of two commercial European modified-live vaccines against porcine reproductive and respiratory syndrome viruses in pregnant gilts. Vet J 172:506–514

Scortti M, Prieto C, Simarro I, Castro JM (2006b) Reproductive performance of gilts following vaccination and subsequent heterologous challenge with European strains of porcine reproductive and respiratory syndrome virus. Theriogenology 66:1884–1893

Shi M, Lam TT, Hon C-C, Murtaugh MP, Davies PR, Hui RK, Li J, Wong LT, Yip CW, Jiang JW, Leung FC-C (2010) A phylogeny-based evolutionary, demographical and geographical dissection of North American type 2 porcine reproductive and respiratory syndrome viruses. J Virol 84(17):8700–8711

Shimizu M, Yamada S, Murakami Y, Morozumi T, Kobayashi H, Mitani K, Ito N, Kubo M, Kimura K, Kobayashi M, Yamamoto K, Miura Y, Yamamoto T, Watanabe K (1994) Isolation of porcine reproductive and respiratory syndrome (Prrs) virus from Heko-Heko disease of pigs. J Vet Med Sci 56:389–391

Shin J, Torrison J, Choi CS (1997) Monitoring of porcine reproductive and respiratory syndrome virus infection in boars. Vet Microbiol 55:337–346

Snijder EJ, Meulenberg JJ (1998) The molecular biology of arteriviruses. J Gen Virol 79(5):961–979

Spagnuolo Weaver M, Walker IW, Campbell ST (2000) Rapid detection of porcine reproductive and respiratory syndrome viral nucleic acid in blood using a fluorimeter based PCR method. Vet Microbiol 76:15–23

Spagnuolo Weaver M, Walker IW, McNeilly F (1998) The reverse transcription polymerase chain reaction for the diagnosis of porcine reproductive and respiratory syndrome: comparison with virus isolation and serology. Vet Microbiol 62:207–215

Stadejek T, Oleksiewicz MB, Potapchuk D, Podgorska K (2006) Porcine reproductive and respiratory syndrome virus strains of exceptional diversity in Eastern Europe support the definition of new genetic subtypes. J Gen Virol 87(Pt 7):1835–1841

Stadejek T, Oleksiewicz MB, Scherbakov AV, Timina AM, Krabbe JS, Chabros K, Potapchuk D (2008) Definition of subtypes in the European genotype of porcine reproductive and respiratory syndrome virus: nucleocapsid characteristics and geographical distribution in Europe. Arch Virol 153:1479–1488

Stevenson GW, Vanalstine WG, Kanitz CL, Keffaber KK (1993) Endemic porcine reproductive and respiratory syndrome virus-infection of nursery pigs in 2 swine herds without current reproductive failure. J Vet Diagn Investig 5:432–434

Suarez P, Zardoya R, Martin MJ, Prieto C, Dopazo J, Solana A, Castro JM (1996) Phylogenetic relationships of European strains of porcine reproductive and respiratory syndrome virus (PRRSV) inferred from DNA sequences of putative ORF-5 and ORF-7 genes. Virus Res 42(1–2):159–165

Suarez P, Zardoya R, Prieto C (1994) Direct detection of the porcine reproductive and respiratory syndrome (PRRS) virus by reverse polymerase chain reaction (RT-PCR). Arch Virol 135:89–99

Takikawa N, Kobayashi S, Ide S (1996) Detection of antibodies against porcine reproductive and respiratory syndrome (PRRS) virus in swine sera by enzyme-linked immunosorbent assay. J Vet Med Sci 58:355–357

Terpstra C, Wensvoort G, Pol JMA (1991) Experimental reproduction of porcine epidemic abortion and respiratory syndrome (mystery swine disease) by infection with Lelystad virus: Koch's postulates fulfilled. Vet Q 13:131–136

Thanawongnuwech R, Amonsin A, Tatsanakit A, Damrongwatanapokin S (2004) Genetics and geographical variation of porcine reproductive and respiratory syndrome virus (PRRSV) in Thailand. Vet Microbiol 101(1):9–21

Thanawongnuwech R, Young TF, Thacker BJ, Thacker EL (2001) Differential production of proinflammatory cytokines: in vitro PRRSV and mycoplasma hyopneumoniae co-infection model. Vet Immunol Immunopathol 79:115–127

Tian K, Yu X, Zhao T, Feng Y, Cao Z, Wang C, Hu Y, Chen X, Hu D, Tian X, Liu D, Zhang S, Deng X, Ding Y, Yang L, Zhang Y, Xiao H, Qiao M, Wang B, Hou L, Wang X, Yang X, Kang L, Sun M, Jin P, Wang S, Kitamura Y, Yan J, Gao GF (2007) Emergence of fatal PRRSV variants: unparalleled outbreaks of atypical PRRS in China and molecular dissection of the unique hallmark. PLoS One 2(6):e526

Tong GZ, Zhou YJ, Hao XF, Tian ZJ, An TQ, Qiu HJ (2007) Highly pathogenic porcine reproductive and respiratory syndrome, China. Emerg Infect Dis 13:1434–1436

Van Reeth K, Labarque G, Nauwynck H (1999) Differential production of proinflammatory cytokines in the pig lung during different respiratory virus infections: correlations with pathogenicity. Res Vet Sci 67:47–52

Van Woensel PA, Liefkens K, Demaret S (1998) Effect on viremia of an American and a European serotype PRRSV vaccine after challenge with European wild-type strain of the virus. Vet Rec 142:510–512

Vu HL, Kwon B, Yoon KJ, Laegreid WW, Pattnaik AK, Osorio FA (2011) Immune evasion of porcine reproductive and respiratory syndrome virus through glycan shielding involves both glycoprotein 5 as well as glycoprotein 3. J Virol 85:5555–5564

Wang X, Eaton M, Mayer M, Li H, He D, Nelson E (2007) Porcine reproductive and respiratory syndrome virus productively infects monocyte-derived dendritic cells and compromises their antigen-presenting ability. Arch Virol 152:289–303

Wang Y, Liang Y, Han J, Burkhart KM, Vaughn EM, Roof MB, Faaberg KS (2008) Attenuation of porcine reproductive and respiratory syndrome virus strain MN184 using chimeric construction with vaccine sequence. Virology 371(2):418–429

Wei X, Decker JM, Wang S, Hui H, Kappes JC, Wu X (2003) Antibody neutralization and escape by HIV-1. Nature 422:307–312

Welch SK, Calvert JG (2010) A brief review of CD163 and its role in PRRSV infection. Virus Res 154:98–103

Wensvoort G, de Kluyver EP, Luijtze EA (1992) Antigenic comparison of Lelystad virus and swine infertility and respiratory syndrome (SIRS) virus. J Vet Diagn Investig 4:134–138

Wensvoort G, Terpstra C, Pol JM, terLaak EA, Bloemraad M, de Kluyver EP, Kragten C, van Buiten L, den Besten A, Wagenaar F, Broekhuijsen JM, Moonen PLJM, Zetstra T, de Boer EA, Tibben HJ, de Jong MF, van 't Veld P, GJR G, van Gennep JA, Voets M, JHM V, Braamskamp J (1991) Mystery swine disease in The Netherlands: the isolation of Lelystad virus. Vet Q 13(3):121–130

Wesley RD, Mengeling WL, Lager KM, Clouser DF, Landgraf JG, Frey ML (1998) Differentiation of a porcine reproductive and respiratory syndrome virus vaccine strain from North American field strains by restriction fragment length polymorphism analysis of ORF 5. J Vet Diagn Investig 10(2):140–144

Wills RW, Doster AR, Galeota JA, Sur JH, Osorio FA (2003) Duration of infection and proportion of pigs persistently infected with porcine reproductive and respiratory syndrome virus. J Clin Microbiol 41:58–62

Wills RW, Zimmerman JJ, Yoon KJ, Swenson SL, McGinley MJ et al (1997) Porcine reproductive and respiratory syndrome virus: a persistent infection. Vet Microbiol 55:231–240

Yoon IJ, Joo HS, Christianson WT (1992a) An indirect fluorescent antibody test for the detection of antibody to swine infertility and respiratory syndrome virus in swine sera. J Vet Diagn Investig 4:144–147

Yoon IJ, Joo HS, Christianson WT (1992b) Isolation of a cytopathic virus from weak pigs on farms with a history of swine infertility and respiratory syndrome. J Vet Diagn Investig 4:139–143

Yoon KJ, Wu LL, Zimmerman JJ, Platt KB (1997) Field isolates of porcine reproductive and respiratory syndrome virus (PRRSV) vary in their susceptibility to antibody dependent enhancement (ADE) of infection. Vet Microbiol 55:277–287

Yoon KJ, Zimmerman JJ, McGinley MJ (1995a) Failure to consider the antigenic diversity of porcine reproductive and respiratory syndrome (PRRS) virus isolates may lead to misdiagnosis. J Vet Diagn Invest 7:386–387

Yoon KJ, Zimmerman JJ, Swenson SL, McGinley MJ, Eernisse KA, Brevik A, Rhinehart LL, Frey

ML, Hill HT, Platt KB (1995) Characterization of the humoral immune response to porcine reproductive and respiratory syndrome (PRRS) virus infection. J Vet Diagn Investig 7:305–312

Zhou E, Zimmerman JJ, Zhou KX .(2001) Development of a blocking ELISA for detection of swine antibodies to PRRSV. Proceedings of the American Association of Veterinary Laboratory Diagnosticians, p 60

Zhou L, Ni YY, Pineyro P, Sanford BJ, Cossaboom CM, Dryman BA, Huang YW, Cao DJ, Meng XJ (2012) DNA shuffling of the GP3 genes of porcine reproductive and respiratory syndrome virus (PRRSV) produces a chimeric virus with an improved cross-neutralizing ability against a heterologous PRRSV strain. Virology 434:96–109

Zhou YJ, Hao XF, Tian ZJ, Tong GZ, Yoo D, An TQ, Zhou T, Li GX, Qiu HJ, Wei TC, Yuan XF (2008) Highly virulent porcine reproductive and respiratory syndrome virus emerged in China. Transbound Emerg Dis 55(3–4):152–164

Zimmerman JJ, Benfield DA, Murtaugh MP, Osorio FA, Stevenson GW, Torremorell M (2006) Porcine reproductive and respiratory syndrome virus (porcine arterivirus). In: Straw BE, Zimmerman JJ, D'Allaire S, Taylor DJ (eds) Diseases of swine, 9th edn. Blackwell Publishing, Ames, IA, pp 387–417

Zuckermann FA, Garcia EA, Luque ID, Christopher-Hennings J, Doster A, Brito M (2007) Assessment of the efficacy of commercial porcine reproductive and respiratory syndrome virus (PRRSV) vaccines based on measurement of serologic response, frequency of gamma-IFN-producing cells and virological parameters of protection upon challenge. Vet Microbiol 123:69–85

第 13 章 小反刍兽疫病毒

Balamurugan Vinayagamurthy, Govindaraj Gurrappa Naidu,
Parimal Roy

13.1 引言

在不发达国家和发展中国家，饲养小反刍动物能够保障生活贫困、没有土地的农民维持生计。绵羊和山羊就是贫穷失地农民的“随时提款机”，在一整年中，为个体经济增加收入做出贡献。PPR又称之为“小反刍瘟疫”或“羊瘟”，在绵羊和山羊中发病率和死亡率非常高，给PPR流行国家的经济发展构成了严重威胁。PPR是一种具有传染性的、急性的跨界动物疾病，主要感染家养绵羊、山羊和野生小反刍动物，给牲畜饲养带来了重大经济损失。PPR每年造成的经济损失为14.5~21亿美元(WOAH和FAO，2015)。PPR最早于1942年在西非科特迪瓦(Gargadennec和Lalanne，1942)被发现，随后传播至非洲其他地区、阿拉伯半岛、中东和亚洲地区(Balamurugan等，2014a；Muthuchelvan等，2015；Parida等，2015b；Baron等，2016)。PPR临床症状主要表现为抑郁、高热、鼻眼分泌物、坏死性口腔溃疡、坏死性胃肠炎和腹泻，可伴有呼吸困难，继而为支气管肺炎。

小反刍动物麻疹病毒(small ruminant morbillivirus，SRMV)，又称小反刍兽疫病毒(peste des petits ruminant virus，PPRV)(Gibbs等，1979)，具有致病性。PPRV基因组为单股负链RNA病毒(Adombi等，2017)，根据核蛋白(N)基因、融合蛋白(F)基因序列和遗传进化分析，PPRV基因可分为四个谱系(Shaila等，1996；Balamurugan等，2010b)。一般情况下，谱系Ⅰ到Ⅲ主要在非洲流行，而谱系Ⅳ(亚洲谱系)则在亚洲大陆流行。最近在非洲国家(摩洛哥、土耳其马尔马拉地区)发现了亚洲谱系Ⅴ(Banyard等，2010；Kwiatek等，2011)，在中国发现了非洲谱系(Zhou等，2018)。近期随着PPR在保加利亚暴发流行(Altan等，2019)，增加了PPRV入侵欧洲的可能性(Kwiatek等，2011；Baazizi等，2017)。不同谱系的PPRV在非洲、亚洲和欧洲国家中广泛传播，这使其成为全球关注的焦点(Kwiatek等，2011；Balamurugan等，2014；Kumar等，2014；Parida等，2015b；Niyokwishmira等，2019)。

据估计，PPRV的跨界传播特性能使病毒扩散到更远区域，使得牲畜生产性能降低，对发展中国家和不发达国家的影响尤为严重。目前为止，非洲和亚洲共有76个国家、17亿只绵羊和山羊受到PPRV的威胁。对于不发达国家和发展中国家来说，绵羊和山羊在粮食安全和社会经济增长方面发挥着重要作用，因此世界粮农组织(FAO)联合WOAH共同发起了一项国际计划，即到2030年实现PPR的控制和根除(WOAH和FAO，2015；Parida等，2019)。

13.2 历史

1942年首次对“小反刍兽疫”进行报道(Gargadennec和Lalanne，1942)，这是一种小反刍动物流行病，与牛瘟相似，但不传染牛。1984年，该疾病已传播到尼日利亚、塞内加尔、加纳、苏丹等非洲国家，随后进一步传播到非洲和亚洲等地区。据报道，印度首次暴发PPR疫情是1987年在印度南部(Shaila等，1989)，随后蔓延至印度北部。到目前为止，该病成为印度地方性动物疫病，全年经常暴发流行(Balamurugan等，2014a)。根据国家动物疫病转诊专家系统的分析报道，1991—2017年PPR主要在印度绵羊和山羊中发生(Balamurugan等，2016)。PPR如同滚雪球般在流行国家不断蔓延，不时也有其他国家确认感染PPRV的报道。在非洲、亚洲和南欧国家暴发的PPRV的谱系不同，2018年6月在保加利亚暴发PPRV，敲响了动物健康的警钟(Kwiatek等，2011；Balamurugan等，2014a；Banyard等，2010)。随着PPR的传播范围不断扩大，在许多国家，PPR已经成为地方性动物疫病。非洲、亚洲和欧洲各国不断有PPR的疫情报道，且各国流行的PPRV谱系不同(Banyard等，

2015)。

13.3 病毒结构

PPRV 是有囊膜的 SRMV,属于副黏病毒科麻疹病毒属,并且是单股负链 RNA 病毒(Gibbs 等,1979)。PPRV 与麻疹病毒属的其他成员具有遗传相似性，是一个新发现的病毒。因此，我们对 PPRV 结构和分子生物学的认识大多是基于与其他麻疹病毒的比较(Baron,2015)。PPRV 能与人麻疹病毒(measles virus,MV),牛瘟病毒(rinderpest virus,RPV),犬瘟热病毒(canine distemper virus,CDV),海豚麻疹病毒(dolphin and porpoise morbillivirus,DMV),以及海豹瘟疫病毒(phocine distemper virus,PDV)。等发生免疫交叉反应(Barrett 等,1993)。PPRV 呈多形性,具有囊膜和核糖核蛋白核心,为单股负链、不分节段的 RNA 病毒(Haffar 等,1999),基因组全长约为 16kb (Chard 等,2008)。囊膜厚度在 8~15nm、带有 8.5~14.5nm 的纤突,类似"人"形结构的核糖核酸蛋白链长 4~23nm。PPRV 基因组符合"六碱基"原则,按照 3'N-P/C/V-M-F-H-L5'的顺序依次编码蛋白(Bailey 等,2005)。根据 PPRV 进化动力学研究结果,推测 PPRV 进化率为 2.61×10^6 变异/位点/日 (Bao 等,2017)。病毒基因组可划分为 6 个转录单元,每个转录本单元被保守的基因三核苷酸分隔,6 个转录单元编码 6 个结构蛋白[融合(F)蛋白、血凝素(H)、基质(M)蛋白、核衣壳(N)蛋白、大(L)蛋白、磷(P)蛋白]和两个非结构蛋白(V 蛋白和 C 蛋白)。N 蛋白和 P 蛋白与 L 蛋白共同组成酶复合体。基于 N 蛋白核苷酸(nt)和氨基酸(aa)序列分析，麻疹病毒通常分为两组：一组包含 CDV 和 PDV；另一组则包含 RPV、MV 和 PPRV(Diallo,1990)。据报道,PPRV 与 DMV 的 N、V 和 H 蛋白的抗原性接近(Bailey 等,2005)。病毒 3'和 5'端的保守序列是互补的,在基因组的复制、转录和包装过程中起着重要的调节作用(Banyard 等,2010)。带有 N 基因 3' UTR 的 PPRV 前导序列能够产生病毒基因组启动子，带有 L 基因 5' UTR 的小序列则生成抑制病毒基因组启动子。

13.3.1 核衣壳(N)蛋白

病毒的 N 蛋白是含量最丰富且保守性较强的免疫原性蛋白，是病毒粒子核衣壳蛋白最主要的组成部分，在病毒的转录和复制过程中起着重要的作用。与其他病毒蛋白相比,N 蛋白是最为保守的免疫原蛋白,且在感染细胞中表达水平高。N 蛋白因其抗原性稳定、型特异性及其存在免疫交叉反应的抗原表位而成为病毒诊断的重要候选靶点(Munir 等,2013)。N 蛋白虽然具有高度免疫原性,但由于其位于病毒内部,因此 N 蛋白产生的抗体不具有免疫保护作用。基于 N 基因序列遗传多样性的特点,常被用于 PPRV 遗传分析,根据 N 基因序列,PPRV 被分为不同的谱系。N 基因位于基因组的 3'端,编码 PPRV 基因组中表达量最多的蛋白,即 N 蛋白。N 蛋白由 525 个氨基酸组成,通过与其他 N 蛋白(N-N)、P 蛋白(N-P)和 L 聚合酶蛋白(P-L)发生相互作用,进而参与病毒复制。虽然已经绘制出 N 蛋白自身组装的必要序列,但证实 N 蛋白两个区域[N-端(位于 1~120aa)]和中心区域(位于 146~241aa)负责 N-N 自身组装,且两个区域间的小段肽(位于 146~241aa)对 N 蛋白的稳定性起着至关重要的作用。2018 年,Ma 等证实了两个跨界 PPRV 的编码 N 蛋白使用同义密码，并且认为宿主 tRNA 变异的丰度可能对 PPRV N 蛋白二级结构的形成有潜在影响。此外,N 蛋白 106~210 位氨基酸序列能有效抑制 IRF3(IFN 调节因子 3)核转运和 IFN-β 的生成,而 IRFβ 区域(140~400aa)是 N 蛋白与 IRF3 相互作用的核心区域(Zhu 等,2019)。PPRV N 蛋白通过与 IRF3 相互作用并将 IRF3 去磷酸化,能够有效抑制 IRF3 功能和减少 IFN Ⅰ类的产生；通过干扰 TBK1-IRF3 络合物生成达到抑制 IFN 形成的目的。因此,PPRV 是极重要的病毒拮抗因子

(Zhu 等,2019)。N 蛋白在病毒复制过程中也扮演重要的角色,在病毒复制的过程中 N 蛋白通过沉默 mRNA 阻断转录,进而阻止 M 蛋白的合成,引发更多的细胞融合,从而抑制病毒粒子的释放。

13.3.2 基质(M)蛋白

M 蛋白是最保守、最小的蛋白质之一,位于病毒囊膜内,本身能够与脂质膜相结合(Haffar 等,1999)。M 蛋白在病毒粒子的组装和释放过程中起着关键作用,新组装的病毒粒子通过出芽方式释放到细胞外,同样在病毒样颗粒组装和释放过程中也起着关键作用(Wang 等,2017)。M 蛋白具有高度保守性,含有 335 个氨基酸,分子量为 37.8 kDa,在子代病毒的形成及其与细胞膜糖蛋白互作中具有重要作用。此外,M 蛋白还能在宿主细胞膜特定区域介导病毒的出芽过程。M 蛋白构成病毒包膜内层,是连接 F 蛋白、H 蛋白和核心 RNP 的桥梁。由于细胞运输需要高浓度肌动蛋白丝,因此 MV 出芽发生在顶端微绒毛。此外,基序(FMYL 50~53aa)则是进行 PPRV 蛋白定位所必需的,就像 PPRV 一样为确保出芽生殖,尼帕病毒也具有同样的基序(FMYL 50~53aa)。然而,在这些病毒中是否具有相同功能尚不明确。M 蛋白缺失能够抑制 MV 和 CDV 产生持续性感染。Liu 等人(2015)展示了在体外抑制病毒复制的过程,即通过干扰小 RNA 沉默 M 蛋白 mRNA,增加了病毒介导的细胞膜融合。

13.3.3 表面糖蛋白

病毒表面糖蛋白(H 蛋白和 F 蛋白)能够促进病毒的吸附和侵入。保守基因 F 编码的蛋白分子量为 137kDa,而具有高免疫原性的 F 蛋白,分子量为 59 kDa (Bailey 等,2005)。F 蛋白可形成纤突,嵌入脂质双层膜(Diallo 等,2007)。病毒的毒力由 F 蛋白裂解特性决定,F 蛋白的前体蛋白(F0),可被水解为 F1 和 F2 两部分,两者由二硫键连接。裂解是病毒具备融合性和感染性的必要条件,且麻疹病毒 F2 亚单位包含一个高度保守的 NXS/T 糖基化位点。虽然具体机制尚不明确,但 PPRV 含有保守基序 RRTRR(位于 104~108aa),这个保守基序与麻疹病毒特征性保守基序RRX1X2R(X1,任一氨基酸;X2,精氨酸或赖氨酸)(Chard等,2008)一致,可被反式高尔基相关的呋喃肽酶识别和切割。PPRV 的 F1 亚单位具有四个保守基序:N 末端融合肽、七倍重复体 1、2 和一个跨膜区。结构相同的七倍重复体具有一个融合机制,该机制在出芽生殖中具有一定的作用。此外,在细胞膜上融合肽区域与七倍重复体的二聚体结合引起了细胞的融合(Rahaman 等,2003)。保守的亮氨酸拉链基序(PPRV 的 459~480aa)可使蛋白质通过一种未知的过程发生聚集和融合。病毒 F 蛋白是病毒进入细胞、在细胞间扩散所必需的,并且在病毒诱导细胞病理、溶血(Devireddy 等,1999),以及细胞融合和半融合过程中起着关键作用(Seth 和 Shaila,2001)。

H 蛋白能够促进病毒与细胞受体 CD150 和 SLAM 结合(Tatsuo 等,2001)。H 蛋白基因 ORF (7376~9152 个核苷酸) 编码了分子量为 67 kDa 的 H 蛋白(血凝素-神经氨酸酶)。虽然 PPRV 与 RPV 的 H 蛋白都只有 609 个氨基酸,但这两个病毒的 H 精氨酸的同源性仅 50%,这反映了两种病毒的细胞偏嗜性和宿主的特异性不同。H 蛋白在病毒感染及其与宿主细胞膜特异性结合,宿主免疫应答产生病毒中和抗体的过程中均发挥重要作用。此外,它也是 MV 中重要的抗原决定簇,是决定细胞偏嗜性的主要因素,也是引起跨种传播的重要原因。在 PPRV 的复制过程中,H 蛋白需要 F 同源蛋白协同发挥作用,且 H 蛋白具有吸附红细胞特性和神经氨酸酶活性(Seth 和 Shaila,2001),而在一些副黏病毒中,其表面蛋白既具有血凝特性也有神经氨酸酶活性。PPRV 能够凝集不同哺乳动物和鸟类宿主的红细胞,包括感染了 PPRV 的细胞培养物,这些细胞很容易吸附鸡红细胞。此外,抗表位及其基序的识别有助于我们更好地理解 PPRV H 蛋白的抗原特性,进而也为基于表位的诊断检测和多重表位疫苗的研究奠定了基础(Yu 等,2017)。

13.3.4 大分子(L)蛋白

RNA 依赖的 RNA 聚合酶 L 蛋白是 PPRV 中最大、最保守的蛋白质,但数量最少。PPRV 的 L 蛋白由 2183 个 aa 构成，分子量与牛瘟病毒 L 蛋白相同,约为 247.3 kDa,(Bailey 等,2005)。L 蛋白是一种负责基因组 RNA 转录和复制的多功能催化蛋白,除了实现病毒 mRNA 帽化、甲基化和病毒 mRNA 的聚腺苷酸化作用外，还具有 RNA 三磷酸酶(RTPase)、尿苷转移酶(GTase)和甲基转移酶活性(Ansari 等,2019)。虽然 L 蛋白三种功能基序已被鉴定，但该蛋白对 PPRV 直接作用的研究尚未开展(Munir 等,2013)。除 PPRV 第一个缬氨酸外,P 蛋白和 L 蛋白相互作用的结合位点序列(9~21 aa)在副黏病毒属中是保守序列(Chard 等,2008)。此外,虽然该蛋白的重要功能仍未确定,但 PPRV 的 L 蛋白的多功能活性或许可通过重组体系得到确定 (Yunus 和 Shaila,2012)。而且,2019 年,Ansari 等在 L 蛋白的 C 末端区域 (1640~1840aa)发现了 RTPase 区,该区域具有 RTPase 及 RTPase 相关的三磷酸酶活性(NTPase)。

13.3.5 磷(P)蛋白

麻疹病毒的 P 蛋白、V 蛋白和 C 蛋白是由 P 基因 ORF 重叠序列编码而成 (Mahapatra 等,2003)。PPRV 的 P 蛋白变异率较高,因富含丝氨酸和苏氨酸，且翻译后经过多次磷酸化修饰呈酸性。在麻疹病毒中,P 蛋白的组成从 506~509 个氨基酸不等,在 PPRV 基因组中 P 蛋白基因最长,其在病毒复制、转录和免疫调节等多层面发挥作用(Mahapatra 等,2003)。P 蛋白的 C 端比 N 端更趋于保守,它参与了 N-P 的相互作用,是控制细胞周期、调控转录和翻译所必需的。PPRV 的 P 蛋白中的保守基序，是 P-N 的相互作用所必需的,是构成聚合酶复合体的重要组成部分，也是决定跨物种感染的关键因素。

13.3.6 辅助蛋白

非结构辅助蛋白——C 蛋白和 V 蛋白是 P 基因的 ORF 在病毒感染细胞中分别利用替代的起始密码子和 RNA 编辑机制合成的。麻疹病毒 C 蛋白的 C 端高度保守(Mahapatra 等,2003),在病毒复制中发挥重要作用,通过阻断 IFN-β 活化转录因子抑制 IFN-β 的产生。这种抑制 RPV 的 C 蛋白分子机制还有待进一步研究。虽然 PPRV C 蛋白的生物功能尚不明确,但在 MV 感染中,它以毒力因子的形式存在。病毒的 V 蛋白通过 RNA 编辑,在磷蛋白的 mRNA 中掺入一个或多个 G 残基翻译产生。麻疹病毒 V 蛋白的氨基酸长度并不相同,V 蛋白的 N 端与 P 蛋白具有相同的氨基酸残基,但在 C 端不同,C 端富含半胱氨酸残基。V 蛋白和 P 蛋白一样，也进行磷酸化,含有 60%的丝氨酸残基。V 蛋白在病毒转录和复制过程中起调节作用,是 IFN 的有效抑制剂,通过干扰和抑制 STAT 介导信号来阻断 IFN 信号(Ma 等,2015)。

13.3.7 感染机制

细胞受体是决定病毒宿主范围及其组织嗜性的首要因素,PPRV 受体是细胞 SLAM 或羊体内的 Nectin-4 分子(Birch 等,2013)。PPRV 可在非洲绿猴肾细胞(LeFevre 和 Diallo,1990)、马莫塞特 B-淋巴细胞-B95a(Sreenivasa 等,2006)和 VeroNectin-4(Fakri 等,2016)细胞等细胞系进行体外传代培养、分离和滴定。2018 年,Pawar 等观察到，在家畜 PBMC 中 PPRV 复制与 SLAM 的 mRNA 水平相关，并且推断，在非自然宿主中 PPRV 利用 SLAM 受体效率并不高。一般情况下,病毒感染细胞后 3~5 天，被感染细胞中会出现典型细胞病变效应,如细胞变圆、葡萄样簇、聚合、空泡状态、颗粒样化和合胞体。此外,感染 PPRV 的

PBMC 出现细胞病变效应,有细胞退化、膨大、变圆、聚合、无合胞体等特点(Mondal 等,2001),但在感染后 2~3 天,B95a 细胞中会发现合胞体(Sreenivasa 等,2006)。一般来说,PPRV 在乙醚中很脆弱、敏感度高,在 pH 值 5.8~10.0 环境中具有抵抗力,在 50℃ 加热 60 分钟可灭活(WOAH,2013)。

病毒复制周期第一步是吸附并融合在细胞膜表面,进而将遗传物质释放到细胞质中。当病毒吸附侵入到子宫内膜上皮细胞后，将对早期细胞基因表达造成严重影响(Yang 等,2018a,b)。病毒 H 蛋白通过受体 SLAM 或 cd150(Pawar 等,2008)或 Nectin-4(Birch 等,2013)实现与细胞膜上唾液酸结合(Munir 等,2013),病毒结合可激活 F 蛋白融合,促进病毒囊膜与细胞膜的融合,释放遗传物质进入细胞中。麻疹病毒的复制只在宿主细胞质中进行。PPRV 基因组由 N 蛋白将其包裹进而形成螺旋状 RNP,基因组从未以裸 RNA 的形式出现。该复合物包含被核衣壳蛋白包裹的 RNA，其与 P 蛋白和 L 蛋白结合形成微型复制单位(Parida 等,2015b)。聚合酶复合物作用于病毒基因组,与基因组启动子结合,启动短前导链 RNA 的转录,然后进行所有基因的转录,从而建立转录梯度。某些位点上,聚合酶复合体从 mRNA 合成转到全基因组正义链 RNA 的合成。虽然具体机制尚不明确,但认为聚合酶复合物与病毒蛋白的积累有关(Parida 等,2015a,b)。全基因组反义 RNA 合成后,聚合酶与反基因组 3’ 启动子结合并生成新的完整正义基因组。病毒组合成后,导致病毒从宿主细胞中逸出。病毒的 M 蛋白负责将新生 RNP 和病毒糖蛋白带到宿主细胞膜上，从而开始新的病毒粒子包装、出芽和释放(Parida 等,2015a,b)。

此外,Chaudhary 等人观察到(2015)受体酪氨酸激酶调节病毒复制，肿瘤相关的标志物（如 PVRL4)具有调控作用,且在发病机制中发挥重要作用,有望用于癌症治疗(Delpeut 等,2014)。此外,Balamurugan 等人证实(2008)阿拉伯金合欢的水提取物对病毒体外增殖具有抑制作用,而 Khandelwal 等于 2015 年报道指出银钠米粒子能够有效抑制病毒增殖。此外,Kumar 等 2019 年通过使用肌质网/内质网的钙镁 ATP 酶特异性抑制剂(毒萝卜素)阻断病毒进入靶细胞和病毒蛋白的合成，证明了该酶对病毒复制具有调控作用。Qi 等人在 2018 年研究了宿主 miRNA 在 PPR 病毒复制及发病机制中的作用。据 Yang 等人(2018a, b）报道，病毒感染山羊子宫内膜上皮细胞后,经 PPRV 的 C 蛋白和 N 蛋白介导作用，能够激活自噬反应，而细胞自噬能够抑制依赖于细胞凋亡蛋白酶的细胞凋亡，从而促进宿主细胞中的病毒复制和成熟。

通过对非洲、亚洲大陆不同国家(Padhi,Ma, 2014;Baron,2015;Banyard 和 Parida,2015)的PPRV 分离株/毒株进行分子生物学特性和遗传进化分析,界定了 PPRV 的四个谱系流行范围(Shaila等, 1996;Dhar 等,2002)。20 世纪 70 年代,谱系Ⅰ型 PPRV 最早从西非的尼日利亚和塞内加尔国家中分离得到,后来在科特迪瓦、几内亚、塞内加尔和布基纳法索被发现。20 世纪 80 年代，谱系Ⅱ型 PPRV 在西非不同国家中分离到（Dundon 等, 2018),目前为止,该非洲谱系病毒还未穿过红海传播到亚洲大陆。此外,美国、埃塞俄比亚、也门、苏丹、阿曼和布隆迪报道了谱系Ⅲ型病毒(Niyokwishimira 等,2019)，而亚洲谱系Ⅳ型 PPRV 在阿拉伯半岛、中东和东南亚部分地区被发现。最近,非洲和南欧也发现了亚洲谱系Ⅳ型 PPRV(Altan 等,2019)。在尼日利亚(Woma 等,2016)和中国(Liu 等,2018)出现不同谱系 PPRV 同时流行的现象,而埃塞俄比亚(Muniraju 等,2014)发现了谱系Ⅳ型 PPRV。据报道,目前在印度发现的 PPRV 只与谱系Ⅳ型 PPRV 毒株/分离株(Balamurugan 等, 2010b);Muthuchelvan 等,2014）和谱系Ⅲ型分离株(PPR TIN/92)有关(Shaila 等,1996)。

13.4　流行病学

世界各国报道，在牛、水牛、骆驼(Govindarajan 等，1997；Abraham 等，2005；Balamurugan 等，2012a，2014b；Sen 等，2014；Omani 等，2019）等小反刍动物身上，检测到了病毒抗体和病毒抗原/基因组(Singh 等，2004a；Balamurugan 等，2011，2012b，2015；Khan 等，2008；Mbyuzi 等，2014；Gari 等，2017；Balamurugan 2017，Ali 等，2019)。在不发达国家和发展中国家，不定期的疫苗接种或疫苗接种受限，导致小反刍动物和大反刍动物(Abubakar 等，2017；Woma 等，2016)存在着 PPRV 的亚临床感染/隐性感染/非致命性感染。此外，成年的小反刍动物体内检测到抗体并不代表患有 PPR。牛和猪的血清是病毒转换器，它们感染既无临床症状，也不具传染性。然而，据 2018 年 Schulz 等的报道，猪和野牛可能是感染源。牛是 PPRV 终末宿主，在流行过程中，不能发挥病毒维持和传播的作用(Couacy-Hymann 等，2019)。这些研究一方面证明 PPR 能够从小反刍动物传播到大反刍动物身上，另一方面也为病毒在体外环境中生存提供了机制。在某些地区(Balamurugan 等，2012a，2014b)，小反刍动物和大反刍动物同时饲养的农业体系中(Balamurugan 等，2014b)，非自然宿主能够抑制 PPRV 传播，这可能是由于病毒的毒力偶然适应变化或改变。但是，这还需要根据病毒的存活情况、宿主的易感性、病毒的突变、疾病严重程度变化等进一步确认。

野生动物对于病毒传播具有重要意义，因而相关研究(Taylor，1984)将野生动物在动物流行病学中的作用列为重点，尤其是对野生动物物种(如易感的野生小型反刍动物 (Marashi 等，2017)、野生有蹄类动物(Rahman 等，2018)、野生及家养动物(Li 等，2017)。野生动物感染可能是由于野生动物与家养动物在同一牧场(Mahapatra 等，2015)进行采食和饮水(Rahman 等，2016)。然而，目前为止，野生动物在 PPRV 流行病学中的作用仍不确定，还有待进一步研究(Banyard 等，2010；Balamurugan 等，2015)。此外，除了绵羊、山羊、自然感染产生抗体外，在牛和野生反刍动物体内也存在抗 PPRV 的抗体(Abraham 等，2005；Balamurugan 等，2015)。然而，除了瞪羚、骆驼 (Khalafallaa 等，2010；Zakian等，2016)、野生岩羊和四角羚(牛亚科成员)(Jaisree 等，2018)外，其他野生动物体内尚未检测到 PPRV。对犬鼻拭子通过微阵列筛选到 PPRV (Ratta 等，2016)，在狮子组织中检测到 PPRV 基因组(Balamurugan 等，2012c)，为 PPRV 跨越种属屏障传播的可能性提供了新认识。鉴于 PPRV 具有危害性，应采取接种疫苗的方法消灭病毒，打破野生小反刍动物与家养小反刍动物之间的恶性循环。

13.5　病毒感染

该病毒主要感染家养绵羊和山羊，很少感染骆驼和其他野生动物。许多研究人员指出，PPRV 虽然主要感染小反刍动物，但受影响最严重的是山羊 (Tripathi 等，1996；Singh 等，2004a；Balamurugan 等，2015)。虽然现在还没有带毒状态证据(Furley 等，1987)，但动物在感染后第 3~26 天甚至 16 周后，仍可通过分泌物和排泄物排出 PPRV (Balamurugan 等，2006，2010a；Liu 等，2014；Wasee Ullah 等，2016)，这是在 PPRV 流行中，造成 PPRV 隐性感染的主要原因。不同物种/品种对 PPRV 易感性不同的原因是 TOLL 样受体和细胞因子在其中发挥作用(Dhanasekaran 等，2015)，通过这些因素及其宿主基因因素分析能够帮助我们进一步了解宿主的多态性及其易感性。动物品种不同，病毒感染情况也可能不同(Lefevre 和 Diallo，1990)。除品种因素外，疾病与动物年龄(6~12 个月)相关，即幼畜感染引起的疾病更严重。1 岁以下的幼畜是农村或牧场环境中 PPR 的重要预测因子(Huyam 等，2014；Gitonaga，2015)，在疫情流行 4~

5个月后，后代的被动免疫就会消失(Balamurugan等，2012d)。2016年，Gowane等报道称母源抗体影响疫苗接种反应，据此推断，通过对疫苗遗传反应的合理推测，山羊在注射疫苗后免疫效果会更好。

由于受到环境和遗传影响，包括MHC Ⅱ类限制的影响，绵羊和山羊对PPRV疫苗的接种反应截然不同(Gowane等，2016、2017、2018)。动物品种、年龄、性别和群体大小都被认定是PPR的易感性因素(Teshale等，2018)。PPR流行的季节性与小反刍动物的活动范围和气候因素也有关。不过，在地方性流行时，PPR多发生在歉收年间(Balamurugan等，2011，2012b，2016)。通常情况下，小反刍动物是由农民在草原、灌木和森林地区自由放养，但在歉收年间，这些动物需要通过长途跋涉寻找食物和水源，导致PPR发病率大幅上升(Balamurugan等，2016)。大多数研究人员认为，将新引入、来路不明的羊混群饲养是PPR暴发的一个重要风险因素，Kardjadj等(2015)报道称，PPR是造成小反刍动物流产及其他相关风险的因素。一般来说，PPR在绵羊和山羊中流行，4个月至1岁的幼畜受到影响的风险最高。

虽然饲养方式被认为是发生传染性疾病的风险因素，但目前还没有证据证明舍饲和放牧的饲养方式有传染疾病的可能(Rahman等，2016)。舍饲的饲养管理方式像动物的购进，都会增加疾病传播的风险。尽管放牧并不是重大风险因素，但由于家畜在牧场和水源地觅食饮水，很有可能导致疾病不断蔓延。Rony等(2017)基于动物医院的病例进行对照研究，明确了PPR流行的决定因素和空间分布。他们指出在冬季、雨季开始之前，在PPR高发地区，尤其是高危种群(4~24月龄的小动物)应优先接种疫苗，从而提高它们在歉收之年的免疫防御能力。有研究认为PPRV的阳性血清检出率与动物品种、性别、年龄、季节和地理位置有关(Abubakar等，2008)。然而，Khan等(2008)却指出，研究发现，相较于雄性动物，雌性动物的阳性血清比例要高得多。这可能与雌雄性之间的生理差异有关，雌性的感染表现在一定程度上是由产奶和妊娠的压力造成的。与雄性动物相比，雌性动物具有更重要的生产潜力，更长生产时间，久而久之，增加了雌性动物暴露在PPRV的可能性。动物迁徙是引起该病传播的主要因素(Almeshay等，2017)。

13.6 病理变化

该病是通过PPRV感染上呼吸道上皮细胞引起的(Parida等，2015a)。Pope等(2013)证明了病毒最初在扁桃体和接种部位的淋巴结复制。他们提出，免疫细胞将PPRV转运至发生病毒复制的场所即淋巴组织后，病毒进入血液循环(最先是病毒血症)，随着病毒在靶细胞/组织中不断增殖，出现临床症状(Truong等，2015)。病毒增殖及其致病性与许多家畜的流行病决定因素有关(Munir等，2013；Balamurugan等，2015)，包括遗传学和非遗传学因素等(Gowane等，2016，2017)。Kumar等(2004)报道了在呼吸道上皮合胞体细胞中检测到了PPRV抗原。在派尔集合淋巴结和肠系膜淋巴结中发现，病毒对肺壁组织细胞和淋巴细胞具有偏嗜性，在回肠上皮细胞中也检测到了以深褐色颗粒形式存在的病毒。许多研究人员已经证实病毒存在于不同的组织和器官中，包括肠隐窝上皮细胞。关于PPRV感染发病机制及其感染期间涉及免疫细胞的相关细节已有报道(Balamurugan，2017)。白细胞短暂增多的变化被认为是一种应激反应和免疫反应，并且B细胞可能在二次病毒血症期间被激活。此外，PPRV感染诱导IFN-β能力较弱且持续时间短，病毒可主动阻断IFN-β诱导的防护作用(Sanz Bernardo等，2017)。对接种疫苗的绵羊和山羊免疫原性进行了研究(Singh等，2004b，2004c；Rajak等，2005)，单次免疫的保护性抗体可维持3~6年(Saravanan等，2010；Zahur

等,2014)。Rojas(2019)等人发现,病毒抗原 F 和 H 是感染过程中天然的抗体依赖细胞介导的细胞毒作用靶点。除此之外,他们发现了宿主能够防御 PPRV 感染的一种新免疫效应机制可能有利于病毒的清除。体液免疫反应和细胞免疫反应均可参与 PPRV 引起的免疫抑制(Rajak 等,2005),但激活宿主视黄酸诱导基因Ⅰ样受体实现精准免疫抑制的机制尚未被阐明。除此之外,Zhu 等(2019)人证实,PPRV 通过抑制 IFN-β 和 IFN 刺激基因表达机制,可显著抑制视黄酸诱导基因Ⅰ样受体途径的激活和Ⅰ型 IFN 的产生。Jagtap 等(2012)报道,与非免疫抑制动物相比,免疫抑制动物的病毒血症期短,病程长,病症重,病死率高。Mondal 等(2001)证实病毒可诱导 PBMC 细胞凋亡,随后 Kumar 等(2002)报道了 PPRV 感染动物 PBMC 淋巴细胞凋亡。通过对宿主和病毒的基因组和转录组分析发现,有 985 个不同表达的基因和转录因子对免疫通路进行调控,包括调控,剪接,并且证实了凋亡途径的异常调节。他们证实 PPRV 诱导产生 miR-21-3p、miR-320a 和 miR-363,这些能够协同下调免疫应答基因,提高病毒发病机制。基于转录组分析和 qRT-PCR 验证,Manjunath 等(2017)预测 IRF 通过与干扰素无关的方式诱导 IFN 刺激基因产生,进而激活免疫应答。Baron 等(2015)表示在山羊发病过程中 CD4+T 细胞不断减少,同年,Truong 等人的研究认为病毒复制的主要场所为淋巴结、淋巴组织和胃肠道。

13.7 传播

由于病毒在体外环境抵抗力低、不稳定,需要通过被感染动物和易感宿主近距离接触实现病毒传播(Balamurugan,2017)。在感染期间,大量的病毒通过被感染动物的排泄物和分泌物被排出体外是病毒感染发生的重要因素(Balamurugan,2017)。病毒传播主要通过与感染动物密切接触,尤其是与周边动物直接接触,吸入具有传染性气溶胶飞沫。这些病毒粒子通过污染水、饲料槽、垫料等进行间接传播。但由于病毒在体外存活时间短,因此间接传播方式似乎对 PPRV 传播意义并不大(Lefevre 和 Diallo,1990)。此外,在当地市场交易中,不同来源动物相互间的密切接触,也可能加剧病毒的传播。除绵羊和山羊外,病毒也能够传播给野生反刍动物和非自然宿主牛、水牛、骆驼、猪等,说明了病毒体外生存的机制(Abraham 等,2005),在非自然宿主和野生反刍动物中存在 PPRV 抗体,表明病毒在这些动物中进行了自然传播(Abraham 等,2005;Balamurugan 等,2012a)。Sevik 和 Oz(2019)对库蠓属在病毒传播中的作用进行了调查研究。尽管在非自然宿主有关于 PPRV 血清阳性的报道,但尚无有关非自然宿主感染 PPRV 的临床疾病/症状的报道(Sen 等,2014)。小反刍动物迁徙过程中经常通过与其他感染动物密切接触(Singh 等,2004a),引起病毒传播。2013 年,Kivaria 等报道,为了更好地理解 PPRV 的传播,发展中国家和不发达国家都需要建立 PPRV 持续感染和传播的流行病学模型研究,并对其进行预测。因此,动物迁移在疾病传播及 PPRV 在环境中的保存方面起着关键作用。Fournie(2018)等通过采用模拟集合种群的模型推断,在疫区,至少 71% 农村的易感动物的免疫率大于 37%时,才能有效防止病毒传播。

13.8 临床症状

小反刍兽疫的主要临床症状为高热、眼鼻口分泌物、口腔坏死性溃疡、胃肠炎、腹泻、呼吸急促和支气管肺炎(Balamurugan,2017)。骆驼的临床症状主要表现为结膜炎和腹部水肿(Omani 等,2019)。根据疾病严重程度,将 PPR 分为超急性、急性、轻型(WOAH,2013;Balamurugan,2017),且 PPR 的临床症状由病毒的毒力(Couacy Hymann 等,2007)、宿主、品种、年龄和动物免疫状态等因素决定。PPR 感染期分为 4~5 个阶段,即:①短潜伏期，3~10 天不等；②发热前驱期/发热期；③黏膜病变，伴有高热、结膜炎、眼和鼻腔分泌物（图 13.1A，B），口腔黏膜糜烂，尤其是舌头、上颚、唇和其他部位（图 13.1C）；④腹泻期（图 13.1D），伴有肺炎、脱水和死亡；⑤恢复阶段，大多为非致死阶段，动物在此阶段获得终身免疫。关于该病临床症状及其并发症的研究已有相关报道（Tripathi 等，1996）。

13.9 诊断

PPR 可根据感染动物的特征性临床症状和病理变化，包括肺实变，脾脏及肠系膜淋巴结肿大，结肠黏膜条纹状出血等进行诊断（图 13.1E，F），但需要通过一系列的实验室检测确诊。通过大量的实验室技术实现了对病毒抗原、抗体或基因组的检测。一般而言，包括金标准病毒分离在内的传统分析/检测技术，具有耗时长，灵敏度低，对专业技术和培养设施要求高等问题，不适合应用于早期诊断。

ELISA 被广泛用于疫病的监测，甚至应用于疾病诊断。随着对病毒学特性和分子生物学技术的深入了解，用于病毒快速检测的分子检测方法（RT-PCR、实时定量 RT-PCR 和 LAMP）被开发出

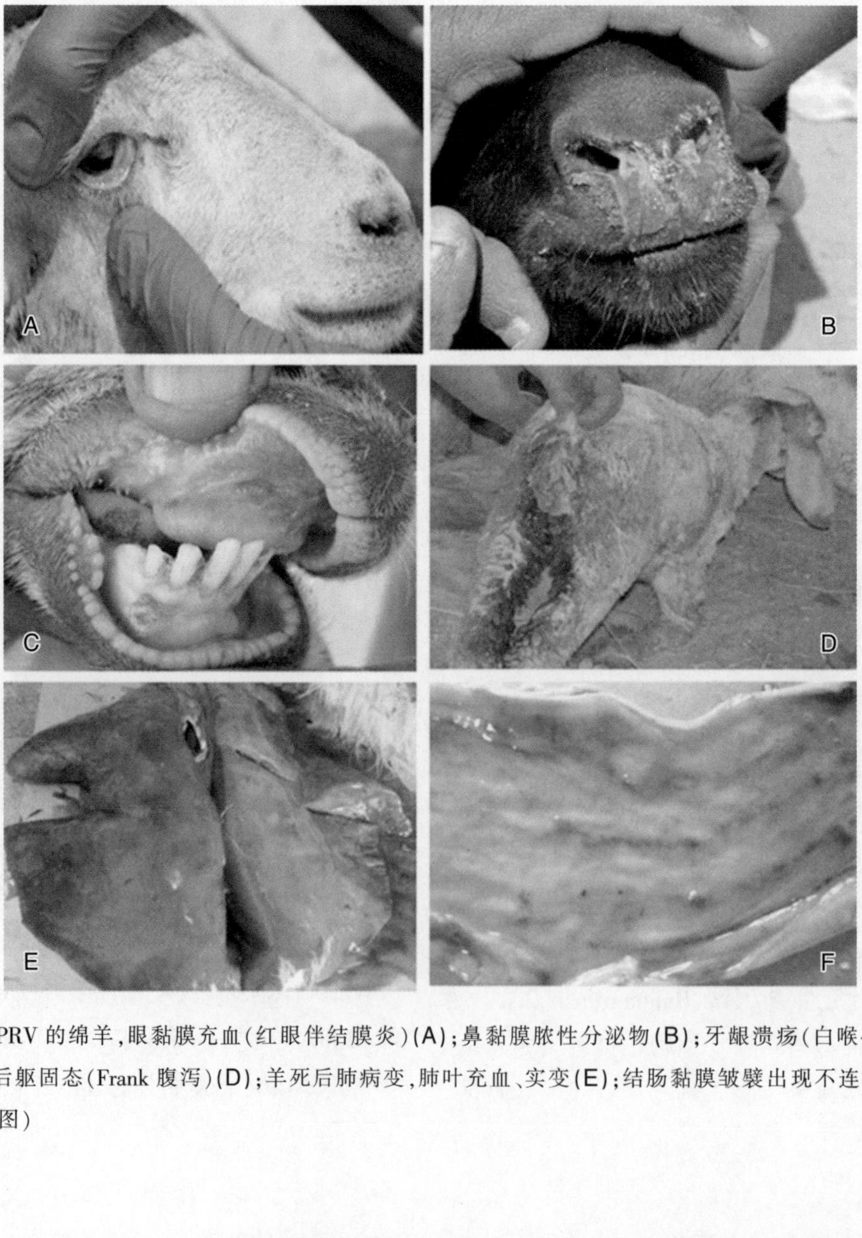

图 13.1 感染 PPRV 的绵羊，眼黏膜充血（红眼伴结膜炎）(A)；鼻黏膜脓性分泌物(B)；牙龈溃疡（白喉-斑块）和坏死斑(C)；严重腹泻，后躯固态（Frank 腹泻）(D)；羊死后肺病变，肺叶充血、实变(E)；结肠黏膜皱襞出现不连续的充血和出血(F)。（扫码看彩图）

来，这些分子生物检测方法具有特异性强、敏感性强的优势。此外，在设备较差的实验室，研制的胶体金卡可用于实时检测或现场检测，但其使用效果未达到预期水平（Raj 等，2008）。此外，运用 DNA 重组技术生产重组病毒蛋白，既安全又简单，能够较好地替代活病毒抗原并用于疫病诊断的免疫测定中。近期，研究人员研发了多种现代诊断技术，用于检测血清中的 PPRV 抗体和分泌物，血液和组织样本（包括牛奶）中的 PPRV 抗原/核酸（Clarke 等，2018），具体汇总见表 13.1。2019 年，Parida 等人报道称，鼻拭子是分子诊断中的最佳样本。鉴于可用的资源和需要筛选的样本数量，这些诊断方法在控制和根除 PPR 的不同阶段都能够得到有效地应用（Libeau，2015；Santhamani 等，2016）。

PPR 要与其他具有相似临床症状的疫病进行鉴别诊断，如传染性胸膜肺炎、巴氏杆菌病、蓝舌病、传染性脓疱、口蹄疫等。PPRV 与其他病毒共

表 13.1 PPR 诊断方法

诊断方法	特点
病毒分离（金标准检测）	主要用牛、羊细胞，Vero 细胞系、狨猴 B 类淋巴母细胞-95a、表达连接素-4 Vero 细胞（Fakri 等，2016）
病毒或血清中和试验	利用细胞培养
琼脂凝胶免疫扩散试验/对流免疫电泳	初步诊断
血凝反应	简易、价廉、几小时内出结果
竞争性 ELISA/阻断 ELISA/表位-阻断 ELISA	H 蛋白/N 蛋白的中和单抗（Singh 等，2004；Bodjo 等，2018）
免疫捕捉 ELISA/夹心 ELISA	应用 N 蛋白单克隆抗体进行快速鉴别诊断（Singh 等，2004b）
间接 ELISA/联合间接 ELISA	PPRV 抗原/多克隆抗体/G 蛋白（Yousuf 等，2015）
SNAP-ELISA	RT-PCR/ELISA 系统
免疫组织化学检测	在组织内检测抗原
免疫过氧化物酶单层试验（IPMA）	IPMA 是 VNT 替代法，通过使用 BHK-21 细胞系稳定表达山羊 SLAM
斑点 ELISA	抗 M 蛋白 MAb/抗 N 蛋白 MAb
免疫渗滤法/抗原竞争 ELISA/层析检测	在临床样本中进行抗原/抗体检测（Raj 等，2008）
以重组抗原为基础的 ELISA	使用 N/H/F 抗原（Balamurugan 2017；Basagoudanavar 等，2018）
核酸分子杂交和 RT-PCR	以 F 基因、N 基因、M 基因为靶基因进行一步法或两步法检测（Forsyth 和 Barrett，1995；Couacy-Hmann 等，2002；Balamurugan 等，2006）
PCR-ELISA	PCR 和 ELISA 相结合（Saravanan 等，2004）
实时 PCR-ELISA	以 SYBR 染色/TaqMan 探针为基础的单重或双重实时定量 RT-PCR 方法，或一步法多重实时定量 RT-PCR
环介导等温扩增（LAMP）	以 N 基因为靶基因
荧光素酶免疫沉淀反应系统（LIPS）	对 PPRV 抗体快速检测和特异鉴别
反转录重组酶聚合酶扩增实验（RT-RPA）	以 N 基因为靶基因的常规和实时 RT-RPA（Zhang 等，2018；Li 等，2018）
实时层析 RT-RPA 检测	以 PPRV N 基因为靶基因的实时荧光检测
免疫层析检测试纸条（LFIAS）	快速、超灵敏试纸条系统，将量子点（QD）与运用重组 PPRV N 的 LFIAS 相结合（Cheng 等，2017）

存/并发、混合感染引起的疫病也应进行鉴别(Balamurugan,2017;Kumar 等,2016;Maan 等,2018;Adedeji 等,2019;Malik 等,2011)。动物贸易能够促进病毒的传播和流行,因此,对这些病毒或病毒混合感染进行调查,在流行病学和经济上都具有重要意义,有助于为动物贸易制订更好的指导方针(Kumar 等,2016)。

13.10 经济损失

在发展中国家和不发达国家,绵羊和山羊为偏远地区的贫困失地小农提供了社会和经济保障。生活在本地的农民由于受到相关社会、经济、气候风险因素影响,生活已经陷入困境,而 PPR 在绵羊和山羊中的流行,更是让他们的生活受到了影响。PPR 在初生小反刍动物的流行,致死率为 50%~80%。高致死率直接导致畜牧业生产受损。此外,被感染的种群中也会出现体重下降、产奶量减少和少数流产病例。为了控制疫病,农民们需要支付治疗费用和受感染动物护理的人工费用。少数情况下,尤其是农场大批动物死亡时,也会廉价出售。在宏观层面上,如果 PPR 在更广范围内暴发,将对整个小反刍动物价值链产生明显的连锁效应,造成家畜相关生产环节上的经济溢出效应。有些影响是直接的,如生产损失,有些是间接的,如处理成本、折价出售和市场限制。有些影响是有形的,易于估计,但许多是无形的,难以衡量。有形影响包括致死率、产奶、产肉量降低、收入减少,而无形影响包括资本累积、投资分散和收入下滑等损失(de Haan 等,2015)。通过分析 PPR 对食品供应、消费变化的影响,感染 PPR 羊群的收入和支出调整及 PPR 对女性参与、就业、资产损失、家庭教育等其他社会变量的影响等,综合评估出 PPR 造成的整体影响。从小佃农的角度看,PPR 感染小反刍动物在农业和服务领域造成的严重影响需要给予适当的干预措施(de Haan 等,2015)。因此,PPR 在农场、区域、国家和全球层面上造成的经济成本是巨大的。PPR 对农民造成了严重的社会经济影响,据报道,每年造成的经济损失为 14.5~21 亿美元(WOAH 和 FAO,2015)。

通过评估不同国家、区域、不同生产环境和管理模式对 PPR 的影响,有利于在国家、地区、区域制订适当的 PPR 控制计划。这将有利于说服各国政府和国际组织提供必要的资金支持 PPR 控制,最终实现 PPR 全球根除计划。据估计,要实现全球 PPR 根除大约需要 3.08 亿美元 (Jones 等,2016),包括全球、区域和国家协调相关费用;机构发展、流行病学监测,诊断实验室建立、疫苗接种实施、培训和研究;社会经济学;应急措施;应急响应。据 Govindaraj 等(2016)估计,在印度每年 10% 的发病率就会造成高达 161.1 亿卢比 (1 卢比≈0.08 元)损失。

13.11 预防与控制

一般而言,PPR 的控制和消除依赖于对易感动物群体进行及时、有效地诊断、监测和免疫接种。现有 PPR 疫苗已经广泛用于印度(Sungri 96 毒株)和世界其他国家(尼日利亚 75/1 毒株)绵羊和山羊的 PPR 防控。然而,这些疫苗的使用不利于区分被感染动物和接种疫苗动物。PPR 严重影响绵羊和山羊生产,限制了疫情流行国家贸易。因此,PPR 预防和控制有重要意义。实施有效的预防措施是疫病控制的当务之急。PPR 的传播地域及其流行病学参数和传播方式可借助诊断设备进行简化处理(Taylor,2016)。在疫区,现有控制措施建议通过疫苗接种进行疫病控制。尽管在疫区对动物进行隔离或限制其活动能够控制疫病传播,但由于不发达国家和发展中国家存在的诸多问题,很难在实地建立起这种机制。此外,在 PPR 流行的国家,卫生、植物检疫控制措施难以有效防控。因此,大规模接种疫苗成了有效控制该病的唯一选择。除了疫苗接种外,对新购买动物实施 2~3 周隔离,了解所购买动物的健康情况和来源,对被感染动物进行持续监测，对被污染地区的设备和衣物等进行消毒，都能够达到控制疫病的目的。目前尚无针对 PPR 的具体治疗方法。受感染动物的治疗方案主要是疫情暴发的第一周使用广谱抗生素加补液疗法，对受感染动物群进行疫苗接种(Abubakar 等,2017)。

13.11.1　疫苗

各国基于国内疫病流行情况制订了不同的控制方案。然而,由于社会经济原因,不发达国家和发展中国家制订的政策受到限制，根除政策在这些国家不具备可行性。疫苗接种可以使该疫病得到有效防控，从而避免农民遭受经济损失(Singh 等,2009;Singh,2011;Balamurugan 等,2016)。在许多国家牛瘟被成功地消灭，增强了人们采取类似方案控制 PPR 的信心。从生产和采购到现场疫苗接种的各个阶段,科技人员,训练有素的兽医人员、技术和兽医辅助人员参与对疫苗接种程序的制订实施都具有重大意义。参与大规模免疫接种计划的兽医和辅助人员的专业认定对疫苗接种计划成功实施至关重要(Singh 等,2009)。随着公私合作关系的建立,尤其是非政府组织、合作社和私人兽医从业人员参与到上述所说的疫病控制方案的实施过程，使得预防性服务正在全球逐步展开(Singh 等,2009)。此外,建立网络数据库能够有效协调疫病控制方案的实施。一些国家通过利用自身资源或在国际机构的帮助下已经启动了 PPR 疫苗接种控制方案，从而增加小反刍动物的产量(Singh 和 Bandyopadhyay,2015)。因此,开展 PPR 防控计划和根除计划在技术上、经济上和实际执行上都是切实可行的(Singh 等,2009)。

鉴于小反刍动物种群动态变化，每年新生绵羊、山羊数量(占 30%~40%),以及农业实践和农业气候条件,需实现更大范围的 PPR 疫苗接种目标,达到 80%的群体免疫(Singh,2011;Woma 等,2016)。因此,在特定时期对特定区域所有种群进行免疫接种，然后再对 3~4 个月的新生动物进行免疫接种(Singh,2011),这是避免新生动物在易感窗口期感染的必要措施(Balamurugan 等,2012d)。同样,疫苗接种也要针对风险性高的仔畜、迁徙动物及市场交易畜群等(Singh,2011);另外,在一个划定区域内执行严格的疫苗接种政策，可在 PPR 流行地区设置“无 PPR 区”。包括疫苗接种策略、防控计划和根除计划在内的 PPR 防控措施(Balamurugan,2016;Balamurugan,2017;Raj 和 Thangavelu,2015)。

13.12　展望

PPR 是小反刍动物跨界动物病毒性疾病的一种,限制了国内和国际贸易和生产,因此实现 PPR 控制对不发达国家和发展中国家消除贫困具有重要意义。家畜流行病对农民、畜牧业和国民经济造成了危害。由于绵羊和山羊种群是动态变化的,只有加强 PPR 的免疫接种才能控制该病。到目前为止，一些发展中国家已经通过使用安全有效的 PPR 疫苗实现了对 PPR 有效控制。然而,病毒在更大范围内的快速传播使相关研究变得有必要,以便更好理解农业气候变化对不同地区疫情发生的影响。研究包括疾病传播机制和传播动力学,分析疫情发生与风险因素的关系，构建合适的疫病预警和预测模型。此外,根据种群动态和当地具体流行病学特点调整疫苗接种策略，有利于实现有限资源的最优分配，提高根除 PPR 的可能性(Fournie 等,2018)。国家和国际资助机构提供的疫苗接种方案显示,如果确定了 PPRV 的传染源,就能在病毒传播高峰期实现对病毒的高效消除(Taylor,2016)。要实现病毒的高效防控、迅速根除,需要进一步研究、开发、应用最新的或下一代的疫苗和病毒诊断技术,更确切地说,还需要谨慎应用流行病学和传统病毒学提供的有限信息,有效控制、迅速根除该病(Baron 等,2017)。在目前全世界根除牛瘟的状况下，由于 PPRV 对大型反刍动物也有感染性,因此,根除 PPRV 尤为重要。

此外，FAO、WOAH 和其他国际机构通过加强合作、增进了解、分享经验、采取措施，PPR 的控制和消除将比牛瘟的控制和消除更为迅速。

参考文献

Abraham G, Sintayehu A, Libeau G, Albina E, Roger F, Laekemariam Y, Abayneh D, Awoke KM (2005) Antibody seroprevalences against Peste des petits ruminants (PPR) virus in camels, cattle, goats and sheep in Ethiopia. Prev Vet Med 70:51–57

Abubakar M, Ali Q, Khan HA (2008) Prevalence and mortality rate of Peste des petits ruminant (PPR): possible association with abortion in goat. Trop Anim Health Prod 40(5):317–321

Abubakar M, Mahapatra M, Muniraju M, Arshed MJ, Khan EH, Banyard AC, Ali Q, Parida S (2017) Serological detection of antibodies to Peste des petits ruminants virus in large ruminants. Transbound Emerg Dis 64(2):513–519

Adedeji AJ, Dashe Y, Akanbi OB, Woma TY, Jambol AR, Adole JA, Bolajoko MB, Chima N, Asala O, Tekki IS, Luka P, Okewole P (2019) Co-infection of Peste des petits ruminants and goatpox in a mixed flock of sheep and goats in Kanam, North Central Nigeria. Vet Med Sci 5(3):412–418. https://doi.org/10.1002/vms3.170

Adombi CM, Waqas A, Dundon WG, Li S, Daojin Y, Kakpo L, Aplogan GL, Diop M, Lo MM, Silber R, Loitsch A, Diallo A (2017) Peste des petits ruminants in Benin: persistence of a single virus genotype in the country for over 42 years. Transbound Emerg Dis 64(4):1037–1044

Ali WH, Osman NA, Asil RM, Mohamed BA, Abdelgadir SO, Mutwakil SM, Mohamed NEB (2019) Serological investigations of Peste des petits ruminants among cattle in Sudan. Trop Anim Health Prod 51(3):655–659

Almeshay MD, Gusbi A, Eldaghayes I, Mansouri R, Bengoumi M, Dayhum AS (2017) An epidemiological study on Peste des petits ruminants in Tripoli Region, Lybia. Vet Ital 53(3):235–242

Altan E, Parida S, Mahapatra M, Turan N, Yilmaz H (2019) Molecular characterization of Peste des petits ruminants viruses in the Marmara Region of Turkey. Transbound Emerg Dis 66(2):865–872

Ansari MY, Singh PK, Rajagopalan D, Shanmugam P, Bellur A, Shaila MS (2019) The large protein ‘L’ of Peste-des-petits-ruminants virus exhibits RNA triphosphatase activity, the first enzyme in mRNA capping pathway. Virus Genes 55(1):68–75

Baazizi R, Mahapatra M, Clarke BD, Ait-Oudhia K, Khelef D, Parida S (2017) Peste des petits ruminants (PPR): a neglected tropical disease in Maghreb region of North Africa and its threat to Europe. PLoS One 12(4):e0175461

Bailey D, Banyard AC, Dash P, Ozkul A, Barrett T (2005) Full genome sequence of peste des petits ruminants virus, a member of the Morbillivirus genus. Virus Res 110:119–124

Balamurugan V (2017) Peste des petits ruminants. In: Bayry J (ed) Emerging and re-emerging infectious diseases of livestock. Springer International Publishing AG, Cham, pp 55–98. ISBN978-3-3-19-47424-3

Balamurugan V, Govindaraj G, Rahman H (2016) Planning, implementation of Peste des petits ruminants control programme and strategies adopted for disease control in India. British J Virol 3:53–62

Balamurugan V, Hemadri D, Gajendragad MR, Singh RK, Rahman H (2014a) Diagnosis and control of peste des petits ruminants: a comprehensive review. Virus 25:39–56

Balamurugan V, Krishnamoorthy P, Raju DSN, Rajak KK, Bhanuprakash V, Pandey AB, Gajendragad MR, Rahman PK (2014b) Prevalence of *Peste-des-petits-ruminant virus* antibodies in cattle, buffaloes, sheep and goats in India. Indian J Virol 25:85–90

Balamurugan V, Krishnamoorthy P, Veeregowda BM, Sen A, Rajak KK, Bhanuprakash V, Gajendragad MR, Prabhudas K (2012a) Seroprevalence of Peste des petits ruminants in cattle and buffaloes from Southern Peninsular India. Trop Anim Health Pro 44:301–306

Balamurugan V, Rahman H, Munir M (2015) Host susceptibility to PPR virus. In: Munir M (ed) Peste des petits ruminants virus. Springer-Verlag Berlin Heidelberg, Berlin, pp 39–50

Balamurugan V, Saravanan P, Sen A, Rajak KK, Bhanuprakash V, Krishnamoorthy P, Singh RK (2011) Sero-epidemiological study of peste des petits ruminants in sheep and goats in India between 2003 and 2009. Rev Sci Tech 30:889–896

Balamurugan V, Saravanan P, Sen A, Rajak KK, Venkatesan G, Krishnamoorthy P, Bhanuprakash V, Singh RK (2012b) Prevalence of peste des petits ruminants among sheep and goats in India. J Vet Sci 13:279–285

Balamurugan V, Sen A, Saravanan P, Bhanuprakash V, Patra RC, Swarup D, Mahesh Kumar MP, Yadav MP, Singh RK (2008) Potential antiviral effect of aqueous extract of *Acacia arabica* leaves on peste des petits ruminants (PPR) virus replication. Pharm Biol 46:171–179

Balamurugan V, Sen A, Saravanan P, Singh RP, Singh RK, Rasool TJ, Bandyopadhyay SK (2006) One-step multiplex RT-PCR assay for the detection of Peste-des-petits-ruminants virus in clinical samples. Vet Res Commun 30:655–666

Balamurugan V, Sen A, Venkatesan G, Bhanot V, Yadav V, Bhanuprakash V, Singh RK (2012c) Peste des petits ruminants virus detected in tissues from an Asiatic lion (*Panthera Leo Persica*) belongs to Asian lineage IV. J Vet Sci 13:203–206

Balamurugan V, Sen A, Venkatesan G, Rajak KK, Bhanuprakash V, Singh RK (2012d) Study on passive immunity: time of vaccination in kids born to goats vaccinated against peste des petits ruminants. Virol Sin 27:228–233

Balamurugan V, Sen A, Venkatesan G, Yadav V, Bhanot V, Bhanuprakash V, Singh RK (2010a) Application of semi-quantitative M gene-based hydrolysis probe (TaqMan) real-time RT-PCR assay for the detection of Peste des petits ruminants virus in the clinical samples for investigation into clinical prevalence of disease. Transbound Emerg Dis 57:383–395

Balamurugan V, Sen A, Venkatesan G, Yadav V, Bhanot V, Bhanuprakash V, Singh RK (2012e) A rapid and sensitive one step-SYBR green based semi quantitative real time RT-PCR for the detection of peste des petits ruminants virus in the clinical samples. Virol Sin 27:1–9

Balamurugan V, Sen A, Venkatesan G, Yadav V, Bhanot V, Riyesh T, Bhanuprakash V, Singh RK (2010b) Sequence and phylogenetic analyses of the structural genes of virulent isolates and vaccine strains of Peste des petites ruminants virus from India. Transbound Emerg Dis 57:352–364

Banyard AC, Parida S (2015) Molecular epidemiology of PPR. In: Munir M (ed) Peste des petits ruminants virus. Springer-Verlag Berlin Heidelberg, Berlin, pp 69–94

Banyard AC, Parida S, Batten C, Oura C, Kwiatek O, Libeau G (2010) Global distribution of peste des petits ruminants virus and prospects for improved diagnosis and control. J Gen Virol 91:2885–2897

Banyard AC, Wang Z, Parida S (2015) Peste des petits ruminants virus, eastern Asia. Emerg Infect Dis 20:2176–2178

Bao J, Li L, Wang Z, Barrett T, Suo L, Zhao W, Liu Y, Liu C, Li J (2008) Development of one-step real-time RT-PCR assay for detection and quantitation of peste des petits ruminants virus. J Virol Methods 148:232–236

Bao J, Wang Q, Li L, Liu C, Zhang Z, Li J, Wang S, Wu X, Wang Z (2017) Evolutionary dynamics of recent peste des petits ruminants virus epidemic in China during 2013–2014. Virology 510:156–164

Baron J, Bin-Tarif A, Herbert R, Frost L, Taylor G, Baron MD (2015) Early changes in cytokine expression in peste des petits ruminants disease. Vet Res 45:22

Baron MD (2015) The molecular biology of PPRV. In: Munir M (ed) Peste des petits ruminants virus. Springer-Verlag Berlin Heidelberg, Berlin, pp 11–38

Baron MD, Diallo A, Lancelot R, Libeau G (2016) Peste des petits ruminants virus. Adv Virus Res 95:1–42

Baron MD, Diop B, Njeumi F, Willett BJ, Bailey D (2017) Future research to underpin successful peste des petits ruminants virus (PPRV) eradication. J Gen Virol 98(11):2635–2644

Barrett T, Amarel-Doel C, Kitching RP, Gusev A (1993) Use of the polymerase chain reaction in differentiating rinderpest field virus and vaccine virus in the same animals. Rev Sci Tech 12:865–872

Basagoudanavar SH, Hosamani M, Muthuchelvan D, Singh RP, Santhamani R, Sreenivasa BP, Saravanan P, Pandey AB, Singh RK, Venkataramanan R (2018) Baculovirus expression and purification of peste-des-petits-ruminants virus nucleocapsid protein and its application in diagnostic assay. Biologicals 55:38–42. https://doi.org/10.1016/j.biologicals.2018.07.0030

Berguido FJ, Bodjo SC, Loitsch A, Diallo A (2016) Specific detection of peste des petits ruminants virus antibodies in sheep and goat sera by the luciferase immunoprecipitation system. J Virol Methods 227:40–46

Birch J, Juleff N, Heaton MP, Kalbfleisch T, Kijas J, Bailey D (2013) Characterization of ovine Nectin-4, a novel peste des petits ruminants virus receptor. J Virol 87:4756–4761

Bodjo SC, Baziki JD, Nwankpa N, Chitsungo E, Koffi YM, Couacy-Hymann E, Diop M, Gizaw D, Tajelser IBA, Lelenta M, Diallo A, Tounkara K (2018) Development and validation of an epitope-blocking ELISA using an anti-haemagglutinin monoclonal antibody for specific detection of antibodies in sheep and goat sera directed against peste des petits ruminants virus. Arch Virol 163(7):1745–1756. https://doi.org/10.1007/s00705-018-3782-1

Chard LS, Bailey DS, Dash P, Banyard AC, Barrett T (2008) Full genome sequences of two virulent strains of peste-des-petits ruminants virus, the Coted'Ivoire 1989 and Nigeria 1976 strains. Virus Res 136:192–197

Chaudhary K, Chaubey KK, Singh SV, Kumar N (2015) Receptor tyrosine kinase signaling regulates replication of the peste des petits ruminants virus. Acta Virol 59:78–83

Cheng S, Sun J, Yang J, Lv J, Wu F, Lin Y, Liao L, Ye Y, Cao C, Fang L, Hua Q (2017) A new immunoassay of serum antibodies against Peste des petits ruminants virus using quantum dots and a lateral-flow test strip. Anal Bioanal Chem 409(1):133–141

Clarke BD, Islam MR, Yusuf MA, Mahapatra M, Parida S (2018) Molecular detection, isolation and characterization of Peste-des-petits ruminants virus from goat milk from outbreaks in Bangladesh and its implication for eradication strategy. Transbound Emerg Dis 65(6):1597–

1604. https://doi.org/10.1111/tbed.12911

Couacy-Hymann E, Bodjo C, Danho T, Libeau G, Diallo A (2007) Evaluation of the virulence of some strains of peste-des-petits-ruminants virus (PPRV) in experimentally infected West African dwarf goats. Vet J 173:178–183

Couacy-Hymann E, Koffi MY, Kouadio VK, Mossoum A, Kouadio L, Kouassi A, Assemian K, Godji PH, Nana P (2019) Experimental infection of cattle with wild type peste-des-petits-ruminants virus—their role in its maintenance and spread. Res Vet Sci 124:118–122

Couacy-Hymann E, Roger F, Hurard C, Guillou JP, Libeau G, Diallo A (2002) Rapid and sensitive detection of peste des petits ruminants virus by a polymerase chain reaction assay. J Virol Methods 100:17–25

de Haan NC, Kimani T, Rushton J, Lubroth J (2015) Why is small ruminant health important–Peste des petits ruminants and its impact on poverty and economics? In: Munir M (ed) Peste des petits ruminants virus. Springer-Verlag Berlin Heidelberg, Berlin, pp 195–226

Delpeut S, Noyce RS, Richardson CD (2014) The tumor-associated marker, PVRL4 (nectin-4), is the epithelial receptor for morbilliviruses. Viruses 6:2268–2286

Devireddy LR, Raghavan R, Ramachandran S, Shaila MS (1999) The fusion protein of peste des petits ruminants virus is a hemolysin. Arch Virol 144:1241–1247

Dhanasekaran S, Biswas M, Vignesh AR, Ramya R, Raj GD, Tirumurugaan KG, Raja A, Kataria RS, Parida S, Elankumaran S (2015) Toll-like receptor responses to Peste des petits ruminants virus in goats and water buffalo. PLoS One 9:e111609

Dhar P, Sreenivasa BP, Barrett T, Corteyn M, Singh RP, Bandyopadhyay SK (2002) Recent epidemiology of peste des petits ruminants virus (PPRV). Vet Microbiol 88:153–159

Diallo A (1990) Morbillivirus group: genome organization and proteins. Vet Microbiol 23:55–163

Diallo A, Minet C, Le Goff C, Berhe G, Albina E, Libeau G, Barrett T (2007) The threat of peste des petits ruminants: progress in vaccine development for disease control. Vaccine 25:5591–5597

Dundon WG, Adombi CM, Kanu S, Loitsch A, Cattoli G, Diallo A (2018) Complete genome sequence of a lineage II Peste des petits ruminants virus from sierra Leone. Genome Announc 6(1):e01417

Fakri F, Elarkam A, Daouam S, Tadlaoui K, Fassi-Fihri O, Richardson CD, Elharrak M (2016) VeroNectin-4 is a highly sensitive cell line that can be used for the isolation and titration of Peste des Petits ruminants virus. J Virol Methods 228:135–139

Forsyth MA, Barrett T (1995) Detection and differentiation of rinderpest and peste des petits ruminants viruses in diagnostic and experimental samples by polymerase chain reaction using P and F gene-specific primers. Virus Res 39:151–163

Fournié G, Waret-Szkuta A, Camacho A, Yigezu LM, Pfeiffer DU, Roger F (2018) A dynamic model of transmission and elimination of peste des petits ruminants in Ethiopia. Proc Natl Acad Sci U S A 115(33):201711646. https://doi.org/10.1073/pnas.1711646115

Furley CW, Taylor WP, Obi TU (1987) An outbreak of *peste des petits ruminants* in a zoological collection. Vet Rec 121:443–447

Gargadennec L, Lalanne A (1942) La peste des petits ruminants. Bull Serve Zootech Epizoot Afr Occid Fr 5:16–21

Gari G, Serda B, Negesa D, Lemma F, Asgedom H (2017) Serological investigation of Peste des petits ruminants in East Shewa and Arsi Zones, Oromia Region, Ethiopia. Vet Med Int 17:9769071

Gibbs EP, Taylor WP, Lawman MJ, Bryant J (1979) Classification of peste des petits ruminants virus as the fourth member of the genus Morbillivirus. Intervirology 11:268–274

Gitonga P (2015) Spatial analysis of risk factors and their effects on peste des petits ruminants control strategies in Kajiado and Marsabit pastoral systems of Kenya. Doctor of Philosophy in Dryland Resource Management. https://doi.org/10.13140/RG.2.1.5043.4164

Govindaraj G, Balamurugan V, Rahman H (2016) Estimation of economic loss of PPR in sheep and goats in India: an annual incidence based analysis. Br J Virol 3(3s):77–85

Govindarajan R, Koteeswaran A, Venugopalan AT, Shyam G, Shaouna S, Shaila MS, Ramachandran S (1997) Isolation of pestes des petits ruminants virus from an outbreak in Indian buffalo *(Bubalus bubalis)*. Vet Rec 141:573–574

Gowane GR, Akram N, Misra SS, Prakash V, Kumar A (2018) Genetic diversity of Cahi DRB and DQB genes of caprine MHC class II in Sirohi goat. J Genet 97(2):483–492

Gowane GR, Akram N, Prince LLL, Prakash V, Kumar A (2017) Ovar-MHC polymorphism in Malpura and Avikalin sheep vaccinated for Peste des petits ruminants (PPR) virus. Anim Biotechnol 28(4):306–314

Gowane GR, Najif Akram SS, Misra VP, Kumar A (2016) Assessment of the antibody response to Peste des petits ruminants (PPR) disease vaccination in a flock of Sirohi goat kids. Small Rumin Res 138:20–24

Haffar A, Libeau G, Moussa A, Cècile M, Diallo A (1999) The matrix protein gene sequence analysis reveals close relationship between PPR virus and dolphin morbillivirus. Virus Res 64:69–75

Huyam AMES, Abdelhamid AMEK, Intisar S, Yahia HA (2014) Seroprevalence and risk factors of

Peste des petits ruminants in sheep and goats in Sudan. J Advan Vet Anim Res 1:42–49. https://doi.org/10.5455/javar.2014.a12

Jagtap SP, Rajak KK, Garg UK, Sen A, Bhanuprakash V, Sudhakar SB, Balamurugan V, Patel A, Ahuja A, Singh RK, Vanamayya PR (2012) Effect of immunosuppression on pathogenesis of peste des petits ruminants (PPR) virus infection in goats. Microb Pathog 52:217–226

Jaisree S, Aravindhbabu RP, Roy P, Jayathangaraj MG (2018) Fatal peste des petits ruminants disease in Chowsingha. Transbound Emerg Dis 65(1):e198–e201

Jones BA, Rich KM, Mariner JC, Anderson J, Jeggo M, Thevasagayam S, Cai Y, Andrew RP, Peter R (2016) The economic impact of eradicating PPR: a benefit-cost analysis. PLoS One 11(2):E0149982

Kardjadj M, Kouidri B, Metref D, Luka PD, Ben-Mahdi MH (2015) Abortion and various associated risk factors in small ruminants in Algeria. Prev Vet Med 123:97–101

Khalafallaa AI, Saeed IK, Ali YH, Abdurrahman MB, Kwiatekc O, Libeau G, Obeidaa AA, Abbas Z (2010) An outbreak of peste des petits ruminants (PPR) in camels in the Sudan. Acta Trop 116:161–165

Khan HA, Siddique M, Rahman S, Abubakar M, Ashraf MJ (2008) The detection of antibody against peste des petits ruminants virus in sheep, goats, cattle and buffaloes. Trop Anim Health Pro 40:521–527

Khandelwal N, Kaur G, Chaubey KK, Singh P, Sharma S, Tiwari A, Singh SV, Kumar N (2015) Silver nanoparticles impair Peste des petits ruminants virus replication. Virus Res 190:1–7

Kivaria FM, Kwiatek O, Kapaga AM, Swai ES, Libeau G, Moshy W, Mbyuzi AO, Gladson J (2013) The incursion, persistence and spread of peste des petits ruminants in Tanzania: epidemiological patterns and predictions. Onderstepoort J Vet Res 80:593

Kumar N, Barua S, Riyesh T, Chaubey KK, Rawat KD, Khandelwal N, Mishra AK, Sharma N, Chandel SS, Sharma S, Singh MK, Sharma DK, Singh SV, Tripathi BN (2016) Complexities in isolation and purification of multiple viruses from mixed viral infections: viral interference, persistence and exclusion. PLoS One 11(5):e0156110

Kumar N, Khandelwal N, Kumar R, Chander Y, Rawat KD, Chaubey KK, Sharma S, Singh SV, Riyesh T, Tripathi BN, Barua S (2019) Inhibitor of sarco/endoplasmic reticulum calcium-ATPase impairs multiple steps of paramyxovirus replication. Front Microbiol 10:209

Kumar N, Maherchandani S, Kashyap SK, Singh SV, Sharma S, Chaubey KK, Ly H (2014) Peste des petits ruminants virus infection of small ruminants: a comprehensive review. Viruses 6(6):2287–2327

Kumar P, Kumar R, Sharma AK, Tripathi BN (2002) Pathology of peste des petits ruminants (PPR) in goats and sheep : spontaneous study. Indian J Vet Pathol 26:15–18

Kumar P, Tripathi BN, Sharma AK, Kumar R, Sreenivasa BP, Singh RP, Dhar P, Bandyopadyaya SK (2004) Pathological and immunohistochemical study of experimental Peste des petits ruminants virus infection in goats. J Veterinary Med Ser B 51:1–7

Kwiatek O, Ali YH, Saeed IK, Khalafalla AI, Mohamed OI, Obeida AA, Rahman MBA, Osman HM, Taha KM, Abbas Z, Harrak ME, Lhor Y, Diallo A, Lancelot R, Albina E, Libeau G (2011) Asian lineage of Peste des petits ruminants virus, Africa. Emerg Infect Dis 17:1223–1231

Lefèvre PC, Diallo A (1990) Peste des petits ruminants virus. Rev Sci Tech 9:951–965

Li J, Li L, Wu X, Liu F, Zou Y, Wang Q, Liu C, Bao J, Wang W, Ma W, Lin H, Huang J, Zheng X, Wang Z (2017) Diagnosis of Peste des petits ruminants in wild and domestic animals in Xinjiang, China, 2013–2016. Transbound Emerg Dis 64(6):e43–e47

Li L, Bao J, Wu X, Wang Z, Wang J, Gong M, Liu C, Li J (2010) Rapid detection of peste des petits ruminants virus by a reverse transcription loop-mediated isothermal amplification assay. J Virol Methods 170:37–41

Li Y, Li L, Fan X, Zou Y, Zhang Y, Wang Q, Sun C, Pan S, Wu X, Wang Z (2018) Development of real-time reverse transcription recombinase polymerase amplification (RPA) for rapid detection of peste des petits ruminants virus in clinical samples and its comparison with real-time PCR test. Sci Rep 8(1):17760

Libeau G (2015) Current advances in serological diagnosis of PPRV. In: Munir M (ed) Peste des petits ruminants virus. Springer-Verlag Berlin Heidelberg, Berlin, pp 133–154

Liu F (2018) Letter to the editor concerning "First report of peste des petits ruminants virus lineage II in Hydropotes inermis, China" by Zhou et al. Transbound Emerg Dis; 2017: https://doi.org/10.1111/tbed.12683. Transbound Emerg Dis 65(4):1125. https://doi.org/10.1111/tbed.12840

Liu F, Wu X, Zou Y, Li L, Liu S, Chi T, Wang Z (2015) Small interfering RNAs targeting peste des petits ruminants virus M mRNA increase virus-mediated fusogenicity and inhibit viral replication in vitro. Antivir Res 123:22–26

Liu W, Wu X, Wang Z, Bao J, Li L, Zhao Y, Li J (2014) Virus excretion and antibody dynamics in goats inoculated with a field isolate of peste des petits ruminants virus. Transbound Emerg Dis 60(suppl 2):63–68

Ma X, Yang X, Nian X, Zhang Z, Dou Y, Zhang X, Luo X, Su J, Zhu Q, Cai X (2015) Identification of amino-acid residues in the V protein of peste des petits ruminants essential for interference and suppression of STAT-mediated interferon signaling. Virology 483:54–63

Ma XX, Wang YN, Cao XA, Li XR, Liu YS, Zhou JH, Cai XP (2018) The effects of codon usage on the formation of secondary structures of nucleocapsid protein of peste des petits ruminants virus. Genes Genomics 40(9):905–912

Maan S, Kumar A, Gupta AK, Dalal A, Chaudhary D, Gupta TK, Bansal N, Kumar V, Batra K, Sindhu N, Kumar A, Mahajan NK, Maan NS, Mertens PPC (2018) Concurrent infection of bluetongue and Peste-des-petits-ruminants virus in small ruminants in Haryana state of India. Transbound Emerg Dis 65(1):235–239

Mahapatra M, Parida S, Egziabher BG, Diallo A, Barrett T (2003) Sequence analysis of the phosphoprotein gene of peste des petits ruminants (PPR) virus: editing of the gene transcript. Virus Res 96(1–2):85–98

Mahapatra M, Sayalel K, Muniraju M, Eblate E, Fyumagwa R, Shilinde L, Mdaki M, Keyyu J, Parida S, Kock R (2015) Spillover of Peste des petits ruminants virus from domestic to wild ruminants in the Serengeti ecosystem, Tanzania. Emerg Infect Dis 21(12):2230–2234

Malik YS, Singh D, Chandrashekar KM, Shukla S, Sharma K, Vaid N, Chakravarti S (2011) Occurrence of dual infection of peste-des-petits-ruminants and goatpox in indigenous goats of Central India. Transbound Emerg Dis 58(3):268–273

Manjunath S, Kumar GR, Mishra BP, Mishra B, Sahoo AP, Joshi CG, Tiwari AK, Rajak KK, Janga SC (2015) Genomic analysis of host—Peste des petits ruminants vaccine viral transcriptome uncovers transcription factors modulating immune regulatory pathways. Vet Res 46:15

Manjunath S, Mishra BP, Mishra B, Sahoo AP, Tiwari AK, Rajak KK, Muthuchelvan D, Saxena S, Santra L, Sahu AR, Wani SA, Singh RP, Singh YP, Pandey A, Kanchan S, Singh RK, Kumar GR, Janga SC (2017) Comparative and temporal transcriptome analysis of peste des petits ruminants virus infected goat peripheral blood mononuclear cells. Virus Res 229:28–40

Marashi M, Masoudi S, Moghadam MK, Modirrousta H, Marashi M, Parvizifar M, Dargi M, Saljooghian M, Homan F, Hoffmann B, Schulz C, Starick E, Beer M, Fereidouni S (2017) Peste des petits ruminants virus in vulnerable wild small ruminants, Iran, 2014–2016. Emerg Infect Dis 23(4):704–706

Mbyuzi AO, Komba EV, Kimera SI, Kambarage DM (2014) Sero-prevalence and associated risk factors of peste des petits ruminants and contagious caprine pleuro-pneumonia in goats and sheep in the southern zone of Tanzania. Prev Vet Med 116:138–144

Mondal B, Sreenivasa BP, Dhar P, Singh RP, Bandyopadhyay SK (2001) Apoptosis induced by *peste des petits* ruminants virus in goat peripheral blood mononuclear cells. Virus Res 73:113–119

Munir M, Zohari S, Berg M (2013) Molecular biology and pathogenesis of Peste des petits ruminants virus. In: Springer Briefs Anim Sci. Springer Science & Business Media, Berlin, pp 23–32

Muniraju M, Munir M, Parthiban AR, Banyard AC, Bao J, Wang Z, Ayebazibwe C, Ayelet G, El Harrak M, Mahapatra M, Libeau G, Batten C, Parida S (2014) Molecular evolution of peste des petits ruminants virus. Emerg Infect Dis 20:2023–2033

Muthuchelvan D, De A, Debnath B, Choudhary D, Venkatesan G, Rajak KK, Sudhakar SB, Hemadri D, Pandey AB, Parida S (2014) Molecular characterization of peste-des-petits ruminants virus (PPRV) isolated from an outbreak in the Indo-Bangladesh border of Tripura state of North-East India. Vet Microbiol 174(3–4):591–595

Muthuchelvan D, Rajak KK, Ramakrishnan MA, Choudhary D, Bhadouriya S, Saravanan P, Pandey AB, Singh RK (2015) Peste-des-petits-ruminants: an Indian perspective. Adv Anim Vet Sci 3:422–429

Niyokwishimira A, de D Baziki J, Dundon WG, Nwankpa N, Njoroge C, Boussini H, Wamwayi H, Jaw B, Cattoli G, Nkundwanayo C, Ntakirutimana D, Balikowa D, Nyabongo L, Zhang Z, Bodjo SC (2019) Detection and molecular characterization of Peste des petits ruminants virus from outbreaks in Burundi, December 2017–January 2018. Transbound Emerg Dis 66(5):2067–2073. https://doi.org/10.1111/tbed.13255

OIE and FAO (2015). Manual on global strategy for the control and eradication of Pesti des petits ruminants. http://www.fao.org/3/a-i4460e.pdf

OIE—World Organisation for Animal Health (2013) Manual of diagnostic tests and vaccines for terrestrial animals. OIE, Paris, pp 1–14

Omani RN, Gitao GC, Gachohi J, Gathumbi PK, Bwihangane BA, Abbey K, Chemweno VJ (2019) Peste des petits ruminants (PPR) in dromedary camels and small ruminants in Mandera and Wajir counties of Kenya. Adv Virol 2019:4028720

Padhi A, Ma L (2014) Genetic and epidemiological insights into the emergence of peste des petits ruminants virus (PPRV) across Asia and Africa. Sci Rep 4:7040

Parida S, Couacy-Hymann E, Pope RA, Mahapatra M, El Harrak M, Brownlie J, Banyard AC (2015a) Pathology of PPR. In: Munir M (ed) Peste des petits ruminants virus. Springer-Verlag Berlin Heidelberg, Berlin, pp 51–68

Parida S, Muniraju M, Mahapatra M, Muthuchelvan D, Buczkowski H, Banyard AC (2015b) Peste des petits ruminants. Vet Microbiol 181:90–106

Parida S, Selvaraj M, Gubbins S, Pope R, Banyard A, Mahapatra M (2019) Quantifying levels of Peste des petits ruminants (PPR) virus in excretions from experimentally infected goats and its importance for nascent PPR eradication programme. Viruses 11(3):E249

Pawar RM, Raj GD, Kumar TM, Raja A, Balachandran C (2008) Effect of siRNA mediated suppression of signaling lymphocyte activation molecule on replication of peste des petits ruminants virus *in vitro*. Virus Res 136:118–123

Polci A, Cosseddu GM, Ancora M, Pinoni C, El Harrak M, Sebhatu TT, Ghebremeskel E, Sghaier S, Lelli R, Monaco F (2015) Development and preliminary evaluation of a new real-time RT-PCR assay for detection of Peste des petits ruminants virus genome. Transbound Emerg Dis 62:332–338

Pope RA, Parida S, Bailey D, Brownlie J, Barrett T, Banyard AC (2013) Early events following experimental infection with Peste-Des-Petits ruminants virus suggest immune cell targeting. PLoS One 8(2):e55830

Qi X, Wang T, Xue Q, Li Z, Yang B, Wang J (2018) MicroRNA expression profiling of goat peripheral blood mononuclear cells in response to peste des petits ruminants virus infection. Vet Res 49(1):62

Rahaman A, Srinivasan N, Shamala N, Shaila MS (2003) The fusion core complex of the peste des petits ruminants virus is a six-helix bundle assembly. Biochemistry 42(4):922–931

Rahman A-U, Wensman JJ, Abubakar M, Shabbir MZ, Rossiter P (2018) Peste des petits ruminants in wild ungulates. Trop Anim Health Prod 50(8):1815–1819. https://doi.org/10.1007/s11250-018-1623-1626

Rahman UA, Abubakar M, Rasool MH, Manzoor S, Saqalein M, Rizwan M, Munir M, Ali Q, Wensman JJ (2016) Evaluation of risk factors for Peste des petits ruminants virus in sheep and goats at the wildlife-livestock interface in Punjab Province, Pakistan. Biomed Res Int 2016:7826245

Raj GD, Thangavelu A, Munir M (2015) Strategies and future of global eradication of PPRV. In: Munir M (ed) Peste des petits ruminants virus. Springer-Verlag Berlin Heidelberg, Berlin, pp 227–254

Raj GD, Rajanathan TM, Kumar CS, Ramathilagam G, Hiremath G, Shaila MS (2008) Detection of peste des petits ruminants virus antigen using immunofiltration and antigen-competition ELISA methods. Vet Microbiol 129(3–4):246–251

Rajak KK, Sreenivasa BP, Hosamani M, Singh RP, Singh SK, Singh RK, Bandyopadhyay SK (2005) Experimental studies on immunosuppressive effects of peste des petits ruminants (PPR) virus in goats. Comp Immunol Microbiol Infect Dis 28:287–296

Ratta B, Pokhriyal M, Singh SK, Kumar A, Saxena M, Sharma B (2016) Detection of Peste des petits ruminants virus (PPRV) genome from nasal swabs of dogs. Curr Microbiol 73(1):99–103

Rojas JM, Rodríguez-Martín D, Avia M, Martín V, Sevilla N (2019) Peste des petits ruminants virus fusion and hemagglutinin proteins trigger antibody-dependent cell-mediated cytotoxicity in infected cells. Front Immunol 9:3172

Rony MS, Rahman AKMA, Alam MM, Dhand N, Ward MP (2017) Peste des petits ruminants risk factors and space-time clusters in Mymensingh, Bangladesh. Transbound Emerg Dis 64(6):2042–2048

Santhamani R, Singh RP, Njeumi F (2016) Peste des petits ruminants diagnosis and diagnostic tools at a glance: perspectives on global control and eradication. Arch Virol 161(11):2953–2967

Sanz Bernardo B, Goodbourn S, Baron MD (2017) Control of the induction of type I interferon by Peste des petits ruminants virus. PLoS One 12(5):e0177300

Saravanan P, Singh RP, Balamurugan V, Dhar P, Sreenivasa BP, Muthuchelvan D, Sen A, Aleyas AG, Singh RK, Bandyopadhyay SK (2004) Development of a N gene-based PCR-ELISA for detection of Peste-des-petits-ruminants virus in clinical samples. Acta Virol 48(4):249–255

Saravanan P, Balamurugan V, Sen A, Sreenivasa BP, Singh RP, Bandyopadhyay SK, Singh RK (2010) Long term immune response of goats to a Vero cell adapted live attenuated homologous PPR vaccine. Indian Vet J 87:1–3

Schulz C, Fast C, Schlottau K, Hoffmann B, Beer M (2018) Neglected hosts of small ruminant Morbillivirus. Emerg Infect Dis 24(12):2334–2337

Sen A, Saravanan P, Balamurugan V, Bhanuprakash V, Venkatesan G, Sarkar J, Rajak KK, Ahuja A, Yadav V, Sudhakar SB, Parida S, Singh RK (2014) Detection of subclinical peste des petits ruminants virus infection in experimental cattle. Virus 25:488–492

Seth S, Shaila MS (2001) The hemagglutinin-neuraminidase protein of peste des petits ruminants virus is biologically active when transiently expressed in mammalian cells. Virus Res 75:169–177

Settypalli TB, Lamien CE, Spergser J, Lelenta M, Wade A, Gelaye E, Loitsch A, Minoungou G, Thiaucourt F, Diallo A (2016) One-step multiplex RT-qPCR assay for the detection of Peste des petits ruminants virus, Capripoxvirus, Pasteurella multocida and mycoplasma capricolum subspecies (ssp.) capripneumoniae. PLoS One 11(4):e0153688

Şevik M, Oz ME (2019) Detection of Peste des petits ruminants virus RNA in Culicoides imicola (Diptera: Ceratopogonidae) in Turkey. Vet Ital 55(2):173–179

Shaila MS, Purushothaman V, Bhavasar D, Venugopal K, Venkatesan RA (1989) Peste des petits ruminants in India. Vet Rec 125:602
Shaila MS, Shamaki D, Forsyth MA, Drallo A, Groatley L, Kitching RP, Barrett T (1996) Geographic distribution and epidemiology of peste des petits ruminants viruses. Virus Res 43:149–153
Shatar M, Khanui B, Purevtseren D, Khishgee B, Loitsch A, Unger H, Settypalli TBK, Cattoli G, Damdinjav B, Dundon WG (2017) First genetic characterization of peste des petits ruminants virus from Mongolia. Arch Virol 162(10):3157–3160
Singh RK, Balamurugan V, Bhanuprakash V, Sen A, Saravanan P, Yadav MP (2009) Control and eradication of peste des petits ruminants in sheep and goats in India: possibility. Vet Ital 45:449–462
Singh RP (2011) Control strategies for peste des petits ruminants in small ruminants of India. Rev Sci Tech 30:879–887
Singh RP, Bandyopadhyay SK (2015) Peste des petits ruminants vaccine and vaccination in India: sharing experience with disease endemic countries. Virus Dis 26(4):215–224
Singh RP, Saravanan P, Sreenivasa BP, Singh RK, Bandyopadhyay SK (2004a) Prevalence and distribution of peste des petits ruminants virus infection in small ruminants in India. Rev Sci Tech 23:807–819
Singh RP, Sreenivasa BP, Dhar P, Bandyopadhyay SK (2004b) A sandwich-ELISA for the diagnosis of Peste des petits ruminants (PPR) infection in small ruminants using anti-nucleocapsid protein monoclonal antibody. Arch Virol 149:2155–2170
Singh RP, Sreenivasa BP, Dhar P, Shah LC, Bandyopadhyay SK (2004c) Development of a monoclonal antibody based competitive-ELISA for detection and titration of antibodies to peste des petits ruminants (PPR) virus. Vet Microbiol 98:3–15
Sreenivasa BP, Singh RP, Mondal B, Dhar P, Bandyopadhyay SK (2006) Marmoset B95a cells: a sensitive system for cultivation of Peste des petits ruminants (PPR) virus. Vet Res Commun 30:103–108
Tatsuo H, Ono N, Yanagi Y (2001) Morbilliviruses use signaling lymphocyte activation molecules (CD150) as cellular receptors. J Virol 75(13):5842–5850
Taylor W (2016) The global eradication of peste des petits ruminants (PPR) within15 years—is this a pipe dream? Trop Anim Health Prod 48(3):559–567
Taylor WP (1984) The distribution and epidemiology of Peste des petits ruminants. Prev Vet Med 2:157–166
Teshale S, Moumin G, Moussa C, Gezahegne M (2018) Seroprevalence and risk factors for peste des petits ruminants in sheep and goats in Djibouti. Rev Sci Tech 37(3):961–969
Tripathi BN, Chhattopadhyay SK, Parihar NS (1996) Clinicopathological studies on an outbreak of peste des petits ruminants in goats and sheep. Indian J Vet Pathol 20:99
Truong T, Boshra H, Embury-Hyatt C, Nfon C, Gerdts V, Tikoo S, Babiuk LA, Kara P, Chetty T, Mather A, Wallace DB, Babiuk S (2015) Peste des petits ruminants virus tissue tropism and pathogenesis in sheep and goats following experimental infection. PLoS One 9:e87145
Wang Q, Ou C, Dou Y, Chen L, Meng X, Liu X, Yu Y, Jiang J, Ma J, Zhang Z, Hu J, Cai X (2017) M protein is sufficient for assembly and release of Peste des petits ruminants virus-like particles. Microb Pathog 107:81–87
Wasee Ullah R, Bin Zahur A, Latif A, Iqbal Dasti J, Irshad H, Afzal M, Rasheed T, Rashid Malik A, Qureshi ZU (2016) Detection of Peste des petits ruminants viral RNA in fecal samples of goats after an outbreak in Punjab Province of Pakistan: a longitudinal study. Biomed Res Int 2016:1486824
Woma TY, Ekong PS, Bwala DG, Ibu JO, Ta'ama L, Dyek DY, Saleh L, Shamaki D, Kalla DJ, Bailey D, Kazeem HM, Quan M (2016) Serosurvey of peste des petits ruminants virus in small ruminants from different agro-ecological zones of Nigeria. Onderstepoort J Vet Res 83(1):1035
Yang B, Qi X, Guo H, Jia P, Chen S, Chen Z, Wang T, Wang J, Xue Q (2018a) Peste des petits ruminants virus enters caprine endometrial epithelial cells via the Caveolae-mediated endocytosis pathway. Front Microbiol 9:210
Yang B, Xue Q, Qi X, Wang X, Jia P, Chen S, Wang T, Xue T, Wang J (2018b) Autophagy enhances the replication of Peste des petits ruminants virus and inhibits caspase-dependent apoptosis in vitro. Virulence 9(1):1176–1194
Yang Y, Qin X, Zhang X, Zhao Z, Zhang W, Zhu X, Cong G, Li Y, Zhang Z (2017) Development of real-time and lateral flow dipstick recombinase polymerase amplification assays for rapid detection of goatpox virus and sheeppox virus. Virol J 14(1):131
Yousuf RW, Sen A, Mondal B, Biswas SK, Chand K, Rajak KK, Gowane GR, Sudhakar SB, Pandey AB, Ramakrishnan MA, Muthuchelvan D (2015) Development of a single-plate combined indirect ELISA (CI-ELISA) for the detection of antibodies against peste-des-petits-ruminants and bluetongue viruses in goats. Small Rumin Res 124:137–139
Yu R, Zhu R, Gao W, Zhang M, Dong S, Chen B, Yu L, Xie C, Jiang F, Li Z (2017) Fine mapping and conservation analysis of linear B-cell epitopes of peste des petits ruminants virus hemagglutinin protein. Vet Microbiol 208:110–117

Yunus M, Shaila MS (2012) Establishment of an in vitro transcription system for Peste des petits ruminant virus. Virol J 9:302

Zahur AB, Irshad H, Ullah A, Afzal M, Latif A, Ullah RW, Farooq U, Samo MH, Jahangir M, Ferrari G, Hussain M, Ahmad MM (2014) Peste des petits ruminants vaccine (Nigerian strain 75/1) confers protection for at least 3 years in sheep and goats. J Biosci Med 2:23–27

Zakian A, Nouri M, Kahroba H, Mohammadian B, Mokhber-Dezfouli MR (2016) The first report of peste des petits ruminants (PPR) in camels (*Camelus dromedarius*) in Iran. Trop Anim Health Prod 48(6):1215–1219

Zhang J, Liu W, Chen W, Li C, Xie M, Bu Z (2016) Development of an immunoperoxidase monolayer assay for the detection of antibodies against Peste des petits ruminants virus based on BHK-21 cell line stably expressing the goat signaling lymphocyte activation molecule. PLoS One 11(10):e0165088

Zhang Y, Wang J, Zhang Z, Mei L, Wang J, Wu S, Lin X (2018) Development of recombinase polymerase amplification assays for the rapid detection of peste des petits ruminants virus. J Virol Methods 254:35–39

Zhou XY, Wang Y, Zhu J, Miao QH, Zhu LQ, Zhan SH, Wang GJ, Liu GQ (2018) First report of peste des petits ruminants virus lineage II in *Hydropotes inermis*, China. Transbound Emerg Dis 65(1):e205–e209

Zhu Z, Li P, Yang F, Cao W, Zhang X, Dang W, Ma X, Tian H, Zhang K, Zhang M, Xue Q, Liu X, Zheng H (2019) Peste des petits ruminants virus nucleocapsid protein inhibits interferon-β production by interacting with IRF3 to block its activation. J Virol 93(16):e00362

第 14 章 萨佩罗病毒

Yashpal Singh Malik, Sudipta Bhat, Anastasia N. Vlasova,
Fun-In Wang, Nadia Touil, Souvik Ghosh, Kuldeep Dhama,
Mahendra Pal Yadav, Raj Kumar Singh

14.1 引言

萨佩罗病毒(sapelovirus,SPV)最早从猪的肠道中分离获得，因此，之前将其称为肠道病毒 8 型,命名为 PEV-8,在猪群广泛存在,PEV-8 感染主要是引发无症状胃肠道感染。后来对其基因分析，发现了其特有基因,PEV-8 也因此被分类为萨佩罗病毒属(Krumbholz 等,2002)。目前已证实该病毒对猪、禽和灵长类动物均具有感染性。相关的 PSV 感染临床症状主要表现为肺炎、生殖缺陷、皮肤损害、肠炎和脑脊髓炎。目前由于对禽萨佩罗病毒(avian sapelovirus,ASV)和猿萨佩罗病毒(simian sapelovirus,SSV)相关研究报道较少,为此本章仅对 PSV 相关研究进行介绍。

14.2 结构与理化特性

PSV 是属于小 RNA 病毒科的萨佩罗病毒属,无囊膜、球形、表面光滑、直径约为 35nm,病毒衣壳呈 20 面体,立体对称结构(Baiet 等,2018)。成熟病毒粒子对外界环境具有抵抗力,耐酸环境,在 60℃加热 10 分钟、脂溶剂和一些消毒剂中仍能保持稳定。但亚氯酸钠或 70%乙醇可有效杀死 SPV (Horak 等,2016)。SPV 在氯化铯中浮力密度为 1.32~1.34g/cm^3。

14.3 分类

SPV 分类是 SPV 病毒研究小组和 ICTV 共同提出的,并经过数据证实,取得专家共识。萨佩罗病毒属包含禽萨佩罗病毒(ASV)、猪萨佩罗病毒(SV-A,PSV)和猿萨佩罗病毒(SV-B,SSV)(ICTV 第 10 届报道)。ASV 和 SV-A 血清抗体并不相同,为此分别称之为 ASV-1 和 PSV-1。而 SSV 由于高度分化，又被分为 SSV-1、SSV-2 和 SSV-3 三个基因亚型。PSV-csh 株于 2011 年在中国首次被分离到(Lan 等,2011)。到目前为止,中国、韩国、日本、德国、法国、印度和英国共公布了 PSV 的 50 个毒株的全基因组序列 (Lan 等,2011;Chen 等,2012;Schock 等,2014;Son 等,2014;Ray 等,2018;Piorkowski 等,2018;Bai 等,2018)。PSV 不同毒株对靶器官都具有不同亲嗜性，例如，中国上海株 PSV-csh 和英国株 PSV-G5 主要侵害神经系统，而 KS0515、KS04105、KS0552179 和 PSV 中国株 YC2011 则会导致腹泻。特别值得关注的是 PSV-csh 株主要侵犯胃肠道，但感染初期也会引起呼吸困难,随后出现脑脊髓灰质炎症状(Lan 等,2011)。

14.4 基因组

PSV 是正链 RNA 病毒,5’端和 3’端分别含有一个基因连接病毒蛋白。PSV 基因组大小为 7.5~8.3kb,编码一个大的多聚蛋白前体,随后被自身编码的蛋白酶裂解成 12 种成熟的功能蛋白，即 1 个前导蛋白(L)、4 个结构蛋白(VP-1-4)和 7 个非结构蛋白(2A-2C,3A-3D)。萨佩罗病毒编码的多聚蛋白含有 2322~2521 个氨基酸。PSV(V13 株）全基因组及每个蛋白的核苷酸长度如图 14.1 所示。PSV 具有独特的蛋白 L 和 2A(Krumbholz 等,2002),虽然这两种蛋白质的功能尚未明确,但推测其可能全是蛋白酶。三个种属的萨佩罗病毒均含有一个内部核糖体结合位点。ASV 的 L 蛋白(451 个氨基酸)比 PSV(84 个氨基酸)和 SSV(88 个氨基酸)的 L 蛋白长。2A 蛋白则非常小,ASV 的 2A 蛋白仅有 12 个氨基酸，而 PSV 和 SSV 2A 蛋白分别含有 226 和 302 个氨基酸。萨佩罗病毒属遵循物种分界标准,其多聚蛋白质中氨基酸 70%以上具有一致性，其中 P1 蛋白(VP4,VP2,VP3,VP1)中氨基酸一致性大于 64%,且 2C+3CD 中氨基酸一致性大于 70%,2C+3CD 基因组碱基组成差异性不超过 1%，且具有相同的基因组结构(ICTV,2018 年第 10 期报道)。

14.5 流行病学

14.5.1 感染宿主

目前，已从家猪、野猪、猴和鸭中分离到了萨佩罗病毒，同时在蝙蝠(多种)、猫、牛、犬、鸽子、鹌鹑和海狮等动物中也发现了类似萨佩罗病毒的基因组(Li 等，2011)。值得注意的是，迄今为止，这些病毒是否会由动物传染给人还未可知。

14.5.2 病毒变异

世界各地的健康猪和腹泻病猪身上均检测到了 PSV 的存在，属于感染猪的一种新的肠道病原体。PSV 免疫原性最好的衣壳蛋白 VP1 常用于不同毒株间的变异分析，基于此，PSV-1 虽然只有一个血清型，但不同地区分离到 PSV-1 毒株之间的相似性较小。根据 VP1 基因序列的遗传进化分析显示，来源于不同国家的 PSV 形成不同的分支。VP1 全长(855bp)和 VP1 部分基因(544bp)均可以作为靶基因对 PSV 进行遗传进化分析，利用 Splits Tree4 软件以 VP1 基因为靶基因进行遗传进化分析表明，PSV 可形成三个分支，分为三个种(图 14.2)，PSV 在世界范围内广泛传播流行，且各分离株间存在明显变异。中国学者的进一步研究表明，PSV 分离株间发生了基因重组，这进一步加强了 PSV 抗原差异性(Yang 等，2017)。近期，法国和意大利学者已经对萨佩罗病毒进行了鉴定和基因组特征分析(Ray 等，2018；Piorkowski 等，2018；Kumari 等，2019)。

14.6 传播与发病机制

该病毒最常见的传播方式为粪-口传播，由于其可在外界环境中稳定存在，所有被病毒粒子污染的物体都可作为传播媒介(Huang 等，1980；Lan 等，2011)。PSV 经口摄入后，主要在肠道中进行复制，所以常用肠上皮细胞对 PSV 致病性与毒性进行研究(Lan 等，2013)。

尽管尚不明确 PSV 受体，但近来一项研究发现，在 GD1A 神经节苷脂细胞表面有一个 2,3 连接的末端唾液酸 SA 可能是介导 PSV 感染的受体(Kim 等，2016)。研究表明，PSV 可利用细胞质膜微囊相关的胞吞途径进入 PK-15 细胞，且低 pH 值环境、发动蛋白、Rab7 和 Rab11 均可影响 PK-15 对 PSV 的胞吞过程(Zhao 等，2019)。此外，PSV 感染也能导致病毒血症，并通过血液循环侵入中枢神经系统，潜伏期一般 5~14 天。

14.7 临床症状

PSV 自然感染和人工感染仔猪时，有的有临床症状，有的无临床症状。一般情况下，PSV 与其他肠道病原体，如猪嵴病毒、猪肠道病毒、猪捷申病毒、轮状病毒 A~C 群、猪札幌病毒、猪诺如病毒、猪传染性胃肠炎病毒、猪流行性腹泻病毒、大肠杆菌、沙门菌、胞内劳森菌、猪痢疾短螺旋体和产气荚膜梭菌形成共感染。症状主要表现为神

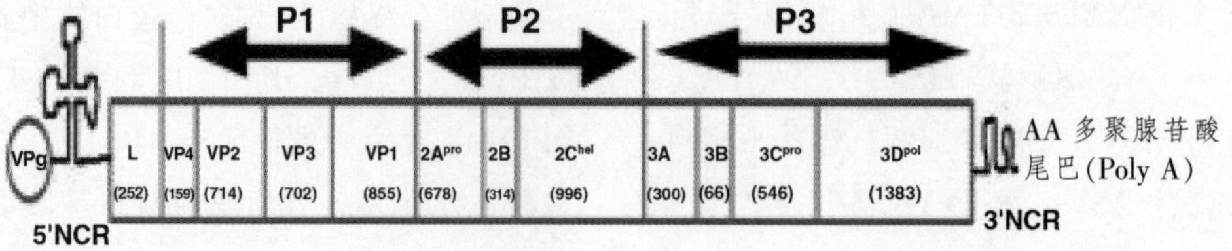

图 14.1 猪萨佩罗病毒基因组示意图[根据第一株分离的 V13 株 PSV(NC 003987)得出不同基因碱基对的核苷酸长度]。图示为基因连接病毒蛋白基因组相关病毒蛋白、NCR 非编码 RNA、前导蛋白、VP 病毒蛋白、P1 结构蛋白、P2-P3 非结构蛋白、pro 蛋白酶、hel 解旋酶和 pol 聚合酶。

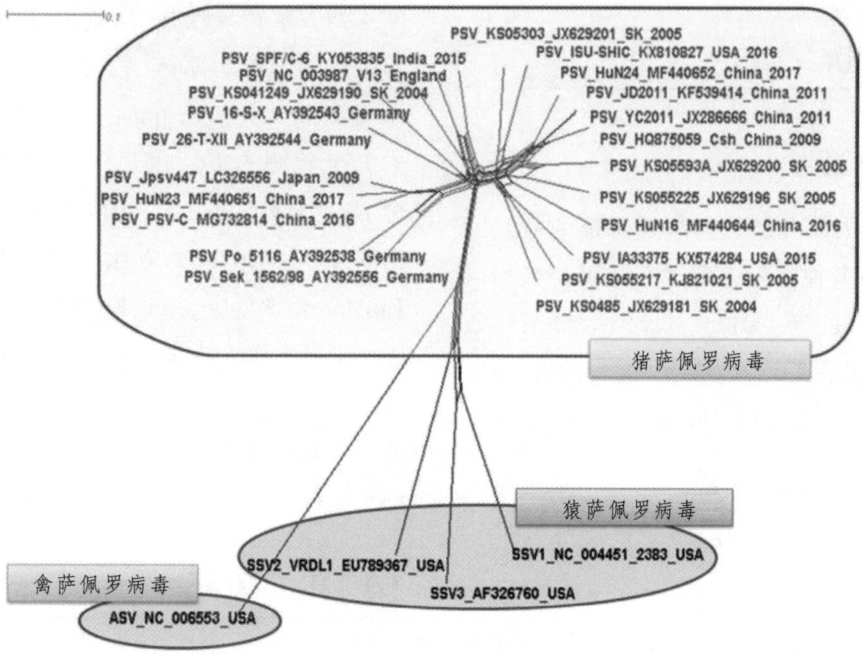

图 14.2　以 VP1(544bp)核苷酸序列为靶基因,对来源于中国、日本、英国、印度、美国、德国和韩国的猪萨佩罗病毒、猿萨佩罗病毒和禽萨佩罗病毒典型代表株的遗传进化分析(Huson 和 Bryant,2006)。

经症状,如脊髓损伤、运动失调、精神迟钝、麻痹,以及对环境刺激的反应迟钝。处于生长期感染的猪表现为脑灰质炎特征,运动失调及下肢轻瘫,在实验室条件下对 50~60 日龄的生长猪进行人工感染,出现类似临床表现,研究也发现,美国新分离到的 PSV 与典型的脑脊髓灰质炎有关(Arruda 等,2016)。PSV 诱发脑脊髓灰质炎,同时可能引起胃肠炎和呼吸困难的症状(Lan 等,2011)。在妊娠至第 15 天的母猪阴道内接种 PSV,会造成早期胚胎死亡和胚胎吸收,而在妊娠至第 30 天进行 PSV 接种时,胎儿的死亡率明显增加(Huang 等,1980)。在印度对死胎猪的肠内容物进行检测,同样也检测到 PSV,这表明该病毒可通过胎盘进行传播。

14.8　发病率和死亡率

目前,PSV 感染造成死亡或发病的具体数据还不完善。实验表明,妊娠第 30 天的母猪通过阴道和子宫内感染 PSV 后胎儿死亡率为 94.4%(Huang 等,1980)。据报道,2016 年美国的疫情发病率和死亡率分别为 20%和 30%(Arruda 等,2016)。

14.8.1　病理学

病变主要见于中枢神经系统,与其他嗜神经病毒感染引起的病理变化一致,临床上表现为亚急性、多病灶和非化脓性脑脊髓灰质炎的病理变化,同时伴有硬脑膜点状出血和充血(Lan 等 2011;Schock 等,2014)。常见症状还有神经元空泡化和周围管套现象(Lan 等,2011)。充血是小肠上最常见的病理损失,病理组织变化可观察到小肠绒毛有明显损失,且有出血症状(Lan 等,2011)。临床上常见肺炎,表现为肺叶实变和多处出血,红细胞渗透到肺间质和肺泡,导致肺泡壁变薄(Lan 等,2011)。在印度,发现自然感染猪的病理变化可见脑膜重度浑浊、增厚和充血,气管内有泡沫渗出物,肠系膜淋巴结充血、肝脏变黄、回肠褶皱和肠道黏膜皱褶增厚(Kumari 等,2019)。

14.9 诊断

14.9.1 临床症状

具有运动失调和肢体瘫痪为特征的神经障碍,无论是否有其他临床特征(腹泻或肺炎),均可疑似感染 PSV(Lan 等,2011)。根据一窝中有少数至几个死胎或出现木乃伊胎的情况,可以怀疑由 PSV 感染诱发的生殖障碍(Huang 等,1980)。

14.9.2 被检样本

活体动物样本包括粪便和血液(血清)样本。为防止样本污染,粪便样本直接从肛门采集,有神经症状动物尸体采集的首选样本是脊髓和大脑。除此之外,肝脏、脾脏、气管、肺和肠的样本也可用于 PSV 感染诊断。虽然还未从死胎或木乃伊胎的组织中分离到 PSV,但已经从死胎病例中检测到 PSV 核酸(Huang 等,1980;Kumari 等,2019)。

14.9.3 诊断

常规 PSV 实验诊断主要包括用细胞对病毒进行分离、培养和病毒基因组鉴定。PSV 可在 EFN、PK-15、IBRS-2 及 LLC-PK 猪肾细胞系内生长,并引起 CPE。此外,在 Vero 细胞和 BHK-21 细胞也可生长,传 3~5 代后,可见细胞收缩、细胞变圆和细胞脱离现象。一种人类肝癌细胞系 PLC/PRF/5 通常用于从猪粪便样本中分离戊型肝炎,但偶然分离到了 PSV 毒株(Bai 等,2018)。人工培养的 PSV 通常运用病毒中和试验(Sozzi 等,2010)和免疫荧光抗体试验进行鉴定。然而,这些方法耗时费力,成本高,且特异性不强。基于此,RT-PCR、套式反转录聚合酶(nRT-PCR)和实时定量 PCR 等多种分子生物学方法被用于 PSV 检测(Zell 等,2000;Palmquist 等,2002;Krumbholz 等,2003)。研究人员正是通过表 14.1 描述的靶基因和引物序列实现了对 PSV 的检测。这些分子生物学检测方法特异性强、敏感性高、速度快且不易与猪疾病常见病原体(类似于小 RNA 病毒等)发生交叉反应(Palmquist 等,2002)。用于野外样本 PSV 检测的实时 PCR,最低可检测到 10^2~10^3 个 DNA 拷贝(Chen 等,2014;Kumari 等,2019)。这两种实时 PCR 检测方法应用了特异性极高的 TaqMan 化学反应。RT-LAMP 则是另一种有效的 PSV 诊断方法,由 Wang 等于 2014 年开发而成,其最低可检测到 10 个 DNA 拷贝。这些简便、快速、灵敏的诊断方法能够帮助我们进一步了解 PSV 流行病学,以期找到更好的预防方法。

14.10 免疫

关于 PSV 免疫相关研究报道较少,主要基于 IgA 体液免疫。通过猪肠上皮细胞感染研究发现,PSV 感染能够改变先天性免疫应答(Lan 等,2013)。研究表明 PSV 感染可引起先天免疫系统分子 TLR3 上调和 TLR4 下调(Lan 等,2013)。研究发现,在胎盘感染情况下,母源抗体起不到保护作用(Huang,1980)。通过阴道和子宫内感染 PSV,血清呈阳性的母猪会导致胚胎和胎儿感染(Huang,1980)。虽然认为初乳中的母源抗体具有保护作用,但断奶后动物体中仍发现了该病毒(Schock 等,2014)。至今尚未研制出 PSV 疫苗。

14.11 展望

目前,尽管 PSV 在世界各地猪群中广泛存在,但有关病毒进化方式、引发变异机制等相关信息仍然很匮乏。然而,有关于 PSV 在猪和野猪中感染的流行情况和病毒的遗传多样性的精确数据只在几个国家进行了报道。此外,PSV 与其他肠道致病菌的共感染关系仍不清楚,因此,对 PSV 的致病潜力了解不深。最近关于 PSV 株间重组事件的发现,更突显了对 PSV 遗传变异研究的重要性。

表 14.1 用于 PSV 检测的 RT-PCR 试验

试验	引物序列(5'-3')	靶基因	参考文献
RT-PCR	FP-CCCTGGGACGAAAGAGCCTG	5'UTR	Zell 等(2000)
	RP-CCTTTAAGTAAGTAGTAAAGGG		
nRT-PCR	FP-CCAAGATTAGAAGTTGATTTG		
	RP-GGGTAGCCTGCTGATGTAGTC		
用于 PTV 和 PSV 的双重 PCR	FP-GTGGCGACAGGGTACAGAAGAG	5'UTR	Palmquist 等(2002)
	RP-GGCCAGCCGCGACCCTGTCAG		
RT-PCR (基因型分型)	FP-AGGATGTGGTGCAAGCAAGCAT	VP1	Son 等(2014)
	RP-AGGCAGCACCGTTCTGGTCAA		
RT-LAMP	FP-CCATACCCTACCCTCCCTTC	5'UTR	Wang 等(2014)
	RP-GCCCATAGTTCACTGCCTAC		
实时 RT-PCR	FP-GGCAGTAGCGTGGCGAGC	5'UTR	Chen 等(2014)
	RP-CTACTCTCCTGTAACCAGT		
	Taqman 探针—CGATAGCCATGTTAGTG		
实时 RT-PCR	FP-GGAAACCTGGACTGGGYCT	5'UTR	Kumari 等(2018)
	RP-ACACGGGCTCTCTGTTTCTT		
	Tagman 探针—CCAGCCGCGACCCTATCAGG		

参考文献

Arruda PH, Arruda BL, Schwartz KJ, Vannucci F, Resende T, Rovira A, Sundberg P, Nietfeld J, Hause BM (2016) Detection of a novel sapelovirus in central nervous tissue of pigs with polioencephalomyelitis in the USA. Transbound Emerg Dis 64(2):311–315

Bai H, Liu J, Fang L, Kataoka M, Takeda N, Wakita T, Li TC (2018) Characterization of porcine sapelovirus isolated from Japanese swine with PLC/PRF/5 cells. Transbound Emerg Dis 65(3):727–734

Chen J, Chen F, Zhou Q, Li W, Chen Y, Song Y, Zhang X, Xue C, Bi Y, Cao Y (2014) Development of a minor groove binder assay for real-time PCR detection of porcine Sapelovirus. J Virol Methods 198:69–74

Chen JW, Chen F, Zhou QF, Li W, Song Y, Pan Y, Zhang X, Xue C, Bi Y, Cao Y (2012) Complete genome sequence of a novel porcine sapelovirus strain YC2011 isolated from piglets with diarrhea. J Virol 86(19):10898–10898

Horak S, Killoran K, Leedom Larson KR (2016) Porcine sapelovirus. Swine Health Information Center and Center for Food Security and Public Health. http://www.cfsph.iastate.edu/pdf/shic-factsheet-porcine-sapelovirus

Huang J, Gentry RF, Zarkower A (1980) Experimental infection of pregnant sows with porcine enteroviruses. Am J Vet Res 41(4):469–473

Huson DH, Bryant D (2006) Application of phylogenetic networks in evolutionary studies. Mol Biol Evol 23(2):254–267

Kim DS, Son KY, Koo KM, Kim JY, Alfajaro MM, Park JG, Hosmillo M, Soliman M, Baek YB, Cho EH, Lee JH (2016) Porcine sapelovirus uses α2, 3-linked sialic acid on GD1a ganglioside as a receptor. J Virol 90(8):4067–4077

Krumbholz A, Dauber M, Henke A, Birch-Hirschfeld E, Knowles NJ, Stelzner A, Zell R (2002) Sequencing of porcine enterovirus groups II and III reveals unique features of both virus groups. J Virol 76(11):5813–5821

Krumbholz A, Wurm R, Scheck O, Birch-Hirschfeld E, Egerer R, Henke A, Wutzler P, Zell R (2003) Detection of porcine teschoviruses and enteroviruses by LightCycler real-time PCR. J Virol Methods 113(1):51–63

Kumari S, Ray PK, Singh R, Desingu PA, Sharma GT, Saikumar G (2018) Development of a Taqman-based real-time PCR assay for detection of porcine sapelovirus infection in pigs. Anim Biotechnol:1–4. https://doi.org/10.1080/10495398.2018.1549561. (Epub ahead of print)

Kumari S, Ray PK, Singh R, Desingu PA, Varshney R, Saikumar G (2019) Pathological and molecular investigation of porcine sapelovirus infection in naturally affected Indian pigs. Microb Pathog 127:320–325

Lan D, Tang C, Yue H, Sun H, Cui L, Hua X, Li J (2013) Microarray analysis of differentially expressed transcripts in porcine intestinal epithelial cells (IPEC-J2) infected with porcine sapelovirus as a model to study innate immune responses to enteric viruses. Arch Virol 158(7):1467–1475

Li L, Shan T, Wang C, Côté C, Kolman J, Onions D, Gulland FM, Delwart E (2011) The fecal viral flora of California sea lions. J Virol 85(19):9909–9917

Palmquist JM, Munir S, Taku A, Kapur V, Goyal SM (2002) Detection of porcine teschovirus and enterovirus type II by reverse transcription-polymerase chain reaction. J Vet Diagn Investig 14(6):476–480

Piorkowski G, Capai L, Falchi A, Casabianca F, Maestrini O, Gallian P, Barthélémy K, Py O, Charrel R, De Lamballerie X (2018) First identification and genomic characterization of a porcine sapelovirus from Corsica, France, 2017. Microbiol Resour Announc 7(11):e01049–e01018

Ray PK, Desingu PA, Kumari S, John JK, Sethi M, Sharma GK, Pattnaik B, Singh RK, Saikumar G (2018) Porcine sapelovirus among diarrhoeic piglets in India. Transbound Emerg Dis 65(1):261–263

Son KY, Kim DS, Kwon J, Choi JS, Kang MI, Belsham GJ, Cho KO (2014) Full-length genomic analysis of Korean porcine sapelovirus strains. PLoS One 9(9):11

Sozzi E, Barbieri I, Lavazza A, Lelli D, Moreno A, Canelli E, Bugnetti M, Cordioli P (2010) Molecular characterization and phylogenetic analysis of VP1 of porcine enteric picornaviruses isolates in Italy. Transbound Emerg Dis 57(6):434–442

Wang CY, Yu DY, Cui L, Hua X, Yuan C, Sun H, Liu Y (2014) Rapid and real-time detection of Porcine Sapelovirus by reverse transcription loop-mediated isothermal amplification assay. J Virol Methods 203:5–8

Yang T, Yu X, Yan M, Luo B, Li R, Qu T, Luo Z, Ge M, Zhao D (2017) Molecular characterization of Porcine sapelovirus in Hunan, China. J Gen Virol 98(11):2738–2747

Zell R, Krumbholz A, Henke A, Birch-Hirschfeld E, Stelzner A, Doherty M, Hoey E, Dauber M, Prager D, Wurm R (2000) Detection of porcine enteroviruses by nRT-PCR: differentiation of CPE groups I-III with specific primer sets. J Virol Methods 88(2):205–218

Zhao T, Cui L, Yu X, Zhang Z, Shen X, Hua X (2019) Porcine Sapelovirus enters PK-15 cells via caveolae-dependent endocytosis and requires Rab7 and Rab11. Virology 529:160–168. https://doi.org/10.1016/j.virol.2019.01.009

第 15 章 戊型肝炎病毒

Harsh Kumar, Nassim Kamar, Gheyath K. Nasrallah,
Dinish Kumar

15.1 引言

据估计,在全球范围内戊型肝炎病毒(hepatitis E virus,HEV)的感染人数已超过 2000 万人,其中 330 万人有症状感染。2015 年有记录的死亡病例有 4.4 万例(Rein 等,2012;WHO,2017)。患有自限性 HEV 感染的患者可自行痊愈,也有少数患者出现腹痛、厌食、发热、黄疸、重症黄疸、病毒性肝炎等症状(Aggarwal,2011;Mirazo 等,2014)。另一方面,老年人、免疫抑制患者、HIV/AIDS 呈阳性的癌症患者或接受过器官移植的患者感染 HEV 的概率更高。慢性戊型肝炎患者死亡率高,需长期住院治疗。妊娠期间,HEV 对母婴安全构成威胁(Navaneethan 等,2008;Poovorawan 等,2014)。

经鉴定,HEV 的四种主要基因型(属同一血清型)能导致人类戊型肝炎的发生(Emerson 和 Purcell,2003)。HEV 基因 1 型和 2 型主要在非洲、亚洲部分地区等低收入国家人群中暴发(Teshale 等,2010b)。在工业化国家和发展中国家,基因 3 型和 4 型通常表现为本土戊型肝炎病例(Dalton 等,2008)。据报道,HEV 可感染鸡、鹿、兔、鼠、家猪和野猪等多种动物(Intharasongkroh 等,2017)。同样,人源 HEV 毒株与猪源 HEV 毒株有相似之处,尤其是基因型 3(Suwannakarne 等,2010;Temmame 等,2013)。据报道,在法国和日本等发达国家,HEV 感染人类并引发人兽共患可能与食用受污染的肉类、猪肉制品,以及生香肠有关(Intharasongkroh 等,2017)。

15.2 病原发现

1980 年,建立了用于检测甲型与乙型肝炎病毒的血清学方法。然而,在对 1955—1956 年在新德里暴发的水源性肝炎样本进行检测时,发现非 A 型和非 B 型肝炎(Viswanathan,1957)。类似的报道还有 1978—1979 年在克什米尔的暴发的肝炎(Khuroo,1980)。两次疫情暴发结果显示,HEV 是引起感染的病原体。这是首次将 HEV 与水源性传染病联系在一起进行调查(Wong 等,1980;Teshale 和 Hu,2011)。1983 年,一个名叫 Mikhail Balayan 的苏联医生调查了苏联士兵在阿富汗战争中非 A 和非 B 型肝炎的病例(Balayan 等,1983),在返回莫斯科的途中,他将被 9 名士兵粪便污染的酸奶样本服下,然后静待症状表现。35 天后,他出现了重型肝炎的症状。在分析了自己的粪便后,他发现了该病毒的新毒株。这些毒株对试验动物的肝脏也能诱发损伤。这是首次检测到 HEV 的基因 1 型(Teshale 和 Hu,2011;Pinto 等,2017)。

其他关于 HEV 感染病例的记载始于 1975 年的哥斯达黎加(Khuroo,1980;Khuroo 等,2016),1988 年发生在非洲索马里的疫情感染多达 11000 人。在拉丁美洲,首次检测到 HEV 基因 2 型的报道是 1986—1987 年在墨西哥暴发的疫情(Rendon 等,2016;Pinto 等,2017)。

15.3 分类与基因组

根据分类学体系,HEV 科被分为两属:鱼戊型肝炎病毒属 (Piscihepevirus genera)(割喉鳟病毒)和正戊型肝炎病毒属(Orthohepevirus genera)(哺乳动物和禽类毒株)(ICTV,2019)。正戊型肝炎病毒属分为四种:正戊型肝炎病毒 A(Orthohepevirus A)、正戊型肝炎病毒 B(Orthohepevirus B)(感染鸟类)、正戊型肝炎病毒 C(Orthohepevirus C)(感染啮齿动物、鼩形目和食肉动物)和正戊型肝炎病毒 D(Orthohepevirus D)(感染蝙蝠)(ICTV,2019)。正戊型肝炎病毒 A 分为 7 个基因型,能够感染人类(HEV1、2、3、4 和 7)、猪(HEV3 和 4)、兔子(HEV3)、野猪(HEV3、4、5 和 6)、猫鼬(HEV3)、鹿(HEV3)、牦牛(HEV4)和骆驼(HEV7)(ICTV 2019)。HEV 无囊膜,二十面体立体对称,直径一般为 27~34nm。HEV 基因组全长约 7.2kb,为单股正链 RNA 病毒,3’端有多聚腺苷 A 尾巴,5’端有甲基鸟嘌呤帽子结构 (Reyes 等,1990;Tam 等,

1991)。HEV 基因组含 3 个ORF,如图 15.1 所示。ORF1 编码多聚功能蛋白（1693 个氨基酸）(Koonin 等,1992),包含半胱氨酸蛋白酶、甲基转移酶、RNA 解旋酶，以及 RNA 依赖的 RNA 聚合酶。ORF 2 编码衣壳蛋白(660 个氨基酸),在病毒组装及其与靶细胞互作，免疫原性方面起着至关重要的作用（Li 等,1997;He 等,2008;Kalia 等,2009;Xing 等,2011)。ORF 2 编码的蛋白有三个线性区域:中心域(M)、外壳域(S)和突出域(P)。HEV 中和抗原表位在 P 域内(aa 455~602)(Meng 等,2001;Guu 等,2009;Yamashita 等,2009;Xing 等,2010;Tang 等,2011)。ORF 3 与ORF 2 重叠编码一个小蛋白(113~114 个氨基酸长度),该蛋白与病毒粒子形态及其释放有关（Graff 等,2006;Yamada 等,2009a;Emerson 等,2010)。研究表明，一些分离于不同地区 HEV 分离株基因组的结构域共有序列相似度很高(Arankalle 等,1999)。属于基因 1 型的 HEV 在非洲和亚洲地区更为普遍。在墨西哥分离到了基因 2 型 HEV,同样在非洲地方病中也分离到了罕见的 HEV 基因 2 型变异株。猪和人 HEV 基因 3 型在工业化国家比较常见,而在东亚,以猪和人 HEV 为代表的基因 4 型则比较常见(Chandra 等,2008)。据报道,禽HEV 与基因 5 型相似,但尚未得到证实(Haqshenas 等,2001;Huang 等,2004)。

据报道,HEV 基因 5 型和 6 型通常会感染野猪(AI-Sadeq 等,2018)。阿联酋发现了一种新的 HEV 基因型并将其命名为 HEV7(Woo 等,2014)。HEV7 主要感染单峰骆驼，在骆驼粪便中首次被发现。然而,从肝炎患者的粪便中也分离出了相同的病毒序列，表明该病毒对人和骆驼都具有感染性(Rasche 等,2016)。在双峰驼中发现了 HEV8,但目前还没有在人体内检测出 HEV8 的报道(Woo 等,2016)。

15.4　感染

一般认为,HEV 感染的严重程度取决于宿主免疫状态,但病毒本身特性也对 HEV 致病机制起着重要作用。因此,HEV 基因型也被认为是影响该病严重程度主要因素,例如,感染 HEV 基因 4 型患者表现的症状要比感染 HEV 基因 3 型的患者严重。然而,HEV 基因 1 型引起的感染最为严重，能够引起暴发性肝功能衰竭和严重的胎盘疾病(Al-Sadeq 等,2017,2018)。HEV 衣壳蛋白与宿主细胞受体结合使病毒进入细胞,而后开始复制。例如,ORF 2 编码的多肽结合实验表明,ORF 2 编码多肽的 C 端区域可通过与宿主细胞表面受体的热休克蛋白同源物(HSC70)结合,介导 HEV 进入细胞(Zhou 和 Emerson,2006)。另外,还发现了存在于宿主细胞表面的多重肝素蛋白多糖(heparan sulphate proteoglycan,HSPG),也可作为辅助受体普遍存在于宿主细胞表面（Kalia 等,2009）(图 15.2)。换言之,HEV 衣壳的截短二聚体(大小为 23nm)蛋白质(HEV239)能够与 HSPG 结合(He

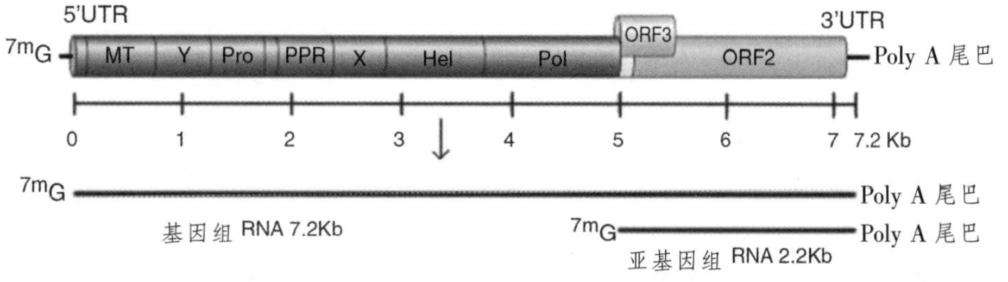

图 15.1　HEV 基因组。RNA 基因组 3'端有 poly A 尾巴,5' 端有 7-甲基鸟嘌呤帽子结构(7mG)。ORF 1 编码非结构蛋白,包括甲基化转移酶、半胱氨酸蛋白酶、RNA 解旋酶、RNA 依赖的 RNA 聚合酶以及 Y 结构域、富含脯氨酸铰链域、X 宏结构域。(Adopted from Kamar et al. 2014.)

等,2008)。

在另一项研究中同样发现截短蛋白能与Grp78的细胞伴侣蛋白也能发生相互作用。综上所述,HEV239衣壳蛋白截短体在HEV与宿主细胞互作、HEV生命周期中都起着至关重要的作用(Yu等,2011)。当病毒进入易感细胞后,HEV通过一种未知机制进行脱壳释放RNA。格尔达霉素是HSP90的抑制剂,能够阻碍HEV239衣壳蛋白的胞内运动,但对HEV239衣壳蛋白截短体与细胞的结合和进入没有任何作用。因此,证实了HSP90对病毒在胞内运输中具有重要的作用(Zheng等,2010)。

病毒基因组脱壳并释放到细胞后,存在于HEV基因组中的5'端非编码区的7-甲基鸟嘌呤引导核糖体亚单位(40S),开始进行病毒基因组帽依赖性翻译。病毒RNA依赖性RdRp产生负链复制RNA中间体。中间体作为模板产生正链病毒基因组。经研究发现,HEV RdRp活性在HEV复制系统中起着重要的作用。换言之,HEV感染的动物组织中检测到了中间体病毒RNA(负链)的存在(Nanda等,1994;Meng等,1998;Agrawal等,2001;Williams等,2001;Graff等,2005a)。顺式反应元件进一步结合带有HEV基因组3' NCR和RdRp揭示了这种交互作用在HEV复制中的重要性(Agrawal等,2001;Emerson等,2001;Graff等,2005a)。顺式反应元件参与HEV基因组铰链区及亚基因组RNA合成的重要性已得到阐述(Graff等,2005b;Cao等,2010)。亚基因组RNA经

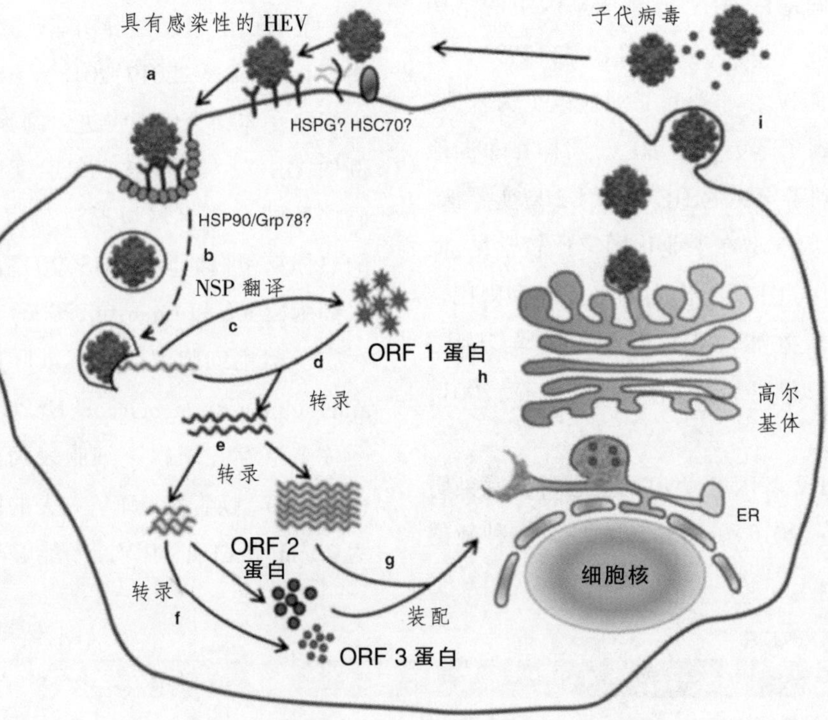

图15.2 HEV生命周期。HEV通过HSPG、HSC70或其他假定的附着受体附着于细胞表面(步骤a)。然后通过未知的特异性细胞受体进入细胞(步骤b)。HEV病毒粒子通过膜进入细胞。HSP90和Grp78也可能参与了此次运输。之后病毒粒子进行脱壳并将正链基因组RNA释放到细胞质中(步骤c)。以正链基因组病毒RNA为模板在细胞质内对ORF 1非结构多聚蛋白进行翻译(步骤d)。病毒RdRp从正链基因组RNA中合成中间体复制负链RNA(步骤e)。正链基因组RNA是生产正链子代病毒基因组的模板。ORF 2和ORF 3蛋白是由亚基因组正链RNA翻译而来(步骤f)。ORF 2衣壳蛋白对病毒基因组RNA进行包装并组装新的病毒粒子(步骤g)。新生的病毒粒子被运输到细胞膜上,ORF 3蛋白加速病毒粒子的运输(步骤h)。新生的病毒分子从感染细胞中释放出来(步骤i)。(Adopted from Cao and Meng 2012.)

过翻译后、编码小蛋白 ORF 3,以及编码 HEV 衣壳蛋白 ORF 2。

最后,虽然对 HEV 的组装和释放机制尚不明确,但 ORF 2 编码蛋白在子代病毒组装和包装过程中起主导作用。经发现,位于 HEV 基因组 5'端的 76nt 结合域与 ORF 2 编码蛋白具有交互作用,这表明该区域在 HEV 基因组包装中具有重要作用(Surjit 等,2004)。ORF 3 编码蛋白也具有类似功能(Yamada 等,2009b;Emerson 等,2010)。ORF 3 编码蛋白含有的基序在膜结合型 HEV 颗粒从感染细胞中释放时发挥着中枢作用(Nagashima 等,2012)。

15.5 流行病学

HEV 通常通过粪-口传播(Khuroo,1991),发展中国家暴发戊型肝炎主要与粪便污染水有关(Khuroo 等,2016b),在未净化水和污染水样本中均发现了与动物和人相关的 HEV 分离株(Vaidya 等,2003;Ippagunta 等,2007;Ishida 等,2012;Masclaux 等,2013)。露天排便是造成农作物、地下水和下水道粪便污染的原因之一(Gurav 等,2007)。例如,在印度有大约 3 亿人露天排便,对河流、井水和溪流等水源造成污染 (Khuroo 和 Khuroo,2015)。除此之外,自来水也会受其污染。印度城市和城镇的水是通过管道输送的,通常情况下,这些供水管道要么与下水道平行分布, 要么与下水道交叉分布。在预定的间歇性供水时期,一旦旧水管出现裂缝和孔洞,便会造成自来水粪便污染。在印度, 大多数情况下发生的流行性戊型肝炎病例与此有关(Khuroo 等,2016a,b)。

HEV3 和 HEV4 引起人兽共患病,在家养猪、梅花鹿和野猪中均有发现(Pavic 等,2015)。感染模式主要有三种,即与污染环境直接相互作用,被感染的动物, 以及食用有人兽共患病的动物源性食品(Yugo,Meng 2013)。食用有人兽共患病的动物源性食品可引起 HEV3 和 HEV4 传播已得到证实 (Izopet 和 Kamar,2008;Yugo 和 Meng,2013),例如,食用了未煮熟的家猪、梅花鹿和野猪可导致戊型肝炎和本地病例的暴发 (Khuroo 和 Khuroo,2008;Miyashita 等,2012), 在欧洲,HEV 广泛传播的一个常见途径是由于食用了科西嘉无花果肠和生吃从超级市场买来的猪肝(Khuroo 等,2016)。

虽然 HEV 主要通过粪-口途径传播, 但最近研究还发现其他传播途径:垂直传播(妊娠期间)和输血传播。2004 年,HEV 被认定为是一种可以通过输血传播的病原体, 对血库供应构成极大威胁。在日本、英国、法国、丹麦和沙特阿拉伯等国家都有 HEV 输血传播病原体的相关记录(Al-sadeq 等,2017)。因此,对于免疫缺陷患者进行输血,可能引起急性或慢性 HEV 感染。尽管如此,目前在世界上许多国家,还没有足够证据证明 HEV 是血液供应的一种新兴血液病原体。现已证明,HEV 垂直传播感染可导致新生儿急性肝炎, 并已经证实 HEV 可在宫内感染, 并有 HEV 宫内感染的相关证据(Khuroo 等,2009)。新生儿 HEV 感染(母婴传播)具有自限性,不会导致持久性病毒血症或临床病程延迟。

2009—2018 年间, 印度综合疾病监测组织共报道了 29 100 例戊型肝炎病例, 其中死亡病例有 164 例。如表 15.1(IDSP,2019)所示,印度还报道了 6897 例疑似肝炎病例,其中死亡病例有 28 例。在所有病例中, 饮用被粪便污染的饮用水是引起发病的主要原因, 在患者血清中检测到了 HEV 的 IgM 抗体。通过研究, 南亚和东南亚国家证实了 HEV 基因 3 型和 4 型的存在。HEV 基因 3 型和 4 型是从动物和动物食品中分离出来的(Kumar 等,2019)。然而,在西亚国家,血源性 HEV 是感染的主要原因 (Hesamizadeh 等,2016;Parsa 等,2016;Nasrallah 等,2017)。在过去的 10 年间(2005—2015 年), 戊型肝炎病例数在欧洲呈指数增长,从 514 例增至 5617 例(Aspinall 等,2017)。HEV 基因 3 型与本地病例有关,在一些病例中,相同的基因

表 15.1 2009—2018 年印度戊型肝炎病毒病例(基于医院数据)

年份	案例数(例)	死亡(例)	疑似病例(例)	死亡(例)
2009	0	0	152	3
2010	153	0	1102	3
2011	2750	4	2246	15
2012	3526	8	1641	6
2013	1153	1	400	0
2014	1097	12	746	0
2015	4460	26	145	0
2016	13997	112	90	0
2017	1611	1	120	1
2018	353	0	255	0

数据来源:http://ww.idsp.nic.in

数据从第 25 周计算

型出现在欧洲猪肉产品里。另一方面,旅行病例与 HEV 基因 1 型或 4 型相关(Aspinall 等,2017)。

在乍得、肯尼亚、索马里、苏丹和乌干达(非洲部分地区),由于居住在避难所和难民营的人们无法获得清洁的水和良好的卫生条件,经常暴发 HEV 感染。由于自身营养不良,较差的生活环境使暴露风险增高,免疫系统受到抑制,这些因素使得那里的人群极易感染 HEV(Teshale 等,2010)。

在阿根廷(HEV3)、巴西(HEV3)、哥伦比亚(HEV3)、墨西哥(HEV2,仅人类)、美国(HEV3)、乌拉圭(HEV1 和 HEV3)和委内瑞拉(HEV1 和 HEV3),人和动物中均发现了 HEV 基因型 1~3(Melgaco 等,2018)。在加拿大零售商店销售的猪肉中发现了基因 3 型,HEV 在猪肉和生猪肝中感染率分别为 47%和 10.5%(Mykytczuk 等,2017)。

15.6 人兽共患病

到目前为止,全球范围内,仅在猪中发现了 HEV 基因 3 型和 4 型。同样,在日本的野猪和家猪身上也发现了 HEV 基因 3 型和 4 型(Takahashi 等,2004,2014)。此外,中国和日本的研究报道也证实了 HEV 基因 3 型和 HEV 基因 4 型的存在。然而美国研究报道指出,在家猪和野猪身上中只发现了 HEV 基因 3 型(Pauli 等,2015)。

此外,英国、美国、德国和日本销售的猪肝及其他产品中也检测到了 HEV(Pauli 等,2015)。从中国、法国、美国等不同国家兔子身上采集到了 HEV 样本,通过进化分析发现,这些 HEV 分离株与 HEV 基因 3 型有很高的相似性。此外,遗传进化分析还表明,兔 HEV(rbHEV)存在感染人类的可能性(Izopet 等,2012)。

在德国从鼠身上获得的类似于 HEV 基因序列,但属于不同的基因型,从遗传进化上看,从 HEV 基因 1 型跨越到 4 型,因此,它被归类到一个单独的群(Johne 等,2010)。此外,通过对美国鼠进行生物研究,发现了类似的 HEV 序列(Purcell 等,2011)。对从不同鼠分离到的毒株进行遗传进化分析发现,HEV 基因 1 型和 4 型具有高度变异性(Mulyanto 等,2013)。这些结果使得大鼠 HEV 被归为正戊型肝炎病毒属或罗卡戊型肝炎病毒属(Johne 等,2014;Smith 等,2014)。

有趣的是,在埃及 13%的马体内检测到 HEV 抗体。此外,有三个序列与 HEV 基因 1 型序列(即埃及人基因型)相似系数最高(Saad 等,2007)。在澳大利亚和美国,有鸡出现肝、脾大的记录,血清学和遗传学分析表明禽 HEV(aHEV)与人 HEV 之间有关联(Payne 等,1999;Haqshenas 等,2001)。此外,BLAST 分析表明日本蝙蝠身上的 HEV 与德国分离株相似系数高。虽然研究人员对不同种蝙蝠 HEV 的地理分布进行了报道,但仍未能将日本和德国存在的类似分离株相联系起来(Kobayashi 等,2018)。

在日本利用 ELISA 和 RT-PCR 方法对猫鼬身上采集血清样本进行 HEV 检测。ELISA 检测结果显示,有 8.3%的猫鼬身上存在 HEV 抗体,但 RT-PCR 检测阴性。该研究表明,猫鼬不应被认为是 HEV 人兽共患病的主要宿主(Li 等,2006)。但后来的研究发现,在野生猫鼬和猫鼬分离到的 HEV 基因 3 型序列,与猪身上分离到 HEV 非常相似(Nidaira 等,2012)。

15.7 诊断

通过演示和诊断程序等方法确认并收集 HEV 相关信息。在日常定期检查中,丙氨酸转氨酶、天冬氨酸转氨酶、胆红素、γ-谷氨酰转肽酶和可溶性磷酸盐升高,认为是肝遭受 HEV 损伤的一些非特异性标志(Aggarwal 和 Jameel,2011)。通过免疫电子显微镜检查证明,在粪便样本中存在着类似 HEV 感染的颗粒,用 HEV 康复者的血清沉淀这些带有天然抗原的颗粒,可分离到 HEV (Dienstag 等,1976)。利用免疫荧光显微法也可以在肝活检组织中检测到 HEV 抗原,荧光素标记的抗 HEV 抗体与 HEV 结合后,在荧光显微镜可观察到抗 HEV 抗体与 HEV 形成的复合体(Krawczynski 和 Bradley,1989)。用于识别 HEV 的方法如图 15.3 所示。

在患者的早期诊断中,粪便或血清样本及时收集和恰当处理直接决定 HEV RNA 检测方法的准确性。通常情况下,对于急性感染 HEV 患者,发病 2~6 周时在血清里就能够检测到 HEV RNA,3~8 周时能够在粪便中检测到 HEV RNA (Alsadeq 等,2017,2018)。核酸扩增技术可以准确识别 HEV 的感染性,因此生物样本中病毒 RNA 的检测是鉴定重型戊型肝炎的黄金标准(Huang 等,2010)。一般情况下,普通的诊断研究中心并不具备以核酸扩增技术为基础的鉴定方法,因为该鉴定方法需要受过专业训练的人员进行样本分析(Mirazo 等,2014)。目前对粪便和血清样本中 HEV 的核酸扩增技术检测,是在反转录基础上,进行的 PCR、环介导等温扩增和实时定量 PCR。通过对核酸扩增技术设计、优化,可以实现对感染人的 HEV 四个基因型的检测 (Jothikumar 等,2006;Gyarmati 等,2007;Lan 等,2009)。

内部测试显示,HEV 核酸在体外检测具有很高的不稳定性(Baylis 等,2011)。采用一步法多重 RT-PCR 检测患者血清中的非甲、非乙、非丙型肝炎病毒和 HEV 的 RNA。该方法是将所有病毒基因组中保守区作为靶序列进行扩增,具有快速、灵敏、可重复的优点(Irshad 等,2013)。

在对抗体与 HEV 结合反应及其结合反应模式认识和研究之前,有关 HEV 感染的免疫学方面信息甚少。用于诊断 HEV 感染的 ELISA 商品化试剂盒的敏感性和特异性见表 15.2。抗 HEV 的抗体 IgM 在重型 HEV 感染患者的黄疸症状首次出现后的 4 天就可被检测到,并可持续存在 5 个月(Favorov 等,1992)。一般情况下,90%的重型戊型肝炎患者中可在感染后 14 天检测到抗 HEV 抗体 IgM。相比之下,IgM 产生后不久,IgG 也可被检测到(Favorov 等,1992)。在重度 HEV 感染患者中,由于两种抗体的同时产生,为诊断带来了很大困难。常见的抗 HEV 的 IgM 检测有效率仅有 90%~97%,假阳性反应高达 10%,有一些感染 HEV 基因 1 型的患者检测结果为假阴性(Herremans 等,2007;Legrand -Abravanel,2009;Drobeniuc 等,2010)。在商品化 HEV 血清学检测方法中,ORF 2、ORF 3 编码的抗原和免疫活性肽可有效地识别抗 HEV 的抗体 (IgA、IgM 和 IgG)(Takahashi 等,

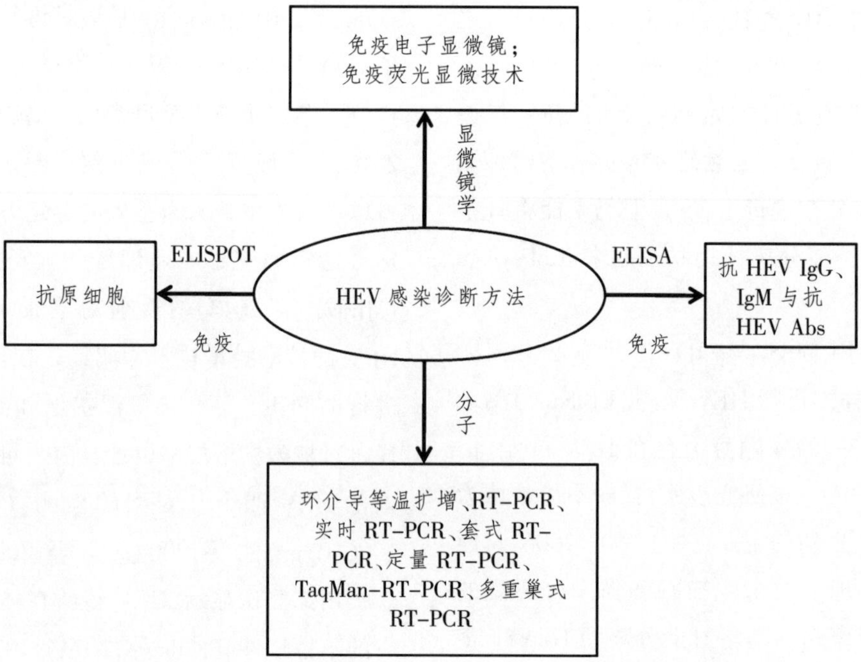

图 15.3 戊型肝炎病毒诊断方法。(Adopted from Kumar et al. 2019)

2005;Zhang 等,2009b)。有时，不同病毒与 HEV 的血清学交叉反应也与肝炎病毒感染有关(Hyams 等,2014)。由于在患者感染期间,HEV 血清抗体产生存在延迟或中止的情况，因而在血清学检测时患者病情是影响 HEV 检测的另一限制因素。在这种情况下,荧光定量分析法可以实现对 HEV 感染快速检测(Brown 等,2016)。本实验使用跨越开放读码框架(ORF 1~3)(即基因型 3a 肽库)的 616 条重叠多肽用于产生 IFN-γ 的 T 细胞 ELISA 斑点分析中。

15.8 预防与疫苗

在 HEV 暴发期间,提供清洁的饮水和改善卫生条件是预防 HEV 最重要的两项措施。但能够及时实施这两项措施具有相当大的挑战性，尤其是在疫情暴发期间(Teshale 和 Hu,2011)。因此,开发一种有效的疫苗尽可能降低 HEV 感染变得尤为重要。两种 HEV 疫苗已应用于临床初期,第一个 rHEV5 疫苗是由比利时葛兰素史克生产(比利时,里克森萨特,GSK)(Li 等,2015)。rHEV5 的免疫接种实验是在尼泊尔军人身上进行，但接种一剂疫苗后,并没有获得免疫性保护,在接种两个剂量疫苗后可以提供保护作用，但该研究未能提供确切的结果(Shrestha 等,2007)。

另一种疫苗 HEV239(改名为 Hecolin9)由中国厦门 Innovax 公司研制 (中国，厦门)(Zhu 等,2009;Zhu 等,2010;Wu 等,2012)。2012 年,HEV 239 被中国政府批准使用,主要应用于 16 岁以上的人群。然而,该疫苗在其他国家的使用仍在审批中(Park,2012;Riedmann,2012)。该重组疫苗利用 HEV 高度保守抗原，且 HEV 仅一个血清型,为此,该疫苗对 HEV 的四个基因型(HEV1~4)都具有较强保护作用(Melgaco 等,2018)。

中国疾病预防控制中心正在通过互联网收集接种后的不良反应,免疫效果看似很好。目前只有中国人接种 HEV 疫苗预防 HEV,且 HEV 感染已成为一个健康问题,为此,接种 HEV 疫苗的成本-效益已经被讨论过多次。厦门 Innovax 公司制成 HEV239 每剂成本为 17.60~41.70 美元，比每剂

表 15.2　检测 HEV 各类商品化的 ELISA 检测方法

商品化分析试剂	生产商	应用于 HEV 实验的抗原	抗体检测	敏感性(%)	特异性(%)	参考文献
Wantai	Wantai Biological Pharmacy Enterprise Co（中国，北京）	IgM：微型孔用抗 μ 链包被	IgM	97.7；75；65.4；87.3；96.7	99.6；>99；NS；100；96.7	Abravanel 等（2013），Pas 等（2013），Avellon 等（2015），Zhou 等（2008），Abravanel 等（2015）
		IgG：重组 HEV 抗原	IgG	93.2；72.5	97.8；NS	
Mikrogen	Mikrogen GmbH（德国，诺伊里德）	HEV 基因 1 型和 3 型 ORF 2 重组抗原	IgM	38；74；75；92	99；99；NS；95.6	Norder 等(2016)，Pas 等(2013)，Avellon 等(2015)，Drobeniuc 等(2010)
			IgG	62；72.5	99；NS	
Euroimmun	Euroimmun Medizinische Labordiagnostika AG（德国，吕贝克）	HEV 基因 1 型和 3 型重组抗原	IgM	24；61.5	100；NS	Drobeniuc 等(2010)，Avellon 等(2015)，Norder 等(2016)
			IgG	42；57.5	99；NS	
MP Diagnostics	MP Biomedicals（新加坡）	ORF 2 编码抗原	IgM	74；88；59.6；67.3；72；80；72.5	84；99.5；NS；89.1；93；86.1；93	Pas 等(2013)，Legrand- Abravanel 等(2009)，Avellon 等(2015)，Zhou 等(2008)，Abravanel 等(2015)，Wu 等(2014)，Khudyakov 和 Kamili (2011)
		3 个 ORF 2 重组抗原	IgG	70；73..3	NS；65.3	
DSI	DSI S.R.l（意大利，萨龙诺）	HEV 基因 1 型、2 型和 3 型 ORF 2 及 ORF 3 重组多肽	IgM	63；71；80.8	99；90；NS	Norder 等(2016)，Pas 等(2013)，Avellon 等(2015)
			IgG	72；75	90；NS	
DS-EIA	RPC Diagnostic Systems（俄罗斯，下诺夫哥里德）	HEV 基因 1 型和 3 型 ORF 2 及 ORF 3 重组多肽	IgM	71；98	90；95.6	Pas 等（2013），Khudyakov 和 Kamili（2011）
			IgG	71	90	

（待续）

表 15.2(续)

商品化分析试剂	生产商	应用于 HEV 实验的抗原	抗体检测	敏感性(%)	特异性(%)	参考文献
DiaPro	Diagnostic Bioprobes S.r.l (意大利,米兰)	HEV 基因 1 型、2 型、3 型和 4 型 ORF 2 和 ORF 3 保守抗原表位	IgM IgG	72;81;59.6 98;77.5	100;98;NS 96;NS	Norder 等(2016),Pas 等(2013),Avellon 等(2015)
Axiom	Axiom Diagnostics, Worms (德国)	HEV 基因 1 型的 ORF 2 羧基端	IgM IgG	29 95	99 98	Norder 等(2016)
Adaltis EIAgen	Adaltis S.r.l., Guidonia Montecelio (意大利)	ORF 2 和 ORF 3 抗原决定簇	IgM IgG	97.7;90;80 81.8;80	100;100;87.4 89.5;62.9	Abravanel 等(2013),Legrand-Abravanel 等 (2009),Wu 等(2014)
Fortress Diagnostics	Fortress Diagnostics (英国,阿尔斯特)	IgM:IgM:微型孔用抗 μ 链包被 IgG:HEV 重组抗原	IgM 和 IgG IgG	92;100;95 99.5;98	88;86.2;97 99.6;NS	Galiana 等 (2008),Schnegg 等(2013),Yan 等 (2008),Bendall 等(2010)
International Immuno Diagnostics	International Immuno-Diagnostics (美国,福斯特)	NM	IgM	82.4	91.7	Khudyakov 和 Kamili (2011)

23.21 美元的甲型肝炎疫苗要便宜得多(Riedmann,2012;Zhao 等,2016)。免疫接种能够有效降低住院和治疗的成本，因此在疫苗市场上推出 HEV 疫苗可能是一种经济有效的措施。然而,某些区域的信息仍然匮乏，如免疫接种的儿童(<16 岁)、老年人(>65 岁)、妊娠女性、隐性肝病患者或免疫抑制病患者。这些人群是感染 HEV 的高风险人群,急需这种疫苗(Li 等,2015)。

15.9 展望

在世界范围内,HEV 被认为是非甲和非乙型肠道传播重症病毒性肝炎的主要病因。在发展中国家，恶劣的卫生条件是造成这种疾病肆虐的主要原因。在发达国家,主要原因则是人兽共患病和输血。在疫情期间,预防 HEV 感染的第一步,就是提供清洁的饮用水,以及对人类垃圾的正确处理。目前已经成功建立了 HEV 检测的血清学方法和分子生物学方法，通过对患者血清中的 HEV 抗原、HEV 抗体或 RNA 检测，从而实现对 HEV 感染有效监测。这些方法都各有利弊,但检测 HEV RNA 仍为标准方法。尽管在疫苗接种和治疗方面取得了很大的进步,但阻碍根除 HEV 的制约因素依然存在。

参考文献

Abravanel F, Chapuy-Regaud S, Lhomme S et al (2013) Performance of anti-HEV assays for diagnosing acute hepatitis E in immunocompromised patients. J Clin Virol 58:624–628

Abravanel F, Lhomme S, Chapuy-Regaud S et al (2015) Performance of a new rapid test for detecting anti-hepatitis E virus immunoglobulin M in immunocompetent and immunocompromised patients. J Clin Virol 70:101–104

Agrawal S, Gupta D, Panda SK (2001) The 3′ end of hepatitis E virus (HEV) genome binds specifically to the viral RNA-dependent RNA polymerase (RdRp). Virology 282:87–101

Aggarwal R (2011) Clinical presentation of hepatitis E. Virus Res 161:15–22

Aggarwal R, Jameel S (2011) Hepatitis E. Hepatology 54:2218–2226

Al-Sadeq DW, Majdalawieh AF, Nasrallah GK (2017) Seroprevalence and incidence of hepatitis E virus among blood donors: a review. Rev Med Virol 2017:e1937

Al-Sadeq DW, Majdalawieh AF, Mesleh AG et al (2018) Laboratory challenges in the diagnosis of hepatitis E virus. J Med Microbiol 67:466–480

Arankalle VA, Paranjape S, Emerson SU et al (1999) Phylogenetic analysis of hepatitis E virus isolates from India (1976–1993). J Gen Virol 80:1691–1700

Aspinall EJ, Couturier E, Faber M et al (2017) Hepatitis E virus infection in Europe: surveillance and descriptive epidemiology of confirmed cases, 2005 to 2015. Euro Surveill 22:30561

Avellon A, Morago L, Garcia-Galera del Carmen M et al (2015) Comparative sensitivity of commercial tests for hepatitis E genotype 3 virus antibody detection. J Med Virol 87:1934–1939

Balayan MS, Andjaparidze AG, Savinskaya SS et al (1983) Evidence for a virus in non-A, non-B hepatitis transmitted via the fecal-oral route. Intervirology 20:23–31

Baylis SA, Hanschmann KM, Blümel J et al (2011) Standardization of hepatitis E virus (HEV) nucleic acid amplification technique (NAT)-based assays: an initial study to evaluate a panel of HEV strains and investigate laboratory. J Clin Microbiol 49:1234–1239

Bendall R, Ellis V, Ijaz S et al (2010) A comparison of two commercially available anti-HEV IgG kits and a re-evaluation of anti-HEV IgG seroprevalence data in developed countries. J Med Virol 82:799–805

Brown A, Halliday JS, Swadling L et al (2016) Characterization of the specificity, functionality, and durability of host T-cell responses against the full-length hepatitis E virus. Hepatology 64:1934–1950

Cao D, Huang YW, Meng XJ (2010) The nucleotides on the stem–loop RNA structure in the junction region of the hepatitis E virus genome are critical for virus replication. J Virol 84:13040–13044

Cao D, Meng XJ (2012) Molecular biology and replication of hepatitis E virus. Emerg Microbes Infect 1:e17

Chandra V, Taneja S, Kalia M et al (2008) Molecular biology and pathogenesis of hepatitis E virus. J Biosci 33:451–464

Dalton HR, Bendall R, Ijaz S et al (2008) Hepatitis E: an emerging infection in developed countries. Lancet Infect Dis 8:698–709

Dienstag JL, Alling DW, Purcell RH (1976) Quantitation of antibody to hepatitis A antigen by immune electron microscopy. Infect Immun 13:1209–1213

Drobeniuc J, Meng J, Reuter G et al (2010) Serologic assays specific to immunoglobulin M antibodies against hepatitis E virus: pangenotypic evaluation of performances. Clin Infect Dis 51:e24–e27

Emerson SU, Nguyen HT, Torian U et al (2010) Release of genotype 1 hepatitis E virus from cultured hepatoma and polarized intestinal cells depends on open reading frame 3 protein and requires an intact PXXP motif. J Virol 84:9059–9069

Emerson SU, Purcell RH (2003) Hepatitis E virus. Rev Med Virol 13:145–154

Emerson SU, Zhang M, Meng XJ et al (2001) Recombinant hepatitis E virus genomes infectious for primates: importance of capping and discovery of a cis-reactive element. Proc Natl Acad Sci U S A 98:15270–15275

Favorov MO, Fields HA, Purdy MA et al (1992) Serologic identification of hepatitis E virus infections in epidemic and endemic settings. J Med Virol 36:246–250

Fujiwara S, Yokokawa Y, Morino K et al (2014) Chronic hepatitis E: a review of the literature. J Viral Hepat 21:78–89

Galiana C, Fern_andez-Barredo S, García A et al (2008) Occupational exposure to hepatitis E virus (HEV) in swine workers. Am J Trop Med Hyg 78:1012–1015

Graff J, Nguyen H, Kasorndorkbua C et al (2005a) In vitro and in vivo mutational analysis of the 39-terminal regions of hepatitis E virus genomes and replicons. J Virol 79:1017–1026

Graff J, Nguyen H, Yu C et al (2005b) The open reading frame 3 gene of hepatitis E virus contains a cis-reactive element and encodes a protein required for infection of macaques. J Virol 79:6680–6689

Graff J, Torian U, Nguyen H et al (2006) A bicistronic subgenomic mRNA encodes both the ORF2 and ORF3 proteins of hepatitis E virus. J Virol 80:5919–5926

Gurav YK, Kakade SV, Kakade RV et al (2007) A study of hepatitis E outbreak in rural area of Western Maharashtra. Indian J Community Med 32:182–184

Guu TS, Liu Z, Ye Q et al (2009) Structure of the hepatitis E virus-like particle suggests mechanisms for virus assembly and receptor binding. Proc Natl Acad Sci U S A 106:12992–12997

Gyarmati P, Mohammed N, Norder H et al (2007) Universal detection of hepatitis E virus by two real-time PCR assays: TaqMan and primer-probe energy transfer. J Virol Methods 146:226–235

Haqshenas G, Shivaprasad HL, Woolcock PR et al (2001) Genetic identification and characterization of a novel virus related to human hepatitis E virus from chickens with hepatitis-splenomegaly syndrome in the United States. J Gen Virol 82:2449–2462

Herremans M, Bakker J, Duizer E et al (2007) Use of serological assays for diagnosis of hepatitis E virus genotype 1 and 3 infections in a setting of low endemicity. Clin Vaccine Immunol 14:562–568

Hesamizadeh K, Sharafi H, Keyvani H et al (2016) Hepatitis A virus and hepatitis E virus seroprevalence among blood donors in Tehran, Iran. Hepat Mon 16:e32215

He S, Miao J, Zheng Z et al (2008) Putative receptor-binding sites of hepatitis E virus. J Gen Virol 89:245–249

Huang FF, Sun ZF, Emerson SU et al (2004) Determination and analysis of the complete genomic sequence of avian hepatitis E virus (avian HEV) and attempts to infect rhesus monkeys with avian HEV. J Gen Virol 85:1609–1618

Huang S, Zhang X, Jiang H et al (2010) Profile of acute infectious markers in sporadic hepatitis E. PLoS One 5:e13560

Hyams C, Mabayoje DA, Copping R et al (2014) Serological cross reactivity to CMV and EBV causes problems in the diagnosis of acute hepatitis E virus infection. J Med Virol 86:478–483

ICTV (2019) International committee on taxonomy of viruses. https://talk.ictvonline.org/ictv-reports/ictv_online_report/positive-sense-rna-viruses/w/hepeviridae. Accessed 25 Mar 2019

IDSP (2019). Weekly outbreaks. http://www.idsp.nic.in. Accessed 1 Mar 2019

Intharasongkroh D, Sa-nguanmoo P, Tuanthap S et al (2017) Hepatitis E virus in pork and variety meats sold in fresh markets. Food Environ Virol 9:45–53

Ippagunta SK, Naik S, Sharma B et al (2007) Presence of hepatitis E virus in sewage in Northern India: frequency and seasonal pattern. J Med Virol 79:1827–1831

Irshad M, Ansari MA, Irshad K et al (2013) Novel single-step multiplex real-time polymerase chain reaction assay for simultaneous quantification of hepatitis virus A, B, C, and E in serum. J Gastroenterol Hepatol 28:1869–1876

Ishida S, Yoshizumi S, Ikeda T et al (2012) Detection and molecular characterization of hepatitis E virus in clinical, environmental and putative animal sources. Arch Virol 157:2363–2368

Izopet J, Dubois M, Bertagnoli S et al (2012) Hepatitis E virus strains in rabbits and evidence of a closely related strain in humans, France. Emerg Infect Dis 18:1274–1281

Izopet J, Kamar N (2008) Hepatitis E: from zoonotic transmission to chronic infection in immunosuppressed patients. Med Sci 24:1023–1025

Johne R, Dremsek P, Reetz J et al (2014) Heperviridae: an expending family of vertebrate viruses. Infect Genet Evol 27:212–229

Johne R, Plenge-Bonig A, Hess M et al (2010) Detection of novel hepatitis E-like virus in faeces of wild rats using a nested broad-spectrum RT-PCR. J Gen Virol 91:750–758

Jothikumar N, Cromeans TL, Robertson BH et al (2006) A broadly reactive one-step real-time RT-PCR assay for rapid and sensitive detection of hepatitis E virus. J Virol Methods 131:65–71

Kalia M, Chandra V, Rahman SA et al (2009) Heparan sulfate proteoglycans are required for cellular binding of the hepatitis E virus ORF2 capsid protein and for viral infection. J Virol 83:12714–12724

Kamar N, Dalton HR, Abravanel F et al (2014) Hepatitis E virus infection. Clin Microbiol Rev 27:116–138

Khudyakov Y, Kamili S (2011) Serological diagnostics of hepatitis E virus infection. Virus Res 161:94–92

Khuroo MS (1980) Study of an epidemic of non-A, non-B hepatitis. Possibility of another human hepatitis virus distinct from post-transfusion non-A, non-B type. Am J Med 68:818–824

Khuroo MS (1991) Hepatitis E: the enterically transmitted non-A, non-B hepatitis. Indian J Gastroenterol 10:96–100

Khuroo MS, Kamili S, Khuroo MS (2009) Clinical course and duration of viremia in vertically transmitted hepatitis E virus (HEV) infection in babies born to HEV-infected mothers. J Viral Hepat 16:519–523

Khuroo MS, Khuroo MS (2008) Hepatitis E virus. Curr Opin Infect Dis 21:539–543

Khuroo MS, Khuroo MS (2015) Sanitation and sewage disposal in India. JK-Practitioner 20:43–46. http://medind.nic.in/jab/t15/i1/jabt15i1p43.pdf. Accessed 1 Mar 2019

Khuroo MS, Khuroo MS, Khuroo NS (2016a) Hepatitis E: discovery, global impact, control and cure. World J Gastroenterol 22:7030–7045

Khuroo MS, Khuroo MS, Khuroo NS (2016b) Transmission of hepatitis E virus in developing countries. Viruses 8:E253

Kobayashi T, Murakami S, Yamamoto T et al (2018) Detection of bat hepatitis E virus RNA in microbats in Japan. Virus Genes 54:599–602

Koonin EV, Gorbalenya AE, Purdy MA et al (1992) Computer-assisted assignment of functional domains in the nonstructural polyprotein of hepatitis E virus: delineation of an additional group of positive-strand RNA plant and animal viruses. Proc Natl Acad Sci U S A 89:8259–8263

Krawczynski K, Bradley DW (1989) Enterically transmitted non-A, non-B hepatitis: identification of virus-associated antigen in experimentally infected cynomolgus macaques. J Infect Dis 159:1042–1049

Kumar H, Kamar N, Kumar D (2019) Hepatitis E: current status in India and other Asian countries. J Pure Appl Microbiol 13:141–159

Lan X, Yang B, Li BY et al (2009) Reverse transcription-loop-mediated isothermal amplification assay for rapid detection of hepatitis E virus. J Clin Microbiol 47:2304–2306

Legrand-Abravanel F, Thevenet I, Mansuy JM et al (2009) Good performance of immunoglobulin M assays in diagnosing genotype 3 hepatitis E virus infections. Clin Vaccine Immunol 16:772–774

Li SW, Zhao Q, Wu T et al (2015) The development of a recombinant hepatitis E vaccine HEV 239. Hum Vaccin Immunother 11:908–914

Li T-C, Saito M, Ogura G et al (2006) Serologic evidence for hepatitis E virus infection in mongoose. Am J Trop Med Hyg 74:932–936

Li TC, Yamakawa Y, Suzuki K et al (1997) Expression and self-assembly of empty virus-like particles of hepatitis E virus. J Virol 71:7207–7213

Masclaux FG, Hotz P, Friedli D et al (2013) High occurrence of hepatitis E virus in samples from wastewater treatment plants in Switzerland and comparison with other enteric viruses. Water Res 47:5101–5109

Melgaço JG, Gardinali NR, de VDM M et al (2018) Hepatitis E: update on prevention and control. Biomed Res Int 2018:5769201

Meng J, Dai X, Chang JC et al (2001) Identification and characterization of the neutralization epitope (s) of the hepatitis E virus. Virology 288:203–211

Meng XJ, Halbur PG, Haynes JS et al (1998) Experimental infection of pigs with the newly identified swine hepatitis E virus (swine HEV), but not with human strains of HEV. Arch Virol 143:1405–1415

Mirazo S, Ramos N, Mainardi V et al (2014) Transmission, diagnosis, and management of hepatitis E: an update. Hepat Med 6:45–59

Miyashita K, Kang JH, Saga A et al (2012) Three cases of acute or fulminant hepatitis E caused by ingestion of pork meat and entrails in Hokkaido, Japan: zoonotic food-borne transmission of hepatitis E virus and public health concerns. Hepatol Res 42:870–878

Mulyanto, Depamede SN, Sriasih M et al (2013) Frequent detection and characterization of hepatitis E virus variants in wild rats (*Rattus rattus*) in Indonesia. Arch Virol 158:87–96

Mykytczuk O, Harlow J, Bidawid S et al (2017) Prevalence and molecular characterization of hepatitis E virus in retail pork products marketed in Canada. Food Environ Virol 9:208–218

Nagashima S, Takahashi M, Jirintai S et al (2012) Tsg101 and the vacuolar protein sorting pathway are required for release of hepatitis E virions. J Gen Virol 92:2838–2248

Nanda SK, Panda SK, Durgapal H et al (1994) Detection of the negative strand of hepatitis E virus RNA in the livers of experimentally infected rhesus monkeys: evidence for viral replication. J Med Virol 42:237–240

Nasrallah GK, Al-Absi ES, Ghandour R et al (2017) Seroprevalence of hepatitis E virus among blood donors in Qatar (2013–2016). Transfusion 57:1801–1807

Navaneethan U, Al Mohajer M, Shata MT (2008) Hepatitis E and pregnancy: understanding the pathogenesis. Liver Int 28:1190–1199

Nidaira M, Takahashi K, Go O et al (2012) Detection and phylogenetic analysis of hepatitis E viruses from mongooses in Okinawa, Japan. J Vet Med Sci 74:1665–1668

Norder H, Karlsson M, Mellgren Å et al (2016) Diagnostic performance of five assays for anti-hepatitis E virus IgG and IgM in a large cohort study. J Clin Microbiol 54:549–555

Park SB (2012) Hepatitis E vaccine debuts. Nature 491:21–22

Parsa R, Adibzadeh S, Behbahani AB et al (2016) Detection of hepatitis E virus genotype 1 among blood donors from Southwest of Iran. Hepat Mon 16:e34202

Pas SD, Streefkerk RH, Pronk M et al (2013) Diagnostic performance of selected commercial HEV IgM and IgG ELISAs for immunocompromised and immunocompetent patients. J Clin Virol 58:629–634

Pauli G, Aepfelbacher M, Bauerfeind U et al (2015) Hepatitis E virus. Transfus Med Hemother 42:247–265

Pavio N, Meng XJ, Doceul V (2015) Zoonotic origin of hepatitis E. Curr Opin Virol 10:34–41

Payne CJ, Ellis TM, Plant SL et al (1999) Sequence data suggests big liver and spleen disease virus (BLSV) is genetically related to hepatitis E virus. Vet Microbiol 68:119–125

Pinto MA, de Oliveira JM, González J (2017) Hepatitis A and E in South America: new challenges toward prevention and control. In: Human virology in Latin America. Springer, Cham, pp 119–138

Poovorawan K, Jitmitrapab S, Treeprasertsuk S et al (2014) Risk factors and molecular characterization of acute sporadic symptomatic hepatitis E virus infection in Thailand. Asian Pac J Trop Med 7:709–714

Purcell RH, Engle RE, Rood MP et al (2011) Hepatitis E virus in rats, Los Angeles, California, USA. Emerg Infect Dis 17:2216–2222

Rasche A, Saqib M, Liljander AM et al (2016) Hepatitis E virus infection in dromedaries, North and East Africa, United Arab Emirates, and Pakistan, 1983–2015. Emerg Infect Dis 22:1249–1252

Rein DB, Stevens GA, Besides TJ (2012) The global burden of hepatitis E virus genotypes 1 and 2 in 2005. Hepatology 55:988–997

Rendon J, Hoyos MC, Di Filippo D et al (2016) Hepatitis E virus genotype 3 in Colombia: survey in patients with clinical diagnosis of viral hepatitis. PLoS One 11:e0148417

Riedmann EM (2012) Chinese biotech partnership brings first hepatitis E vaccine to the market. Hum Vaccin Immunother 8:1743–1744

Reyes GR, Purdy MA, Kim JP et al (1990) Isolation of a cDNA from the virus responsible for enterically transmitted non-A, non-B hepatitis. Science 247:1335–1339

Saad MD, Hussein HA, Bashandy MM et al (2007) Hepatitis E virus infection in work horses in Egypt. Infect Genet Evol 7:368–373

Schnegg A, Bürgisser P, André C et al (2013) An analysis of the benefit of using HEV genotype 3 antigens in detecting anti-HEV IgG in a European population. PLoS One 8:e62980

Surjit M, Jameel S, Lal SK (2004) The ORF2 protein of hepatitis E virus binds the 5′ region of viral RNA. J Virol 78:320–328

Suwannakarn K, Tongmee C, Theamboonlers A et al (2010) Swine as the possible source of hepatitis E virus transmission to humans in Thailand. Arch Virol 155:1697–1699

Shrestha MP, Scott RM, Joshi DM et al (2007) Safety and efficacy of a recombinant hepatitis E vaccine. N Engl J Med 356:895–903

Smith DB, Simmonds P, Jameel S et al (2014) Consensus proposals for classification of the family Hepeviridae. J Gen Virol 95:2223–2232

Takahashi K, Kitajima N, Abe N et al (2004) Complete or near-complete nucleotide sequences of hepatitis E virus genome recovered from a wild boar, a deer, and four patients who ate the deer. Virology 330:501–505

Takahashi M, Kusakai S, Mizuo H et al (2005) Simultaneous detection of immunoglobulin A (IgA) and IgM antibodies against hepatitis E virus (HEV) is highly specific for diagnosis of acute HEV infection. J Clin Microbiol 43:49–56

Takahashi M, Nishizawa T, Nagashima S et al (2014) Molecular characterization of a novel hepatitis E virus (HEV) strain obtained from wild boar in Japan that is highly divergent from previously recognized HEV strains. Virus Res 180:59–69

Tam AW, Smith MM, Guerra ME et al (1991) Hepatitis E virus (HEV): molecular cloning and sequencing of the full-length viral genome. Virology 185:120–131

Tang X, Yang C, Gu Y et al (2011) Structural basis for the neutralization and genotype specificity of hepatitis E virus. Proc Natl Acad Sci U S A 108:10266–10271

Temmam S, Besnard L, Andriamandimby SF et al (2013) High prevalence of hepatitis E in humans and pigs and evidence of genotype-3 virus in swine, Madagascar. Am J Trop Med Hyg 88:329–338
Teshale EH, Hu DJ (2011) Hepatitis E: epidemiology and prevention. World J Hepatol 3:285–291
Teshale EH, Howard CM, Grytdal SP et al (2010a) Hepatitis E epidemic, Uganda. Emerg Infect Dis 16:126–129
Teshale EH, Hu DJ, Holmberg SD (2010b) The two faces of hepatitis E virus. Clin Infect Dis 51:328–334
Vaidya SR, Tilekar BN, Walimbe AM et al (2003) Increased risk of hepatitis E in sewage workers from India. J Occup Environ Med 45:1167–1170
Viswanathan R (1957) Infectious hepatitis in Delhi (1955–56): a critical study-epidemiology 1957. Natl Med J India 26:362–377
WHO (2017) The global hepatitis report, 2017. https://apps.who.int/iris/bitstream/handle/10665/255016/9789241565455-eng.pdf;jsessionid=09784545BEEE72E45785F98132749D65?sequence=1. Accessed 1 Mar 2019
Williams TP, Kasorndorkbua C, Halbur PG et al (2001) Evidence of extrahepatic sites of replication of the hepatitis E virus in a swine model. J Clin Microbiol 39:3040–3046
Wong DC, Purcell RH, Sreenivasan MA et al (1980) Epidemic and endemic hepatitis in India: evidence for a non-A, non-B hepatitis virus aetiology. Lancet 2:876–879
Woo PC, Lau SK, Teng JL et al (2014) New hepatitis E virus genotype in camels, the Middle East. Emerg Infect Dis 20:1044–1048
Woo PC, Lau SK, Teng JL et al (2016) New hepatitis E virus genotype in Bactrian camels, Xinjiang, China, 2013. Emerg Infect Dis 22:2219–2221
Wu T, Li SW, Zhang J et al (2012) Hepatitis E vaccine development: a 14year odyssey. Hum Vaccin Immunother 8:823–827
Wu WC, Su CW, Yang JY et al (2014) Application of serologic assays for diagnosing acute hepatitis E in national surveillance of a nonendemic area. J Med Virol 86:720–728
Xing L, Li TC, Mayazaki N et al (2010) Structure of hepatitis E virion-sized particle reveals an RNA-dependent viral assembly pathway. J Biol Chem 285:33175–33183
Xing L, Wang JC, Li TC et al (2011) Spatial configuration of hepatitis E virus antigenic domain. J Virol 85:1117–11124
Yamada K, Takahashi M, Hoshino Y et al (2009a) ORF3 protein of hepatitis E virus is essential for virion release from infected cells. J Gen Virol 90:1880–1891
Yamada K, Takahashi M, Hoshino Y et al (2009b) Construction of an infectious cDNA clone of hepatitis E virus strain JE03-1760F that can propagate efficiently in cultured cells. J Gen Virol 90:457–462
Yamashita T, Mori Y, Miyazaki N et al (2009) Biological and immunological characteristics of hepatitis E virus-like particles based on the crystal structure. Proc Natl Acad Sci U S A 106:12986–12991
Yan Q, Du HL, Wang YB et al (2008) Comparison of two diagnostic reagents to detect anti-hepatitis E virus IgG antibodies. Chin J Zoon 24:1087–1089
Yugo DM, Meng XJ (2013) Hepatitis E virus: foodborne, waterborne and zoonotic transmission. Int J Environ Res Public Health 10:4507–4533
Yu H, Li S, Yang C et al (2011) Homology model and potential virus-capsid binding site of a putative HEV receptor Grp78. J Mol Model 17:987–995
Zhang J, Liu CB, Li RC et al (2009a) Randomized-controlled phase II clinical trial of a bacterially expressed recombinant hepatitis E vaccine. Vaccine 27:1869–1874
Zhang S, Tian D, Zhang Z et al (2009b) Clinical significance of anti-HEV IgA in diagnosis of acute genotype 4 hepatitis E virus infection negative for anti-HEV IgM. Dig Dis Sci 54:2512–2518
Zhao Y, Zhang X, Zhu F et al (2016) A preliminary cost-effectiveness analysis of hepatitis E vaccination among pregnant women in epidemic regions. Hum Vaccin Immunother 12:2003–2009
Zheng ZZ, Miao J, Zhao M et al (2010) Role of heat-shock protein 90 in hepatitis E virus capsid trafficking. J Gen Virol 91:1728–1736
Zhou H, Jiang CW, Li LP et al (2008) Comparison of the reliability of two ELISA kits for detecting IgM antibody against hepatitis E virus. Chin J Prev Med 42:667–671
Zhou Y, Emerson SU (2006) P.302 Heat shock cognate protein 70 may mediate the entry of hepatitis E virus into host cells. J Clin Virol 36:S155
Zhu FC, Zhang J, Zhang XF et al (2010) Efficacy and safety of a recombinant hepatitis E vaccine in healthy adults: a large-scale, randomised, double-blind placebo-controlled, phase 3 trial. Lancet 376:895–902

《新发和跨界动物病毒》:世界病毒学学会组织出版

关于世界病毒学学会

世界病毒学学会(world society for virology,WSV)是一个非营利性组织(501c3-ID No. 001303257),成立于2017年,旨在加强对人类、动物和植物等不同病毒性疫病的病毒学研究。

WSV的主要宗旨包括但不限于:

1. 世界各地的病毒学家共同组成学会,且不收取会员费(这是许多国家的一些病毒学家所面临的主要问题),但尽可能为大家提供帮助。

2. 为全球病毒学家建立一个相互合作的科学网络。

3. 为世界各地的病毒学实验室搭建国际桥梁。

4. 协助世界各地的病毒学家发展他们的事业,并给予奖励。

5. 向所有会员提供免费的教育资源。

6. 为全球病毒学家申请奖学金和职位提供帮助。

7. 根据病毒学家的专业领域建立病毒学家数据库,以便在暴发任何疫病时能够及时进行远程协助和指导。

详情请访问:www.ws-virology.org

索 引